Inorganic Compounds with Unusual Properties—II

Inorganic Compounds with Unusual Properties—II

R. Bruce King, EDITOR

The University of Georgia

A symposium sponsored by the Division of Inorganic Chemistry at the Inorganic Chemistry Symposium, Athens, Georgia, February 1–3, 1978.

ADVANCES IN CHEMISTRY SERIES **173**

AMERICAN CHEMICAL SOCIETY
WASHINGTON, D. C.   1979

Library of Congress CIP Data

Inorganic compounds with unusual properties—II.

(Advances in chemistry series; v. 173 ISSN 0065-2393)

Bibliography: p.
Includes index.
1. Chemistry, Inorganic—Congresses.
I. King, R. Bruce. II. American Chemical Society.
Division of Inorganic Chemistry. III. Series.

QD1.A355 no. 173 [QD146] 540'.8s [546] 78-31900
ISBN 0-8412-0429-2 ADCSAJ 173 1–418 1979

Copyright © 1979

American Chemical Society

All Rights Reserved. The appearance of the code at the bottom of the first page of each article in this volume indicates the copyright owner's consent that reprographic copies of the article may be made for personal or internal use or for the personal or internal use of specific clients. This consent is given on the condition, however, that the copier pay the stated per copy fee through the Copyright Clearance Center, Inc. for copying beyond that permitted by Sections 107 or 108 of the U.S. Copyright Law. This consent does not extend to copying or transmission by any means—graphic or electronic—for any other purpose, such as for general distribution, for advertising or promotional purposes, for creating new collective works, for resale, or for information storage and retrieval systems.

The citation of trade names and/or names of manufacturers in this publication is not to be construed as an endorsement or as approval by ACS of the commercial products or services referenced herein; nor should the mere reference herein to any drawing, specification, chemical process, or other data be regarded as a license or as a conveyance of any right or permission, to the holder, reader, or any other person or corporation, to manufacture, reproduce, use, or sell any patented invention or copyrighted work that may in any way be related thereto.

PRINTED IN THE UNITED STATES OF AMERICA

Advances in Chemistry Series

Robert F. Gould, *Editor*

Advisory Board

Kenneth B. Bischoff

Donald G. Crosby

Robert E. Feeney

Jeremiah P. Freeman

E. Desmond Goddard

Jack Halpern

Robert A. Hofstader

James D. Idol, Jr.

James P. Lodge

John L. Margrave

Leon Petrakis

F. Sherwood Rowland

Alan C. Sartorelli

Raymond B. Seymour

Aaron Wold

Gunter Zweig

FOREWORD

ADVANCES IN CHEMISTRY SERIES was founded in 1949 by the American Chemical Society as an outlet for symposia and collections of data in special areas of topical interest that could not be accommodated in the Society's journals. It provides a medium for symposia that would otherwise be fragmented, their papers, distributed among several journals or not published at all. Papers are reviewed critically according to ACS editorial standards and receive the careful attention and processing characteristic of ACS publications. Volumes in the ADVANCES IN CHEMISTRY SERIES maintain the integrity of the symposia on which they are based; however, verbatim reproductions of previously published papers are not accepted. Papers may include reports of research as well as reviews since symposia may embrace both types of presentation.

CONTENTS

Preface xi

1. Selective Hydrogenation of Polyunsaturated Olefins 1
   John C. Bailar, Jr.

2. Some Aspects of the Coordination Chemistry and Catalytic Properties of Cationic Rhodium–Phosphine Complexes 16
   J. Halpern, A. S. C. Chan, D. P. Riley, and J. J. Pluth

3. Rhodium–Phosphine Complexes as Homogeneous Catalysts. Hydrogenation of Aromatic Nitro Compounds 26
   Pál Kvintovics, Bálint Heil, and László Markó

4. Pentamethylcyclopentadienyl–Rhodium and –Iridium Complexes as Catalysts for Olefin and Arene Hydrogenation 31
   Peter M. Maitlis

5. Catalytic Homogeneous Hydrogenation Using Micellar and Phase Transfer Reaction Conditions 43
   Daniel L. Reger and M. M. Habib

6. Asymmetric Hydrosilylation 50
   H. B. Kagan, J. F. Peyronel, and T. Yamagishi

7. Activation of C–H Bonds by Bidentate Phosphorus Ligand Complexes of Iron 67
   S. D. Ittel, C. A. Tolman, A. D. English, and J. P. Jesson

8. Homogeneous Catalysis of the Water Gas Shift Reaction by Metal Carbonyls 81
   Peter C. Ford, Robert G. Rinker, Richard M. Laine, Charles Ungermann, Vincent Landis, and Sergio A. Moya

9. Homogeneous Catalysis of the Water Gas Shift Reaction: Pentacarbonyliron and the Metal Hexacarbonyls as Active Catalyst Precursors 94
   C. C. Frazier, R. Hanes, A. D. King, Jr., and R. B. King

10. Oxygen-Exchange and Ligand Substitution Reactions in $Cr(CO)_6$ and $\mu\text{-H}[Cr(CO)_5]_2^-$, and the Water Gas Shift Reaction 106
    Donald J. Darensbourg, Marcetta Y. Darensbourg, Robert R. Burch, Jr., Joseph A. Froelich, and Michael J. Incorvia

11. Catalytic Reductions Using Carbon Monoxide and Water in Place of Hydrogen 121
    R. Pettit, K. Cann, T. Cole, C. H. Mauldin, and W. Slegeir

12. Mechanistic Studies Related to the Metal-Catalyzed Reduction of Carbon Monoxide to Hydrocarbons 131
    Charles P. Casey and Stephen M. Neumann

13. Carbon Monoxide–Metal Oxide Interactions. Surface Site Requirements for Electron Transfer Processes 140
    Kenneth J. Klabunde, Richard A. Kaba, and Russell M. Morris

14. Structure and Reactivity of Ni(II)-d^8 Complexes with Monodentate Tertiary Phosphine:CO Fixation 152
    C. Saint-Joly, M. Dartiguenave, and Y. Darteguenave

15. Binuclear Metal Complexes of Cofacial Diporphyrins 162
    C. K. Chang

16. Homogeneous Oxidative Coupling Catalysts. Products of the Oxidation of Copper(I) Chloride by Oxygen in Polar, Aprotic Media 178
    Geoffrey Davies, Mohamed F. El-Shazly, Deborah R. Kozlowski, Charles E. Kramer, Martin W. Rupich, and Robert W. Slaven

17. Novel Cleavage and Oligomerization Reactions of Nickel(0) Complexes. Applications to Homogeneous Deoxygenation and Desulfurization 195
    John J. Eisch and Kyoung R. Im

18. Gaseous Evolution of Molecular Hydrogen and Oxygen in Photochemical Splitting of Water by Platinized Chlorophyll a Dihydrate Polycrystals. Laboratory Simulation of the Primary Light Reaction in Plant Photosynthesis 210
    L. Galloway, D. R. Fruge, and F. K. Fong

19. Further Studies of the Spectroscopic Properties and Photochemistry of Binuclear Rhodium(I) Isocyanide Complexes 225
    Kent R. Mann and Harry B. Gray

20. Scope and Applications of Light-Induced Electron-Transfer Reactions of Metal Complexes for Energy Conversion and Storage . 236
    Patricia J. DeLaive, David G. Whitten, and Charles Giannotti

21. Tungsten(IV) Chelates—Potential Energy Transfer Complexes ... 252
    Ronald D. Archer, Craig J. Donahue, William H. Batschelet, and David R. Whitcomb

22. Stereochemical Aspects of Expanded Coordination Spheres: Seven-Coordinate Tungsten Complexes 263
    Joseph L. Templeton

23. Effect of Aluminum Additions on the Thermodynamic and Structural Properties of $LaNi_{5-x}Al_x$ Hydrides 279
    Marshall H. Mendelsohn, Dieter M. Gruen, and Austin E. Dwight

24. Sunlight Engineering Efficiency of Thin-Layer Iron–Thiazine Photogalvanic Cells. Evidence That Surface-Induced Back Reaction Is a Key Limiting Factor 296
    Norman N. Lichtin, Peter D. Wildes, Terry L. Osif, and Dale E. Hall

25. Nickel(0) Catalyzed Reactions of Strained Ring Systems 307
    R. Noyori

26. Sensitization of Olefin Photoreactions by Copper(I) Compounds .. 325
    Charles Kutal and Paul A. Grutsch

27. Catalysts for the Isomerization of Quadricyclane to
    Norbornadiene in a Photochemical Energy Storage System 344
    E. M. Sweet, R. B. King, R. M. Hanes, and S. Ikai

28. An Entatic State for Copper in Redox Enzymes? 358
    Robert H. Lane, Nantelle S. Pantaleo, James K. Farr,
    William M. Coney, and M. Gary Newton

29. Metal Tetrathiolenes: Chemistry, Stereochemistry,
    Electrochemistry, and Semiconductivity 364
    Boon-Keng Teo

30. Templates in Zeolite Crystallization 387
    Louis D. Rollmann

31. Reaction Schemes for Dinuclear Compounds Containing
    Metal–Metal Triple Bonds Illustrated by Recent Findings
    in the Chemistry of Molybdenum and Tungsten 396
    Malcolm H. Chisholm

Index 411

PREFACE

The purpose of the symposium, "Inorganic Compounds with Unusual Properties. II. Molecular Catalysis and the Conversion, Production, and Storage of Energy," was to stimulate communication between scientists concerned with the synthesis, characterization, and reactivity of inorganic, coordination, and organometallic compounds and scientists concerned with applications of these compounds in molecular catalysis and in the conversion, production, and storage of energy. This symposium was a sequel to a symposium entitled "Inorganic Compounds with Unusual Properties" held in January, 1975, at the University of Georgia and published in *Advances in Chemistry Series* (1976) 150.

The program of the 1978 symposium included 33 papers of which 31 are included in this volume. Seventeen of the 33 papers presented at the symposium were by persons invited by the Symposium Planning Committee. The remaining 16 papers were selected by the Symposium Planning Committee from those submitted for presentation at the Symposium. The 33 speakers at the Symposium included four persons from industry, one person from a government laboratory (Argonne), and 28 persons from universities and colleges. Two speakers were from France, one each from Japan, Hungary, and England, and the remaining 28 from the United States.

The papers emphasized applications of inorganic chemistry to molecular catalysis and energy. Areas of molecular catalysis receiving particular attention include homogeneous hydrogenation and the water gas shift reaction. Other areas of catalysis covered in the program include oxygenation, deoxygenation, desulfurization, and asymmetric hydrosilylation. In addition, several papers dealt with aspects of photochemistry and molecular catalysis pertaining to solar energy storage. Examples of other related topics discussed in this symposium include the fundamental chemistry of porphyrins, redox enzymes, zeolites, carbon–hydrogen bond activation, metal–metal triple bonds, and metal clusters.

I would like to acknowledge the tremendous help of my University of Georgia colleagues, John K. Ruff, Charles R. Kutal, and Robert H. Lane, who served on the Symposium Planning Committee in connection with the planning and arranging of both the scientific and nonscientific program of the Symposium. We thank the U.S. Army Research Office (Durham), the Petroleum Research Fund of the American Chemical

Society, and the U.S. Department of Energy (Energy Research and Development Administration, Division of Energy Storage Systems) for sponsorship of the Symposium through major financial contributions. Finally, I am pleased to acknowledge assistance from both Helen Mills of the Georgia Center for Continuing Education and my wife, Jane K. King, with the nonscientific program of the symposium.

The University of Georgia R. BRUCE KING
Athens, Georgia
March 1978

Selective Hydrogenation of Polyunsaturated Olefins

JOHN C. BAILAR, JR.

University of Illinois, Urbana, IL 61801

The selective hydrogenation of soybean methyl ester, short chain dienes, and cyclooctadiene to the monoene stage under the catalytic influence of $[Pt(PO_3)_2(SnCl_3)Cl]$ is described. In this catalyst, the platinum can be replaced by palladium, the phosphorus by arsenic, antimony, sulfur, or selenium, the phenyl groups by other aromatic, aliphatic, or ester groups, the tin by lead or germanium, and the chlorine by bromine, iodine, or cyanogen. Terminal double bonds, even in monoenes, are hydrogenated. Isomerization to conjugation apparently precedes hydrogenation. Under mild conditions only isomerization is observed. The isomerized products are largely in the trans form. The catalyst, when made heterogeneous by attaching it to cross-linked polystyrene, still retains its ability to hydrogenate polyunsaturated molecules selectively.

The term, "selective hydrogenation," as it is used in this discussion, refers to the hydrogenation of some of the double bonds in a molecule, leaving other, similar bonds unattacked. Our research on this subject began with the hydrogenation of soybean methyl ester, which, in its original form, is a mixture of the glycerine esters of linolenic, linoleic, oleic, stearic, and palmitic acids. For our studies, the glycerol ester has been converted to the methyl ester (shown in Table I).

Most of the soybean oil of commerce is used in the manufacture of oleomargarine, salad oils, salad dressings, and other foods. Unfortunately, linolenic ester has a poor flavor, so it is desirable to hydrogenate one of the double bonds, leaving the others intact. Ideally, the ethylenic bond in the 15-position would be reduced, and all of the double bonds remaining would retain their positions and their cis configurations.

Table I. Major Constituents of Soybean Oil[a]

$\overset{15}{C}CC=\overset{12}{C}CC=\overset{9}{C}CC=CCCCCCCCOOR$	linolenic	9%
CCCCCC=CCC=CCCCCCCCOOR	linoleic	50%
CCCCCCCC=CCCCCCCCOOR	oleic	27%
CCCCCCCCCCCCCCCCCOOR	stearic	4%
CCCCCCCCCCCCCCCCOOR	palmitic	10%

[a] All double bonds are cis.

A great deal of work has been done on the selective hydrogenation of polyolefinic materials, including natural oils as well as hydrocarbons, and a large number of catalysts has been used (1–22). Some of these are effective only on conjugated systems, some bring about migration of olefinic bonds to conjugation and then reduction, and still others bring about hydrogenation of terminal double and triple bonds selectively, though the selectivity is not always complete (23). It was reported by Abley and McQuillan (24) that isomerization of octene-1 takes place only in the presence of hydrogen, but Bailar and Itatani found that soybean ester isomerizes in the absence of hydrogen (25).

The two classes of polyolefinic compounds that have received the greatest amount of attention in regard to selective hydrogenation are natural oils (e.g., soybean, cottonseed) and hydrocarbons. In both cases, a variety of catalysts and solvents has been used. Candlin and Oldham have reviewed this subject and, in the same article, have discussed their own work (26). They hydrogenated a number of polyolefinic alkenes and alkynes using the Wilkinson catalyst, $RhCl(PO_3)_3$. They found that terminal alkenes are hydrogenated more slowly than unsubstituted ones (as had been found by other investigators) and cyclic alkenes more slowly than terminal alkenes. They noted that the rate of hydrogenation is dependent on the coordinating ability of the substrate, and that substrates containing electron withdrawing groups hydrogenate rapidly. For the hydrogenation of hydrocarbons, Takegami et al. used $FeCl_3$ and

LiAlH$_4$ in THF (27). Tikhomirov et al. used a mixture of Cr(acac)$_3$ and Al(*i*-bu)$_3$ in decalin (28), and Kwiatek and his colleagues used K$_3$Co(CN)$_5$ (4, 5, 6, 7). Tajima and Kunioka used a variety of organometallic compounds (e.g., Cp$_2$VCl$_2$, Cp$_2$ZrCl$_2$, CpCo(CO)$_2$, *n*-C$_4$H$_9$Li, C$_6$H$_5$MgBr, Bu$_3$Al) with moderate success (29). For the selective reduction of cottonseed oil, Kaliev and Bizhanov used a mixture of nickel and molybdenum (9:1) without a solvent (30), but Zueva and Potselueva preferred a 1:1 mixture of palladium and nickel in absolute ethanol (31). A catalyst consisting of Al (48%), Ni (40%), Cu (10%), and Cr (2%) was found to be much superior to the same mixture with the chromium left out, in relation to the velocity of hydrogenation, to the formation of trans acids, and to the degree of isomerization (32). In recent years, at least, the selective hydrogenation of soybean ester has received far more attention than that of other vegetable oils. Frankel and his co-workers have had excellent success with metal carbonyls (33–38). For the most part, they used iron and chromium carbonyls, which gave good selectivity. They found that cobalt carbonyl was much inferior. Emken, Frankel, and Butterfield also used the acetylacetonates of Ni(III), Co(III), Cu(II), and Fe(III) (45). [Ni(III)acac$_3$] was the most active and the most selective of this group. [Cuacac$_2$] showed little selectivity. They found that no hydrogenation took place in the absence of solvents, and that methanol was a better solvent than either acetic acid or dimethylformamide. The chief products were monoenes, as long as some triene remained. They reported that the nature of the solvent had little effect on the rate of catalysis, except that when the solvent contained pyridine the reaction was extremely slow. Bailar and Tayim (46), however, found that chlorinated hydrocarbons gave a much faster reaction than did a benzene–methanol mixture and that the presence of pyridine or thiophene completely blocked the reaction.

The very high selectivity of H$_2$PtCl$_4$–SnCl$_2$ and H$_2$PtCl$_6$–SnCl$_2$ mixtures was first observed by Bailar and Itatani (47). This was based on the observation of Cramer, Jenner, Lindsey, and Stolberg (48) that this catalyst was extremely effective for the hydrogenation of ethylene. Lindsey's group has published extensively on the Pt–Sn chloride complex, but not in terms of its catalytic selectivity (49, 50, 51). van Bekkum and his co-workers (20), Bond and Hellier (53), and van't Hof and Linsen have used this catalyst for the hydrogenation of polyolefins but not for vegetable oil esters (54). Several investigators have reported that the mole ratio of platinum to tin is important, but their estimates of the optimum ratio vary from 1:5 to 1:10.

Bailar and Itatani later adopted [Pt(PO$_3$)$_2$Cl$_2$] + SnCl$_2$ as the catalyst for the hydrogenation of soybean ester. This complex seems to have the highest selectivity of any that have been studied. Frankel and

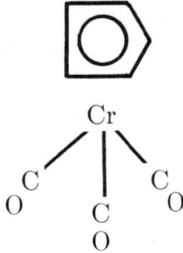

his associates (55, 56, 57) have recently found that cyclopentadiene chromium tricarbonyl is selective and gives cis products. It is a superior catalyst for this purpose.

We were asked several years ago by the Northern Utilization Research and Development Division of the U.S. Department of Agriculture to study this selective hydrogenation. While our efforts have not yet led to complete success, they have led to some interesting and useful results. In addition to soybean methyl ester, we have examined some short chain diolefins and some cyclic compounds such as cyclooctadiene. The catalyst that we first used was suggested by the work of Cramer, Lindsey, Prewitt, and Stolberg. It is a mixture of chloroplatinic acid(IV) and tin(II) chloride. We found that it is also a selective catalyst for the hydrogenation of soybean ester—selective in the sense that it leaves one double bond unhydrogenated.

We soon modified this catalyst to gain greater selectivity, and throughout most of our work we have used $[Pt(PO_3)_2Cl_2] + SnCl_2$. This, however, is only one of a large family of materials that have the same catalytic properties, some to a greater degree and some to a lesser degree. Instead of platinum, palladium can be used. It gives a much more active catalyst than platinum. In fact, in many cases the palladium catalyst is effective without any addition of tin chloride. Nickel can also be used, but the properties of the resulting catalyst are somewhat different than those of the platinum catalyst. The phosphorus can be replaced by arsenic, antimony, sulfur, or selenium. Arsenic gives a more active

Table II. Hydrogenation of Propylene at 44°C and 500 psi of Hydrogen in Chloroform with $[PtCl_2L_2] + SnCl_2 \cdot 2H_2O$

L	% Propane after 3 hr
$P\varnothing_3$	24.5
$P(p\text{-}CH_3\varnothing)_3$	24.4
$P(CH_3)\varnothing_2$	76.9
$P(CH_3)_2\varnothing$	91.8
$P(CH_3)_3$	6.0
$P(CH_2CH_2CH_2CH_3)_3$	7.0

catalyst than phosphorus; the others are less active. The phenyl groups of the triphenyl phosphine can be replaced by other aromatic groups, phenoxy groups, or, in part, by aliphatic groups. The presence of one or two aliphatic groups increases the catalytic activity, but trialkyl phosphines give poor catalysts. The phenyl group evidently takes part in the reaction. Table II illustrates the relative activities of some catalysts containing different phosphines. The halides in the $SnCl_3$ group can be chloride, bromide, iodide, or cyanide. If $[Pt(PR_3)_2I_2]$ or $[Pt(PO_3)_2(CN)_2]$ is used it is not necessary to add a tin halide, for iodide and cyanide, like the $SnCl_3$ group, are σ donors as well as π acceptors. Iodide is also a π donor.

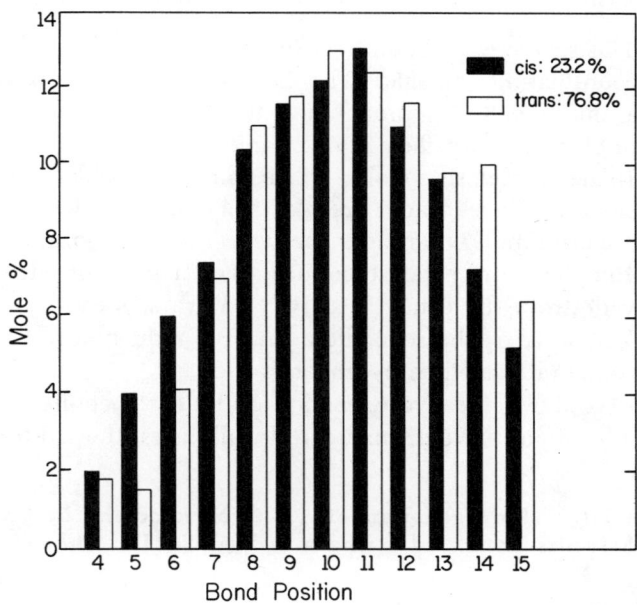

Journal of Organic Chemistry

Figure 1. Distribution of the ethylenic bonds after selective hydrogenation (14)

The properties of this group of catalysts can be summarized as follows:

(1) Under suitable conditions, they leave one double bond unhydrogenated.

(2) They catalyze the migration of the double bonds along the carbon chain. Frankel and Emken (*14*) observed the distribution of the one unhydrogenated double bond in soybean ester to be as shown in Figure 1. Under suitable conditions, this isomerization can be effected without hydrogenation.

(3) The double bond in the esters that are formed are mostly of the trans configuration (*see* Figure 1).

(4) With short-chain diolefins, at least, terminal double bonds are reduced even if there are no other double bonds present.

(5) Short-chain conjugated diolefins, such as butadiene and isoprene, not only are not reduced, but they form stable complexes with the catalyst and thus destroy its catalytic property.

The nature of the solvent is very important. In our early work we used a 3:2 mixture of benzene and methanol, which dissolves both the substrate and the catalyst. However, the realization that methanol is a moderately good complexing agent (it probably forms a stronger bond to Pt(II) than does $\diagdown_{\diagup}C{=}C\diagup_{\diagdown}$) caused us to search for a noncoordinating catalyst. This search was not entirely successful, but it led to experiments with noncoordinating solvents. The best solvent that was found was methylene chloride. In all probability, the olefin and the heavy metal form a complex that is sufficiently "organic" to dissolve in this solvent. Dichloroethane is almost equally active, and even chloroform can be used. Acetone and acetic acid are also suitable and allow the reaction to proceed more rapidly than does the benzene–methanol mixture. Pyridine and thiophene are catalyst poisons; even a trace of either of them completely destroys the catalytic activity. With soybean ester, at least, methanol can serve as the reducing agent, the reduction taking place at about the same rate as when hydrogen is used.

Table III shows the hydrogenation of 1,5-cyclooctadiene in several different solvents or solvent mixtures. In all cases, the 1,5-hydrocarbon

Table III. Hydrogenation of 1,5-Cyclooctadiene in Various Solvents with and without Added Sn(II) Chloride[a]

Solvent	$SnCl_2$ Added	Composition of Product		
		1,3-Cyclo-octadiene	Cyclo-octene	Cyclo-octane
CH_2Cl_2	—	40	58	2
CH_2Cl_2	$SnCl_2 \cdot 2H_2O$	82	18	0
$CH_2Cl_2 + CH_3OH$	—	4	93	3
$CH_2Cl_2 + CH_3OH$ (4:1)	$SnCl_2 \cdot 2H_2O$	93	7	0
CH_3COOH	—	0	81	19
CH_3COOH	$SnCl_2$	90	10	0
$C_6H_6 + CH_3OH$	—	0	81	19
$C_6H_6 + CH_3OH$ (3:2)	$SnCl_2$	12	88	0

[a] Reaction time, 5 hr; $[Pd(PO_3)_2Cl_2]$: 90°C.

Table IV. Electronic Structure of Pt(II) and Its Complexes

	e^-	5d	6s	6d
Pt°	78	⊙ ⊙ ⊙ ⊙ ⊙	⊙	
Pt$^{(II)}$	76	⊙ ⊙ ⊙ ⊙		
[PtX$_4$]$^{n+}$	84	⊙ ⊙ ⊙ ⊙ ⊗	⊗ ⊗ ⊗	
Rn	86	⊙ ⊙ ⊙ ⊙ ⊙	⊙	⊙ ⊙ ⊙ ⊙ ⊙

was isomerized to the 1,3-configuration before any samples were removed for analysis, and the 1,3-isomer was partially or entirely reduced to cyclooctene or cyclooctane. It should be noticed also that in all cases, in the presence of tin chloride, there has been no reduction to cyclooctane (58).

The mechanism of the catalytic reaction is not fully known. Four-covalent complexes of Pt(II), Pd(II), and Ni(II) lack two electrons of the next rare gas structure, so they have a vacant space in the coordination sphere. This is shown for platinum in Table IV. There is some evidence that the early part of the reaction proceeds through the steps:

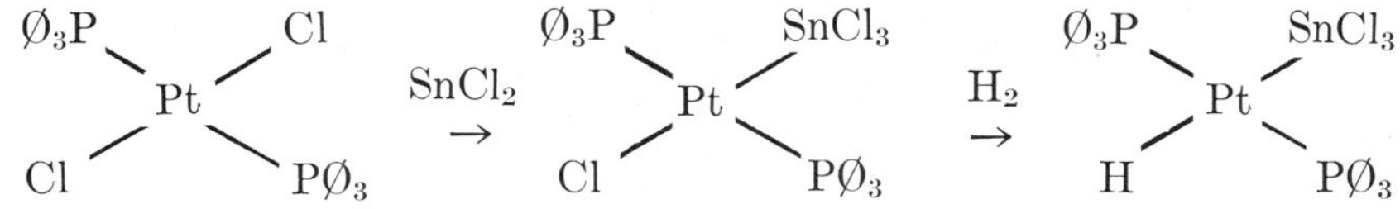

$$\underset{Cl}{\overset{\varnothing_3P}{\diagdown}}Pt\underset{P\varnothing_3}{\overset{Cl}{\diagup}} \xrightarrow{SnCl_2} \underset{Cl}{\overset{\varnothing_3P}{\diagdown}}Pt\underset{P\varnothing_3}{\overset{SnCl_3}{\diagup}} \xrightarrow{H_2} \underset{H}{\overset{\varnothing_3P}{\diagdown}}Pt\underset{P\varnothing_3}{\overset{SnCl_3}{\diagup}}$$

and that the hydride is the actual catalyst. This compound has not been isolated, but addition of triethylamine allowed the isolation of triethylamine hydrochloride (59).

If we abbreviate the formula for [(Pt or Pd)(PO$_3$)$_2$(SnCl$_3$)H] to MH, the reaction sequence may be something like that shown below. (No experiments have been done with labeled hydrogen, but some of the hydrogen atoms in this figure have been marked for identification.) Reaction 1 shows the formation of a π bond between the catalyst and one of the ethylenic bonds, and Reaction 2 shows the conversion of this bond into two σ bonds. Both of these reactions are reversible. However, Reaction 3 can take place as readily as Reaction 2; this accounts for the migration of the double bond along the hydrocarbon chain. Reaction 4 illustrates the formation of an additional π bond and the formation of a cycle. Upon hydrogenation, the carbon–metal σ bond is broken, with one hydrogen atom going to the carbon atom and the other to the catalyst. Hydrogenation takes place only when this cycle is present. If the polyolefin contains another double bond, Reactions 1, 2, and 3 can proceed until the two remaining double bonds are close enough together to allow

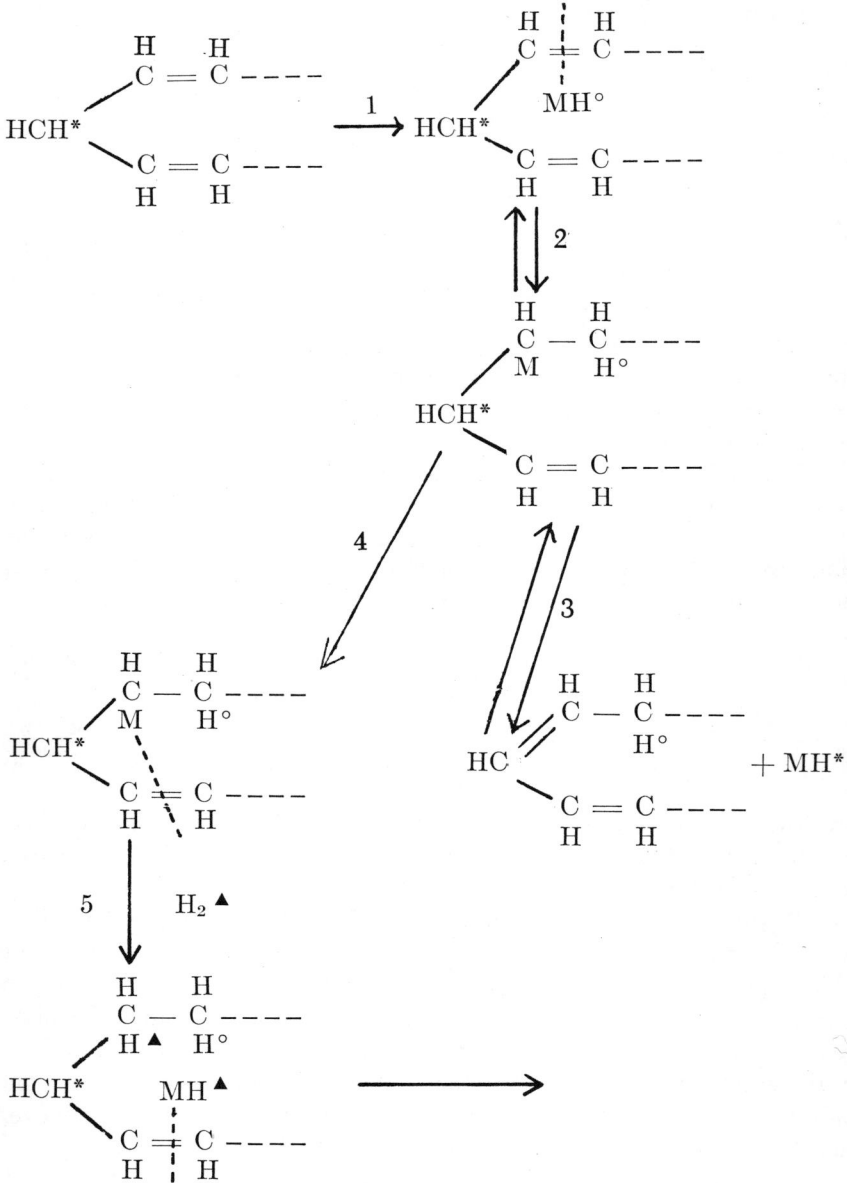

the formation of a new cycle, when Reactions 4 and 5 take place. Since the formation of the cycle must involve two double bonds, hydrogenation stops when only one remains. The isolation of an allylic intermediate in the hydrogenation of 1,5-cyclooctadiene (Figure 2) is in line with this mechanism (58).

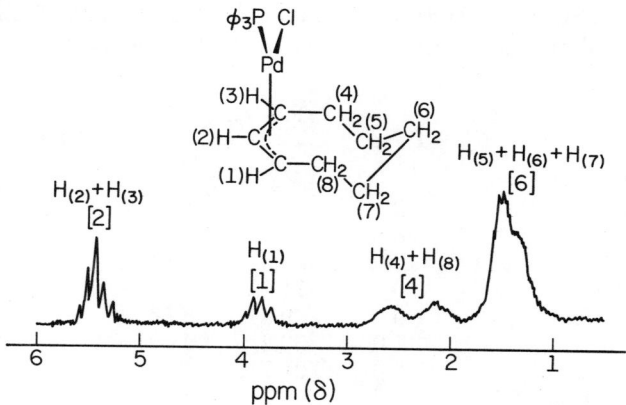

Figure 2. PMR spectrum of [PdCl(C_8H_{13})(PO_3)] in $CDCl_3$. [ ] represents the number of protons. $H_{(1)}$: δ = 3.85 ppm, $J_{1,2}$ = ca. 6.9 Hz, $J_{1,8}$ = ca. 6.8 Hz; $H_{(2)}$: δ = 5.41 ppm, $J_{2,3}$ = ca. 6.9 Hz; $H_{(3)}$: δ = 5.39 ppm, $J_{3,p}$ = ca. 9.6 Hz, $J_{3,4}$ = ca. 6.9 Hz.

Under mild conditions, isomerization can take place without hydrogenation. This, of course, is most evident when hydrogen gas is not used; under that condition, the reaction has been studied in some detail (Figures 3 and 4). It is probable that isomerization to conjugation precedes hydrogenation in all cases. We thought at one time that this is not the case, for 2,3,3-trimethyl-pentadiene (1–22, 25), which cannot form conjugate bonds, is readily hydrogenated (Table V). We have learned since, however, that terminal double bonds hydrogenate even when no other ethylenic groups are present. This is shown in Table VI. This shows the

Table V. Hydrogenation of 2,3,3-Trimethyl-1,4-pentene

Table VI. Catalytic Hydrogenation of Monoenes with $PtCl_2(PO_3)_2$ and $SnCl_2 \cdot 2H_2O$ in Benzene–Methanol (3 hr)

	1-Isomer	2-Isomer cis	2-Isomer trans	3-Isomer	Saturated Hydrocarbon
Ethylene	0				100
Propylene	66				34
1-Butene	12.0	30.3	46.4		11.3
cis-2-Butene	0	71.1	27.3		1.6
trans-2-Butene	0	9.8	89.6		0.6
1-Pentene	11.6	30.2	47.2		11.6
2-Pentene (cis 47.9; trans 52.1)	3.0	27.7	67.5		1.8
1-Hexene	12.5	28.5	41.6	5.6	12.0
2-Hexene[a]					
3-Hexene[a]					

[a] No observable hydrogenation.

products of hydrogenation of some short chain hydrocarbons after 3 hr of hydrogenation. In the case of ethylene, hydrogenation was complete in less than 1 hr. As the chain on one side of the double bond is lengthened by one or two methyl groups, the rate of hydrogenation is decreased. Further lengthening seems to have little effect. If the chain is lengthened on both sides of the ethylenic bond, however, very little hydrogenation takes place, and that little may be preceded by migration of the double bond into a terminal position. Figures 3 and 4 compare the conditions under which isomerization and hydrogenation of 1,5-cyclo-

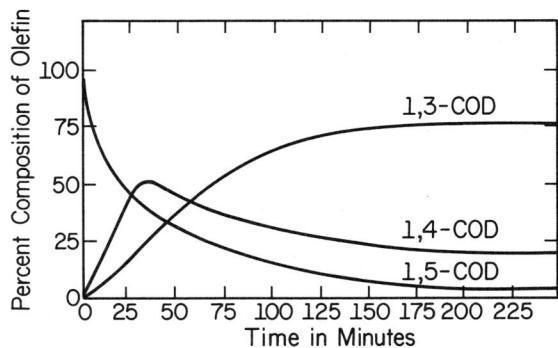

Journal of the American Chemical Society

Figure 3. Catalytic isomerization of 1,5-cyclo-octadiene in C_6H_6–CH_3OH solution under 1 atm N_2 at 60°C (15)

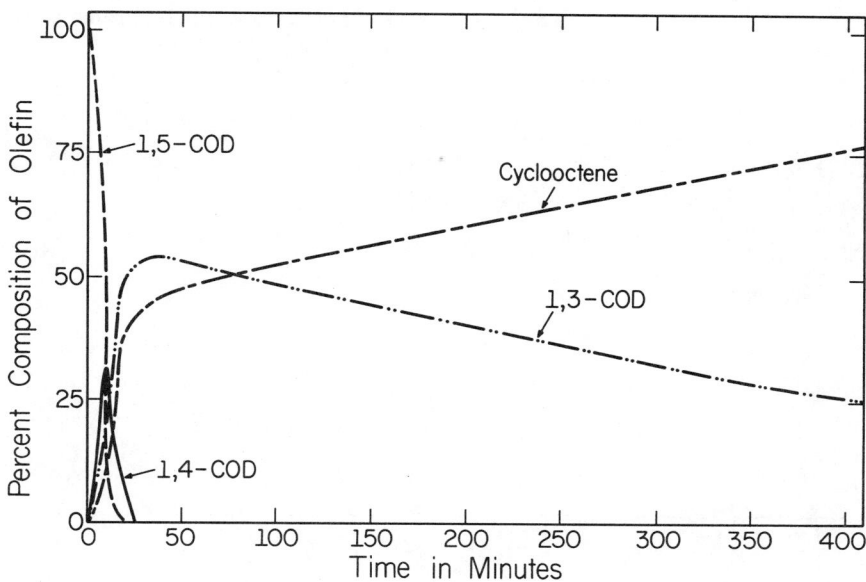

Journal of the American Chemical Society

Figure 4. Catalytic hydrogenation of 1,5-cyclooctadiene in methylene chloride under 500 psi H_2 at 105°C (16)

octadiene take place. It is evident from Figure 4 that the isomerization is reversible and reaches an equilibrium after about 200 min. Short chain dienes often show interesting and instructive results. Table VII shows the results that were obtained with isoprene with different catalysts and solvents (13).

Near the end of our work with soybean methyl ester, we fixed the catalyst on polystyrene cross-linked with divinyl benzene, thus making it a heterogeneous catalyst (Table VIII). Whether the heavy metal further cross-links the polymer or is attached to phosphorus atoms on a single polystyrene chain is not known. In any event, the heterogeneous catalyst has the same sort of selectivity as the homogeneous one. In no case did we succeed in getting all of the phosphorus atoms in the polymer attached to the heavy metal of the catalyst. The first entry in Table IX, for example, indicates that for each unit of the polymer, 0.505 molecules of $PtCl_2$ were attached. In the cases in which palladium chloride was used, no tin chloride was added—the palladium chloride is sufficiently active without it. It is evident that it is possible to destroy the triene without greatly increasing the amount of saturation in the oil. We are still working on this catalyst system, and hope to have more results to report before long.

Table VII. Hydrogenation of Isoprene with

Catalyst	Solvent
$Pt(PO_3)_2Cl_2 + SnCl_2$	$CH_3COOH + CH_3COCH_3$ (4:1)
$Pd(PO_3)_2Cl_2 + SnCl_2$	CH_3COOH
$Pd(PO_3)_2(CN)_2$	$C_6H_6 + CH_3OH$ (3:2)
$Ni(PO_3)_2I_2$	CH_3OH

[a] 55 atm H_2; 5 hr; 100°C.

Table VIII. Polymeric Catalysts[a]

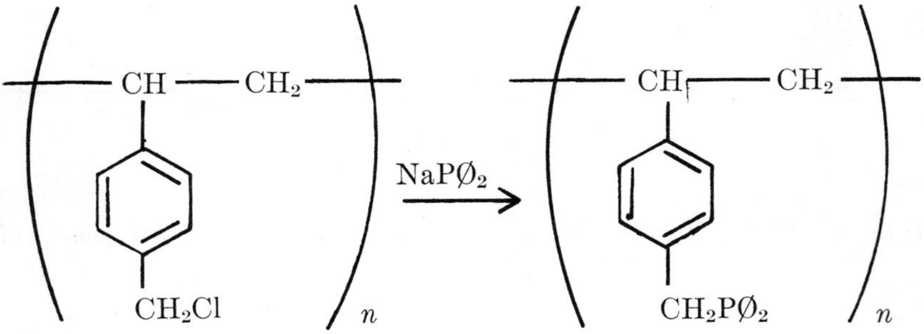

$$\left(\!\!-CH-CH_2-\!\!\right)_n \xrightarrow{NaP\varnothing_2} \left(\!\!-CH-CH_2-\!\!\right)_n$$

(left: with CH_2Cl para-substituent on phenyl; right: with $CH_2P\varnothing_2$ para-substituent on phenyl)

[a] The structure on the left is actually a copolymer of chloromethylstyrene and divinyl benzene.

Table IX. Hydrogenation of Soybean

Catalyst	Solvent	Metal/Olefin Ratio $\times 10^3$
Original substrate		
[Pol. $(PtCl_2)_{0.505}$] 150°C 600 psi H_2	CH_2Cl_2	0.62
[Pol. $(PdCl_2)_{0.851}$] 25°C 600 psi H_2	CH_2Cl_2	1.02
[Pol. $(PdCl_2)_{0.851}$] 25°C 1 atm H_2	$\begin{cases} CH_3OH \\ C_6H_6 \end{cases}$	0.48
[Pol. $(PtCl_2)_{0.505}$] 150°C + $SnCl_2$	CH_2Cl_2	
[Pol. $(PdCl_2)_{0.507}$] 70°C		

Different Catalysts and Different Solvents[a]

Product (%)				
C=C(C)-C=C	C-C(C)=C-C	C=C(C)-C-C	C-C(C)-C=C	C-C(C)-C-C
1	63	8	0	28
30	20	1	49	—
12	64	11	12	1
100	—	—	—	—

Ester on the Heterogeneous Catalyst

Time (hr)	Composition of Product				
	Saturate	Monoene	Diene	Conj. Diene	Triene
	14.2	22.3	56.2		7.0
6	14.5	37.8	28.3	20.1	—
4	15.1	60.4	24.3	—	—
8	14.3	62.4	23.4	—	—
6	13.8	26.6	39.2	18.4	2.0
3	14.8	35.2	45.9	—	4.3

Literature Cited

1. Slaugh, L. H., *Tetrahedron* (1966) **22**, 1741.
2. Slaugh, L. H., *J. Org. Chem.* (1967) **32**, 108.
3. Tajima, Y., Kunioka, E., *J. Org. Chem.* (1968) **33**, 1689.
4. Kwiatek, J., *Catal. Rev.* (1967) **1**, 37.
5. Kwiatek, J., *J. Organomet. Chem.* (1965) **3**, 421.
6. Kwiatek, J., Seyler, J. K., ADV. CHEM. SER. (1968) **70**, 207.
7. Kwiatek, J., Mador, I. L., Seyler, J. K., ADV. CHEM. SER. (1963) **37**, 201.
8. Burnett, M. G., Connolly, P. J., Kemball, C., *J. Chem. Soc. A* (1968) 991.
9. Hallman, P. J., Evans, D., Osborn, J. A., Wilkinson, G., *Chem. Commun.* (1967) 305.
10. Osborn, J. A., Jardine, F. H., Young, J. F., Wilkinson, G., *J. Chem. Soc. A* (1966) 1711.
11. Jardine, F. H., Osborn, J. A., Wilkinson, G., *J. Chem. Soc. A* (1967) 1574.
12. Bailar, J. C., Jr., Itatani, H., *J. Am. Chem. Soc.* (1967) **89**, 1592.
13. Itatani, H., Bailar, J. C., Jr., *J. Am. Chem. Soc.* (1967) **89**, 1600.
14. Frankel, E. N., Emken, E. A., Itatani, H., Bailar, J.C., Jr., *J. Org. Chem.* (1967) **32**, 1447.
15. Tayim, H. A., Bailar, J. C., Jr., *J. Am. Chem. Soc.* (1967) **89**, 3420.
16. Ibid (1967) **89**, 4330.
17. Bailar, J. C., Jr., Itatani, H., Tayim, H., *J. Jpn. Chem.* (1968) **22**, 41.
18. Adams, R. W., Batley, G. E., Bailar, J. C., Jr., *J. Am. Chem. Soc.* (1968) **90**, 6051.
19. Adams, R. W., Batley, G. E., Bailar, J. C., Jr., *Inorg. Nucl. Chem. Lett.* (1968) **4**, 455.
20. Bailar, J. C., Jr., *Platinum Met. Rev.* (1971) **15**, 2.
21. Frankel, E. N., Itatani, H., Bailar, J. C., Jr., *J. Am. Oil Chem. Soc.* (1972) **49**, 132.
22. Itatani, H., Bailar, J. C., Jr., *IEC Prod. Res. Dev.* (1972) **11**, 146.
23. Bond, G. C., Hellier, M., *J. Catal.* (1967) **7**, 217.
24. Abley, P., McQuillin, F. J., *Discuss. Faraday Soc.* (1968) **46**, 31.
25. Bailar, J. C., Jr., Itatani, H., *J. Am. Chem. Soc.* (1967) **89**, 1592.
26. Candlin, J. P., Oldham, A. R., *Discuss. Faraday Soc.* (1968) **46**, 60.
27. Takegami, Y., Ueno, T., Fujii, T., *Bull. Chem. Soc. Jpn.* (1965) **38**, 1279.
28. Tikhomirov, B. I., Klopotova, I. A., Yakabchik, A. I., *Vestn. Leningr. Univ.,* **22**(22), *Fiz. Khim.* (1967) **4**, 147; *Chem. Abstr.* (1968) **68**, 59020.
29. Tajimia, Y., Kunioka, E., *J. Org. Chem.* (1968) **33**, 1689.
30. Kaliev, S. P., Bizhanov, F. B., *Khim. Khim. Technol. (Alma-Ata)* (1971) **2**, 65; *Chem. Abstr.* (1974) **80**, 13779n.
31. Zueva, L. I., Potselueva, L. B., *Tr. Inst. Org. Katal. Elektrokhim., Akad. Nauk Kaz. SSR* (1973) **5**, 82; *Chem. Abstr.* (1974) **80**, 131770a.
32. Nazarova, I. P., Kantsepol'skaya, F. M., Glushenkova, A. I., Markham, A. L., *Maslo-Zhir. Promst.* (1974) **2**, 16; *Chem. Abstr.* (1974) **80**, 131756a.
33. Frankel, E. N., Jones, E. P., Glass, C. A., *J. Am. Oil Chem. Soc.* (1964) **41**, 392.
34. Frankel, E. N., Peters, H. M., Jones, E. P., Dutton, H. J., *J. Am. Oil Chem. Soc.* (1964) **41**, 186.
35. Frankel, E. N., Jones, E. P., Davison, V. L., Emken, E., Dutton, H. J., *J. Am. Oil Chem. Soc.* (1965) **42**, 130.
36. Frankel, E. N., Emken, E. A., Davison, V. L., *J. Am. Oil Chem. Soc.* (1966) **43**, 307.
37. Frankel, E. N., Little, F. L., *J. Am. Oil Chem. Soc.* (1969) **46**, 256.
38. Frankel, E. N., *J. Am. Oil Chem. Soc.* (1970) **47**, 33.
39. Ibid (1970) **47**, 11.

40. Frankel, E. N., Metlin, S., Rohwedder, W. K., Wender, I., *J. Am. Oil Chem. Soc.* (1969) **46,** 133.
41. Frankel, E. N., Emken, E. A., Peters, H. M., Davison, V. L., Butterfield, R. D., *J. Org. Chem.* (1964) **29,** 3292.
42. Frankel, E. N., Emken, E. A., Davison, V. L., *J. Org. Chem.* (1965) **30,** 2739.
43. Frankel, E. N., Selke, E., Glass, C. A., *J. Org. Chem.* (1969) **34,** 3936.
44. Frankel, E. N., Mounts, T. L., Butterfield, R. O., Dutton, H. J., ADV. CHEM. SER. (1968) **70,** 177.
45. Emken, E. A., Frankel, E. N., Butterfield, R. O., *J. Am. Oil Chem. Soc.* (1966) **43,** 14.
46. Tayim, H. A., Bailar, J. C., Jr., *J. Am. Chem. Soc.* (1967) **89,** 4330.
47. Bailar, J. C., Jr., Itatani, H., *Proc. Symp. Coord. Chem., Hungary, 1964,* **67.**
48. Cramer, R. D., Jenner, E. L., Lindsey, R. V., Jr., Stolberg, U. G., *J. Am. Chem. Soc.* (1963) **85,** 1691.
49. Cramer, R. D., Lindsey, R. V., Jr., Prewitt, C. T., Stolberg, U. G., *J. Am. Chem. Soc.* (1965) **87,** 658.
50. Lindsey, R. V., Jr., Parshall, G. W., Stolberg, U. G., *J. Am. Chem. Soc.* (1965) **87,** 658.
51. Parshall, G. W., *J. Am. Chem. Soc.* (1966) **88,** 3534.
52. van Bekkum, H., van Gogh, J., van Minnen-Pathius, G., *J. Catal.* (1967) **7,** 292.
53. Bond, G. C., Hellier, M., *Chem. Ind. (London)* (1965) 35.
54. van't Hof, L. P., Linsen, B. G., *J. Catal.* (1967) **7,** 295.
55. Frankel, E. N., Butterfield, R. O., *J. Org. Chem.* (1969) **34,** 3930.
56. Frankel, E. N., Selke, E., Grass, C. A., *J. Org. Chem.* (1969) **34,** 3936.
57. Frankel, E. N., *J. Org. Chem.* (1972) **38,** 1549.
58. Fujii, Y., Bailar, J. C., Jr., *J. Catal.*, in press.
59. Vassilian, A., experiment in the author's laboratory.

RECEIVED February 22, 1978.

2

Some Aspects of the Coordination Chemistry and Catalytic Properties of Cationic Rhodium–Phosphine Complexes

JACK HALPERN, A. S. C. CHAN, D. P. RILEY, and J. J. PLUTH

Department of Chemistry, University of Chicago, Chicago, IL 60637

> *The chemistry of cationic rhodium complexes containing chelating diphosphine ligands was found to differ significantly from that of corresponding complexes of monodentate phosphine ligands. The characterization of [Rh(DIPHOS)]$^+$ and of its alkene and arene adducts is described, together with studies on the kinetics of the catalytic hydrogenation of alkenes in which such adducts are intermediates. The results of these studies are pertinent to the role of such complexes as catalysts for the asymmetric hydrogenation of prochiral olefins.*

Considerable interest has recently been focused on cationic rhodium(I) complexes containing tertiary phosphine ligands, particularly in the context of such complexes as highly effective asymmetric hydrogenation catalysts (1, 2). While the most extensive studies on the coordination chemistry and catalytic properties relate to such complexes containing monodentate tertiary phosphine ligands, for example, those derived from [Rh(PR$_3$)$_2$(DIENE)]$^+$ (where DIENE = norbornadiene (NOR) or 1,5-cyclooctadiene) (3–13), the highest optical yields to date (> 95% enantiomeric excess in the hydrogenation of prochiral α-acetamidoacrylic acids) have been achieved with cationic rhodium catalysts containing chiral chelating diphosphine ligands, notably 1,2-bis(o-anisylphenylphosphino)ethane (DIPAMP) (14, 15) and 2,3-bis(diphenylphosphino)butane (CHIRAPHOS) (16). Accordingly, we have undertaken an examination of the basic coordination chemistry and catalytic properties of such cationic rhodium diphosphine chelate complexes, notably of [Rh(DIPHOS)(NOR)]$^+$ (1), where DIPHOS = 1,2-bis(diphenylphosphino)ethane and of various other cationic rhodium–DIPHOS complexes derived therefrom by hydrogenation. Unexpectedly, the chemistry of these complexes was found to differ in several important respects, including those bearing on their activity as hydrogenation catalysts, from that of the corresponding complexes containing monodentate phosphine ligands, e.g., $[Rh(PPh_3)_2(NOR)]^+$ (17).

Reaction of $[Rh(DIPHOS)(NOR)]^+$ with Hydrogen

In methanolic solution, $[Rh(DIPHOS)(NOR)]^+$ was found to react rapidly with precisely 2.0 mol of H_2 per Rh (confirmed by spectral titration, e.g., Figure 1) according to the stoichiometry of Reaction 1, quantitatively yielding norbornane (confirmed by NMR) and a cationic

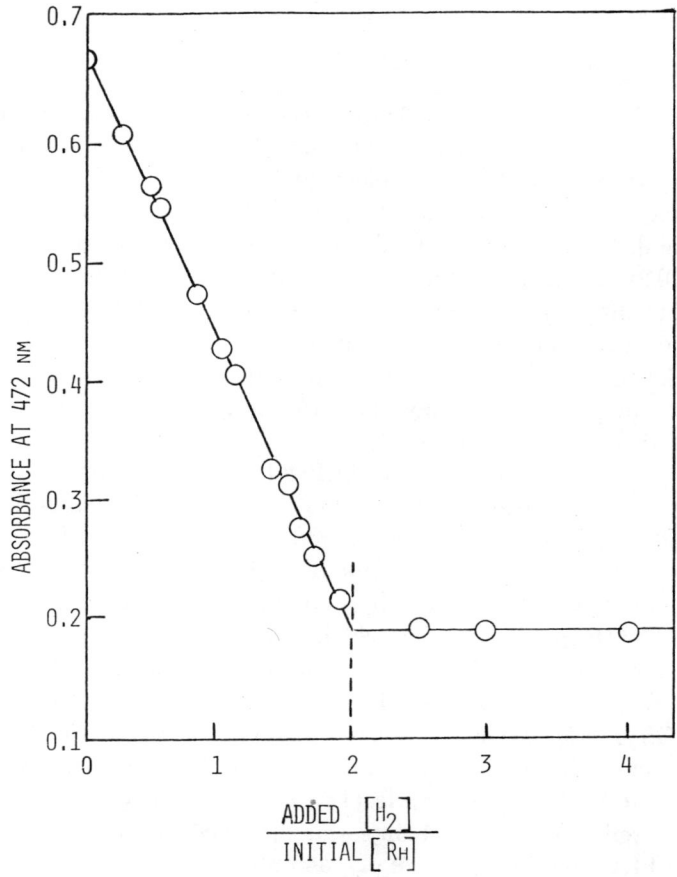

Figure 1. Spectral titration of 2.38×10^{-4}M $[Rh(DIPHOS)-(NOR)]^+$ with hydrogen in methanol

Rh(I) complex of composition (apart from possible solvent coordination) [Rh(DIPHOS)]$^+$, (**2**, λ_{max} 432 nm; ϵ_{max} 1.49 $\times$ 10^3 M^{-1} cm^{-1}). No further uptake of hydrogen nor formation of a hydride complex was detectable (e.g., by NMR). This result is in marked contrast to that reported (8, 10, 11) for [Rh(PPh$_3$)$_2$(NOR)]$^+$ which reacts with 3 mol of H$_2$ under the same conditions to form the Rh(III) hydride complex (**3**) according to Reaction 2.

$$[Rh(DIPHOS)(NOR)]^+ + 2H_2 \rightarrow [Rh(DIPHOS)]^+ + \text{Norbornane} \quad (1)$$
$$\quad\quad\quad\quad \textbf{1} \quad\quad\quad\quad\quad\quad\quad\quad\quad\quad\quad\quad \textbf{2}$$

$$[Rh(PPh_3)_2(NOR)]^+ + 3H_2 \rightarrow$$
$$\quad\quad\quad\quad\quad\quad [RhH_2(PPh_3)_2(\text{Solvent})_2]^+ + \text{Norbornane} \quad (2)$$
$$\quad\quad\quad\quad\quad\quad\quad\quad\quad\quad \textbf{3}$$

Characterization of [Rh(DIPHOS)]$^+$

[Rh(DIPHOS)]$^+$ was isolated as the BF$_4^-$ salt, containing no methanol, and shown by single crystal x-ray diffraction to have a structure corresponding to discrete binuclear [Rh$_2$(DIPHOS)$_2$]$^{2+}$ ions in which each rhodium atom is bonded to two phosphorus atoms of a DIPHOS ligand and, through symmetrical π-arene coordination, to a phenyl ring of the DIPHOS ligand of the second rhodium atom (17). Each rhodium atom thus has an "18-electron valence shell" and the 4.28 Å Rh–Rh separation lies well outside the range of significant metal–metal interaction. There are several precedents for π-arene bonding in other cationic rhodium complexes, including the structurally characterized compound Rh[P(OMe)$_2$]$_2$BPh$_4$ (5, 9, 18, 19).

In methanolic solution, [Rh$_2$(DIPHOS)$_2$][BF$_4$]$_2$ apparently dissociates into mononuclear [Rh(DIPHOS)]$^+$ ions (presumably containing coordinated solvent), as demonstrated by: (i) electrical conductance measurements which yielded a slope of $-$ 270 ohm^{-1} $M^{-0.5}$, corresponding to a 1:1 electrolyte (20) for a plot of equivalent conductance vs. $\sqrt{\text{concn}}$; (ii) ^{31}P NMR measurements which revealed only a single P signal (d, 2P, δ 80, J_{Rh-P} = 203 Hz); and (iii) measurements on the equilibria for the formation of various 1:1 alkene and arene adducts of [Rh(DIPHOS)]$^+$ (*see* below). When base (OMe$^-$ or a sterically hindered amine such as triethylamine) was added to a methanolic solution of [Rh(DIPHOS)]$^+$, an irreversible (i.e., not reversed by addition of acid) yellow to red-brown color change was observed, to yield a new species, [Rh$_3$(DIPHOS)$_3$(OMe)$_2$]$^+$ (**4**, λ_{max} 445 nm, ϵ_{max} 3.3 $\times$ 10^3 M^{-1} cm^{-1}; ^{31}P NMR, d, 6P, δ 76, J_{Rh-P} = 201 Hz), according to Reaction 3, the stoichiometry of which was established by spectral titration. The structure of

4, as deduced from preliminary single crystal x-ray diffraction data for the PF_6 salt, corresponds to a regular triangular array of rhodium atoms, separated by bonding distances of 3.06 Å. Each bidentate DIPHOS ligand is coordinated to one rhodium atom ($r_{Rh-P} = 2.2$ Å) with the P–Rh–P plane perpendicular to the Rh_3 plane. One triply bridging OMe^- ion is symmetrically located on each side of the Rh_3 plane ($r_{Rh-O} = 2.15$ Å). The $[Rh_3P_6O_2]$ framework thus has D_{3h} symmetry.

$$3[Rh(DIPHOS)]^+ + 2OMe^- \rightarrow [Rh_3(DIPHOS)_3(OMe)_2]^+ \quad (3)$$
$$\mathbf{4}$$

Arene and Alkene Adducts of [Rh(DIPHOS)]⁺

In methanol solution, $[Rh(DIPHOS)]^+$ formed 1:1 adducts with a variety of unsaturated substrates (UNSAT) including alkenes and arenes, according to Reaction 4. Reaction 4 could readily be monitored, and the equilibrium constant K_4 ($= [Rh(DIPHOS)(UNSAT)]^+/[Rh(DIPHOS)]^+\cdot[UNSAT]$) determined from the spectral changes accompanying the addition of successive increments of UNSAT (17). In the case of benzene, the composition of the adduct was confirmed by isolating the salt $[Rh(DIPHOS)(C_6H_6)]BF_4 \cdot C_6H_6$ which dissolved in CD_2Cl_2 to yield a solution whose ^1H NMR spectrum contained two sharp singlets of equal intensity (6H) corresponding to free (7.4 δ) and coordinated (6.36 δ) C_6H_6. Values of K_4 in methanol, determined for selected substrates, are: benzene (18 M^{-1}); toluene (97 M^{-1}); o-, m-, or p-xylene (ca. 500 M^{-1}); 1-hexene (2 M^{-1}); styrene (20 M^{-1}); methyl acrylate (3 M^{-1}). The binding constants of arenes are significantly higher than those of simple alkenes and the binding of styrene is clearly attributable primarily to the phenyl ring (also reflected in the similarities of the spectra of the benzene and styrene adducts).

$$[Rh(DIPHOS)]^+ + UNSAT \overset{K_4}{\rightleftarrows} [Rh(DIPHOS)(UNSAT)]^+ \quad (4)$$

[Rh(DIPHOS)]⁺-Catalyzed Hydrogenation

$[Rh(DIPHOS)]^+$ was found to be an effective catalyst for the hydrogenation of simple alkenes as well as various alkene derivatives (styrene, acrylic acid, amidoacrylic acids, etc.). Kinetic measurements on the hydrogenation of 1-hexene (in which the hydrogen uptake was monitored), in conjunction with the equilibrium measurements of the type cited earlier, support the mechanistic scheme of Reactions 5 and 6 which yields the observed rate law, Equations 7 and 8, where $[Rh]_{Tot} = [Rh$-

(DIPHOS)]$^+$ + [Rh(DIPHOS)($\text{C}=\text{C}$)]$^+$. The linear plot of (Rate)$^{-1}$ vs. [1-hexene]$^{-1}$ in Figure 2 is consistent with Equation 8. The values of k_6 and K_4 for 1-hexene in methanol, derived from the slopes and intercepts of such plots, are 0.18 atm^{-1} and 1.6 M^{-1}, respectively. The latter value is in good agreement with the spectrophotometric value (*see* above). Kinetic studies on other substrates are in progress.

$$[\text{Rh(DIPHOS)}]^+ + \overset{}{\underset{}{\text{C}}}=\overset{}{\underset{}{\text{C}}} \underset{\rightleftarrows}{\overset{K_4}{}}$$

$$[\text{Rh(DIPHOS)}(\text{C}=\text{C})]^+ \quad (\text{Rapid equilibrium}) \quad (5)$$

$$[\text{Rh(DIPHOS)}(\text{C}=\text{C})]^+ + \text{H}_2 \xrightarrow{k_6}$$

$$[\text{Rh(DIPHOS)}]^+ + \text{H}-\overset{}{\underset{}{\text{C}}}-\overset{}{\underset{}{\text{C}}}-\text{H} \quad (\text{Rate determining}) \quad (6)$$

$$\text{Rate} = \frac{-d[\text{C}=\text{C}]}{dt} = \frac{k_6 K_4 [\text{Rh}]_{\text{Tot}}[\text{C}=\text{C}][\text{H}_2]}{1 + K_4 [\text{C}=\text{C}]} \quad (7)$$

$$\frac{[\text{Rh}]_{\text{Tot}}[\text{H}_2]}{[\text{Rate}]} = \frac{1}{k_{\text{obs}}} = \frac{1}{k_6} + \frac{1}{k_6 K_4 [\text{C}=\text{C}]} \quad (8)$$

The different reactivities of [Rh(DIPHOS)(NOR)]$^+$ and [Rh(PPh$_3$)$_2$(NOR)]$^+$ toward hydrogen, reflected in Reactions 1 and 2, are intriguing as well as being relevant to the mechanistic features of the catalytic hydrogenation reactions of the two species. A possible explanation of this difference is that whereas [Rh(PPh$_3$)$_2$]$^+$ can form a dihydrogen adduct of Structure 3 in which neither hydrogen ligand is trans to a phosphine ligand (*8, 11*), this is not possible (assuming cis disposition of the two hydrogen atoms) in the case of a chelating diphosphine ligand in which the two phosphorus atoms are constrained to being in mutually cis

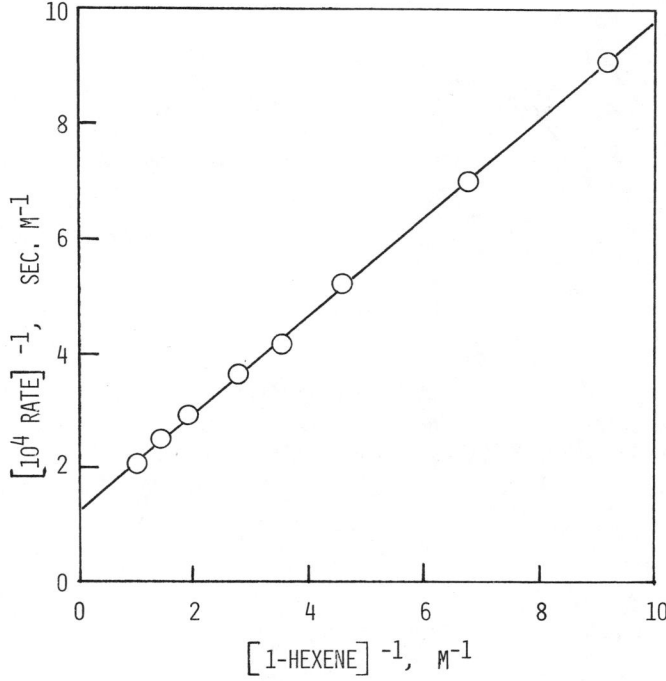

Figure 2. Plot of rate^{-1} vs. [1-hexene]$^{-1}$ for the [Rh-(DIPHOS)]$^+$-catalyzed hydrogenation of 1-hexene in methanol at 25°C (4.3 × 10^{-4}M [Rh(DIPHOS)]$^+$; 1 atm H$_2$)

positions. This is expected to contribute to the instability of the hydrogen adduct of [Rh(DIPHOS)]$^+$ and to result in a considerably reduced equilibrium constant for the oxidative addition of hydrogen to **2**, apparently to the point where the hydride cannot be detected. This reasoning suggests that [Rh(DIPHOS)]$^+$ should, however, be capable of the facile oxidative addition of one hydrogen ligand, i.e., of H$^+$. In accord with this expectation it was found that the addition of a noncoordinating acid such as HBF$_4$, HBF$_6$, or HClO$_4$ to a methanol or acetonitrile solution of [Rh(DIPHOS)]BF$_4$ reversibly discharged the color of the [Rh-(DIPHOS)]$^+$ ion, the spectral changes being quantitatively identifiable with the reversible equilibrium of Reaction 9, with K_9 (MeOH) = 11 ± 2 M^{-1}. The ^{1}H NMR spectrum of [HRh(DIPHOS)]$^{2+}$ in acetonitrile clearly revealed the hydride ligand coupled to the rhodium atom and to two equivalent phosphorus atoms (δ − 15.7, J_{Rh-H} = 12.1 Hz, J_{P-H} = 17.2 Hz, also confirmed by ^{31}P NMR), in accord with Structure **5**.

$$[\text{Rh}(\text{DIPHOS})]^+ + \text{H}^+ \underset{}{\overset{K_9}{\rightleftarrows}} [\text{HRh}(\text{DIPHOS})]^{2+} \quad (9)$$

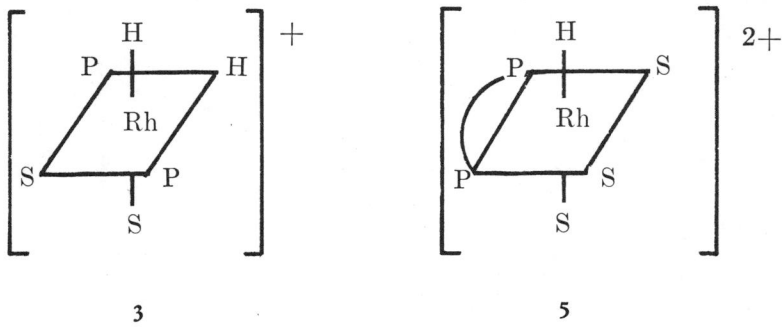

3   5

It should be noted that our mechanism for the [Rh(DIPHOS)]$^+$-catalyzed hydrogenation of alkenes departs significantly from that invoked for the corresponding [Rh(PPh$_3$)$_2$]$^+$-catalyzed reaction in which a principal pathway involves the hydrido-complex, [RhH$_2$(PPh$_3$)$_2$(Solvent)$_n$]$^+$ (*11*).

Characterization of Catalyst–Substrate Intermediates

These studies reveal a number of previously unrecognized features of the coordination chemistry and catalytic activity of cationic rhodium complexes containing chelating diphosphine ligands which differ strikingly from the chemistry of the corresponding monodentate phosphine complexes. The chemistry of these complexes in relatively poorly coordinating solvents such as methanol, which are typically used for catalytic hydrogenation, appears to be dominated by their "ligand deficiency" as reflected in the formation of unusual polynuclear species such as [Rh$_2$-(DIPHOS)$_2$]$^{2+}$ and [Rh$_3$(DIPHOS)$_3$(OMe)$_2$]$^+$, and in the strong binding of typically poor ligands such as arenes (*16*). It seems likely that the striking stereoselectivity which these catalysts exhibit in the asymmetric catalytic hydrogenation of prochiral olefins such as amidoacrylic and amidocinnamic acids reflects the strong tendency of the functional groups typically present in such substrates (C$_6$H$_5$, COOR, NHCOR, etc.) to "coordinate" to the Rh (as has been demonstrated in the comparison of styrene and 1-hexene) and thereby to exert a pronounced "orienting" influence. Our identification of Reaction 4 opens up the possibility of the direct systematic investigation of the effects of various substituents of olefinic substrates both on the equilibrium constants for the binding of the substrate (K_4) and on the structural features (potentially susceptible to elucidation both by NMR and by x-ray diffraction) of the resulting

[Rh(DIPHOS)(C=C)]⁺ adducts which are key intermediates in the catalytic hydrogenation. Such an adduct of [Rh(DIPHOS)]⁺ has been isolated in the case of Z-methyl-α-acetamidocinnamate and its structure, determined by single crystal x-ray diffraction analysis, is depicted in Figure 3. A notable feature of the structure, confirming earlier suggestions (2, 21), is the binding of the substrate to the rhodium atom through the amido-oxygen atom as well as through the C=C bond. The determination of the structure of the corresponding chiral adduct of [Rh-(DIPAMP)]⁺ is in progress. Preliminary results, including comparisons of various spectral features of the DIPHOS and DIPAMP adducts, and the kinetics of the corresponding hydrogenation reactions, suggest that the structural features of the two adducts and the mechanistic features of their hydrogenation reactions are similar.

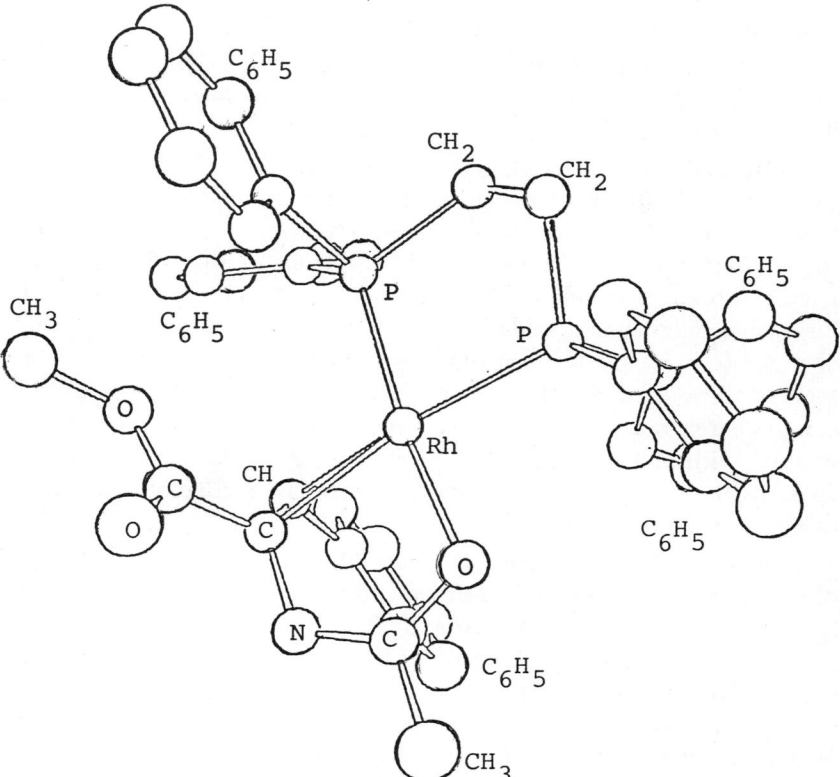

Figure 3. Structure of [Rh(DIPHOS)(Z-methyl-α-acetamidocinnamate)]⁺

Experimental

[Rh(DIPHOS)(NOR)]BF$_4$. This was prepared by adaptation of the procedure used by Schrock and Osborn (*10*) to prepare the corresponding perchlorate salt.

[Rh$_2$(DIPHOS)$_2$][BF$_4$]$_2$. A sample of 0.5 g [Rh(DIPHOS)(NOR)][BF$_4$] was dissolved in 150 mL of methanol. Hydrogen was bubbled through the solution for about 10 min. The resulting orange solution was evaporated under reduced pressure to about 20 mL, and 50 mL of degassed diethyl ether was added to precipitate the crude product which was recrystallized from methanol ether, washed with ether, and dried in vacuo (yield ca. 55%). Single crystals of this compound were obtained and subjected to x-ray structure analysis (*17*).

[Rh(DIPHOS)(BENZENE)]BF$_4$. A sample of 0.3 g of [Rh$_2$(DIPHOS)$_2$]BF$_4$ was dissolved in 50 mL warm methanol and the solution was evaporated by pumping to saturation (ca. 20–30 mL). 100 mL benzene was added and the resulting solution was stirred for about 20 min and evaporated to about 50 mL. Methanol was added to redissolve some oily material which had separated, and 40 mL of diethyl ether was added slowly with stirring. Upon standing for 2 hr, orange crystals of the product precipitated, which were filtered, washed with ether, and dried in vacuo at 50°C (yield ca. 80%).

[Rh$_3$(DIPHOS)$_3$(OCH$_3$)$_2$]PF$_6$. A sample of 0.4 g of [Rh(DIPHOS)(NOR)]PF$_6$ was dissolved in 100 mL of degassed methanol under nitrogen in a Schlenk apparatus. Hydrogen was bubbled through the solution for several minutes until the color of the solution changed from red-orange to yellow. The solution was filtered and 0.126 mL of dicyclohexylamine was added to the filtrate. The color of the solution changed from yellow to deep red-brown, and a red-brown product began to crystallize immediately. After stirring for a few minutes to ensure complete reaction, the red-brown crystals were collected by filtration, washed with methanol and diethyl ether, and dried in vacuo (yield ca. 79%). Single crystals were grown from acetone-methanol and subjected to x-ray structure analysis (*17*).

[Rh(DIPHOS)(Z-methyl-α-acetamidocinnamate)]BF$_4$. A sample of 0.2 g of [Rh$_2$(DIPHOS)$_2$][BF$_4$]$_2$ and 0.1 g of Z-methyl-α-acetamidocinnamate were stirred in ca. 10 mL of warm methanol until a clear red solution was obtained. After cooling to room temperature, diethyl ether was added slowly until the solution turned slightly cloudy, followed by addition of just enough methanol to clarify the solution. Red crystals of [Rh(DIPHOS)(Z-methyl-α-acetamidocinnamate)][BF$_4$] were grown by diffusing diethyl ether into the solution over a period of ca. 20 hr and subjected to x-ray structure analysis.

Measurements. UV-visible spectra were recorded with a Cary 14 spectrophotometer. ^{1}H NMR spectra were determined with a Bruker HS-270 spectrometer and ^{31}P NMR spectra with a Bruker HFX-90 spectrometer. The kinetics of hydrogenation were determined by measuring the rate of uptake of hydrogen gas volumetrically at constant pressure. Data for the single crystal x-ray structure determinations were collected on a Syntex P_{21} diffractometer.

Acknowledgment

We are grateful to the National Science Foundation for support of this research and to W. S. Knowles for generous gifts of samples of DIPAMP and of Z-methyl-α-acetamidocinnamate. The NMR facilities used in this research were supported in part through the University of Chicago Cancer Center Grant NIH-CA-14599.

Literature Cited

1. Knowles, W. S., Sabacky, M. J., Vineyard, B. D., ADV. CHEM. SER. (1974) 132, 274.
2. Kagan, H. B., *Pure Appl. Chem.* (1975) 43, 401 and references therein.
3. Haines, L. M., *Inorg. Nucl. Chem. Lett.* (1969) 5, 399.
4. Haines, L. M., *Inorg. Chem.* (1970) 9, 1517.
5. Haines, L. M., *Inorg. Chem.* (1971) 10, 1685.
6. Johnson, B. F. G., Lewis, J., White, D. A., *J. Am. Chem. Soc.* (1969) 91, 2816.
7. Schrock, R. R., Ph.D. thesis, Harvard University (1971).
8. Shapley, J. R., Schrock, R. R., Osborn, J. A., *J. Am. Chem. Soc.* (1969) 91, 2816.
9. Schrock, R. R., Osborn, J. A., *Inorg. Chem.* (1970) 9, 2339.
10. Schrock, R. R., Osborn, J. A., *J. Am. Chem. Soc.* (1971) 93, 2397.
11. Schrock, R. R., Osborn, J. A., *J. Am. Chem. Soc.* (1976) 98, 2134.
12. Ibid., 2143.
13. Ibid., 4450.
14. Knowles, W. S., Sabacky, M. J., Vineyard, B. D., Weinkauff, D. J., *J. Am. Chem. Soc.* (1975) 97, 2567.
15. Vineyard, B. D., Knowles, W. S., Sabacky, M. J., Bachman, G. L., Weinkauff, D. J., *J. Am. Chem. Soc.* (1977) 99, 5946.
16. Fryzuk, M. D., Bosnich, B., *J. Am. Chem. Soc.* (1977) 99, 6262.
17. Halpern, J., Riley, D. P., Chan, A. S. C., Pluth, J. J., *J. Am. Chem. Soc.* (1977) 99, 8055.
18. Nolte, N. J., Gainer, G., Haines, L. M., *Chem. Commun.* (1969) 1406.
19. Green, M., Kuc, T. A., *J. Chem. Soc., Dalton Trans.* (1972) 832.
20. Feltham, R. D., Hayter, R. G., *J. Chem. Soc.* (1964) 4587.
21. Brown, J. M., Chalmer, P. A., *J. Chem. Soc., Chem. Commun.* (1978) 321.

RECEIVED June 2, 1978.

3

Rhodium–Phosphine Complexes as Homogeneous Catalysts. Hydrogenation of Aromatic Nitro Compounds[1]

PÁL KVINTOVICS, BÁLINT HEIL, and LÁSZLÓ MARKÓ

Department of Organic Chemistry, University of Chemical Engineering, H-8200 Veszprém, Hungary

> *Addition of triethyl amine to $Rh(PPh_3)_3Cl$ or to complexes formed from $[Rh(1,5\text{-}hexadiene)Cl]_2$ and phosphines under hydrogen yields very active catalysts for the hydrogenation of aromatic nitro compounds to amines. The dark brown homogeneous catalyst solutions show highest activity at molar ratios of $Rh/PR_3/Et_3N = 1:1.2:3$. Turnovers above 1 mol H_2/mol Rh min are achieved.*

Only a few homogeneous catalysts have been found to be active for the hydrogenation of the nitro group (2) and also with these, only low reaction rates have been observed in most cases. Perhaps the best homogeneous catalyst described until now for the hydrogenation of nitro compounds is $RuCl_2(PPh_2)_3$ but even this requires rather drastic conditions (3). There is no report about the application of the most widely used homogeneous hydrogenation catalyst $Rh(PPh_3)_3Cl$ for this purpose.

Results and Discussion

The hydrogenation of nitrobenzene with $Rh(PPh_3)_3Cl$ as catalyst was found to be impracticably slow with turnovers below 0.1 min^{-1}. The addition of triethyl amine, however, considerably increased the reaction rate, in accordance with our earlier findings about the favorable effect

---

[1] This is Part 6 of a series. For Part 5 see Ref. 1.

Table I. Hydrogenation of Nitrobenzene and Its Possible Reduction Intermediates with the Rh(PPh$_3$)$_3$Cl + Et$_3$N Catalyst System[a]

Substrate	Solvent	Et$_3$N/Rh	ITO(min^{-1})[b]
Nitrobenzene	benzene	0:1	0.02
Nitrobenzene	benzene/MeOH (1:1)	0:1	0.08
Nitrobenzene	benzene/MeOH (1:1)	3:1	0.41
Nitrobenzene	benzene/MeOH (1:1)	30:1	0.48
Nitrobenzene	benzene/MeOH (1:1)	150:1	0.77
Nitrosobenzene	benzene/MeOH (1:1)	3:1	0.02
N-Phenyl hydroxyl amine	benzene/MeOH (1:1)	3:1	0.05
Azobenzene	benzene/MeOH (1:1)	3:1	0.57
Hydrazobenzene	benzene/MeOH (1:1)	3:1	0.35

[a] 0.025 mmol Rh(PPh$_3$)$_3$Cl, 2.5 mmol substrate, 6 mL solvent, 50°C, and 1 bar H$_2$.
[b] Initial turnover (mol H$_2$/mol Rh min).

of triethyl amine on the activity of rhodium phosphine complex catalysts (1) (Table I).

Aniline was found by GLC to be the only reaction product. The well-known possible intermediates nitrosobenzene, N-phenyl hydroxylamine, azobenzene, and hydrazobenzene could all be hydrogenated by the same catalyst combination to aniline. Comparing the rates of hydrogenations (Table I), the only conclusion one can reach about the mechanism of nitrobenzene hydrogenation is that nitrosobenzene and N-phenyl hydroxylamine (and thus probably also azobenzene and hydrazobenzene) are not intermediates of the reaction.

Even more active was the 'in situ" catalyst prepared from [Rh(1,5-hexadiene)Cl]$_2$ and PPh$_3$ with a low ratio of P/Rh = 1.2, and the use of a p-xylene/methanol solvent instead of benzene/methanol resulted in a further increase of activity (Table II). Both effects already have been observed (1, 4) which shows that these may be characteristic for such rhodium, phosphine, and amine catalyst combinations.

The homogeneous character of the catalyst was checked in several ways: no precipitate or deposition could be observed visually; when filtrating a catalytically active solution, no residue remained on the filter; if no PPh$_3$ was added, a visually well observable, grey precipitate appeared (probably rhodium metal) which proved to be much less active

Table II. Hydrogenation of Nitrobenzene with in situ Rh +

No.	Phosphine	Amine	Rh/P/N
1	PPh$_3$	Et$_3$N	1:1.2:3
2	PPh$_3$	Et$_3$N	1:1.2:3
3	PPh$_3$	Et$_3$N	1:1.2:3
4	PPh$_3$	Et$_3$N	1:1.2:3
5	PPh$_3$	Et$_3$N	1:2.2:3
6	PPh$_3$	Et$_3$N	1:3.2:3
7	—	Et$_3$N	1:0:3
8	PPh$_3$	Et$_3$N	1:1.2:3
9	PPh$_3$	Et$_2$NH	1:1.2:3
10	PPh$_3$	pyridine	1:1.2:3
11	PPh$_3$	pyperidine	1:1.2:3
12	PBu$_3$	Et$_3$N	1:1.2:3
13	Ph$_2$PCH$_2$CH$_2$PPh$_2$	Et$_3$N	1:1.2:3
14	P(OMe)$_3$	Et$_3$N	1:1.2:3

[a] 0.0125 mmol [Rh(1,5-hexadiene)Cl]$_2$, 2.5 mmol nitrobenzene, 6 mL solvent, 50°C, and 1 bar H$_2$.
[b] Initial turnover (mol H$_2$/mol Rh min).
[c] Time necessary for the consumption of 3.75 mmol H$_2$ (50% conversion or turnover number 150).

as a hydrogenation catalyst than the homogeneous system (compare experiments 2 and 7 of Table II).

Some additional information on the nature of the active catalyst was furnished by the experiments at different rhodium concentrations (Table II, No. 2, 3, and 4). As can be seen, the reaction rate showed a small fractional order with respect to catalyst concentration within the range investigated. This suggests that most of the rhodium is present in the form of a metal cluster complex and the active catalyst is a mononuclear species present in small concentration and in equilibrium with

Table III. Hydrogenation of Some

Substrate	ITO(min^{-1})[b]	t$_{0.5}$(min)[c]
Nitrobenzene	3.4	60
1-Nitro naphthalene	0.80	450
o-Nitro phenol	0.90	250
p-Nitro chloro benzene	1.4	230
m-Dinitro benzene	1.5	410

[a] 0.0125 mmol [Rh(1,5-hexadiene)Cl]$_2$, 0.03 mmol PPh$_3$, 0.075 mmol Et$_3$N, 2.5 mmol substrate, 3 mL benzene, 4 mL methanol, 50°C, and 1 bar H$_2$.
[b] Initial turnover (mol H$_2$/mol Rh min).
[c] Time necessary for the consumption of 3.75 mmol H$_2$ (50% conversion of one nitro group or turnover number 150).

Phosphine + Amine Catalysts. Effect of Reaction Parameters[a]

Solvent	$ITO(min^{-1})$[b]	$t_{0.5}(min)$[c]
benzene	1.6	80
benzene/MeOH (1:1)	3.4	70
benzene/MeOH (1:1)	3.8[d]	60
benzene/MeOH (1:1)	4.5[e]	40
benzene/MeOH (1:1)	0.57	—
benzene/MeOH (1:1)	0.38	—
benzene/MeOH (1:1)	0.35[f]	—
p-xylene/MeOH (1:1)	6.7	30
benzene/MeOH (1:1)	3.1	70
benzene/MeOH (1:1)	0.70	310
benzene/MeOH (1:1)	2.7	70
benzene/MeOH (1:1)	0.16	—
benzene/MeOH (1:1)	1.5	120
benzene/MeOH (1:1)	4.1	50

[d] $Rh/PhNO_2 = 1:50$.
[e] $Rh/PhNO_2 = 1:33$.
[f] Rhodium metal precipitated from the solution.

the polynuclear complex. Such a system would at the same time also explain the dark color of the reaction mixture.

Triethyl amine performed as the best amine component, and PBu_3 and $Ph_2PCH_2CH_2PPh_2$ (diphos) were less useful than PPh_3. Trimethyl phosphite gave, however, a rather active catalyst—an observation also not without precedent (1).

The dark brown "in situ" catalyst solution was successfully applied for the hydrogenation of other aromatic nitro compounds too (Table III). The formation of aniline as the main product from p-chloro nitro-

Aromatic Nitro Compounds[a]

H_2 Consumed, mol/mol Substrate	Product	Yield (%)[d]
3.0	aniline	90
1.9[e]	1-amino-naphthalene	52
1.8[e]	o-amino phenol	55
2.2[e]	p-chloro aniline	9
	aniline	48
1.8[e]	m-nitro aniline	57
	m-phenylene diamine	1

[d] Based on total amount of substrate.
[e] Experiment stopped before complete conversion.

benzene deserves mentioning: the HCl acceptor necessary for hydrodehalogenation (4) is apparently the aromatic amine furnished by the reduction of the nitro group, thus enabling the combination of two functions of the same catalyst.

Experimental

Five and one half mg (0.0125 mmol) [Rh(1,5-hexadiene)Cl]$_2$, 7.9 mg (0.03 mmol) PPh$_3$, and 10.6 μL (0.075 mmol) Et$_3$N were dissolved at 50°C in 3.0 mL methanol under H$_2$ in a thermostated reaction flask connected to a thermostated gas burette and equipped with a magnetic stirrer and a silicone rubber cap. This catalyst solution was prehydrogenated for 30 minutes. Following this the substrate (2.5 mmol) was added with a syringe and the reaction was followed by measuring hydrogen consumption. The reaction product was analyzed by GLC.

Literature Cited

1. Nagy-Magos, Z., Vastag, S., Heil, B., Markó, L., *Transition Met. Chem.* (1978) **3**, 123.
2. James, B. R., "Homogeneous Hydrogenation," Wiley, New York, 1973.
3. Knifton, J. F., *J. Org. Chem.* (1976) **41**, 1200.
4. Kvintovics, P., Heil, B., Palágyi, J., Markó, L., *J. Organomet. Chem.* (1978) **148**, 311.

RECEIVED February 22, 1978.

Pentamethylcyclopentadienyl–Rhodium and –Iridium Complexes as Catalysts for Olefin and Arene Hydrogenation

PETER M. MAITLIS

Department of Chemistry, The University, Sheffield, S3 7HF, England

> *A new series of hydrogenation catalysts based on pentamethylcyclopentadienyl-rhodium and -iridium complexes are described. These are derived from $[M(C_5Me_5)Cl_2]_2$ (1a, M = Rh; 1b, M = Ir) which are obtained from hexamethyl Dewar benzene in a two-step reaction. These complexes homogeneously hydrogenate simple olefins at 1 atm and 20°C in the presence of base; polar noncoordinating solvents give the best rates. Benzene and substituted benzenes are also hydrogenated to cyclohexanes by 1a under more vigorous conditions. The mechanism of the olefin hydrogenation reaction has been investigated and shown to involve mononuclear species such as $[M(C_5Me_5)H_2(solv)]$. The binuclear μ-hydrido complexes $[HM_2(C_5Me_5)_2Cl_3]$, $[H_2Ir_2(C_5Me_5)_2Cl_2]$, and $[H_3Ir_2(C_5Me_5)_2]^+$ are much poorer olefin hydrogenation catalysts than 1 and the reasons for this are discussed.*

For a metal complex to be an effective catalyst under homogeneous conditions it must be able to survive cycles of changes of both coordination number and oxidation state without formation of metal occurring. To ensure a reasonable rate of reaction, the complex must have easily accessible vacant coordination sites, which implies that at least some of the ligands must be readily removed; on the other hand, some ligands must remain firmly bound since otherwise, especially under reducing conditions, metal will be formed. Triorgano-phosphine or -phosphite ligands have most commonly been used to stabilize complexes (especially of group VIII metals) as homogeneous catalysts. A good example of such a compound is the Wilkinson hydrogenation catalyst, [Rh(PPh$_3$)$_3$Cl] (e.g., see Ref. 1 and references therein). While such catalysts work well in many cases, they suffer from some disadvantages; for example, since these types of ligands are themselves highly reactive (towards electrophiles, acids, and oxygen), any reactions wherein ligands are displaced, such as occur in catalytic cycles, can lead to side products and eventually to catalyst deactivation.

We have, over the past few years, been developing a new series of homogeneous catalysts that do not contain such phosphine or phosphite ligands (and that are indeed deactivated by them), based on pentamethylcyclopentadienyl-rhodium and -iridium compounds. The parent complexes [M(C$_5$Me$_5$)Cl$_2$]$_2$ (1, M = Rh or Ir) are readily obtained from hexamethyl Dewar benzene and the appropriate hydrated metal chloride in a two-step reaction (2).

1a, M = Rh

1b, M = Ir

These compounds are air stable, diamagnetic (M(III), low spin, d^6), crystalline orange-red (Rh) or yellow-orange (Ir) solids and have good solubility in more polar organic solvents. Their most interesting characteristic is the extreme inertness of the C$_5$Me$_5$–M bond and the high reactivity of the other ligands, which allows them to undergo a variety of reactions (under acidic and basic, reducing and oxidizing and metathesis conditions) while retaining the C$_5$Me$_5$ ligand and without reduction to metal occurring. Examples of such reactions are given in Scheme 1. We ascribe the inertness of the C$_5$Me$_5$–M bond to a combination of electronic and steric effects. The methyls are electron releasing and thus can be expected to stabilize the + III oxidation state particularly well; they will also shield the metal from attack better than hydrogens can and for both reasons, a C$_5$Me$_5$ can be expected to stabilize a complex better than a C$_5$H$_5$ ligand. This is found in practice, and [Rh(C$_5$H$_5$)Cl$_2$]$_n$, as well as being insoluble and amorphous, is also very much more reactive than 1a; for example, it easily gives metal on exposure to H$_2$.

4. MAITLIS *Olefin and Arene Hydrogenation*

Scheme 1

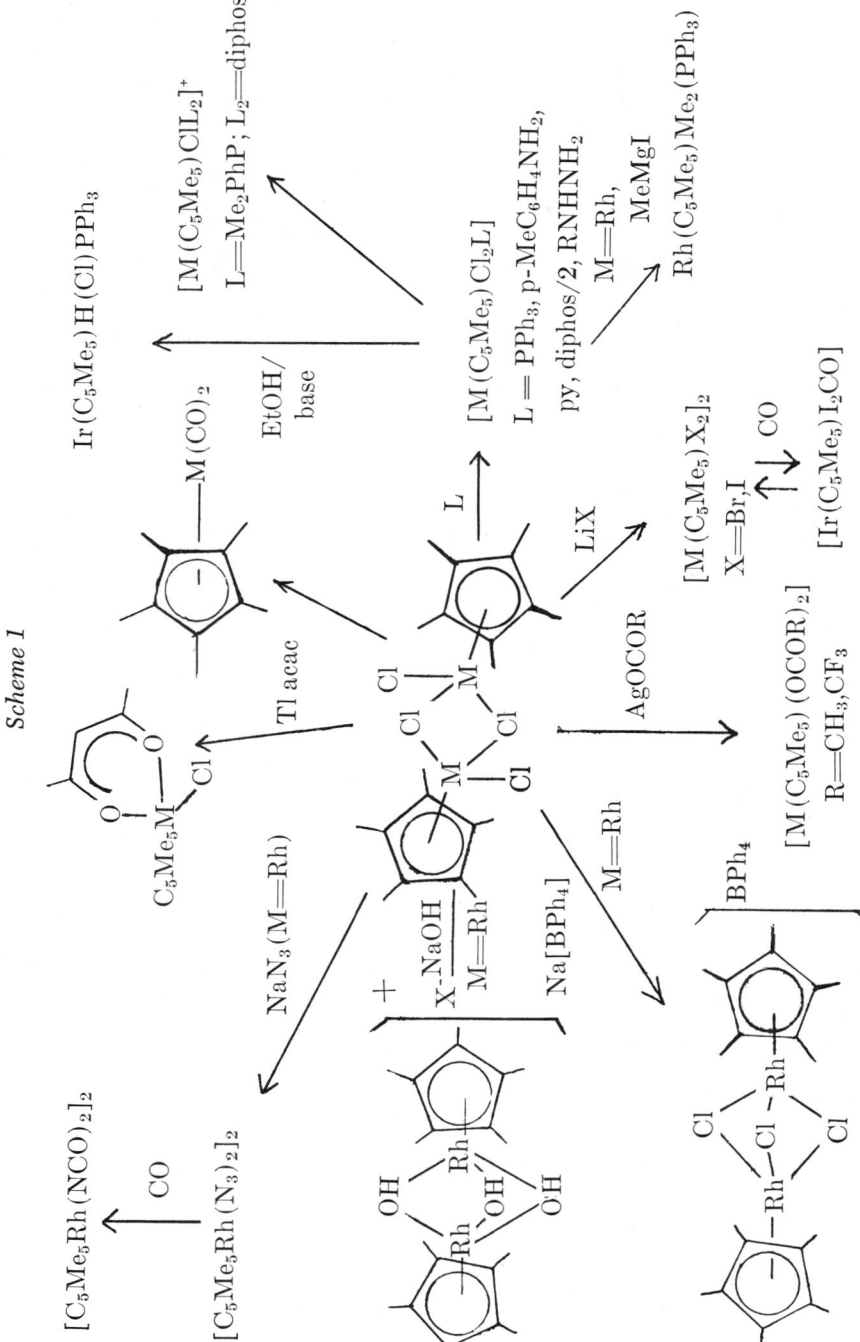

Hydrogen-Transfer Reactions and Metal Hydrides

The first indication that the compounds (1) were able to promote hydrogen-transfer reactions came from our observation that, in alcohols in the presence of base, they reacted with dienes to give enyl (allylic) complexes (3), e.g.,

$$[M(C_5Me_5)Cl_2]_2 + 2CH_2=CHCH=CH_2 + 2C_2H_5OH + \text{base} \rightarrow$$

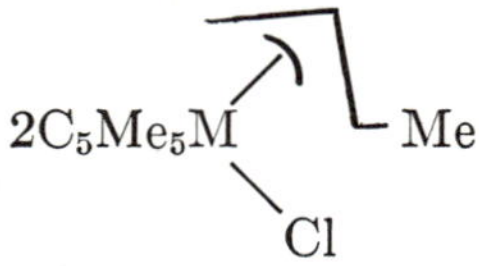

$$2C_5Me_5M\begin{pmatrix} \nearrow \\ \diagdown \\ Cl \end{pmatrix}Me + 2CH_3CHO + 2[\text{base H}]Cl$$

Since this reaction effectively involved the addition of "H–M(C_5Me_5)-Cl" to $\diagup C = C \diagdown$, we then sought to prepare such hydride complexes. A number of such hydrides have now been obtained (4, 5); for rhodium they include the following:

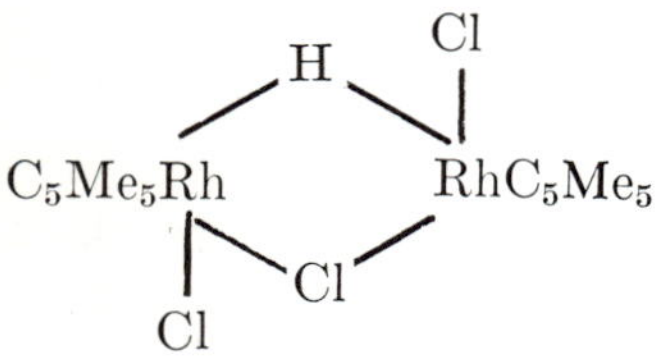

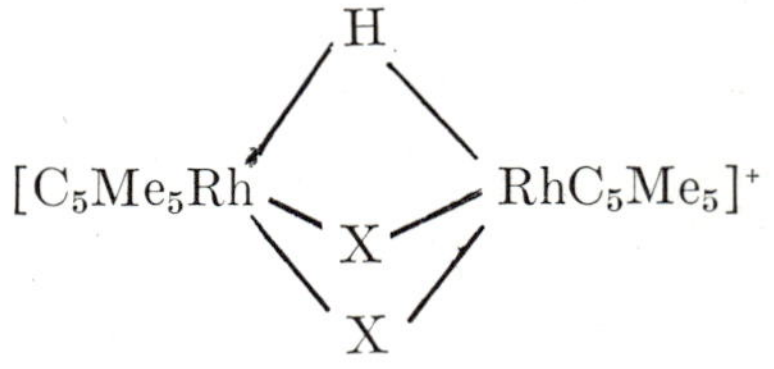

$$C_5Me_5Rh \begin{matrix} H \\ \diagup \diagdown \\ Cl \end{matrix} \begin{matrix} Cl \\ | \\ RhC_5Me_5 \\ Cl \end{matrix} \qquad [C_5Me_5Rh \begin{matrix} H \\ \diagup \diagdown \\ X \\ X \end{matrix} RhC_5Me_5]^+$$

$$(X = CH_3CO_2 \text{ or } CF_3CO_2)$$

2a

while a more extensive series is known for iridium, e.g.,

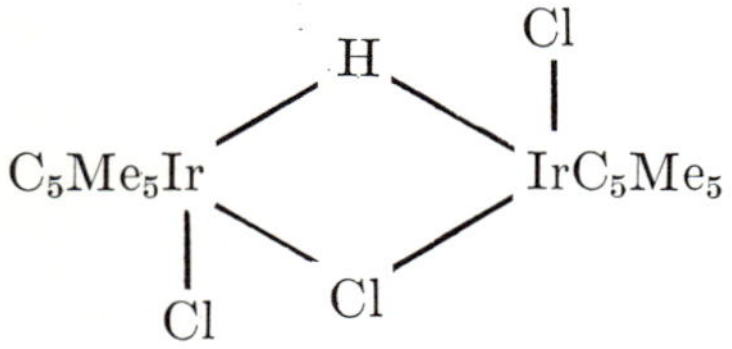

2b 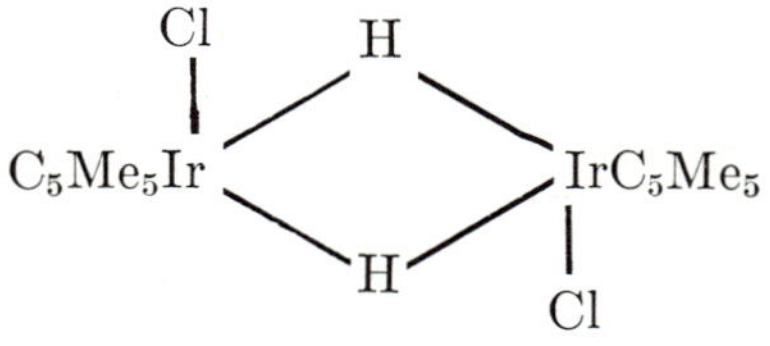   **3**

4. MAITLIS Olefin and Arene Hydrogenation

$$[C_5Me_5Ir\underset{H}{\overset{H}{\diamond}}IrC_5Me_5]^+ \qquad [C_5Me_5Ir\underset{Y}{\overset{H}{\diamond}}IrC_5Me_5]^+$$

$$(X = Y = OCOCH_3 \text{ or } OCOCF_3)$$

4 $\qquad X = H; Y = OCOCH_3 \text{ or } OCOCF_3)$

A variety of methods can be used to prepare these hydrides; they can be formed by the action of H_2/base on the chlorides (*1*) or by reaction with BH_4^- or KOH/2-propanol (*4*). It can be noted that the formation of the hydrides from molecular H_2 overall involves a heterolytic activation and that the role of the base is to combine with the acid

$$[MC_5Me_5Cl_2]_2 + H_2 \rightleftharpoons [H(MC_5Me_5)_2Cl_3] + HCl$$

formed to prevent the back reaction. The acetates activate hydrogen very much more easily and do not require the presence of base since the weaker acetic acid formed.

$$\frac{1}{n}[MC_5Me_5(OAc)_2]_n + H_2 \rightleftharpoons [H(MC_5Me_5)_2(OAc)_2]^+H(OAc)_2^-$$

These complexes are air stable as solids and have all been characterized spectroscopically; four have also had their structures determined by x-ray diffraction (*6, 7, 8, 9, 10*). A characteristic of these binuclear complexes is that the hydride always occupies a bridging position between the two metals atoms.

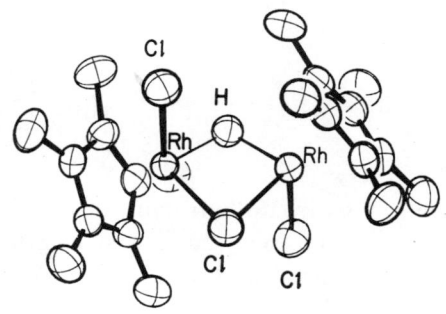

Journal of the American Chemical Society

Figure 1. A general view of [H{Rh(η^5-C_5Me_5$)}_2Cl_3$] (2a) with methyl hydrogens omitted (6)

Stoichiometric Reactions of Hydrides

The hydrides (2) are moderately reactive towards 1,3-diolefins and the following stoichiometric reaction occurs readily in CH_2Cl_2 at $+20°C$:

$$[H(MC_5Me_5)_2Cl_3] + \text{diene} \rightarrow C_5Me_5M(\text{enyl})Cl + \tfrac{1}{2}[C_5Me_5MCl_2]_2$$

Kinetic studies indicated that the reaction was first order in the hydride but zero order in the diene; further, a variety of dienes were found to react at the same rate (11, 12). The absence of an isotope effect for reactions with $[D(MC_5Me_5)_2Cl_3]$, coupled with the observation that the di-μ-hydride (3) did not react with dienes in CH_2Cl_2 at 20°C (5), and the kinetic data, indicated the following points: (i) the rate-determining step is a reorganization of the complex (2) to create a vacant site at which reaction can occur; and (ii) this reorganization does not involve cleavage of either a M–H–M bridge or ionization of Cl^- to create the vacant site, otherwise the $[H_2(IrC_5Me_5)_2Cl_2]$ would be expected to react at a similar rate to $[H(IrC_5Me_5)_2Cl_3]$.

The simplest mechanism that can be put forward which covers these points is one where the rate-determining slow step is the opening of a chloride bridge bond. Coordination of a diene at the vacant site thus created and the subsequent H-transfer reaction to give the enyl complex are then expected to be very fast.

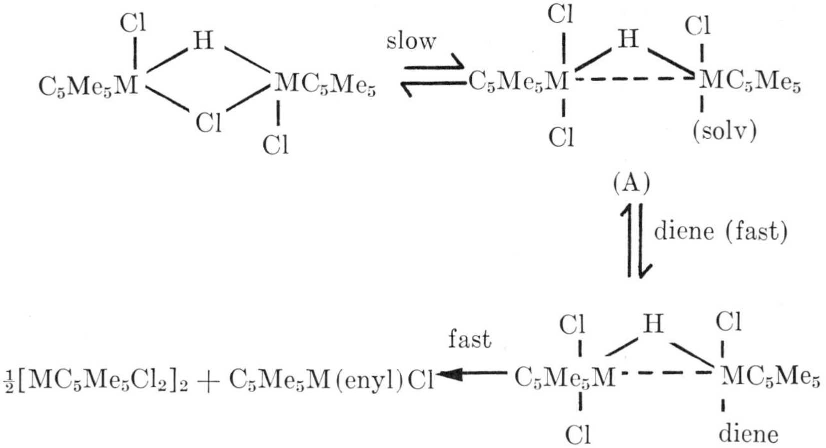

We are therefore faced by the intriguing result that it is apparently easier to break a two-electron, two-center M–Cl bridge than a M–H–M bridge that contains a two-electron, three-center bond. Indeed the transition state (A) contains two large $\{M(C_5Me_5)ClX\}$ entities joined only by one M–H–M bond, and clearly such bonds must be reinforced by M–M bonding. This point is also well illustrated by the very short Ir–Ir distance of 2.455 Å found in the tri-μ-hydrido complex $[H_3(IrC_5Me_5)_2]^+$ (10) which has been likened to a "triply protonated Ir≡Ir triple bond."

Olefin Hydrogenation Catalysts

Both the rhodium and iridium complexes are active catalysts for homogeneous hydrogenation of olefins in polar noncoordinating solvents (e.g., 2-propanol) in the presence of base at 1 atm H_2 and 20°C (*13*). Good coordinating solvents and ligands (R_3P, $(RO)_3P$, R_2S, etc.) deactivate the catalysts. It was originally expected that the most reactive compounds would be the hydrido complexes, but this was found not to be the case. Under the same conditions [for the hydrogenation of cyclohexene to cyclohexane: cyclohexene (2 mL), 2-propanol (18 mL), catalyst (0.05 mmol), H_2 (1 atm and 24°C), and Et_3N (0.22 mmol for the iridium and 4.5 mmol for the rhodium catalysts)], the reactivity measured by take-up of H_2 (mL min^{-1}) was: $[IrC_5Me_5Cl_2]_2$ (27.3) > $[H(IrC_5Me_5)_2Cl_3]$ (15.3) > $[H_2(IrC_5Me_5)_2Cl_2]$ (13.2) >> $[H_3(IrC_5Me_5)_2]^+$ (0); and $[RhC_5Me_5Cl_2]_2$ (9.8) > $[H(RhC_5Me_5)_2Cl_3]$ (5.1). By comparison, the rate of hydrogenation using $[Rh(PPh_3)_3Cl]$ as catalyst (in 2-propanol but no base) was 4.3 mL min^{-1}.

An interpretation of these at-first surprising results is that the chloride complexes (**1**) undergo bridge splitting to monomers more easily than do the μ-hydrido complexes (**2**, **3**, or **4**), and that it is a mononuclear system that is catalytically most active. This is consistent with our suggestion concerning the rate-determining step in the stoichiometric reaction of **2** with dienes and reinforces the idea that such bridging M–H–M bonds are strong and that the more there are, the more inert the complex becomes. This suggestion is complemented by the x-ray data which show that the metal–metal distances decrease from 3.719 Å (**1a**) or 3.769 Å (**1b**) (*7,8*) in the di-μ-chloro complexes to 2.90 Å (*6,7,8*) in the mono-μ-hydrido-μ-chloro complexes (**2a** and **2b**) and to 2.455 Å (*10*) in $[H_3(IrC_5Me_5)_2]^+$. Since the C_5Me_5 ligand shields the reaction sites moderately well (*see* below), the closer the two metals are pulled together, the more shielded and hence more deactivated will the metals become.

The kinetics of cyclohexane hydrogenation catalyzed by **1**, **2**, and **3** have been determined (*13*) and have been shown to follow the general form:

$$-\frac{d[H_2]}{dt} = \frac{k[\text{catalyst}]^a[p(H_2)]^b[\text{olefin}]}{1 + [\text{olefin}]}$$

where $a = 0.5$ for $[(MC_5Me_5)_2Cl_4]$ and $[H_2(IrC_5Me_5)_2Cl_2]$, $a = 1$ for $[H(MC_5Me_5)_2Cl_3]$; and $b = 1$ for $[H(MC_5Me_5)_2Cl_3]$ and $[H_2(IrC_5Me_5)_2Cl_2]$, $b = 2$ for $[(MC_5Me_5)_2Cl_4]$.

The results imply that Complexes **1** and **3** undergo cleavage and react as mononuclear complexes while the mono-μ-hydrides (**2**) react as binuclear complexes, at least during the initial stages of the reactions when measurements were made.

For the reactions of 1 and 3, the following reaction scheme can therefore be proposed (where $m = MC_5Me_5$ and (s) = solvent or vacant site).

Initiation:

$$[m_2Cl_4] \rightleftharpoons 2mCl_2(s)$$
$$mCl_2(s) + H_2 \rightleftharpoons HmCl(s) + HCl$$
$$HmCl(s) + H_2 \rightleftharpoons H_2m(s) + HCl$$
$$[H_2(IrC_5Me_5)_2Cl_2] \rightleftharpoons 2H(IrC_5Me_5)Cl(s)$$

Propagation:

$$H_2m(s) + \text{olefin} \underset{}{\overset{\text{slow}}{\rightleftharpoons}} H_2m(\text{olefin})$$
$$H_2m(\text{olefin}) \overset{\text{fast}}{\rightarrow} m(s)_n + \text{alkane}$$
$$m(s)_n + H_2 \rightleftharpoons H_2m(s)$$

or
$$m(s)_n + HCl \rightleftharpoons HmCl(s)$$

This scheme indicates that while the initiation steps involve heterolytic cleavage of H_2, during the actual hydrogenation and propagation steps the major reactions involve a reductive elimination $[H_2m^{III}(\text{olefin}) \rightarrow m^I(s)_n + \text{alkane}]$ and a subsequent (homolytic) oxidative addition of H_2 to one metal atom. The latter is, of course, the more usual process which has been established for $[Rh(PPh_3)_3Cl]$ catalyzed hydrogenation (*14, 15, 16*).

One chain termination reaction is also clear. It is found that the iridium catalysts are slowly converted into the catalytically inactive $[H_3(IrC_5Me_5)_2]^+$, (4), a process which is assisted by more strongly basic conditions. The complex (4) can then be regarded as formed by reactions such as:

$$H_2(IrC_5Me_5)(s) + H(IrC_5Me_5)Cl(s) \rightarrow [H_3(IrC_5Me_5)_2]^+$$

The formation of this tri-μ-hydrido complex also partly explains why the iridium complexes show maximum activity with ca. 4–5 equiv of base (Et_3N) per metal atom, after which the activity falls. In contrast, the rhodium complexes show a steady increase in activity with increase in base concentration until a plateau is reached at a ratio of ca. 10 equiv of base; this presumably reflects the instability of $[H_3(RhC_5Me_5)_2]^+$ since attempts to prepare it have failed.

The path followed by the hydrogenation reactions catalyzed by the mono-μ-hydrides (**2a** and **2b**) is not so clear, but the kinetic data strongly suggest that the binuclear grouping is retained, at least during the early stages of reaction. A possible first step is the formation of a rearranged complex, such as (A), proposed for the stoichiometric reaction with dienes.

A variety of monoolefins can be hydrogenated by these catalysts; at 20°C and 1 atm H_2, relative rates for catalysis by $[H(RhC_5Me_5)_2Cl_3]$ are found to be:

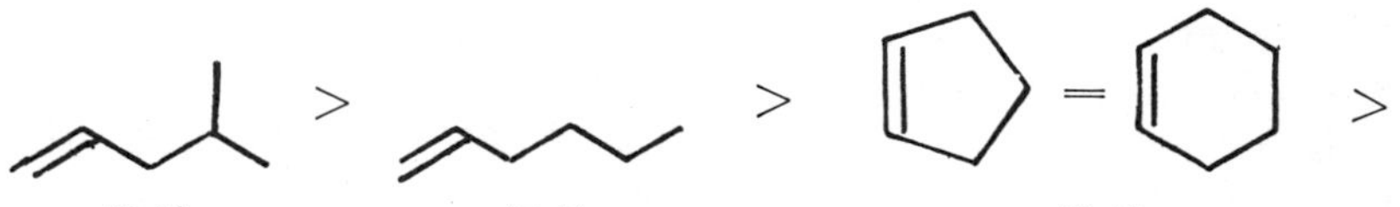

 isobutylene (9.1) > 1-pentene (7.1) > cyclopentene (5.1) = cyclohexene (5.1) >

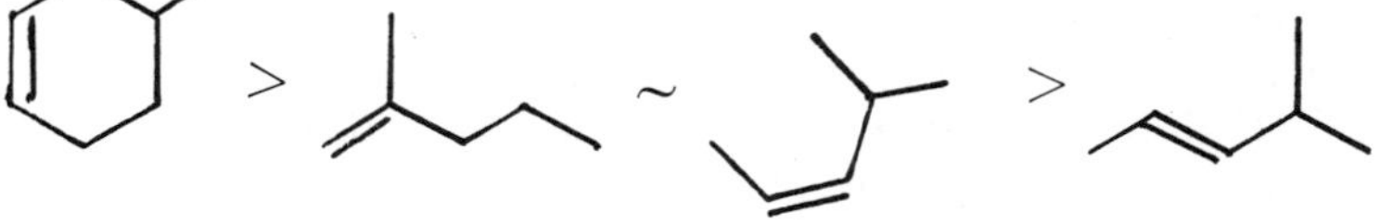

 cyclohexadiene (3.5) > 2-methyl-1-butene (2.8) ~ 2-methyl-2-butene (2.6) > 2-methyl-2-butene (1.4)

Functional groups tend to deactivate the catalysts but reduction can be achieved by working at 100 atm H_2 and 20°C; under these conditions, diolefins (e.g., vinylcyclohexene, 1,5-hexadiene) are reduced to paraffins (ethylcyclohexane, hexane), and unsaturated ketones or esters (mesityl oxide, vinyl acetate) are reduced (to methyl(isobutyl)ketone or ethyl acetate).

As indicated above, the iridium complexes are, in general, more active catalysts than their rhodium analogs; however, isomerization of terminal-to-internal olefins occurs rapidly, and relative rates are therefore rather meaningless. The mono-μ-hydrido iridium complex (**2b**) indeed will isomerize olefins even in the absence of hydrogen.

Arene Hydrogenation

While the homogeneous hydrogenation of olefins catalyzed by group VIII metal complexes is today a rather commonplace reaction and the C_5Me_5–M complexes offer mainly rather higher rates by comparison with the Wilkinson catalyst, very few homogeneous catalysts so far reported are active for the hydrogenation of arenes.

The best of these appear to be the ones developed by Muetterties and his group (*17*, *18*), such as [Co(η^3 − C$_3$H$_5$){P(OMe)$_3$}$_3$], which shows very high stereospecificity for all-cis hydrogenation. Unfortunately, it suffers from the disadvantages of low turnover numbers and short life (typically fewer than 20 moles of arene are hydrogenated per mole of catalyst) and is not easy to prepare.

We find that the pentamethylcyclopentadienyl-rhodium compounds, and particularly the chloride (**1a**), are useful catalysts for arene hydrogenation at 50°C and 50 atm H$_2$ (*19*). Again, base (Et$_3$N) is a necessary co-catalyst and the reactions proceed well in 2-propanol as solvent, where turnover numbers in excess of 200 per rhodium can easily be obtained for benzene-to-cyclohexane hydrogenation. The reactions also proceed in benzene, where turnover numbers in excess of 800 per rhodium are found, and even in the presence of water. No metal is formed under these conditions.

Alkylbenzenes are also hydrogenated to alkylcyclohexanes, the relative rates decreasing with increasing substitution at the ring; for example, benzene (100) ∼ toluene (94) > *o*-xylene (70) > ethylbenzene (58) > mesitylene (35) > *s*-butylbenzene (25) >> durene, hexamethylbenzene (0). These reactions are also stereoselective and, for example, give predominantly the cis-dimethylcyclohexanes from xylenes (e.g., *o*-xylene: cis/trans, 6.2; *m*-xylene: cis/trans, 3.8; and *p*-xylene: cis/trans, 2).

Our preliminary studies show that the catalyst system can tolerate at least some functionalities on the arene; thus, while substrates containing unprotected -OH or CO$_2$H groups are hydrogenated only to a small degree or not at all, aryl-ethers (anisole), -esters (methyl benzoate), -ketones (acetophenone, benzophenone), and *N*,*N*-dimethylaniline are all hydrogenated. In general, no products corresponding to states of intermediate reduction (cyclohexenes or cyclohexadienes) are detected.

It is also noteworthy that the iridium complex (**1b**) is a very much poorer catalyst than the rhodium analog. This point and the fact that the more highly alkyl-substituted benzenes are only hydrogenated to a small degree, even by [RhC$_5$Me$_5$Cl$_2$]$_2$, can be explained by the observations that: (i) iridium forms more stable complexes of the type [M(C$_5$Me$_5$)-(arene)]$^{2+}$ than rhodium does; and (ii) within a series of alkylbenzenes, the more highly substituted ones form the more stable arene complexes for both Rh and Ir (*20*, *21*). Presumably the more stable complexes are reduced less easily.

While the mechanism of the arene hydrogenation has not yet been elucidated, it is interesting to speculate that a path rather analogous to that suggested for olefin hydrogenation may be operative. In this case, the most active species is again probably [H$_2$(RhC$_5$Me$_5$)(*s*)] which can then coordinate (η^2-) arene at the vacant site; *cis*-hydrogen transfer (via

η^3-cyclohexenyl and η^4-cyclohexadiene species) can then take place, followed by further activation (homolytic) of H_2 and further cis-hydrogen transfer reactions. The predominance of the cis isomers in the xylene hydrogenation products suggests that the arene largely remains coordinated to the metal until it is finally expelled as the cyclohexane, but the presence of trans isomers implies that some exchange does occur during hydrogenation.

If this view is correct then the question arises, why is the Wilkinson catalyst [Rh(PPh$_3$)$_3$Cl] also not active for this reaction? A partial answer can be found in the size of the cone angle of the RhC$_5$Me$_5$ unit, which is estimated at 185°, while that of the grouping fac-{(Ph$_3$P)$_2$Cl}Rh (which is the minimum size of the probable active species in the Wilkinson catalyst) is about 230°. This suggests that the active site in the C$_5$Me$_5$Rh catalyst may be more "open" than the corresponding site in the Wilkinson catalyst and that the greater room on C$_5$Me$_5$Rh allows easier coordination of the arene. Such an argument would also explain the lower degree of discrimination observed for the hydrogenation of substituted olefins catalyzed by the C$_5$Me$_5$Rh and -Ir complexes; for example, [Rh(PPh$_3$)$_3$Cl] catalyzes the hydrogenation of cyclohexene about 50 times as fast as 1-methylcyclohexene whereas the corresponding factors are about three (for **1a**) and seven (for **1b**).

Acknowledgment

My warmest thanks go to C. White, J. W. Kang, K. Moseley, A. J. Oliver, H. B. Lee, D. S. Gill, S. J. Thompson, and M. J. H. Russell for their work in developing the chemistry of these complexes and their catalytic reactions.

Literature Cited

1. James, B. R., "Homogeneous Hydrogenation," J. Wiley, New York, 1973.
2. Kang, J. W., Moseley, K., Maitlis, P. M., *J. Am. Chem. Soc.* (1969) **91**, 5970.
3. Moseley, K., Kang, J. W., Maitlis, P. M., *J. Chem. Soc. A* (1970) 2875.
4. White, C., Oliver, A. J., Maitlis, P. M., *J. Chem. Soc., Dalton Trans.* (1973) 1901.
5. Gill, D. S., Maitlis, P. M., *J. Organomet. Chem.* (1975) **87**, 359.
6. Churchill, M. R., Ni, S. W., *J. Am. Chem. Soc.* (1973) **95**, 2150.
7. Churchill, M. R., Julis, S. A., *Inorg. Chem.* (1977) **16**, 1488.
8. Churchill, M. R., Julis, S. A., Rotella, F. J., *Inorg. Chem.* (1977) **16**, 1137.
9. Bailey, P. M., Maitlis, P. M., unpublished results.
10. Bau, R., Carroll, W. E., Teller, R. G., Koetzle, T. F., *J. Am. Chem. Soc.* (1977) **99**, 3872.
11. Lee, H. B., Maitlis, P. M., *J. Chem. Soc., Dalton Trans.* (1975) 2316.
12. Lee, H. B., Moseley, K., White, C., Maitlis, P. M., *J. Chem. Soc., Dalton Trans.* (1975) 2322.

13. Gill, D. S., White, C., Maitlis, P. M., *J. Chem. Soc., Dalton Trans.* (1978) 617.
14. Osborne, J. A., Jardine, F. H., Young, J. F., Wilkinson, G., *J. Chem. Soc. A* (1966) 1711.
15. Halpern, J., Wong, C. S., *J. Chem. Soc., Chem. Commun.* (1973) 629.
16. Halpern, J., Okamoto, T., Zakhariev, A., *J. Mol. Catal.* (1977) **2**, 65.
17. Rakowski, M. C., Hirsekorn, F. J., Stuhl, L. S., Muetterties, E. L., *Inorg. Chem.* (1976) **15**, 2379.
18. Stuhl, L. S., Rakowski, M. C., DuBois, Hirsekorn, F. J., Bloeke, J. R., Stevens, A. E., Muetterties, E. L., *J. Am. Chem. Soc.* (1978) **100**, 2405.
19. Russell, M. J. H., White, C., Maitlis, P. M., *J. Chem. Soc., Chem. Commun.* (1977) 427.
20. White, C., Maitlis, P. M., *J. Chem. Soc. A* (1971) 3322.
21. White, C., Thompson, S. J., Maitlis, P. M., *J. Chem. Soc., Dalton Trans.* (1977) 1654.

RECEIVED February 22, 1978.

5

Catalytic Homogeneous Hydrogenations Using Micellar and Phase Transfer Reaction Conditions

DANIEL L. REGER and M. M. HABIB

University of South Carolina, Columbia, SC 29208

Product distributions can be altered by the addition of neutral micelle-forming surfactants to the $K_3[Co(CN)_5H]$-catalyzed hydrogenation of 2-methyl-1,3-butadiene, 1,3-pentadiene, and 2,3-dimethyl-1,3-butadiene. Although rate accelerations are modest, the micelles significantly prevent decomposition of the catalyst and solubilize the organic substrates in the aqueous medium. Phase transfer reaction conditions have proved to be even more successful for these metal-catalyzed reactions. The reactions are inexpensive, easy to set up, generally regioselective and, because of substantial rate accelerations and stabilization of the catalyst, large amounts of substrate are rapidly converted to product.

There is a considerable research effort aimed at developing new homogeneous catalysts derived from transition metals (1, 2). Although many very useful systems of this type are being developed, the impact of new catalysts in synthetic organic chemistry has been somewhat limited because many of these compounds are either relatively difficult to prepare (3, 4) or use expensive starting materials (5). An alternative approach for developing new catalytic systems useful in organic synthesis is to use known catalysts that are inexpensive and are easy to prepare under unusual experimental conditions. To this end, we have been investigating the hydrogenation of water-insoluble substrates, using the water-soluble catalyst $K_3[Co(CN)_5H]$ (6). This catalyst is easily prepared by simply mixing KCN and $CoCl_2$ in a hydrogen atmosphere. In hydrogenation reactions, it generally only hydrogenates conjugated C–C double bonds. Thus, dienes are hydrogenated to monoenes. Although studied extensively (6, 7, 8), this catalyst has not generally been useful for organic reactions. The reactions are generally not regioselective. For example, 2-methyl-1,3-butadiene (isoprene) is hydrogenated to all three possible monoenes as shown below.

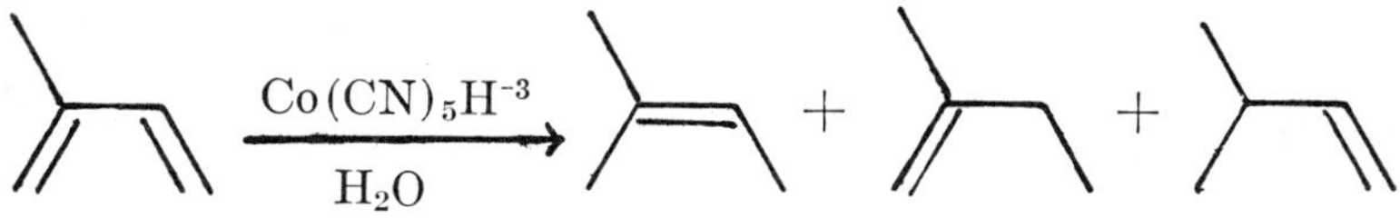

$$\text{isoprene} \xrightarrow[H_2O]{Co(CN)_5H^{-3}} \text{2-methyl-2-butene} + \text{3-methyl-1-butene} + \text{2-methyl-1-butene}$$

Some control over the products can be gained by varying the cyanide-to-cobalt ratio (7). Other problems are that the catalyst is inhibited by excess substrate and that the lifetime of the catalyst is limited, especially if a stoichiometric amount of KCN is added.

This summary covers results on the use of the $K_3[Co(CN)_5H]$ catalyst to hydrogenate dienes under two sets of unusual reaction conditions. The first is to carry out the reactions in the presence of fairly high concentrations of neutral micelle-forming surfactants. The second is to carry out the reactions under phase-transfer reaction conditions.

Reactions in Micellar Solution

The kinetics and mechanisms of many organic and inorganic reactions are generally altered by the presence of micelles (9). For example, a rate acceleration in the inorganic reaction of Hg^{+2}-induced aquation of $[Co(NH_3)_5Cl]^{+2}$ of factors up to 140,000 has been observed (10). It was expected that similar effects would be observed if transition metal catalyzed hydrogenations were carried out in the presence of micelles.

The conditions that were chosen (a full account of this phase of this research has been published in Ref. 11) for the hydrogenation reactions, using the $K_3[Co(CN)_5H]$ catalyst (4.4 mmol) were 40 mL of water, 1 atm H_2, repeated small (0.1 mL) injections of substrate, constant ionic strength, and reaction times never > 24 hr. The catalyst is more stable under a hydrogen atmosphere, and the problems described by others that were caused by aging reactions (12) were not encountered.

Tables I, II, and III show the effect of the neutral surfactant Brij 35 ($C_{12}H_{25}(OCH_2CH_2)_{23}OH$, concentration $= 5.0 \times 10^{-2}$ M) on the hydrogenation of three dienes at various CN:Co ratios (11). Rate accelerations measured by hydrogen uptake were moderate, less than twofold. For isoprene (Table I), the micelles favor the production of 2-methyl-2-butene, making the reaction virtually regiospecific. For 2,3-dimethyl-1,3-butadiene (Table II), the micelles favor production of 2,3-dimethyl-1-butene.

Table I. Product Distribution of $K_3[Co(CN)_5H]$-Catalyzed Hydrogenation of 2-Methyl-1,3-butadiene

CN:Co Ratio	Surfactant	Product Percentages		
4.3	—	3	7	90
4.3	Brij 35	1	2	97
5.2	—	3	9	88
5.2	Brij 35	2	5	93
6.0	—	43	22	35
6.0	Brij 35	29	19	52

This material can be produced 100% selectively at a CN:Co ratio of seven. The presence of micelles on the hydrogenation of 1,3-pentadiene (Table III) causes more 2-pentene to be produced but complicates the reaction somewhat because a larger amount of *cis*-2-pentene is formed.

In addition to these changes in product distributions, the micelles affect the reaction in other ways. As mentioned earlier, the catalyst is not very stable, especially at CN:Co ratios < 6. Thus, the reactions shown in the tables at a CN:Co ratio of 4.3 and 5.2 are not synthetically feasible in aqueous solution because the catalyst decomposes in a few hours under these conditions. On the other hand, the surfactant substantially inhibits this decomposition of the catalyst. Thus, for example, the reaction shown in Table I at a CN:Co ratio of 4.3 in the presence of Brij 35, which is quite regioselective, can be carried out conveniently on a synthetic scale. In addition to preventing decomposition of the catalyst,

Table II. Product Distribution of $K_3[Co(CN)_5H]$-Catalyzed Hydrogenation of 2,3-Dimethyl-1,3-butadiene

CN:Co Ratio	Surfactant	Product Percentages	
4.3	—	38	62
4.3	Brij 35	44	56
5.2	—	68	32
5.2	Brij 35	91	9
7.0	—	88	12
7.0	Brij 35	100	—

Table III. Product Distribution of $K_3[Co(CN)_5H]$-Catalyzed Hydrogenation of 1,3-Pentadiene

CN:Co Ratio	Surfactant	Product Percentages		
4.3	—	2	98	—
4.3	Brij 35	—	95	5
6.0	—	12	80	8
6.0	Brij 35	4	70	26

the micelles solubilize the substrate and reduce inhibition on the catalyst. Thus, as shown in Table IV, larger injections of substrates are possible in the presence of micelles, making these hydrogenations feasible on a synthetic scale. Isolated yields of 74% have been obtained.

Table IV. Hydrogenations Using Excess Substrate

2-Methyl-butadiene Addition[a] (mL)	Surfactant	Product Percentages				Starting Material
0.6	—	3	7	72		18
0.6	Brij 35	2	4	94		—
1.2	Brij 35	2	5	93		—
2.0	—	2	6	68		24
2.0	Brij 35	2	4	90		4

2,3-Dimethyl-butadiene Addition[b] (mL)	Surfactant					
2.0	—	85		7		8
2.0	Brij 35	96		4		—

[a] CN:Co ratio of 4.3.
[b] CN:Co ratio of 7.0.

Phase Transfer Reaction Conditions

Because of the success experienced in organic synthesis by using phase transfer reaction conditions (*13*), it was expected that this would be an excellent method for hydrogenating organic soluble dienes by using the water soluble catalyst $K_3[Co(CN)_5H]$. Others recently have carried out homogeneous catalysis processes under phase-transfer reaction conditions (*14, 15, 16*).

In a typical reaction, $CoCl_2 \cdot 6H_2O$ (0.52 g, 2.2 mmol), NaOH (0.05 g, 1.25 mmol), $(CH_3)_4NCl$ (0.24 g, 2.2 mmol), and 20 mL of degassed water were placed in a 100 mL three-neck flask fitted with a pressure-equalizing dropping funnel, an air tight septum, and a gas inlet tube leading to both a vacuum source and a hydrogen source controlled by a gas buret. A mixture of KCN (0.74 g, 11.44 mmol) and KCl (0.36 g, 4.8 mmol) dissolved in 20 mL of degassed water was added to the dropping funnel. The system was repeatedly evacuated and refilled with hydrogen. Maintaining the hydrogen atmosphere, the cyanide solution was added rapidly to the stirred cobaltous solution. After allowing 15 minutes for the catalyst to completely form, 20 mL of degassed benzene was added followed by 2 mL (1.36 g, 20.0 mmol) of isoprene. The reaction was monitored by hydrogen uptake. As shown by gas chromatography, the substrate was completely hydrogenated in four hours, yielding 2-methyl-2-butene contaminated by 2% of 2-methyl-1-butene. No starting material remained. With benzyltriethylammonium chloride as the phase transfer reagent, the reaction was completely regioselective, yielding 2-methyl-2-butene. In an experiment carried out using 5 mL of substrate, an isolated yield of 87% was obtained. With the amounts of reagents listed above, 6 mL of isoprene could be completely converted to products in 15 hr. Tripling the amount of catalyst tripled the rate of hydrogenation and allowed 15 mL of substrate to be hydrogenated in 13 hr using the amounts of the other components of the system as listed above. If the reaction outlined above was carried out without the addition of a phase transfer reagent, the reaction needed 48 hr to go to completion, at which time the catalyst had decomposed. In the presence of a phase transfer reagent, the catalyst is stable for weeks.

Under similar conditions, 1,3-pentadiene is hydrogenated exclusively to *trans*-2-pentene. Although only one product is obtained, deuterium labeling studies indicate that the reaction is not specific. The hydrogenation of 1,3-pentadiene-1,1-d_2 can yield two *trans*-2-pentene products, one arising from overall 1-4 addition and one arising from 1-2 addition of hydrogen to the external double bond as shown below.

```
                          1-4 addition
                        ─────────────────→      [trans-2-pentene, 55%]
                    ┌
  D─⟨═⟩              
    D               └
                          1-2 addition
                        ─────────────────→      [1-pentene isotopomer, 45%]
```

The percentages shown were determined by analyzing the ^{2}H NMR of the *trans*-2-pentene produced in this hydrogenation. The substrate 2,3-dimethyl-1,3-butadiene is hydrogenated under these conditions to a mixture containing 80% 2,3-dimethyl-2-butene and 20% 2,3-dimethyl-1-butene. Note that because the hydrogenation carried out in micellar solution yields exclusively 2,3-dimethyl-1-butene (Table II), either of the possible hydrogenation products can be obtained by proper choice of reaction conditions.

Conclusion

The two types of unusual reaction conditions, micellar solutions and phase transfer reactions, have proved useful for the $[Co(CN)_5H]^{-3}$-catalyzed hydrogenation of dienes into monoenes. The micelles inhibit decomposition of the catalyst and solubilize the organic substrates, making the reactions possible on a synthetic scale. Even more useful for synthetic scale reactions are the phase-transfer conditions. In this case, large rate accelerations are obtained as well as substantial increases in the lifetime of the catalyst, allowing large amounts of substrate to be hydrogenated. Further studies are in progress using the phase transfer reaction conditions to hydrogenate other dienes as well as C–C double bonds conjugated to other functionality.

Acknowledgment

Acknowledgment is made to the donors of The Petroleum Research Fund, administered by the American Chemical Society, for the support of this research.

Literature Cited

1. Cotton, F. A., Wilkinson, G., "Advanced Inorganic Chemistry," Chap. 24, Interscience, New York, 1972.
2. James, B. R., "Homogeneous Hydrogenation," John Wiley and Sons, New York, 1973.
3. Deeming, A. J., Underhill, M., *J. Chem. Soc., Dalton Trans.* (1973) 2727.
4. Ibid. (1974) 1415.
5. Demitras, G. C., Muetterties, E. L., *J. Am. Chem. Soc.* (1977) **99,** 2796.
6. Kwiatek, J., Mador, I. L., Seyler, J. K., ADV. CHEM. SER. (1963) **37,** 201.
7. Kwiatek, J., Seyler, J. K., ADV. CHEM. SER. (1968) **70,** 207.
8. Murakami, M., Kang, J. W., *Bull. Chem. Soc. Jpn.* (1963) **36,** 763.
9. Cordes, E. H., Dunlap, R. B., *Acc. Chem. Res.* (1969) **2,** 329.
10. Cho, J. R., Morawetz, H., *J. Am. Chem. Soc.* (1972) **94,** 375.
11. Reger, D. L., Habib, M. M., *J. Mol. Cat.* (1978) **4,** 315.
12. King, N. R., Winfield, M. E., *J. Am. Chem. Soc.* (1961) **83,** 3366.
13. Dehmlow, E. V., *Angew. Chem., Int. Ed.* (1977) **16,** 493.
14. Abbayes, H., Alper, H., *J. Am. Chem. Soc.* (1977) **99,** 98.
15. Hai, K., Shaw, B. L., *J. Organomet. Chem.* (1977) **124,** 262.
16. Cassar, L., Foa, M., *J. Organomet. Chem.* (1977) **134,** 615.

RECEIVED February 22, 1978.

6

Asymmetric Hydrosilylation

H. B. KAGAN, J. F. PEYRONEL, and T. YAMAGISHI[1]

Laboratoire de Synthèse Asymétrique, (Associé au CNRS LA n° 040255-02), Université Paris-Sud, 91405-Orsay, France

> *A catalyst for asymmetric hydrosilylation of ketones was prepared from $[Rh(COD)Cl]_2$ and DIOP. The hydrosilylation of acetophenone yields 1-phenylethanol after hydrolysis. The optical yield strongly depends on the nature of the silane $RR'SiH_2$ that was used. Several types of ketones were asymmetrically reduced into chiral alcohols, the highest asymmetric induction being observed from some α-chloro ketones or α-ketoesters. It was remarkable that prochiral benzophenones such as $p\text{-}OMeC_6H_4COC_6H_5$ could be reduced with up to 26% e.e. Mechanism of asymmetric hydrosilylation was discussed in relation with some spin trap experiments. Studies were made on supported rhodium–DIOP catalysts. Some rhodium leaching from support was demonstrated by a three-phase test.*

The stereocontrolled synthesis of a given enantiomer is a key operation in many processes. Very often, one enantiomer rather than the racemic mixture is needed because of its properties. In fragrances, food additives, or pharmaceutical drugs, many such cases can be found. For example, α-aminoacids which enter as components of polypeptides or drugs are always used with a specific absolute configuration. The importance of chiral substances is understandable if it is realized that living systems are essentially chiral themselves and able to differentiate enantiomeric substrates at active sites of enzymes or at biological receptors. One classical preparation of an enantiomer (D) is to resolve a racemic mixture (D, L). This is a time- and energy-consuming process because the undesired enantiomer (L) must be separated by chemical and physical operations and racemized if possible for recycling. It is necessary to use stoichiometric amounts of a chiral auxiliary compound Z* (Figure 1)

[1] Current address: Department of Industrial Chemistry, Tokyo Metropolitan University, Fukazawa, Setagayaku, Tokyo, 158 Japan.

to obtain the resolution through formation and separation of diastereomeric products. Each diastereomer is then destroyed and the desired enantiomer (D) is recovered. The chiral economy is obvious if Z* can be a chiral reagent controlling the direct formation of D; the process is now an asymmetric synthesis. A further improvement occurs when Z* represents a chiral catalyst because a small amount of Z* should be able to control direct production of a large amount of the desired enantiomer. Figure 1 is a schematic comparison of resolution and asymmetric synthesis in the case where the reaction giving rise to a racemic mixture (D + L) is transformed into an asymmetric catalytic process leading to D.

Asymmetric catalysis as a synthetic tool is relatively new (if enzymatic reactions are not considered); its development began 10 years ago, mainly because of the advances in coordination chemistry. Asymmetric hydrogenation started by modifying the Wilkinson catalyst (1). The early results (2, 3, 4) were encouraging enough to initiate a very large amount of research (5, 6). Asymmetric C–C bond formation in olefin co-dimerization was observed for the first time by Wilke and his co-workers (7). Asymmetric hydroformylation (8) as well as several new asymmetric alkylation reactions appeared in the last five years (9, 10). Asymmetric epoxidations were described in 1977 (11, 12).

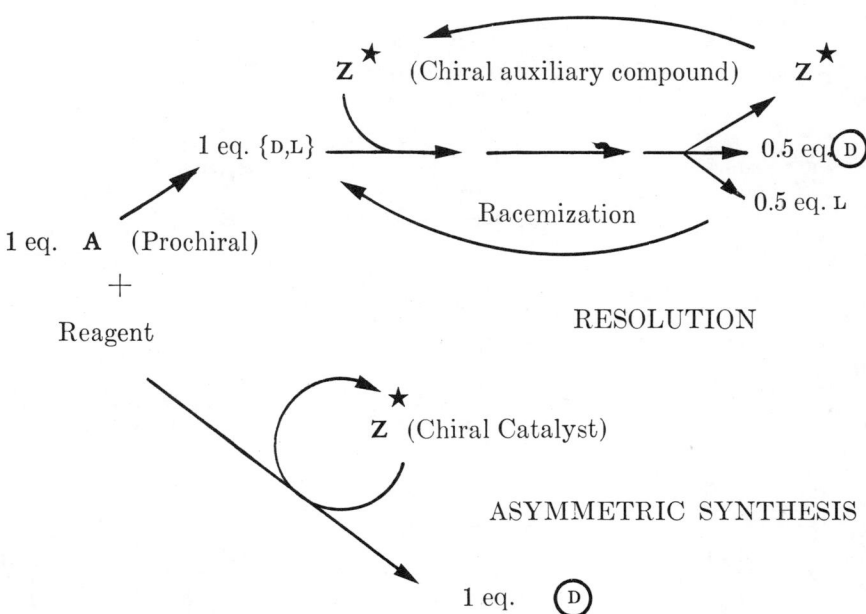

Figure 1. Asymmetric synthesis vs. resolution, an "energy saving" process

In principle, any catalyzed reaction where the ligands of the complex are easily modified could be investigated for asymmetric catalysis. To date, enantioselectivities higher than 95% have been attained in a few cases (13, 14), giving hope that further studies should define a variety of specific chiral catalysts for the production of various chiral compounds.

The creation of an asymmetric center by C–H bond formation is a very common process which can involve several types of reactions. Hydrogenation of prochiral olefins is often used with the rhodium catalysts of the Wilkinson type (5). These catalysts were shown to be inactive for ketone or imine reduction except in some cases (15). It was then interesting to develop an alternate method for asymmetric synthesis of chiral alcohols or amines. Since it was found that $RhCl(PPh_3)_3$ was able to catalyze silane additions to ketones (16, 17) or imines (18), preparation of chiral alcohols or amines by asymmetric hydrosilylation could be envisaged (Figure 2). The 1,4-addition of silanes to conjugated

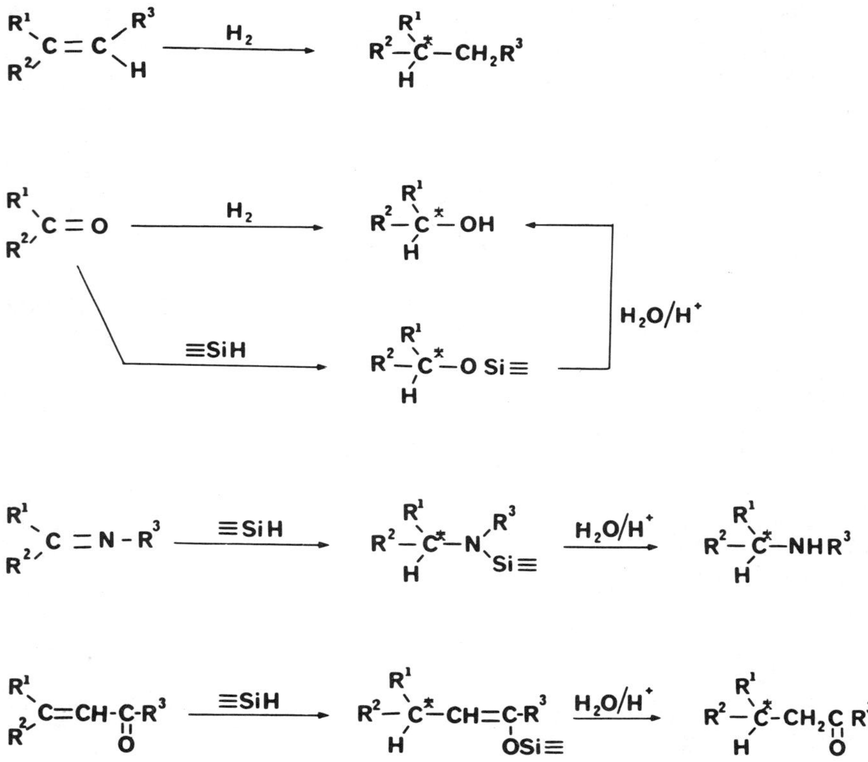

Figure 2. Examples of creation of an asymmetric center by C–H bond formation

esters or ketones (19) is another possibility for asymmetric hydrosilylation and creation of an asymmetric center in the β position relative to a carbonyl group.

We chose to study asymmetric hydrosilylation for the preparation of alcohols (20) and amines (18) because the system permits structural modifications of both the chiral ligands and the silane. The possibility arose of a good matching with the substrate, leading to high stereoselectivity in the formation of the chiral product. We will only consider here the asymmetric synthesis of alcohols.

Synthesis of Chiral Ligands

Phosphines are the usual ligands of rhodium catalysts for hydrosilylation. A classification of chiral phosphines according to their structure is presented in Table I.

Very early we examined the synthesis and behavior of phosphines R*–PPh$_2$ easily prepared from natural products (21). We then introduced, in 1971 (4), the use of chiral chelating diphosphines (phosphines of

Table I. A Structural Classification of Chiral Phosphines[a,b]

Type I: $R_2 \!\!>\!\! P^*$ with R_1, R_3

Type II: $R^* \!\!-\!\! PPh_2$

Type III: $R^* \!\!<\!\! ^{PPh_2}_{PPh_2}$

Type IV: $(CH_2)_n \!\!<\!\! ^{P^*<^{R_1}_{R_2}}_{P^*<^{R_1}_{R_2}}$

Type V: $^{R^*}_{R_1} \!\!>\!\! P^* \!\!-\!\! R_2$

Type VI: $R^* \!\!<\!\! ^{P^*<^{R_1}_{R_2}}_{P^*<^{R_3}_{R_4}}$

[a] This classification gives only the main types of chiral mono- and diphosphines. Many other structures are also possible, allowing, in principle, the synthesis of a large number of new ligands. R* and P* symbolize chiral groups.
[b] For reviews on chiral phosphines prepared for homogeneous asymmetric catalysis, see Ref. 5, 6, and 26.

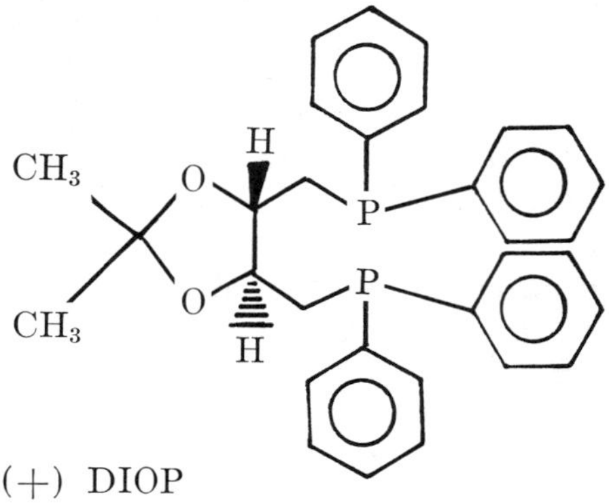

1   (+) DIOP

Type III). Since good results in asymmetric hydrogenation were obtained (22) with DIOP 1, it was then used in asymmetric hydrosilylation (20). Results with DIOP and some other diphosphines will be presented here.

Influence of the Structure of Silanes upon Optical Yield

The standard procedure that was adopted for asymmetric hydrosilylation of ketones was the following. A solution of 1 mmol of ketone and 1.1 mmol of silane in 3 mL benzene in presence of RhCl(DIOP) (0.2% with respect to the ketone) was stirred under nitrogen for a few hours. The catalyst was preformed by mixing [Rh(COD)Cl]$_2$ and DIOP in benzene and stirring 15 min under nitrogen. All of these operations were performed at 20°C.

The complex [Rh(COD)(DIOP)]$^+$ClO$_4^-$ which can be isolated gives no better results than the neutral catalyst prepared in situ. After reaction the solvent was evaporated and the residue hydrolyzed under acidic conditions. The product was recovered by distillation or chromatography. In general, yields were excellent (90–100%).

Dihydrosilanes RR'SiH$_2$ were selected in almost all of our work because of their good reactivity. In Table II some representative dihydrosilanes and their efficiencies in the hydrosilylation of acetophenone are indicated. There are variations in optical yields according to the structure of dihydrosilanes. Steric considerations do not seem to be able to give a simple explanation. For example, Ph(cyclohexyl)SiH$_2$ is more efficient than Ph$_2$SiH$_2$, but (cyclohexyl)$_2$SiH$_2$ is a poor reagent. Similarly, (α-naphthyl)PhSiH$_2$ is better than Ph$_2$SiH$_2$ but not different from (α-naphthyl)$_2$SiH$_2$.

Use of a chiral complex in hydrosilylation of a ketone can provide a route to chiral silanes. This method with RhCl(DIOP) as catalyst was investigated by Corriu et al. (26, 31). A prochiral silane such as (α-Np)-

Table II. Hydrosilylation of Ph–CO–CH$_3$ by RR'SiH$_2$ Catalyzed by RhCl(−)DIOP (23, 24)

Silane[a]		PhCHOHCH$_3$	
R	R'	Absolute Configuration	Optical Yield (%)
Cyclohexyl	cyclohexyl	S	1[b]
Ph	mesityl	R	6[b]
Ph	CH$_3$	R	13
Ph	Ph	R	24
m-Me-Ph	m-Me-Ph	R	30
α-Naphthyl	cyclohexyl	R	32
Ph	o-Me-Ph	R	35
Ph	cyclohexyl	R	49
α-Naphthyl	α-naphthyl	R	50
Ph	α-naphthyl	R	52

[a] See "Influence of the Structure of Silanes upon Optical Yield" for experimental details.
[b] The enol silylether of acetophenone is simultaneously formed in approximately equal amounts as the silylether of 1-phenylethanol.

PhSiH$_2$, after reaction on a symmetrical ketone, is transformed into (α-Np)PhSi̇-(OR), in which the silicon atom is the only source of chirality. Optical yields in the range of 50% were observed.

Relation between Optical Yield and Structure of the Substrates

Often acetophenone is the prochiral ketone which is first tested when a new asymmetric reducing agent has to be evaluated. This was done in asymmetric hydrosilylation by us (20) and others (26). Fortunately, catalytic hydrosilylation is not limited to this case. In Table III, some representative results with α-naphthyl phenyl silane (α-NpPhSiH$_2$) are summarized. Substitution on the methyl group of acetophenone strongly influences the stereospecificity. When there is a free OH group, it is first silylated, followed by an internal hydrosilylation leading to a cyclic silyl diether. Hydrolysis gives the phenyl glycol (Figure 3). Contrary to expectation, this intramolecular process decreases the enantiospecificity.

α-Chloro or bromo ketones were smoothly hydrosilylated with good optical yields (with respect to the unsubstituted ketone). Basic hydrolysis of the halogeno silyl ether gave directly a chiral epoxide. α-Ketoesters gave α-hydroxyesters of high optical purity when RhCl(DIOP) was used as catalyst. Ojima (27) obtained propyl lactate with 85% e.e. by reducing propyl pyruvate in presence of RhCl(DIOP) and also observed high

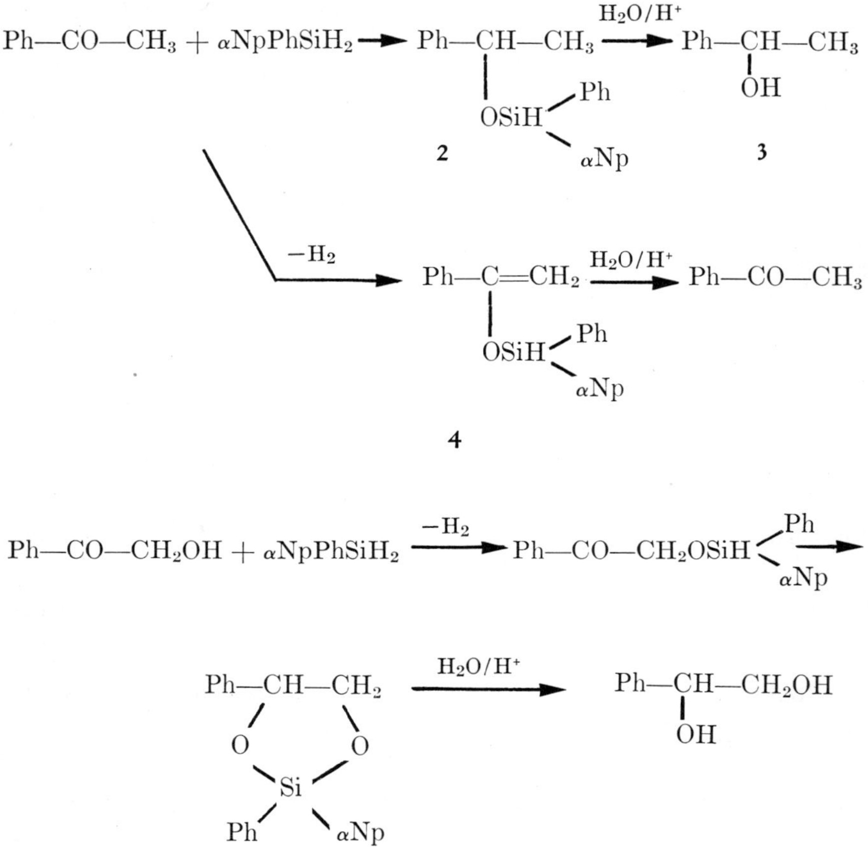

$$Ph-CO-CH_3 + \alpha NpPhSiH_2 \longrightarrow Ph-CH-CH_3 \xrightarrow{H_2O/H^+} Ph-CH-CH_3$$
$$\underset{2}{\overset{|}{OSiH}}\overset{Ph}{\underset{\alpha Np}{\diagdown}} \qquad \underset{3}{\overset{|}{OH}}$$

$$\xrightarrow{-H_2} Ph-C=CH_2 \xrightarrow{H_2O/H^+} Ph-CO-CH_3$$
$$\underset{4}{\overset{|}{OSiH}}\overset{Ph}{\underset{\alpha Np}{\diagdown}}$$

$$Ph-CO-CH_2OH + \alpha NpPhSiH_2 \xrightarrow{-H_2} Ph-CO-CH_2OSiH\overset{Ph}{\underset{\alpha Np}{\diagdown}} \longrightarrow$$

$$Ph-\underset{O\diagdown\underset{Ph}{\overset{Si}{}}\diagup O}{\overset{\diagup\diagdown}{CH-CH_2}} \xrightarrow{H_2O/H^+} Ph-\underset{OH}{\overset{|}{CH}}-CH_2OH$$

Figure 3. Hydrosilylation of acetophenone and PhCOCH$_2$OH

asymmetric inductions in hydrosilylation of γ-ketoesters, presumably because of a labile interaction of the ester group with the rhodium atom.

We have investigated (23, 24) the potential of asymmetric hydrosilylation of various ketones. Methyl alkyl ketones (for example, 2-octanone) gave about 44% e.e. if phenyl cyclohexyl silane was the hydrosilylating reagent (Table III).

An interesting case is the reduction of prochiral benzophenones with one parasubstituted phenyl ring. It is surprising that optical yields as great as 26% could be attained, allowing the preparation of chiral benzhydrols (23, 24). Any of the proposed schemes of asymmetric induction with the rhodium–DIOP catalyst (27, 28) can hardly explain such results. Obviously very subtle interactions between the ketone and the complex are involved. Maybe charge transfer interactions between aromatic

Table III. Asymmetric Hydrosilylation by α-NpPhSiH$_2$ of Some Representative Ketones Catalyzed by RhCl(−)DIOP

Ketone[a]	Alcohol	Absolute Config.	Optical[b] Yield (%)
Ph COCH$_3$	Ph CHOHCH$_3$	R	53
Ph COCH$_2$OH	Ph CHOHCH$_2$OH	S	7
Ph-CO-CH$_2$OCO-Ph	Ph CHOHCH$_2$OCO-Ph	S	34
Ph COCH$_2$Cl	Ph CHOHCH$_2$Cl	S	63
n C$_5$H$_{11}$COCO$_2$nPr	n C$_5$H$_{11}$CHOHCO$_2$nPr	R	66[c]
n C$_6$H$_{13}$COCH$_3$[d]	n C$_6$H$_{13}$CHOHCH$_3$	R	44
pOMe-Ph-CO-Ph	p OMe-Ph-CHOH-Ph	S	26[e]

[a] [Silane]/[ketone] = 1.1%; [Rh]/[ketone] = 0.2%; 30 hr in benzene at 20°C, Yields in product are 90–95%.
[b] Calculated by reference to maximum specific rotation of the alcohol.
[c] Measured on n-C$_5$H$_{11}$CHOHCH$_2$OH after the reaction product was treated with LAH.
[d] Phenylcyclohexylsilane is used instead of α-NpPhSiH$_2$.
[e] Calculated by reference to maximum specific rotation (48).

groups could mediate competitive transition states, as was observed in the asymmetric reduction of prochiral benzophenones by chiral Grignard reagents (29).

Mechanism of Asymmetric Hydrosilylation

The basic features of the mechanism are known (Figure 4). It is quite probable that the reaction starts by an oxidative addition of Si–H bond to a Rh(I) complex. Such reactions are known and adducts have been isolated (30). What is less known are the details of the subsequent steps. The ketone is necessarily coordinated to the complex before the successive transfers of the hydrogen and silicon groups. The chronology of the transfers was discussed by Ojima (26, 31). As the first step, he suggested the formation of an α-silyloxy alkyl-rhodium species (path (a) of Figure 4), followed by a reductive elimination of the product. This hypothesis was mainly based on a detailed analysis of the steric course of reduction of various ketones. Another problem is to decide what is the origin of asymmetric induction. Some models of asymmetric induction were proposed when DIOP was the chiral ligand (27, 28). In these models, the various interactions between the prochiral ketone and the groups that are part of the complex were considered. These models remain very hypothetical because neither the actual structure of the various complexes involved in the catalytic cycle nor the relative rates of the reactions are known. X-ray data on some DIOP complexes (32) or other chiral diphosphines are now available and should be helpful in deciding the conformation of the chiral ligand.

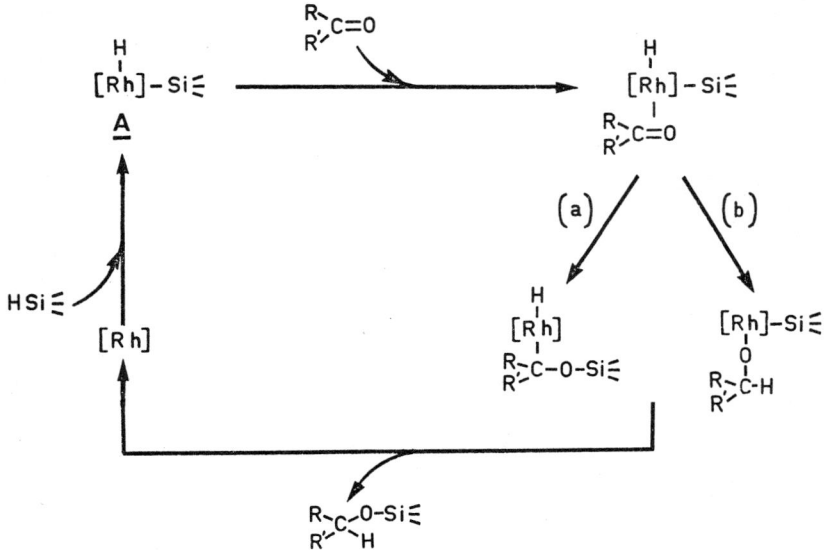

Figure 4. *Tentative mechanisms for hydrosilylation of ketones*

In hydrosilylation it is possible that part of the reaction goes through a radical mechanism. This was postulated to explain some results in hydrosilylation of conjugated esters (*19*). If this mechanism is involved in hydrosilylation of ketones, it should decrease the optical yield. Recently (*33*) we investigated the hydrosilylation of acetophenone by phenyl α-naphthyl silane catalyzed by RhCl(−)DIOP in presence of spin traps (two equivalents with respect to the complex). In the absence of spin traps, there are no ESR signals; when a spin trap such as nitrosodurene is used, a nice signal is observed (Figure 5) that we assigned (*23*) to the nitroxide formed in significant amount (1% of total rhodium) by formal addition of $CH_3-\underset{.}{C}(Ph)-OSiHPh(\alpha-Np)$ to the spin trap. If the same experiment is performed in the absence of acetophenone, another signal is observed that we assigned to $ArN(\overset{O^{\cdot}}{|})-SiHPh(\alpha-Np)$ where Ar = 2,3,5,6-tetramethylphenyl. Since no significant amount of pinacol was observed in hydrosilylation of acetophenone, we do not believe that $CH_3-\underset{.}{C}(Ph)OSiHPh(\alpha-Np)$ represents a reaction intermediate. We propose that the formation of the corresponding addition product on nitrosodurene is related to the presence of spin trap which should induce the decomposition of the alkyl rhodium complex in path (a) (Figure 4). This interpretation is supported by analogous observations (*34, 35*),

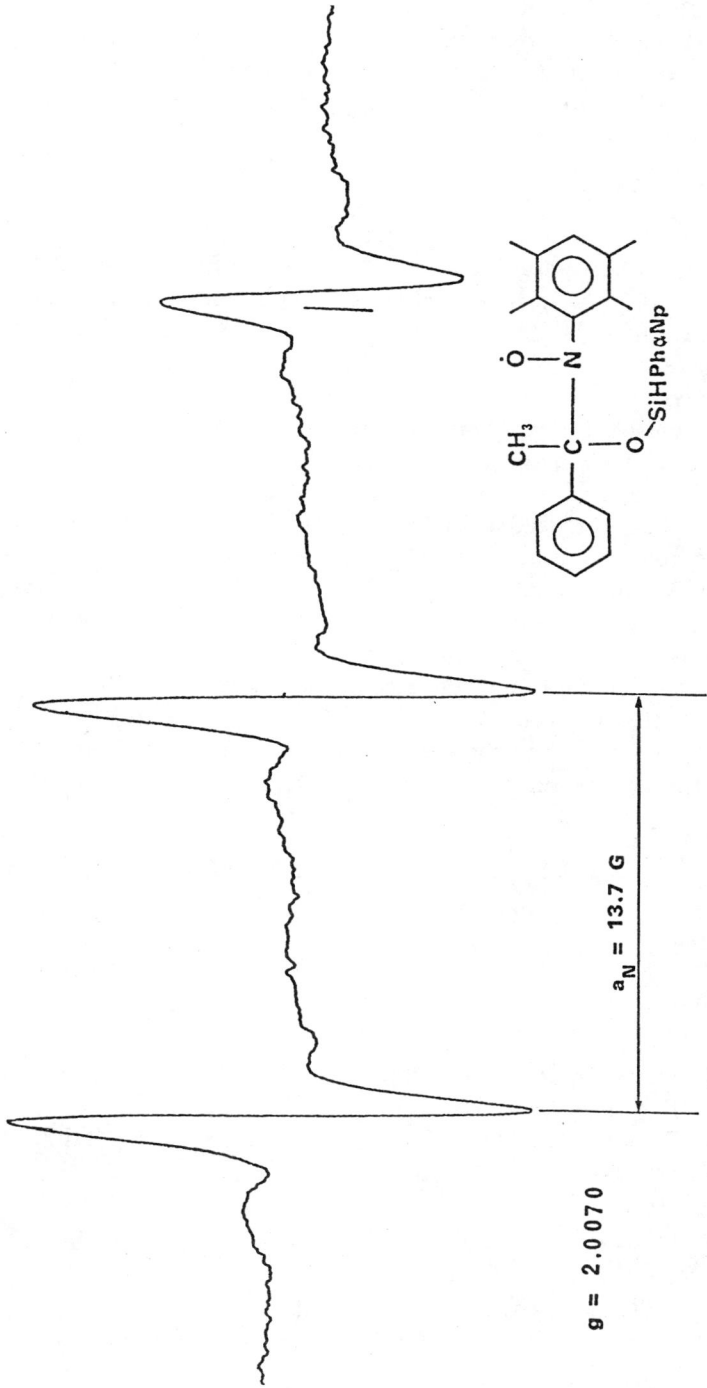

Figure 5. *ESR spectra of a mixture of acetophenone and phenyl α-naphthylsilane in presence of catalytic amounts of RhCl(DIOP) and nitrosodurene*

especially on alkyl palladium complexes. There is then a quite direct proof that path (a) is of importance in hydrosilylation as previously assumed (*31*). Of course it gives no information about the contribution of path (b) through formation of an alkoxyrhodium complex.

Supported Chiral Catalysts

Many attempts were made to prepare supported catalysts derived from soluble transition metal complexes in order to have a recyclable catalyst (*36, 37, 38, 39, 40*). This technique should prove to be useful if it appreciably increases the lifetime of the catalytic site (compared with the homogeneous catalyst). There are no definitive data for many of the published systems; however there are indications that there is some leaching of the metal from the support to the solution. In the special case of catalytic hydrosilylation, some loss of rhodium from the support (*41*) has been observed. We want to describe an example of a three-phase method for studying the possible appearance of minute amounts of soluble, phosphine-free rhodium complexes during heterogeneous catalysis. All experiments involved the rhodium-catalyzed transformation of acetophenone into the silyl ether **2** by reaction with α-naphthylphenylsilane (α-NpPhSiH$_2$). Hydrolysis of **2** gave 1-phenylethanol **3**. A side reaction in the presence of the catalyst is the formation of the silyl enolether of acetophenone **4**, which is then hydrolyzed into acetophenone (Figure 3). In discussing our results in Table II we will first consider the conversion percentage of acetophenone and the selec-

Table IV. Hydrosilylation of Ph–CO–CH$_3$ by

	Supported Catalyst	Soluble Rh Precursor[a]	Second Resin[b]
1	ⓟ/\/DIOP RhCl	(RhCODCl)$_2$	—
2	(ⓟ/\/PPh$_2$)$_2$RhCl	(RhCODCl)$_2$	—
3	ⓟ/\/DIOP RhCl (ⓟ/\/PPh$_2$)$_2$RhCl	(RhCODCl)$_2$	—
4	(ⓟ/\/PPh$_2$)$_2$RhCl	(RhCODCl)$_2$	ⓟ/\/DIOP ⓟ/\/PPh$_2$
5	ⓟ/\/DIOP RhCl	(RhCODCl)$_2$	ⓟ/\/PPh$_2$
6	ⓟ/\/DIOP RhCl	[Rh(C$_2$H$_4$)$_2$Cl]$_2$	—
7	ⓟ/\/DIOP RhCl	[Rh(C$_2$H$_4$)$_2$Cl]$_2$	ⓟ/\/PPh$_2$
8	ⓟ/\/DIOP RhCl	[Rh(C$_2$H$_4$)$_2$Cl]$_2$	ⓟ/\/PPh$_2$

[a] [Silane]/[acetophenone] = 2; [acetophenone]/[Rh] = 50.
[b] Phosphinated resin in equivalent quantity to ⓟ/\/DIOPRhCl/\/ or (ⓟ/\/PPh$_2$)$_2$RhCl.

tivity percentage (yield in 1-phenylethanol with respect to consumed acetophenone).

The initial experiments involved control reactions with the soluble in situ chiral catalyst RhCl(DIOP) (22), with the chiral supported catalysts ⓟ/\/DIOP RhCl that we previously described (20) and with the classical (36) achiral polystyrenic catalyst (ⓟ/\/PPh$_2$)$_2$RhCl. In both cases, ⓟ/\/ represents the polystyrene resin (2% cross linked). The results are reported in Table III and they are in agreement with previous work (20, 22). In a second set of experiments we prepared, as usual, the supported catalysts ⓟ/\/DIOP RhCl and (ⓟ/\/PPh$_2$)$_2$-RhCl. We used each one in the presence of an equal amount of the phosphinated resins, ⓟ/\/PPh$_2$ and ⓟ/\/DIOP, respectively. Clearly the optical purity of the synthesized 1-phenylethanol depended on the presence of a metal-free resin (Table IV). For example, the achiral catalyst (ⓟ/\/PPh$_2$)$_2$RhCl yielded racemic 1-phenylethanol, but in presence of ⓟ/\/DIOP, (S)-1-phenylethanol with 12% e.e. was recovered. The most obvious interpretation of these data would involve a transfer of rhodium from one resin to the other. This situation is typical of the three-phase test which was successfully used (42, 43) to demonstrate transient existence of unstable species. Quantitative information cannot easily be derived from our experiments since the two heterogeneous catalysts have different reactivities, as demonstrated by using an equimolar mixture of them to catalyze the hydrosilylation of acetophenone. The (S)-1-phenylethanol was recovered with 6% e.e. In addition, soluble complexes may display some catalytic activity which

Supported Catalysts in Presence of Phosphinated Resin

Reaction Time (hr)	Acetophenone Conversion (%)	Selectivity in PhCHOHCH$_3$ (%)	Optical Yield[c] (%)
48	99	91	49
48	99	87	0
24	99	79	6
48	92	94	12
260	93	92	34
40	88	97	53
80	93	95	50
50	65	69	17[d]

[c] Optical yield was calculated with respect to the value for optically pure 1-phenyl ethanol. $[\alpha]_D^{23} = -52.5°$ (c = 2.2M CH$_2$Cl$_2$) (47). The recovered alcohol has always the (S) configuration.
[d] In presence of added COD([COD]/[Rh] = 10).

is difficult to evaluate. It was also of interest to know more about the soluble transient complex which carried the rhodium from one resin bead to another. The available data indicates that it must be a phosphine-free complex. The two supported catalysts were prepared by decomposing a soluble complex such as [RhClCOD]$_2$ (COD = 1,5-cyclooctadiene) or [RhCl(C$_2$H$_4$)$_2$]$_2$, using the phosphinated resins Ⓟ/\/DIOP or Ⓟ/\/PPh$_2$. There is some question whether COD or ethylene could stay as ligand in the heterogeneous rhodium complex. If this were the case, an interaction between several sites on the polymer could partly regenerate the starting soluble rhodium complexes. Another possibility is that ketone, silane, or the reaction products are involved in the formation of the soluble rhodium complex. It is interesting to point out that the nature of the precursor complex used for the preparation of the supported catalyst also plays an important role. If Ⓟ/\/DIOP is stirred with a solution of [Rh(C$_2$H$_4$)$_2$Cl]$_2$, the chiral catalyst which is formed is insensitive to the presence of Ⓟ/\/PPh$_2$ (Table IV, entries 6, 7). When COD is added to the solution (entry 8), rhodium transfer occurs giving a strong decrease in the optical yield. The presence of COD in the system seems necessary to promote the formation or to stabilize by coordination the soluble species.

To know more about the species involved and to demonstrate its existence, a transfer experiment was devised using two identical samples of phosphinated polystyrene. Ⓟ/\/PPh$_2$ suspended in benzene was separated from preformed (Ⓟ/\/PPh$_2$)$_2$RhCl by a nylon sack having a porosity of 250μ. After stirring for a long time, the two resins were recovered and analyzed for their rhodium content. There was no transfer of rhodium between them. The same result occured when acetophenone was added to the benzene medium. However, rhodium was detected on Ⓟ/\/PPh$_2$ when α-naphthylphenylsilane was added. Accordingly we must then formulate the soluble complex which carries rhodium as a complex involving silicon; for example, (RhCl(α-NpPhSiH)$_2$)$_2$. It could, too, involve one COD molecule as in RhClCOD(α-NpPhSiH)$_2$. Silyl-stabilized rhodium complexes with similar structures have been isolated (*44, 45*) after reaction with silanes. Because of the small concentration of the soluble complex, we have not been able to isolate and characterize it.

Our results demonstrate that during the course of a heterogeneous hydrosilylation, rhodium can move on the support (if free phosphino groups are still available on the support). It is very probable that part of the reaction could occur with soluble species. In the case of asymmetric catalysis, they are achiral and will lower the optical yield. It also is not clear at present if in homogeneous hydrosilylation similar phenomena do occur, which would increase the number of catalytically actives species.

Conclusion

Asymmetric hydrosilylation is becoming a useful tool for the synthesis of chiral compounds. It also is an interesting method which complements asymmetric hydrogenation and is capable of tolerating many functional groups. A complete review of all aspects of asymmetric hydrosilylation up to 1976 has recently appeared (*26*). It was written by Ojima, Yamamoto, and Kumada, who are important contributors in this field. One attractive aspect of hydrosilylation is the possibility of modifying both the silane and the chiral ligand, thereby increasing the chances of finding the most efficient catalytic system as far as stereoselectivity is concerned. In some cases we have demonstrated the sensitivity of the optical yield with respect to the structure of the dihydrosilane which is chosen (*23, 24*). We have tried a large number of modifications on DIOP itself or we have used other types of ligands. Until now we could not find chiral ligands superior to DIOP. More mechanistic studies are needed to devise proper ligands which will yield the best stereochemical control in hydrosilylation.

Experimental

Rhodium microanalyses were performed by Service Central de Microanalyses du CNRS. Optical rotations were measured on a Perkin–Elmer 141 automatic polarimeter. NMR spectra were performed using a Perkin–Elmer R 32 90 MHz spectrometer and GLC analyses with a Carlo–Erba Fractovap GI chromatograph equipped with a flame ionization detector.

Chemicals and Solvents. The insoluble chiral phosphines used in the hydrosilylation of acetophenone were prepared using the procedure described in (*20*) (Merrifield resin: 200–400 mesh, 2% divinylbenzene). The chiral phosphine is a (+)DIOP analog with a phosphorus content of 0.9 mequiv/g. The insoluble phosphine ⓟ/\/PPh$_2$ which was used in the rhodium transfer experiments was prepared according to the literature (*36*). Ten grams of polystyrene beads (Bio-Beads SM-2, 20–40 mesh) suspended in chloroform was reacted with 20 mL of CH_3OCH_2Cl in the presence of $SnCl_4$ for 12 hr at room temperature. The resin was filtered, washed with a chloroform–methanol mixture (2:3, 4:1, 9:1) and finally with chloroform, and dried at 90°C in vacuo for 10 hr. The chlorine content in the resin was 1.7 mequiv/g. Chloromethylated resin (9 g) was treated with excess $LiP(C_6H_5)_2$ in THF for 40 hr at room temperature under nitrogen. The resin was filtered, washed (THF, ethanol, benzene, hexane), and dried as usual. The phosphorus content was 1.54 mequiv/g.

The ketones were distilled before use. α-Naphthylphenylsilane was prepared according to the literature (*46*). Benzene was purified by first passing it through a basic alumina column followed by distillation over sodium hydride. It was stored under nitrogen.

Hydrosilylation in Homogeneous Media. Details of these experiments can be found in "Influence of the Structure of Silanes upon Optical Yield," and in Ref. *20*, *23*, and *24* and results in Tables II and III.

Hydrosilylation by Supported Rhodium Catalyst. The supported catalyst was prepared directly in a reaction flask stoppered with a serum cap, allowing addition by injection with syringes. The phosphinated resin of 180 mg (200–400 mesh) was stirred at room temperature for 20 hr in a solution containing $[Rh(COD)Cl]_2$ or $[Rh(C_2H_4)_2Cl]_2$ (Rh/P = 1) in 4 mL benzene. The resin was filtered, thoroughly washed by benzene (5 mL × 10) to completely eliminate the rhodium complex adsorbed on the resin surface, and dried. All manipulations were carried out under nitrogen. Then another sample of the phosphinated resin (180 mg) was added swiftly, and the flask was maintained under a nitrogen atmosphere. A mixture of acetophenone and α-naphthylphenylsilane in benzene was syringed into the flask, and the reaction proceeded maintaining the nitrogen atmosphere. After a suitable time the reaction mixture was treated as usual. The conversion and selectivity of the reaction were determined by VPC (Carbowax 20M, 3m, 180°C) and by NMR. These results are given in Table IV.

Rhodium Transfer Experiments. The supported catalyst was prepared in a manner similar to that described (*20*) using 600 mg of the phosphinated resin (20–40 mesh) and equimolar amounts of $[RhCODCl]_2$ or $[Rh(C_2H_4)_2Cl]_2$ in benzene under nitrogen. To the flask, the second resin sample (600 mg) in a nylon sack with a porosity of 250 μ was added swiftly, and the flask was again purged with nitrogen. The resins suspended in benzene were stirred for various times in the presence or absence of additives (acetophenone and/or α-naphthylphenylsilane). The two resins were then separated and analyzed for their rhodium content. Up to 10% of the rhodium present initially could be transferred from one resin to the other. Control experiments demonstrated that the nylon sack did not inhibit the migration of the soluble complexes.

Acknowledgment

One of us (T.Y.) thanks the Ministry of Education of Japan and the French Institute of Petroleum for fellowships. We also thank CNRS and French Institute of Petroleum for a financial support and the Compagnie des Métaux Précieux for a loan of rhodium chloride.

Literature Cited

1. Osborn, J. A., Jardine, F. E., Young, J. F., Wilkinson, G., *J. Chem. Soc.* (1966) 1711.
2. Horner, L., Siegel, H., Buthe, H., *Angew. Chem., Int. Ed., Engl.* (1968) **7**, 941.
3. Knowles, W. S., Sabacky, M. J., *Chem. Commun.* (1968) 1445.
4. Dang, T. P., Kagan, H. B., *Chem. Commun.* (1971) 481.
5. Morrison, J. D., Masler, W. F., Hattaway, S., "Catalysis in Organic Synthesis," Academic, New York, 1976.
6. Kagan, H. B., Fiaud, J. C., *Top. Stereochem.*, E. L. Eliel, A. L. Allinger, Ed. (1978) **10**, 175.
7. Bogdanovic, B., Henc, B., Meister, B., Pauling, H., Wilke, G., *Angew. Chem., Int. Ed., Engl.* (1972) **11**, 1023.
8. Pino, P., Consiglio, G., Botteghi, G., Salomon, C., *Adv. Chem. Ser.* (1974) **132**, 295.
9. Hayashi, T., Tajika, M., Tamao, K., Kumada, M., *J. Am. Chem. Soc.* (1975) **98**, 3718.
10. Trost, B. M., Strege, P. E., *J. Am. Chem. Soc.* (1977) **99**, 1649.
11. Michaelson, R. C., Palermo, R. E., Sharpless, K. B., *J. Am. Chem. Soc.* (1977) **99**, 1290.
12. Yamada, S. I., Mashiko, T., Terashima, S., *J. Am. Chem. Soc.* (1977) **99**, 1988.
13. Vineyard, B. D., Knowles, W. S., Sabacky, M. S., Bachman, G. L., Weinkauff, D. J., *J. Am. Chem. Soc.* (1977) **99**, 1990.
14. Fryzuk, M. D., Bosnich, B., *J. Am. Chem. Soc.* (1977) **99**, 6262.
15. Heil, B., Törös, S., Vastag, S., Markö, L., *J. Organomet. Chem.* (1975) **94**, C47.
16. Ojima, I., Nihonyanagi, M., Nagai, Y., *Chem. Commun.* (1972) 938.
17. Corriu, R. J. P., Moreau, J. J. E., *Chem. Commun.* (1973) 38.
18. Kagan, H. B., Langlois, N., Dang, T. P., *J. Organomet. Chem.* (1975) **91**, 353.
19. Ojima, I., Kumagai, M., Nagai, Y., *J. Organomet. Chem.* (1976) **111**, 43.
20. Dumont, W., Poulin, J. C., Dang, T. P., Kagan, H. B., *J. Am. Chem. Soc.* (1973) **95**, 8295.
21. Dang, T. P., Thèse Orsay (1972).
22. Kagan, H. B., Dang, T. P., *J. Am. Chem. Soc.* (1972) **94**, 6429.
23. Peyronel, J. F., Thèse Orsay (1977).
24. Peyronel, J. F., Fiaud, J. C., Kagan, H. B., unpublished data.
25. Corriu, R. J. P., Moreau, J. J. E., *J. Organomet. Chem.* (1975) **85**, 19.
26. Ojima, I., Yamamoto, K., Kumada, M., "Aspects of Homogeneous Catalysis," Vol. 3, R. Ugo, Ed., D. Reidel, 1977.
27. Ojima, I., Kogure, T., Kumagai, M., *J. Org. Chem.* (1977) **42**, 1671.
28. Glaser, R., *Tetrahedron Lett.* (1975) 2127.
29. Guetté, J. P., Capillon, J., Perlat, M., Guetté, M., *Tetrahedron Lett.* (1974) 2409.
30. Muir, K. W., Ibers, J. A., *Inorg. Chem.* (1970) **9**, 440.
31. Ojima, I., Nithoyanagi, M., Nagai, Y., *Bull. Chem. Soc., Jpn.* (1972) **45**, 3722.
32. Brunie, S., Mazan, J., Langlois, N., Kagan, H. B., *J. Organomet. Chem.* (1976) **114**, 225.
33. Kagan, H. B., Peyronel, J. F., *Nouveau J. Chim.* (1978) **2**, 211.
34. Wong, P. K., Lau, K. S. Y., Stille, J. K., *J. Am. Chem. Soc.* (1974) **96**, 5956.
35. Whitesides, G. M., Bergreiter, D. E., Kindall, P. E., *J. Am. Chem. Soc.* (1974) **96**, 2806.
36. Michalska, Z. M., Webster, D. E., *Chem. Technol.* (1975) 188.

37. Bursian, M., Pracejus, H., *Chem. Abstr.* (1973) **78**, 72591.
38. Takaishi, N., Imai, H., Bertelo, C. A., Stille, J. K., *J. Am. Chem. Soc.* (1976) **98**, 5401.
39. Capka, M., *Collect. Czech. Chem. Commun.* (1977) **42**, 3410.
40. Kolb, I., Cerny, M., Hetflejs, J., *React. Kinet. Catal. Lett.* (1977) **7**, 199.
41. Dietzmann, I., Tomanova, D., Hetflejs, J., *Collect. Czech. Chem. Commun.* (1974) **39**, 123.
42. Rebek, J., Gavina, F., *J. Am. Chem. Soc.* (1975) **97**, 1591.
43. Rebek, J., Gavina, F., Navarro, C., *Tetrahedron Lett.* (1977) 302.
44. Glocking, F., Hill, G. C., *J. Chem. Soc. A* (1971) 2138.
45. Oliver, A. T., Graham, W. A. J., *Inorg. Chem.* (1971) **10**, 1.
46. Corriu, R. J. P., Lanneau, G. F., *Bull. Soc. Chim. Fr.* (1968) 459.
47. Nogai, U., Shishido, T., Shiba, R., Mitsuhashi, H., *Tetrahedron* (1965) **21**, 1701.
48. Guetté, J. P., Perlat, M., Capillon, J., Boucherot, D., *Tetrahedron Lett.* (1974) 2477.

RECEIVED March 13, 1978.

Activation of C–H Bonds by Bidentate Phosphorus Ligand Complexes of Iron

S. D. ITTEL, C. A. TOLMAN, A. D. ENGLISH, and J. P. JESSON

E. I. du Pont de Nemours and Co., Inc., Central Research and Development Dept., Experimental Station, Wilmington, DE 19898

> *Iron complexes containing bidentate alkyl and aryl phosphorus ligands cleave a variety of C–H bonds under mild conditions. Hydrido acetylide complexes were prepared by oxidative addition of primary acetylenes in the Fe(DPPE)$_2$ and the Fe(DMPE)$_2$ systems [DPPE = bis(diphenylphosphino)ethane, DMPE = bis(dimethylphosphino)ethane]. The Fe(DMPE)$_2$ system also cleaves C–H bonds of activated methyl groups, aromatic compounds, and certain other sp^2 hybridized molecules. The C–H cleavage reactions are reversible, resulting in equilibrium mixtures of isomeric products in many cases. Studies of substituted benzenes show that while product stability is favored by electron withdrawing substituents, steric effects play a predominant role in the determination of product distribution.*

One of the basic goals of industrial catalysis is the activation of C–H bonds in a variety of transformations. In heterogeneous catalysis, this reaction is well established (1, 2, 3) but often suffers from indiscriminate bond activation. More recently it has become clear that C–H activation can be accomplished by homogeneous catalytic systems involving transition metal complexes (4, 5, 6). The H–D exchange reaction has been catalyzed by a variety of high oxidation state cyclopentadienyl polyhydride complexes (7, 8, 9, 10) and electrophiles such as Pd^{2+} and Pt^{2+} (2, 11, 12). These simple exchange reactions point the way toward more fruitful synthetic applications.

The exchange reactions by polyhydrides and low-valent species are generally agreed (10) to involve oxidative addition of the C–H bond to form C–M and M–H bonds, exchange of D for H, and finally, reductive elimination yielding the exchanged product. The intermediate aryl hydrido species generally were not detected but were postulated on the precedent of irreversible addition of aryl C–H bonds to the highly reactive transient intermediates $M(C_5H_5)_2$, M = molybdenum (13) and tungsten (14), or $Ru(DMPE)_2$ (15, 16) [DMPE = bis(dimethylphosphino)-ethane].

As part of our continuing study involving the chemistry of low-valent transition metal complexes bearing phosphorus ligands (17), we were intrigued by a literature report of the zero-valent, four-coordinate complex $Fe(DMPE)_2$ (18). It had already been shown that zero-valent complexes of iron are more electron rich than their similarly ligated nickel analogs; $[Fe(P(OMe)_3)_5]$ is protonated by the weak acid methanol (19) but strong acids are required to protonate $Ni[P(OMe)_3)_4]$ (20). It appeared that a coordinatively unsaturated, electron-rich species such as $Fe(DMPE)_2$ should have a wealth of interesting chemistry. The highlights of that chemistry (21, 22, 23, 24) and the chemistry of closely related species (25) are reported here.

Reduction of $Fe(DMPE)_2Cl_2$

The reduction (22) of $Fe(DMPE)_2Cl_2$ was carried out in THF using two equivalents of sodium naphthalenide. The expected product, $Fe(DMPE)_2$, was not obtained for reasons which will soon become apparent. Instead, two other products were obtained. The first was $Fe(DMPE)_2H(C_{10}H_7)$, analogous to previously reported ruthenium and osmium (15, 16) compounds, except that it exists in solution as an equilibrium mixture of cis and trans isomers. The second product was a zero-valent complex, but instead of being four coordinate it is five coordinate, having the structure shown in 2 and displaying an AB_4 $^{31}P\{^1H\}$

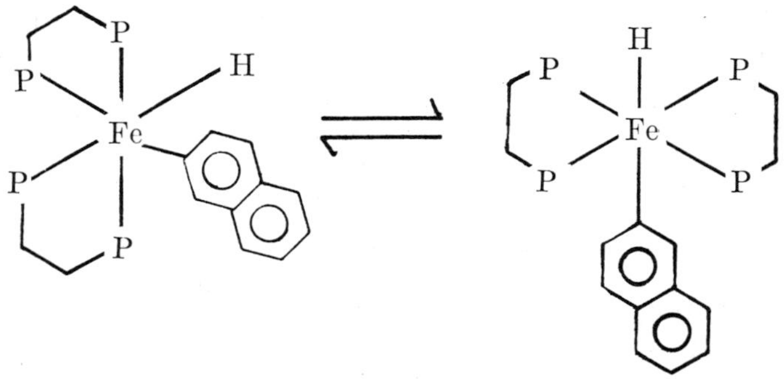

cis-1   trans-1

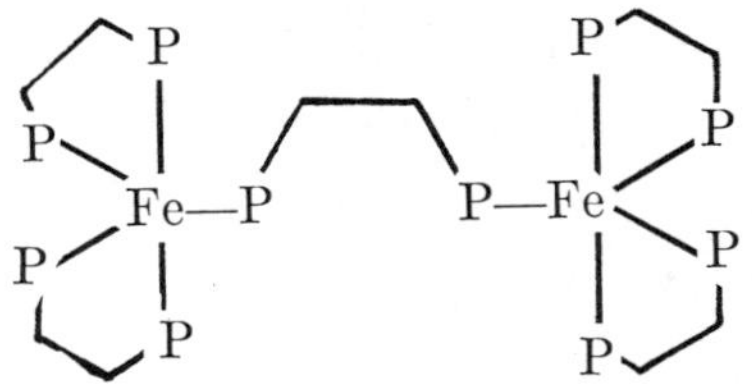

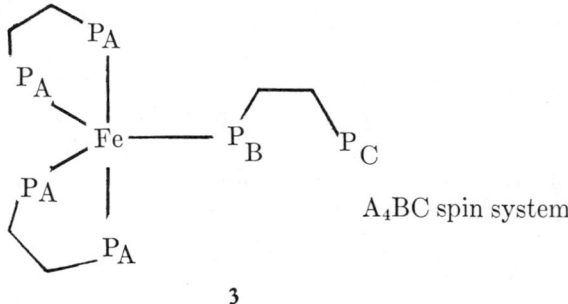

2

NMR spectrum. The origin of this second complex is readily ascertained by reacting one equivalent of DMPE with two equivalents of **1**. The reaction, which takes place in about one hour is conveniently followed by $^{31}P\{^1H\}$ NMR. Resonances attributable to **1** disappear from the spectrum as resonances attributable to **3** appear. Finally, **1** and **3** are completely consumed, giving **2**.

A_4BC spin system

3

The formation of **3** and **2** by reaction of DMPE with **1** explains the source of **2** in the preliminary reduction. More significantly, it demonstrates that **1** can reductively eliminate a C–H bond to give free naphthalene and the undetected intermediate species "Fe(DMPE)$_2$." This reaction is the first step for almost all of the following chemistry. The ruthenium and osmium analogs of **1** have been in the literature for many years (15, 16) but because they are much less prone to eliminate naphthalene, they have shown little of the chemistry described below for the iron complexes.

Formation of Five-Coordinate, Zero-Valent Species

These reactions are outside the scope of C–H activation but are mentioned here because they illustrate some of the properties of the Fe(DMPE)$_2$ system. The "Fe(DMPE)$_2$" generated from **1** reacts with CO to form Fe(DMPE)$_2$CO (22). The carbonyl stretching frequency of Fe(DMPE)$_2$CO is observed at 1812 cm^{-1}, exceptionally low for a non-bridging carbonyl, indicating a very high electron density on the metal.

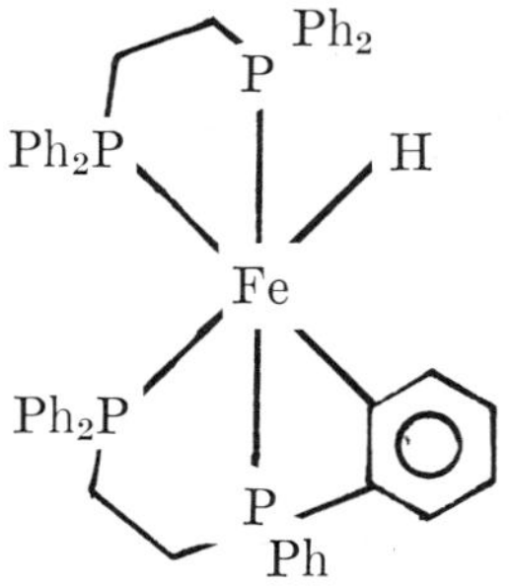

6

Phosphorus ligands also form Fe(DMPE)$_2$L complexes (22), limited only by steric interactions. Ligands ranging in cone angle (26) from P(OMe)$_3$ ($\theta = 107°$) to P(O-o-tolyl)$_3$ ($\theta = 141°$) form the five-coordinate complex, but PPh$_3$ ($\theta = 145°$) does not coordinate through phosphorus; rather, it behaves as a substituted benzene (vida infra).

Phosphorus ligands and CO also react with a variety of Fe(DPPE)$_2$ (25) complexes [DPPE = bis(diphenylphosphino)ethane] such as Fe(DPPE)$_2$(C$_2$H$_4$) (4) and Fe(DPPE)$_2$H$_2$ (5) to form Fe(DPPE)$_2$L complexes, but the fastest reaction is observed with the ortho-metalated species, **6**. Steric limitations are much more severe in the Fe(DPPE)$_2$ systems; the largest phosphorus ligand coordinated was P(OMe)$_2$Ph, having a cone angle of 116°.

A variety of unsaturated molecules react with **1** to form π complexes (22). Ethylene forms the simplest complex of this type:

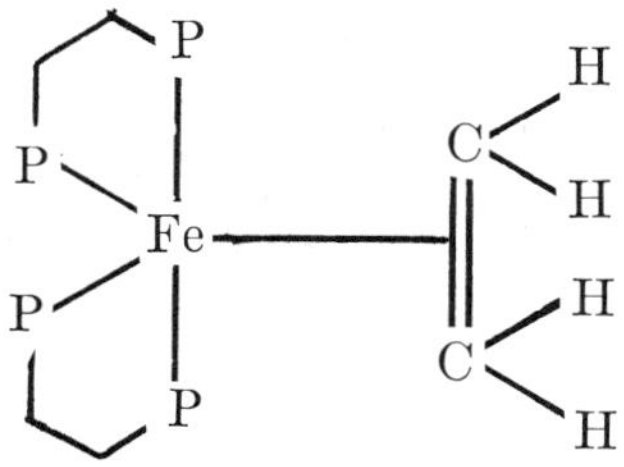

The ^{1}H NMR resonances of the ethylene are shifted upfield to 0.59 ppm below Me$_4$Si. This shift, one of the largest ever reported (27), is another indication of the high electron density on the iron. The ethylene complex and complexes of other unactivated olefins such as propylene and butadiene (which is η^2) undergo a fluxional process on the NMR time scale which equilibrates the four DMPE phosphorus nuclei. Complexes of activated olefins such as acrylonitrile are stereochemically rigid as a result of stronger bonding to the olefin (28, 29). The π complexes of disubstituted acetylenes show marked shifts in $v_{C≡C}$; diphenylacetylene shows a C≡C stretching frequency at 1720 cm^{-1}, 502 cm^{-1} lower than the free acetylene. Azobenzene also forms a π complex, adding one more metal to those observed to form π complexes with diazenes (30, 31).

Anthracene forms an unusual η^4 complex in which one of the DMPE molecules is monodentate (22);

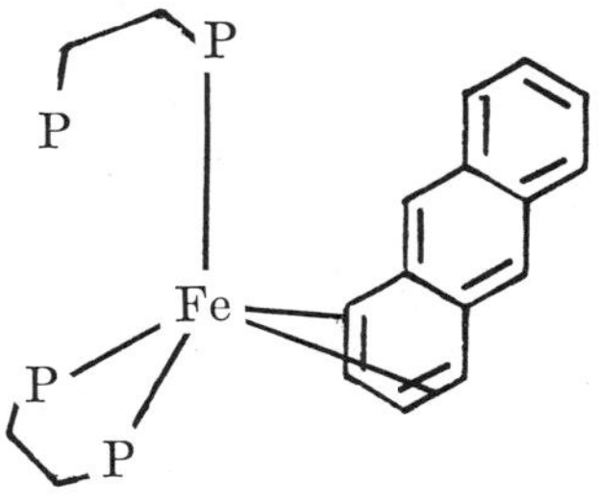

1,3-cyclohexadiene forms an analogous complex in contrast to the η^2 complex formed with butadiene. Apparently the inability of the cyclic species to adopt a transoid geometry favors the more common η^4 coordination enough to overcome the chelate effect of the DMPE.

Activation of sp^3 C–H Bonds

Early in our investigation of **1**, acetonitrile was used as a solvent for NMR studies. Surprisingly, this solvent reacted with **1** to form a trans-hydrido cyanomethyl complex of the type:

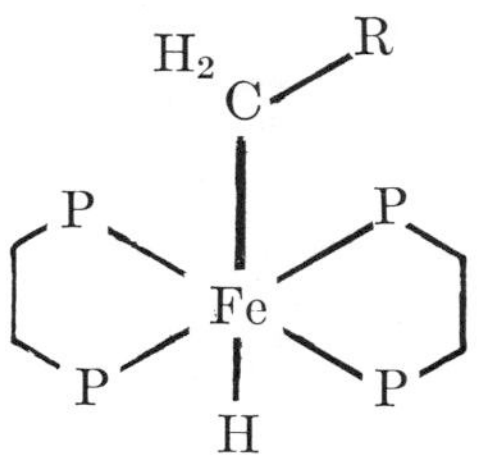

Later, many compounds having activated methyl groups were found to undergo similar reactions; ethyl acetate, dimethyl sulfoxide, ethylcyanoacetate, and malononitrile are just a few of the compounds studied (23). Acetone forms a similar complex but if a deficiency of acetone is used, the dimeric species forms. Similar observations were made in the related systems [Ir(PMe$_3$)$_4$]$^+$ and [Ir(DMPE)$_2$]$^+$ by Herskovitz (32, 33).

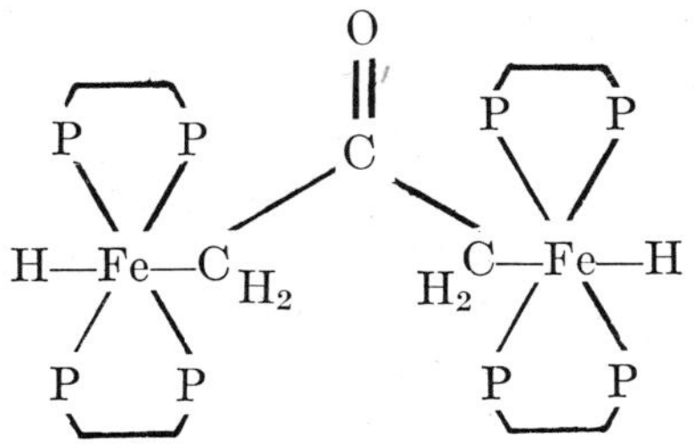

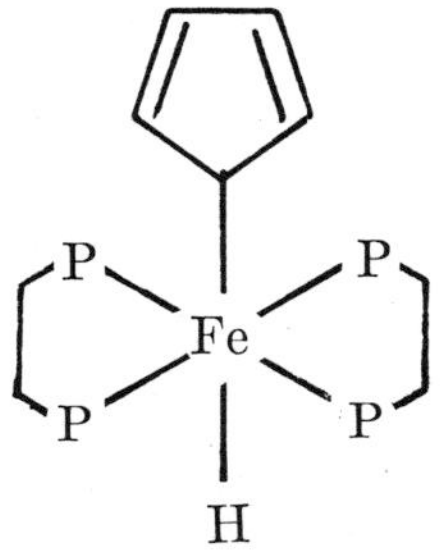

$$\text{H-Fe-C}_{H_2}\overset{O}{\underset{\|}{C}}\text{C-Fe-H}$$
(with P P / P P chelating ligands on each Fe)

[Structure: CpFe(H)(P-P)₂ with four P donors and hydride]

The reaction of 1 with cyclopentadiene (23) yields a complex in which the η^1-cyclopentadienyl group is nonfluxional (34). In contrast, the reaction of 4 with cyclopentadiene yields (25) the known (35) compound:

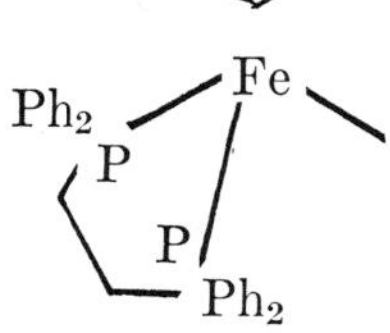

[Structure: CpFe(H)(Ph₂PCH₂CH₂PPh₂)]

After oxidative addition of the C–H bond, steric crowding is relieved by dissociation of one DPPE and subsequent η^5-cyclopentadienyl coordination.

The reaction of compounds having activated methyl groups with 4, 5, or 6 fails to yield hydrido alkyl complexes. Reaction of 4 with β-diketones (36) presumably proceeds through an intermediate similar to the cyclopentadienyl complex, but the final overall reaction is:

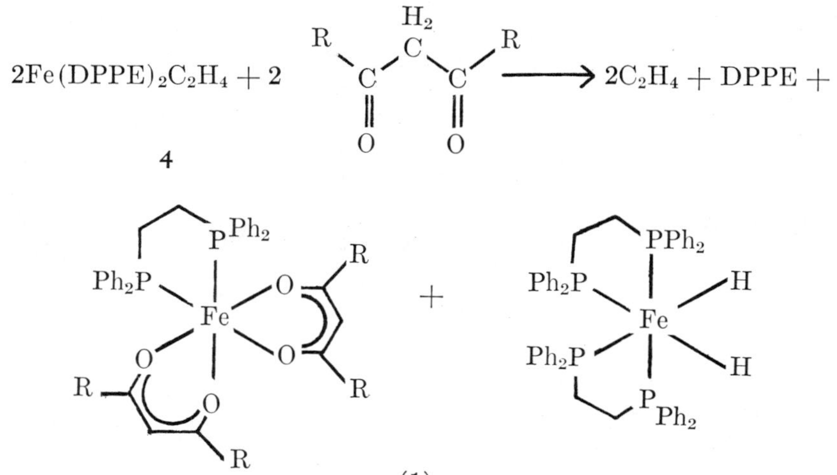

7. ITTEL ET AL. *Bidentate Phosphorus Ligand Complexes*

$$2Fe(DPPE)_2C_2H_4 + 2 \; R{-}C(={O}){-}CH_2{-}C(={O}){-}R \longrightarrow 2C_2H_4 + DPPE +$$

(1)

Kinetic measurements indicate that loss of naphthalene from **1** to yield "Fe(DMPE)$_2$" is the rate-limiting step for the formation of these DMPE complexes. The reactions to give five-coordinate, zero-valent species and the reactions to give new hydrido alkyl species all take place on approximately the same time scale. In addition, the trans-hydrido alkyl complexes represent only the thermodynamic products of the C–H cleavage reactions. When the reaction is followed by ^{31}P and ^1H NMR, a cis intermediate is observed. As an example, the reaction of **1** with acetonitrile followed as a function of time by 100 MHz hydride NMR spectra is shown in Figure 1. An initial pattern at -12.5 ppm attributable to the cis isomer appears initially and then disappears as the trans isomer at -23.2 ppm appears. Hydride resonances from the cis- and trans-hydrido aryl species are absent because the spectra were run in C_6D_6 (vide infra); but it is also probable that the reductive elimination of arene takes place through the cis isomers. These observations and analogous results in other systems strongly suggest, though do not necessitate, a concerted three-center intermediate for the oxidative addition.

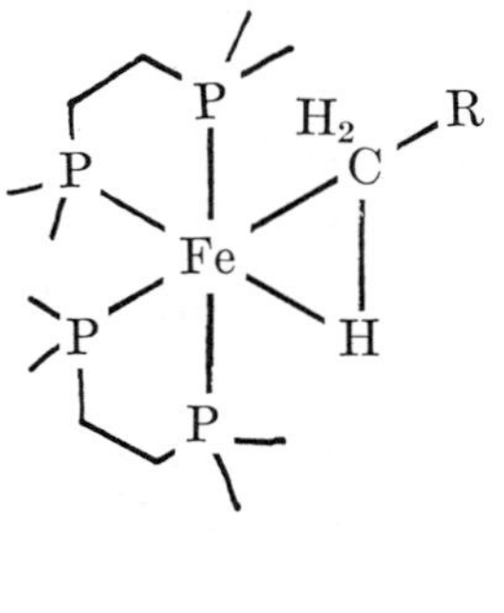

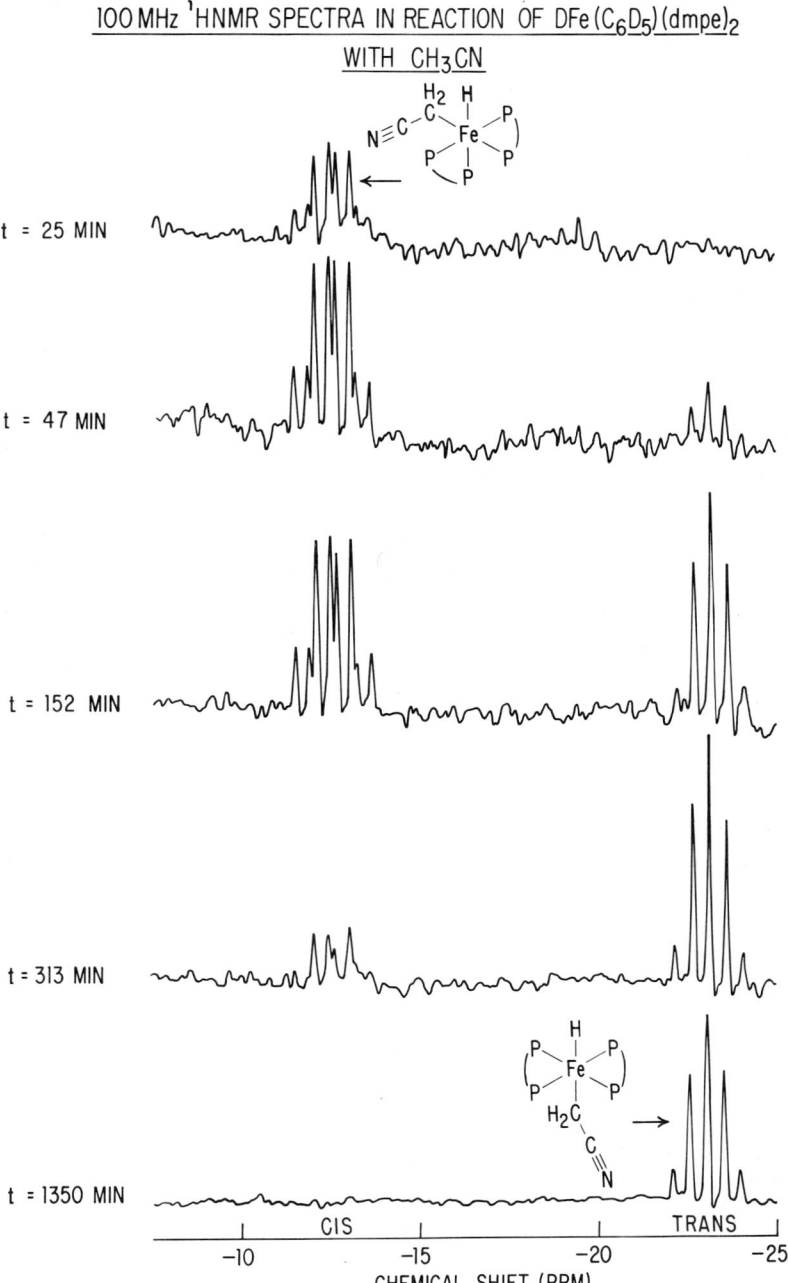

Figure 1. Hydride NMR spectra (100 MHz) of the reaction of acetonitrile with $Fe(DMPE)_2D(C_6D_5)$ (1 dissolved in C_6D_6) taken as a function of time

Activation of sp C–H Bonds

Earlier we noted that disubstituted acetylenes react with 1 to give π-bonded complexes. Acetylene and monosubstituted acetylenes react to give trans-hydrido acetylide complexes (23). The DPPE complexes, 3, 4, and 5, also react with acetylenes to give trans-hydrido acetylide complexes (25, 37). Iso-electronic hydrido cyanide complexes are obtained upon reaction with hydrogen cyanide. The reaction of HCN with $Fe(DMPE)_2H(C_{10}H_7)$ is the first reaction reported here (23) that proceeds by a mechanism different from those already discussed. The reaction, which is over essentially upon mixing, seems to involve electrophilic attack prior to loss of naphthalene.

Activation of Aromatic C–H Bonds (24)

Dissolution of 1 in benzene-d_6 forms an equilibrium mixture of four species given by:

$$\begin{array}{c} \quad\quad\quad K \\ cis\text{-}1 + C_6D_6 \rightleftharpoons cis\text{-}Fe(DMPE)_2D(C_6D_5) + C_{10}H_8 \\ \updownarrow \quad\quad\quad\quad\quad\quad \updownarrow \\ trans\text{-}1 \quad\quad\quad trans\text{-}Fe(DMPE)_2D(C_6D_5) \end{array} \quad (2)$$

The $^{31}P\{^1H\}$ NMR singlet of trans-1 decreases in intensity, the ABCD spin system attributable to cis-1 appears to loose resolution, and a 1:1:1 triplet attributable to the trans deuteride appears. The apparent loss of resolution of the cis isomers is attributable to superposition of the normal ABCD spin system and to another one having an additional 1:1:1 splitting from the deuterium coupling. The observed intensities indicate an equilibrium constant for Reaction 2 of 0.08; $Fe(DMPE)_2$ has an affinity for naphthalene more than one order of magnitude greater than that for benzene.

The reaction of toluene with 1 could conceivably lead to four different trans-isomeric products by attack at the ortho, meta, para, or alpha positions. The carbanion chemistry of toluene would predict attack at the methyl group but, as in other transition metal arene activation studies (8, 9, 10), this is not found. The ^{31}P NMR spectrum of the toluene adducts displays two singlets attributable to trans-tolyl adducts in an intensity ratio of 1:1.6 in addition to resonances of cis-tolyl adducts. Selective mono-deuteration experiments show that the weaker downfield resonance is attributable to the para isomer while the stronger one is attributable to the meta isomer; no ortho isomer is observed. If one corrects for statistical effects, the preference for para, meta, and ortho isomers is 1:0.8:0. We attribute this distribution of isomers primarily to steric effects, a view which is supported by the observation that 1 dissolves in *p*-xylene and mesitylene with no reaction; all of the aromatic C–H bonds are ortho to a methyl group. These results are also consistent with the observation of only the 2-naphthyl isomer of 1; a 1-naphthyl isomer would be ortho substituted and therefore sterically hindered.

Substitution of trifluoromethyl for the methyl of toluene enhances the selectivity for the para position over meta, and again no ortho is observed. Competitive studies show that a trifluoromethyl group activates benzene by almost four orders of magnitude, and methyl deactivates it by almost one order of magnitude. This is a marked change from prior C–H activation studies in which there was less than one order of magnitude change between the most activating and deactivating substituents (*8, 9, 10*), but it should be noted that the earlier studies were kinetic distribu-

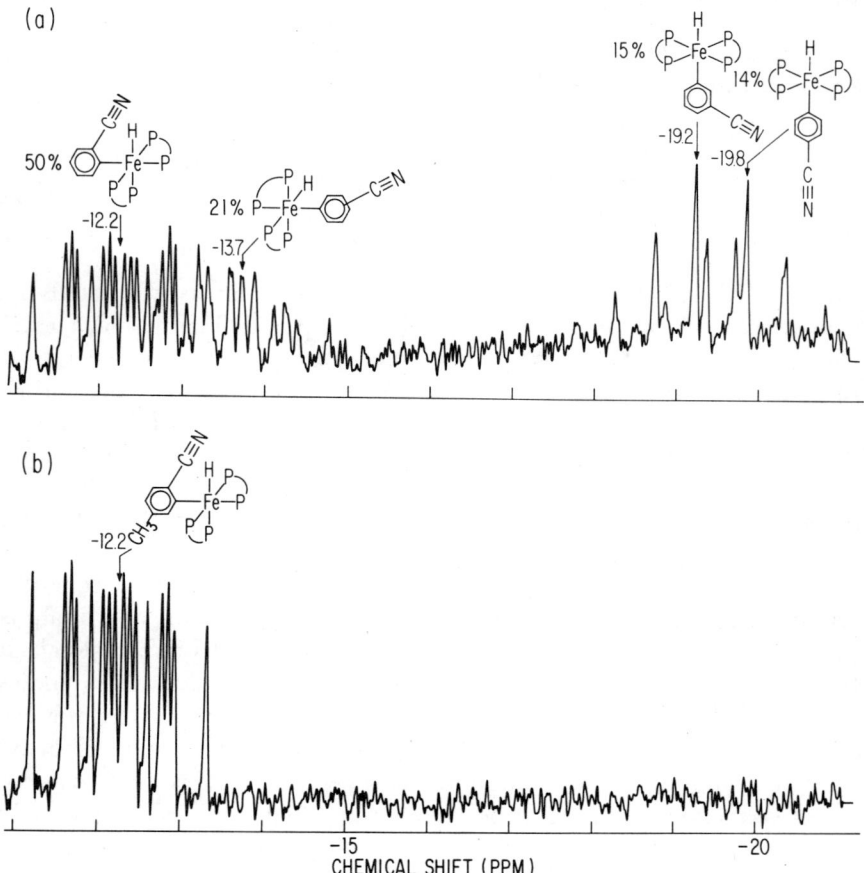

Figure 2. Hydride NMR spectra (100 MHz) of the (a) benzonitrile and (b) para-tolunitrile adducts to Fe(DMPE)$_2$

tions while we are measuring thermodynamic distributions. Fluorine substitution surprisingly yields only one product, the trans-ortho adduct. This observation is a result of the limited steric effect of a fluoro substituent combined with its power of activation. Electron withdrawal typically activates the entire ring with ortho > para > meta.

As a further example of these effects, benzonitrile with its strongly activating cyano group, shows the greatest selectivity for para over meta for trans isomers. The situation is complicated by the observation of two normal and one somewhat abnormal cis isomers (Figure 2). The normal isomers are attributable to meta and para species; the unusual one is ortho. When the reaction is carried out with p-tolunitrile, only the cis-ortho isomer is observed. A space-filling molecular model of cis-orthobenzonitrile can be constructed, but it is not possible to build a model of the trans-ortho isomer.

Activation of Non-Aromatic sp^2 C–H Bonds

Benzaldehyde reacts with 1 to give:

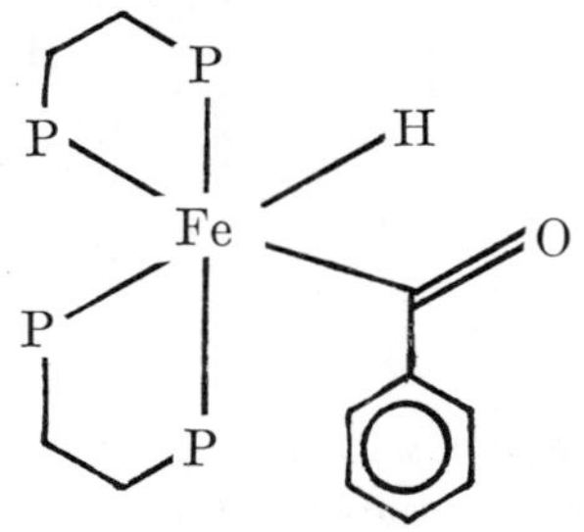

This complex does not extrude CO nor does $Fe(DMPE)_2H(C_6H_5)$ insert CO to yield the above. Propionaldehyde reacts with 1 to give a related acyl hydride, but $Fe(DMPE)_2CO$ is also formed. Such complexes have been proposed as intermediates in the decarbonylation of aldehydes by $RhCl(PPh_3)_3$ (38) but never before observed.

As we have already pointed out, simple olefins form π-bonded complexes with "$Fe(DMPE)_2$." Stimulated by recent work with ruthenium (39), we found that reaction of ethyl methacrylate with 1 forms the vinyl hydride complex:

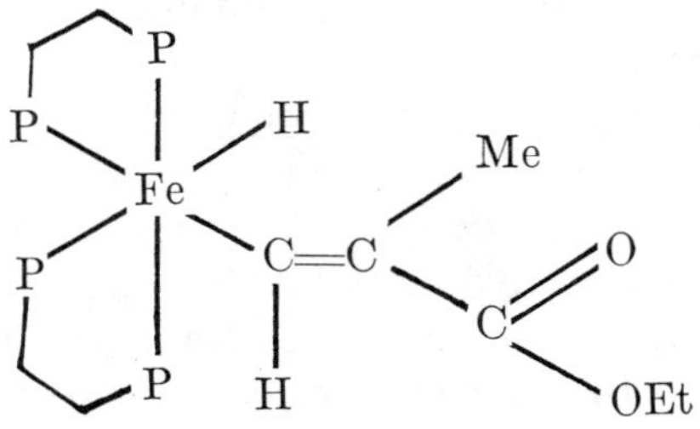

although it was not demonstrated conclusively that the methyl group is cis to the metal. An attempt to prepare the analogous complex with methacrylonitrile unexpectedly gave a π-olefin complex.

Other Reactions of Interest

We have already pointed out that 1 reacts with ethyl acetate at a methyl C–H bond and with ethyl benzoate at the meta and para positions. It was therefore surprising that similar reactions with the corresponding methyl esters gave products resulting from cleavage of the oxygen methyl bonds. In addition to MeC = O and PhC = O, the substituent R can be Ph and Me(MeO)P = O. The only apparent explanation for this difference in behavior between methyl and ethyl is steric; the methyl cone angle (90°) is smaller than that of an ethyl (102°) (*26*).

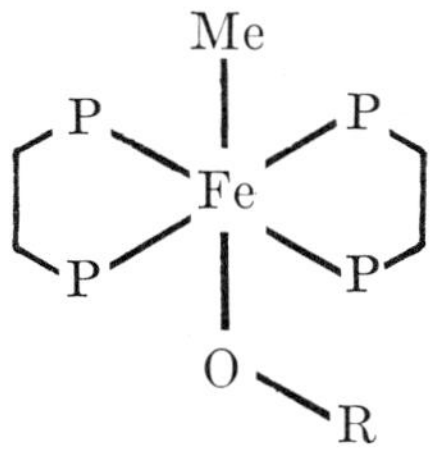

Trimethylsilane reacts with 1 to give a mixture of the cis- and trans-hydrido trimethylsilyl complexes. This reaction is of particular interest because trimethylsilane represents the only case in which we can be certain that there is no form of precoordination before attack to give the hydrido species. In all other cases—arenes, acetylenes, activated methyl groups, and others—there is some conceivable mode of coordination prior to the attack on the C–H bond

Kinetics and Thermodynamics of C–H Activation

The discussion in the preceeding sections dealt with systems at equilibrium, but just as in the earlier example involving acetonitrile, it is possible to monitor the progress of these reactions by a combination of spectroscopic techniques. An interesting example is the reaction of 1 with acetophenone. One of a series of hydride NMR spectra taken as a function of time is shown in Figure 3. The first products observed were the meta- and para-aryl hydride adducts. Later the cis-methyl adduct developed, and finally, the trans-methyl adduct (barely visible in the figure) developed as the system approached equilibrium.

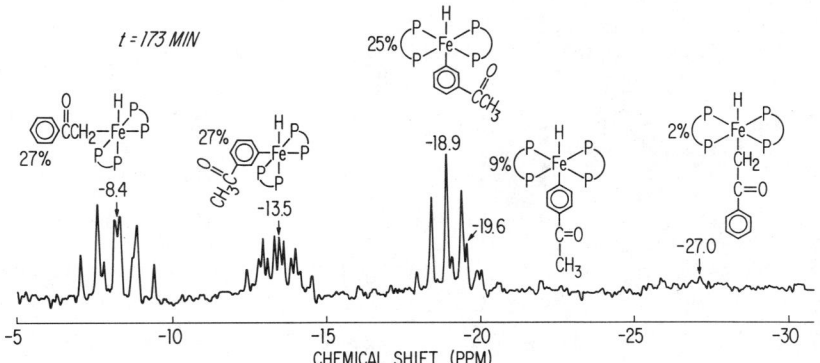

Figure 3. Hydride NMR spectrum (100 MHz) of the reaction of acetophenone with $Fe(DMPE)_2D(C_6D_5)$ (1 dissolved in C_6D_6) before equilibrium

These results point out the following general observations. (1) Aromatic C–H cleavage is faster than sp^3 CH cleavage, but (2) sp^3 C–H bond cleavage results in somewhat more stable products. (3) Isomerization between cis, trans, and positional isomers is relatively fast for aromatic adducts while sp^3 cis-trans isomerization is relatively slow. (4) Reactivity toward C–H cleavage is greatly enhanced by electron-withdrawing substituents on aromatic molecules; and (5) the positional preference increases in the order meta < para < ortho when steric effects do not dictate otherwise.

Literature Cited

1. Farkas, A., Farkas, L., *Trans. Faraday Soc.* (1937) **33**, 827.
2. Garnett, J. L., *Catal. Rev.* (1971) **5**, 229.
3. Somorjai, G. A., *Acc. Chem. Res.* (1976) **9**, 248.
4. Parshall, G. W., *Acc. Chem. Res.* (1970) **3**, 139.
5. Parshall, G. W., *Acc. Chem. Res.* (1975) **8**, 113.
6. Parshall, G. W., *Catalysis (London)* (1977) **1**, 335.
7. Chatt, J., Coffey, R. S., *J. Chem. Soc. A* (1969) 1963.
8. Barefield, E. K., Parshall, G. W., Tebbe, F. N., *J. Am. Chem. Soc.* (1970) **92**, 5234.
9. Tebbe, F. N., Parshall, G. W., *J. Am. Chem. Soc.* (1971) **93**, 3793.
10. Klabunde, U., Parshall, G. W., *J. Am. Chem. Soc.* (1972) **94**, 9081.
11. Hodges, R. J., Webster, D. E., Wells, P. B., *J. Chem. Soc. A* (1971) 3230.
12. Tyabin, M. B., Shilov, A. E., Shteinman, A. A., *Dolk. Akad. Nauk. SSSR* (1971) **198**, 381.
13. Wong, K. L. T., Thomas, J. L., Brintzinger, H. H., *J. Am. Chem. Soc.* (1974) **96**, 3694.
14. Green, M. L. H., Knowles, P. J., *J. Chem. Soc. A* (1971) 1508.
15. Chatt, J., Davidson, J. M., *J. Chem. Soc.* (1965) 843.
16. Iberkne, S. D., Kilbourn, B. T., Raeburn, V. A., Russell, D. R., *J. Chem. Soc. A* (1971) 1118.

17. English, A. D., Ittel, S. D., Tolman, C. A., Meakin, P., Jesson, J. P., *J. Am. Chem. Soc.* (1977) **99**, 117.
18. Chatt, J., Watson, H. R., *J. Chem. Soc.* (1962) 2545.
19. Muetterties, E. L., Rathke, J. W., *Chem. Commun.* (1974) 850.
20. Tolman, C. A., *Inorg. Chem.* (1972) **11**, 3128.
21. Ittel, S. D., Tolman, C. A., English, A. D., Jesson, J. P., *J. Am. Chem. Soc.* (1976) **98**, 6073.
22. Tolman, C. A., Ittel, S. D., English, A. D., Jesson, J. P., *J. Am. Chem. Soc.* (1978) **100**, 4080.
23. Ittel, S. D., Tolman, C. A., English, A. D., Jesson, J. P., *J. Am. Chem. Soc.* (1978) unpublished data.
24. Tolman, C. A., Ittel, S. D., English, A. D., Jesson, J. P., *J. Am. Chem. Soc.* (1978) unpublished data.
25. Ittel, S. D., Tolman, C. A., Krusic, P. J., English, A. D., Jesson, J. P., *Inorg. Chem.* (1978) **17**.
26. Tolman, C. A., *Chem. Rev.* (1977) **77**, 313.
27. Tolman, C. A., English, A. D., Manzer, L. E., *Inorg. Chem.* (1975) **14**, 2353.
28. Ittel, S. D., *Inorg. Chem.* (1977) **16**, 2589.
29. Tolman, C. A., *J. Am. Chem. Soc.* (1974) **96**, 2780.
30. Ittel, S. D., Ibers, J. A., *J. Organomet. Chem.* (1973) **57**, 389.
31. Ittel, S. D., Ibers, J. A., *Inorg. Chem.* (1975) **14**, 1183.
32. English, A. D., Herskovitz, T., *J. Am. Chem. Soc.* (1977) **99**, 1648.
33. Herskovitz, T., private communication.
34. Bennett, M. J., Cotton, T. A., Davison, A., Faller, J. W., Lipparo, J. J., Morehouse, S. M., *J. Am. Chem. Soc.* (1966) **88**, 4371.
35. Mays, M. J., Sears, P. L., *J. Chem. Soc., Dalton Trans.* (1973) 1873.
36. Ittel, S. D., *Inorg. Chem.* (1977) **16**, 1245.
37. Ikariya, T., Yamamoto, A., *J. Organomet. Chem.* (1976) **118**, 65.
38. Lochon, C. F., Miller, R. G., *J. Am. Chem. Soc.* (1976) **98**, 1281.
39. Komiya, S., Ito, T., Cowie, M., Yamamoto, A., Ibers, J. A., *J. Am. Chem. Soc.* (1976) **98**, 3874.

RECEIVED February 22, 1978.

Homogeneous Catalysis of the Water Gas Shift Reaction by Metal Carbonyls

PETER C. FORD[1], CHARLES UNGERMANN, VINCENT LANDIS, and SERGIO A. MOYA—Department of Chemistry, University of California, Santa Barbara, CA 93106

ROBERT C. RINKER—Department of Chemical Engineering, University of California, Santa Barbara, CA 93106

RICHARD M. LAINE—SRI International, Menlo Park, CA 94025

> *Summarized are our recent studies of the homogeneous catalysis of the water gas shift reaction. Characterization of the previously reported catalyst based on $Ru_3(CO)_{12}$ in alkaline, aqueous ethoxyethanol solution indicates that the principal ruthenium components are tetraruthenium carbonyl hydride anions. A mechanism is proposed involving the attack of OH^- or H_2O on coordinated CO to give, after loss of CO_2, a dihydride metal species MH_2. Reductive elimination of hydrogen and coordination of another CO regenerates the original metal carbonyl MCO. A number of other metal carbonyls proved to form active catalysts under analogous conditions. Active catalysts are also formed from $H_4Ru_4(CO)_{12}$ in acidic aqueous diglyme solution and from $H_4Ru_4(CO)_{12}$ or $H_4Ru_4(CO)_{12}/Fe(CO)_5$ mixtures in organic amine solutions.*

An important route for the production of hydrogen from water is the water gas shift reaction (Reaction 1). Since water gas (a mixture of CO_2, CO, H_2O, and H_2) can be obtained from the reaction of steam with hot coke, it will play an important role in methods of improving

$$H_2O(g) + CO(g) \rightleftarrows CO_2(g) + H_2(g) \qquad (1)$$

[1] Address correspondence to this author at the Department of Chemistry, University of California, Santa Barbara, CA 93106.

the usefulness of coal as a primary fuel source in this country. Schemes for the gasification and liquification of coal (i.e., the production of relatively light hydrocarbons) require copious, readily available hydrogen in addition to that already produced. The enormous quantities of industrial hydrogen now used (1976 production in the United States was $\sim 10^{11}$ standard cubic meters (1, 2)) are derived largely by the steam reforming or partial oxidation of hydrocarbons. Thus, even without new, hydrogen-dependent methods for the synthesis of hydrocarbons, a major saving of this valuable resource would be one benefit from the better use of other methods for the production of hydrogen, e.g., the water–gas shift reaction.

Commercial methods for carrying out the shift reaction involve heterogeneous metal oxide catalysts at elevated temperatures (3, 4). According to the thermodynamics of Reaction 1, $\Delta G° = -6.81$ kcal/mol (-4.76), $\Delta H° = -9.83$ kcal/mol (0.68), and $\Delta S° = -10.1$ eu (18.3) [figures in parentheses are for $H_2O(l)$ rather than $H_2O(g)$], greater production efficiency should be realized by carrying out this exothermic reversible process at lower temperatures.

Considerable precedent exists for the use of H_2O/CO mixtures as the source of hydrogen in the homogeneous reductions of various organic compounds. The hydro-hydroxymethylation of olefins (5, 6) is one such example (Reaction 2) where the equivalent of one CO and two hydrogens are added to the olefin in a process which consumes three moles of

$$RCH=CH_2 \xrightarrow[Fe(CO)_5/NR_3]{CO/H_2O} RCH_2CH_2CH_2OH + RCH\!-\!CH_3 \quad (2)$$
$$\qquad\qquad\qquad\qquad\qquad\qquad\qquad\qquad\qquad\qquad\quad |$$
$$\qquad\qquad\qquad\qquad\qquad\qquad\qquad\qquad\qquad\quad CH_2OH$$

CO and two of H_2O and produces two moles of CO_2. Homogeneous catalysis of this reaction can be accomplished with an iron carbonyl in conjunction with a Bronsted acid or base. Thus it appears that conditions might be found where the shift reaction itself can be effected homogeneously using metal complex catalysts. To this end, we have been examining the activity of various homogeneous catalysts for the shift reaction, and our investigations of metal carbonyl cluster complexes are summarized here.

Catalysis by Ruthenium Carbonyl in Alkaline Solution

In a preliminary study (7) we reported that catalysis of the shift reaction is accomplished by a homogeneous solution prepared from $Ru_3(CO)_{12}$. For the initial experiments, the catalysis solution typically contained the following components in 15 mL of purified ethoxyethanol solvent: $Ru_3(CO)_{12}$ (0.126 g, 2×10^{-4} mol), KOH (0.5 g, 0.01 mol), and

H_2O (1.0 g, 0.06 mol). At 100°C under about 1 atm CO (occasionally recharged), this solution produced about 3×10^{-2} mol of H_2 and approximately an equivalent amount of CO_2 over a period of 30 days. Notably, this quantity represents a ratio of 150 moles of H_2 per mole of $Ru_3(CO)_{12}$ added or three moles of H_2 per mole of KOH added. Thus the system is catalytic both in ruthenium and in base.

When this reaction was carried out in a solution prepared from the deuterated solvent $CH_3CH_2OCH_2CH_2OD$ plus D_2O, D_2 (> 90%) was the hydrogen product as measured by mass spectrometry (small amounts of HD and H_2 found could be attributed to the isotopic impurity of the solvent mixture). Thus, the source of the dihydrogen product is water or water-exchangeable hydrogens in the solution (7). The homogeneity of the reaction solution is indicated by its clarity when examined with a strong light and by the fact that an active catalyst solution displayed the same rate of hydrogen production at 110°C before and after filtration through a Fluoropore filter (FHLP, 0.5 μ pore size) under an inert atmosphere. Additional qualitative support for homogeneity of the active components derives from the fact that the catalyst solutions show relatively high reproducibility of activity when prepared a number of different times by different individuals in our laboratories.

Examination of the active ruthenium carbonyl catalyst solution by IR and NMR spectroscopy and by ion exchange chromatography indicate that the solution contains at least three ionic ruthenium components, two of which are major. In addition, traces of a species with an IR spectrum consistent with that of $Ru_3(CO)_{12}$ is seen. The major ionic species are the trihydrido tetraruthenium dodecacarbonyl anion, $H_3Ru_4(CO)_{12}^-$ (> 50%) and a component (X^-) having spectral properties different from those of known ruthenium carbonyl species. IR and ^1H NMR spectra indicate the Et_4N^+ salt of X to be carbonyl hydride (v_{co} = 2072w, 2012s, 1987s, 1950s, br, 1732w in THF; hydride resonance at 22.53 τ in d^6-acetone). The chemical behavior of this material suggests that X^- may be $HRu_4(CO)_{13}^-$. Although purification problems have prevented our obtaining a good elemental analysis of the Et_4N^+ salt of X^-, the fact that the major species formed by the H_2SO_4 neutralization of X^- is $H_2Ru_4(CO)_{13}$ is consistent with this assignment.

When the active catalyst solution is neutralized with H_2SO_4 before isolating the components, three known ruthenium species are found in the reaction mixture: $Ru_3(CO)_{12}$ (20–30%), $H_2Ru_4(CO)_{13}$ (10%), and $H_4Ru_4(CO)_{12}$ (~ 60%). The formation of these species is certainly not unexpected under the reaction conditions given the observations by Lewis et al. (8) that the reactions of $Ru_3(CO)_{12}$ with water leads to $H_4Ru_4(CO)_{12}$ and $H_2Ru_4(CO)_{13}$ and by Kaesz and co-workers (9) that the former species can be deprotonated by base to give $H_3Ru_4(CO)_{12}^-$.

In this context, it is particularly interesting to note that catalysis runs under the same conditions, but using $H_4Ru_4(CO)_{12}$ as the initial source, have activity indistinguishable from that of runs starting with $Ru_3(CO)_{12}$. In addition, the spectral properties of the active solutions from these two sources are indistinguishable. Lastly it is also notable that comparable activity is seen with solutions prepared with ruthenium trichloride ($RuCl_3 \cdot nH_2O$) as the initial ruthenium source.

Closer examination of the reaction solution reveals that the nature of the base species changes markedly in the early stages of the catalysis runs. Carbon monoxide is known to react with aqueous alkali hydroxide to form the analogous alkali formate (*10, 11*). This reaction occurs under the initial solution conditions of the catalyst runs described here and is relatively fast compared with the shift reaction (*12*). Notably, the presence of the ruthenium catalyst has little if any effect on the rate of formate formation. Thus, for the typical catalysis runs at 100°C, titrimetric studies indicate that within a period of hours virtually all the KOH first added has been consumed and that > 90% of the base equivalents are present in the guise of potassium formate. The remaining base equivalents are largely a mixture of potassium bicarbonate and potassium carbonate. In this context, it is notable that $Ru_3(CO)_{12}$ and $H_4Ru_4(CO)_{12}$ solutions prepared using K_2CO_3 plus $KHCO_3$ as the initial bases give comparable catalytic activity to those prepared using KOH (Table I).

Possible Mechanisms for Catalysis

At this point it can be of value to speculate on the mechanisms which might be catalyzing the shift reaction. One mechanism is described in Scheme 1. In this scheme initial activation of carbon monoxide involves nucleophilic attack of OH^- or H_2O on M–CO to form the hydroxycarbonyl complex $M–CO_2H^-$. Ample precedent exists for this reaction. For example, several metal carbonyl complexes have been reported to undergo oxygen exchange with ^{18}O-labeled water in solution (*13, 14*), and the reversible formation of hydroxycarbonyl species has been postulated in logical mechanisms for this exchange. Such species have indeed been isolated, e.g., Reaction 3 (*15*).

$$IrCl_2(PhPMe_2)_2(CO)_2^+ \underset{\text{dry HCl}}{\overset{H_2O}{\rightleftharpoons}} IrCl_2(PhPMe_2)_2(CO)(-CO_2H) \quad (3)$$

The decarboxylation of the hydroxycarbonyl species $HM–CO_2H$ should be facile and has been proposed for other conversions of coordinated carbonyl to hydride, e.g., Reaction 4 (*16*). Subsequent reductive elimination of dihydrogen from MH_2 has precedence for a number of

Scheme 1

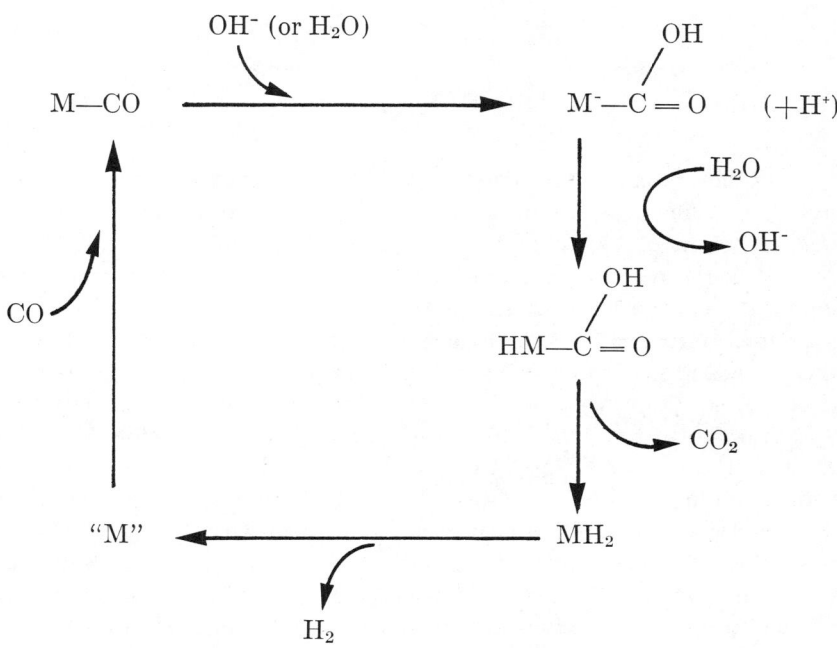

$$\text{trans-Pt(Et}_3\text{P)(CO)Cl}^+ + \text{OH}^- \rightarrow \text{Pt(Et}_3\text{P)}_2\text{ClH} + \text{CO}_2 \quad (4)$$

mononuclear and polynuclear systems (17). This last step forms the coordinatively unsaturated species "M" which undergoes rapid reaction with free CO to reform M–CO. (A similar cyclic scheme has also been proposed recently by Eisenberg (18)). A modification of this scheme would be to have CO assist in the dehydrogenation step in some manner, perhaps by coordination with MH_2 prior to loss of hydrogen.

The composition of the ruthenium catalyst solution is consistent with the steps proposed in Scheme 1, if we visualize M–CO as $HRu_4(CO)_{13}^-$ (or $H_2Ru_4(CO)_{13}$) and MH_2 as $H_3Ru_4(CO)_{12}^-$ (or $H_4Ru_4(CO)_{12}$). Considerably more information (kinetics, etc.) needs accumulation to support or disprove this mechanism, but at present it serves as a reasonable working hypothesis. With regard to the details of Scheme 1, it is notable that nucleophilic attack on coordinated carbonyls is often quite facile (13–16, 19, 20). In addition, we have found the reaction of $Ru_3(CO)_{12}$ with base in aqueous alcoholic solution to occur at lower temperature and more rapidly than the shift reaction catalysis. Thus it appears that CO activation is not rate limiting but another step, perhaps the reductive elimination of hydrogen, is.

Scheme 2

$$CO + OH^- \text{ (or } H_2O) \rightarrow HCO_2^- \text{ (or HCOOH)} \tag{5}$$

$$HCO_2^- \text{ (or HCOOH)} + H_2O \xrightarrow{\text{catalyst}} H_2 + CO_2 + OH^- \text{ (or } H_2O) \tag{6}$$

Another possible mechanism for the shift reaction would be one involving formate or formic acid as an intermediate (Scheme 2). The facile reaction of CO with strong base and the resulting presence of formate in the reaction solution initially prepared with KOH make compelling the consideration of this mechanism. Other platinum metal catalysts have been reported for Reaction 6 (21). However, we and others have demonstrated catalysis of the shift reaction in acidic solution (vide infra). In addition, we have found that adding significant concentrations of sodium formate to an active ruthenium catalyst solution had little effect on the rate of H_2 and CO_2 production. These observations argue against the importance of Scheme 2 for this system. (Our earlier report (7) that HCO_2^- is rapidly decomposed to H_2 plus CO_2 by the ruthenium catalyst in alkaline solution is apparently incorrect. In that study, formate was added as formic acid in quantities sufficient to acidify the solution to give conditions under which the system does decompose formate. We are evaluating the reaction in acid solution further.)

Other Catalysts in Alkaline Alcoholic Solution

We also have studied other metal carbonyl complexes in alkaline ethoxyethanol to survey the generality of the shift-reaction catalysis. Under conditions (0.9 atm CO, 100°C) comparable with those used for the ruthenium catalyst described above, iron, rhodium, osmium, and iridium carbonyls all proved active but rhenium carbonyl did not. For systems starting with the listed complexes, the normalized catalytic activities (*see* Table I; normalized activity is based on the number of

Table I. Activities of Various Catalysts for the Water–Gas Shift Reaction

Initial Complex	Initial Solution	Activity[a]
A. Alkaline solutions (low pressure)[b]		
$H_2FeRu_3(CO)_{13}$	c	10.3
$Ir_4(CO)_{12}$	c	5.3
$H_4Ru_4(CO)_{12}$	c	2.5
	d	3.3
	e	2.5

Table I. Continued

Initial Complex	Initial Solution	Activity[a]
$Ru_3(CO)_{12}$	c	2.3
	d	2.6
	e	~1
	f	3.2
$Fe(CO)_5$	d	~1
$Rh_6(CO)_{16}$	c	3.4[g]
$Ru_6C(CO)_{17}$	c	1.5
$H_3Re_3(CO)_{12}$	c	~0.15
$Re_2(CO)_{10}$	c	<0.1
$Ru_3(CO)_{12}/Fe(CO)_5$	h	7.4[h]

B. Alkaline solutions (high pressure)

$Rh_6(CO)_{16}$	i	110
$Ru_3(CO)_{12}$	i	53
$Ru_3(CO)_{12}/Fe_3(CO)_{12}$[j]	i	55
$Fe_3(CO)_{12}$	i	52[k]
$Ir_4(CO)_{12}$	i	10–20[k]
$Os_3(CO)_{12}$	i	10–17
$(Ir(CO)_3Cl)_2$	i	10
$Re_2(CO)_{10}$	i	~1

C. Acidic solutions (low pressure)[b]

$Ru_3(CO)_{12}$	l	~2.2[m]
$H_4Ru_4(CO)_{12}$	n	0.9
	l	~18[m]
	o	32
$Ru_6C(CO)_{17}$	l	~9[m]
$Fe(CO)_5$	o	~0

[a] Moles of H_2 produced per mole of catalyst per 24-hr period.
[b] $P_{co} = 0.9$ atm, $T = 100°C$ except where noted.
[c] 0.04 mmol complex, 2 mmol KOH, 0.02 mol H_2O, 3 mL ethoxyethanol.
[d] 0.06 mmol complex, 3 mmol KOH, 0.03 mol H_2O, 4.5 mL ethoxyethanol.
[e] 0.06 mmol complex, 0.015 mmol $KHCO_3$, 0.025 mol H_2O, 5 mL ethoxyethanol.
[f] 0.06 mmol complex, 0.15 mmol K_2CO_3, 0.15 mmol $KHCO_3$, 0.025 mol H_2O, 5 mL ethoxyethanol.
[g] Solution may have been heterogeneous.
[h] 0.039 mmol $Ru_3(CO)_{12}$, 0.064 mmol $Fe(CO)_5$, 2 mmol KOH, 0.02 mol H_2O, 3.4 mL ethoxyethanol. Based on total moles of $Fe(CO)_5$ plus $Ru_3(CO)_{12}$. Normalized activity equals 4.2 mol H_2/day/gram atom of metal (Fe + Ru).
[i] 0.1 mmol metal complex, 0.3 mmol KOH, 1.0 mL H_2O, and 6.0 mL CH_3OH, $P_{co} = 75$ atm, $T = 135°C$.
[j] Fe:Ru molar ratio = 3:1.
[k] Significant loss in activity when system is recharged and run a second day.
[l] 0.042 mmol complex, 0.18 mmol H_2SO_4, 20 mmol H_2O, 3 mL ethoxyethanol.
[m] Catalyst dies after several days owing to sublimation of $Ru_3(CO)_{12}$ out of the solution.
[n] 90°C.
[o] 0.042 mmol complex, 1.8 mmol H_2SO_4, 20 mmol H_2O, 3 mL diglyme.

moles of hydrogen produced per gram-atom of metal initially added to the system) followed the order: $H_2FeRu_3(CO)_{13}$ > $Ir_4(CO)_{12}$ > $H_4Ru_4(CO)_{12}$ ≅ $Ru_3(CO)_{12}$ ≅ $Fe(CO)_5$ > $Rh_6(CO)_{16}$ > $Ru_6C(CO)_{17}$ >> $H_3Re_3(CO)_{12}$ > $Re_2(CO)_{10}$. A somewhat different order was seen for the catalysts studied under conditions where CO pressure and temperature were much higher and methanol was the principal solvent (Table I). However, it should be noted that the latter data are largely from single day runs in pressurized bombs in contrast to the continuously monitored, periodically flushed runs in the low pressure glass vessels. It is particularly notable that under the higher pressure and temperature conditions the catalyst solutions based on $Ru_3(CO)_{12}$ and $Rh_6(CO)_{16}$ are quite active for the hydroformylation and hydrohydroxymethylation, respectively, of olefins by CO plus H_2O (22).

A particularly interesting observation is the high activity seen for $H_2FeRu_3(CO)_{13}$. Since this is considerably higher than that seen for either ruthenium carbonyl or iron carbonyl alone, these metals apparently act in a synergistic manner. Similar enhancements of activity are noted when the iron and ruthenium are added together in the forms $Fe(CO)_5$ and $Ru_3(CO)_{12}$ respectively (Table I). Optimal catalytic activity (normalized) was observed when the $Fe(CO)_5$:$Ru_3(CO)_{12}$ ratio is about 2:1 (a Fe:Ru ratio of 2:3). The ^1H NMR and IR spectra of these iron/ruthenium reaction solutions indicated the presence of a number of species. Among these, $Fe(CO)_5$ and $H_3Ru_4(CO)_{12}^-$ could be identified. Neutralization of the reaction solution with sulfuric acid followed by silica gel chromatography led to the identification of $H_4Ru_4(CO)_{12}$ and $H_2FeRu_3(CO)_{13}$ as the major cluster species as well as minor amounts of the triangular carbonyls $Ru_3(CO)_{12}$, $Ru_2Fe(CO)_{12}$, $RuFe_2(CO)_{12}$, and possibly $Fe_3(CO)_{12}$.

Our observations and those of others have led us to view these solutions in the following manner. First, it appears that under the reaction conditions (alkaline solution at 100°C under an atmosphere containing both H_2 and CO) interconversion between the cluster species is relatively facile. Thus, within hours or perhaps less, a solution prepared by mixing $Fe(CO)_5$ and $Ru_3(CO)_{12}$ in a 1:1 molar ratio may not be distinguishable from one prepared from $H_2FeRu_3(CO)_{13}$. Cluster species, specifically hydrido carbonyl anions, are the prominent reservoirs of the metals in the solution, but our mechanistic studies are too undeveloped to establish whether the clusters are the probable catalysts. However, if we follow the course of the earlier mechanistic speculation based on Scheme 1, then $H_4FeRu_3(CO)_{12}$ and $H_2FeRu_3(CO)_{13}$ or their deprotonated analogs may be MH_2 and M–CO respectively in the mixed metal catalysts. Kaesz (23) has reported the synthesis of $H_4FeRu_3(CO)_{12}$, but this species is relatively unstable in solution, decomposing to give the more stable $H_2FeRu_3(CO)_{13}$ among other species. This observation tempts one to suggest that the synergistic effect of the two metal catalysts might be attributed to the instability of the proposed MH_2 intermediate toward reductive elimination (Reaction 7) (8).

$$MH_2 \rightarrow M + H_2 \qquad (7)$$

Catalysis in Acidic Solution

Our initial forays into the shift-reaction catalysis (7) focussed on alkaline conditions owing to the prejudice that initial activation of coordinated CO would be particularly facile via hydroxide attack on CO (Reaction 8). However, water itself may be active (Reaction 9). To

$$M-CO + OH^- \rightleftarrows M-CO_2H^- \qquad (8)$$

$$M-CO + H_2O \rightleftarrows HM-CO_2H \qquad (9)$$

evaluate this possibility, we also have studied the ruthenium clusters in acidic solution. Notably Eisenberg (18) has recently reported a shift-reaction catalyst based on the Rh(I) complex $(Rh(CO)_2Cl)_2$ in an aqueous acetic acid/HCl/NaI medium, confirming the viability of such an approach.

When the reaction was run using $0.1N$ H_2SO_4 in aqueous ethoxyethanol as the solvent and $H_4Ru_4(CO)_{12}$ as the initial metallic species, the activity first seen was a factor of six higher than that found in the alkaline solution (Table I). However, after several days the activity decreased markedly, owing to the sublimation of the ruthenium from the solution into the cooler neck of the all-glass reaction vessel. The orange solid collecting at this location was identified as $Ru_3(CO)_{12}$ by its IR spectrum but may have contained traces of $H_2Ru_4(CO)_{13}$. When attempts were made to effect the catalysis with $Ru_3(CO)_{12}$ itself in the same acidic medium, little reaction was seen owing to the relatively rapid sublimation of this material from the solution. Similarly the carbide cluster $Ru_6C(CO)_{17}$ displayed initial activity much higher in acidic than in alkaline solution but again was unstable toward sublimation of $Ru_3(CO)_{12}$.

Despite the instability of these ruthenium carbonyl solutions in acid, the high initial activity encouraged the search for other solvents in which $Ru_3(CO)_{12}$ may prove more soluble. Diglyme meets this criterion, and preliminary data indicate that $H_4Ru_4(CO)_{12}$ in $1N$ H_2SO_4 aqueous diglyme is roughly one order of magnitude more active as a shift-reaction catalyst than in alkaline ethoxyethanol (Table I).

Various rationale can be offered for the enhanced catalytic activity in acid. The key steps in Scheme 1 are likely to be the activation of CO by nucleophilic attack on M–CO and the reductive elimination of H_2 from MH_2. If M–CO is either $HRu_4(CO)_{13}^-$ or $H_2Ru_4(CO)_{13}$, the latter species (which is favored by lowering the pH) should be the more susceptible one to nucleophilic attack. In acidic solution the more reactive nucleophile OH^- is at an inconsequential concentration, but it is worthwhile to remember that the concentration of H_2O ($6M$ in these experiments) is always much larger than (OH^-) regardless of the pH. With regard to the reductive elimination of hydrogen (Reaction 7), it is entirely possible that a neutral species (e.g., $H_4Ru_4(CO)_{12}$) can be more reactive than its deprotonated analog (e.g., $H_3Ru_4(CO)_{12}^-$). These are mechanistic aspects of these systems in need of greater exploration and under study in our laboratory.

Catalysis in Amine Solution

There is considerable precedent for the reactions of organic amines with metal carbonyls, in particular with iron carbonyls. Not only do amines react directly with $Fe(CO)_5$ to form various products including hydrogen (6, 19), but they are components of the iron carbonyl containing catalysts for the hydroformylation of olefins with CO and H_2O (5, 6, 24). Thus, catalysis of the shift reaction under analogous conditions is not unexpected, especially in the context of the activity displayed by the metal carbonyls in alkaline solutions (above, (7)). Indeed, Imyanitov et al. (25), in a study of the hydrogenation and hydrocarboxylation of dienes with CO and H_2O catalyzed by $Co_2(CO)_8$ and by $Rh_6(CO)_{16}$ plus organic amines, noted the formation of hydrogen and CO_2. In the absence of olefins, $Rh_6(CO)_{16}$ plus pyridine apparently catalyzed the shift reaction under high pressure (250 atm of CO) and elevated temperature (210°C); however, reduction of the pyridine to piperidine is a serious side reaction. Various amines including pyridine also have been reported as components in the purported homogeneous catalysis of the shift reaction by systems containing group VIII metal salts (26), and very recently there has been a report (27) of both the hydroformylation of olefins and the water gas shift reaction catalyzed by several metal carbonyls plus trimethyl amine in pressurized autoclaves.

Our studies have focussed largely on the catalysis of the shift reaction by ruthenium carbonyl and by the ruthenium carbonyl/iron carbonyl mixtures in the presence of organic amines under low pressures of CO. Representative studies are indicated in Table II where it is notable that ruthenium alone is a considerably better catalyst than is iron alone. Among the ruthenium systems, pyridine solutions are somewhat more

Table II. Catalysis in Amine Solutions (Low Pressure)[a]

Complex	Amine	Solvent	Activity[b]
$H_4Ru_4(CO)_{12}$ (0.04 mmol)	pyridine	c	13
$H_4Ru_4(CO)_{12}$ (0.02 mmol)	piperidine	d	8
$Fe(CO)_5$ (0.16 mmol)	piperidine	d	0.9[e]
$H_4Ru_4(CO)_{12}:Fe(CO)_5$ (0.02 mmol:0.08 mmol)	piperidine	d	10[f]

[a] $P_{co} = 0.9$ atm, $T = 100°C$, initial solutions contained 0.022 mol H_2O.
[b] Activity is number of moles of H_2 produced per day per mole of the initial complex added ($H_4Ru_4(CO)_{12}$ plus $Fe(CO)_5$).
[c] Pyridine (3 mL).
[d] Piperidine (1.5 mL):ethoxyethanol (2.8 mL).
[e] Low $CO_2:H_2$ ratios.
[f] Normalized activity is 12 mol H_2/day/g-atcm of metal added (Ru + Fe).

active than piperidine solutions. However, the most active systems among these are derived from iron/ruthenium mixtures, which are much more active than catalysts prepared under similar conditions from the individual metal carbonyls. (If the normalized catalytic activities are compared for piperidine solutions, the $Fe(CO)_5/H_4Ru_4(CO)_{12}$ system is a factor of six more active than that of $H_4Ru_4(CO)_{12}$ and a factor of 13 more active than that based on $Fe(CO)_5$ alone).

The enhanced activity of the ruthenium and the ruthenium/iron catalysts in the amine solutions over those in alkali base solutions may be the result of several perturbations. Certainly, the solvent effects alone can play a role in this case given that the amine concentrations are sufficient to change markedly the properties of the medium. Thus solvation effects on a rate-determining step or key equilibrium in a cycle such as Scheme 1 would have major consequences on the catalytic activity. If CO activation in steps such as Reactions 8 or 9 can be affected instead by other nucleophiles (for example, Reactions 10, 11, 12), then the high concentrations of the amines, higher even than [H_2O] under these conditions, plus the relative nucleophilicity of these species may accelerate the CO activation step.

$$M\text{–}CO + B \rightleftarrows M^-\text{–}\underset{B^+}{C} = O \tag{10}$$

$$M^-\text{–}\underset{B^+}{C}=O + H_2O \rightleftarrows HM\text{–}\underset{B^+}{C}=O \tag{11}$$

$$\text{HM-C}(=\text{O})(\text{B}^+) + \text{H}_2\text{O} \rightleftarrows \text{HMCO}_2\text{H} + \text{H}^+ + \text{B} \qquad (12)$$

$$\text{or} \rightleftarrows \text{MH}_2 + \text{B}^+\text{--CO}_2^- \rightarrow \text{B} + \text{CO}_2$$

Concluding Remarks

In this chapter we have demonstrated that metal carbonyl complexes can be active catalysts for the water–gas shift reaction under a variety of conditions. For example, $H_4Ru_4(CO)_{12}$ forms active catalysts in either acidic or basic solution and for the latter the base can be an alkali hydroxide or carbonate or an organic amine. Some differences between different metal carbonyls are apparent; however, in basic solution the most active catalysts are those prepared from the mixed metal Fe/Ru systems either by starting with the $H_2FeRu_3(CO)_{12}$ or with $Ru_3(CO)_{12}$ (or $H_4Ru_4(CO)_{12}$)/$Fe(CO)_5$ mixtures. A logical mechanism for the catalysis would involve the activation of CO by nucleophilic attack on the coordinated carbon monoxide followed by hydrolytic steps leading to formation of a metal dihydride. A key and perhaps rate-limiting step would be the reductive elimination of dihydrogen from this species. Mechanistic studies currently in progress in these laboratories are directed toward the evaluation of this and other possible catalytic cycles and toward the optimization of catalytic activity.

Experimental Procedures

Catalysis runs under low (0.9 atm) CO pressures were carried out in glass reactors. The catalysis solutions were prepared by dissolving the appropriate components in the solvent under an inert atmosphere at ambient temperature. The resulting solutions in the reaction vessels were then degassed by a freeze/thaw technique and the desired pressure of CO gas containing 6% methane as an inert marker gas was introduced to the bulb at ambient temperature. The glass reactors were then suspended in an oil bath and heated at a constant temperature. In this configuration the reaction solution was agitated vigorously by a magnetic stirring bar. The gas phase above the catalyst solution was sampled periodically by gas syringe, and the gas samples were analyzed by high-resolution gas chromatography on a calibrated Hewlett Packard 5830A programmable GC. Quantities of CO consumed and of CO_2 and H_2 produced were determined by comparison with the marker gas signal. The reaction vessels were periodically recharged by freeze/thaw degassing followed by refilling with the CO/marker gas mixture. Some runs under higher CO pressures were carried out in Parr stainless steel bombs equipped with Teflon liners.

Acknowledgment

This work was supported by the Department of Energy, Division of Basic Energy Sciences. Howard Walker and Ralph G. Pearson contributed significantly to the discussion and interpretation of these results.

Literature Cited

1. Farbsman, G. H., NASA-CR-134918, JA 76.
2. K. E. Cox, K. D. W. Williamson, Jr., Eds., "Hydrogen Production Technology," Vol. I, CRC Press, 1977.
3. Thomas, C. L., "Catalytic Processes and Proven Catalysts," p. 104, Academic, New York, 1970.
4. Aldridge, C. L., U.S. Patent **3,850,840** (1974).
5. Reppe, W., Vetter, H., *Justus Liebigs Ann. Chem.* (1953) **582**, 133.
6. von Kutepow, N., Kindler, H., *Angew. Chem.* (1960) **72**, 802.
7. Laine, R. M., Rinker, R. G., Ford, P. C., *J. Am. Chem. Soc.* (1977) **99**, 252.
8. Eady, C. R., Johnson, B. F. G., Lewis, J., *J. Chem. Soc., Dalton Trans.* (1977) 838.
9. Koepke, J. W., Johnson, J. R., Knox, S. A. R., Kaesz, H. D., *J. Am. Chem. Soc.* (1975) **97**, 3947.
10. "Encyclopedia of Chemical Technology," 2nd ed., Vol. 10, p. 101, Wiley, New York, 1972.
11. Iwata, M., *Nagaoka Kogyo Tanki Daigaku Koto Semmon Gakko Kenkyu Kiyo* (1968) **4**, 307; *Chem. Abstr.* (1969) **70**, 76989.
12. Walker, H., unpublished results.
13. Meutteries, E. L., *Inorg. Chem.* (1965) **4**, 1841.
14. Darensbourg, D. J., Froelich, J. A., *J. Am. Chem. Soc.* (1977) **99**, 5940.
15. Deeming, A. J., Shaw, B. L., *J. Chem. Soc. A* (1969) 443.
16. Clark, H. C., Dixon, K. R., Jacobs, W. J., *J. Am. Chem. Soc.* (1969) **91**, 1346.
17. Mays, M. J., Simpson, R. N. F., Stefanini, F. P., *J. Chem. Soc. A* (1970) 300.
18. Cheng, C. H., Hendriksen, D. E., Eisenberg, R., *J. Am. Chem. Soc.* (1977) **99**, 2791.
19. Edgell, W. F., Yang, M. T., Bulkin, B. J., Bayer, R., Korzumi, N., *J. Am. Chem. Soc.* (1965) **87**, 3080.
20. Casey, C. P., Neumann, S. M., *J. Am. Chem. Soc.* (1976) **98**, 5395.
21. Coffey, R. S., *J. Chem. Soc., Chem. Commun.* (1967) 923.
22. Laine, R. M., unpublished data.
23. Knox, S. A. R., Koepke, J.W., Andrews, M. A., Kaesz, H. D., *J. Am. Chem. Soc.* (1975) **97**, 3942.
24. Hieber, W., Vetter, H., *Z. Anorg. Allg. Chem.* (1933) **212**, 145.
25. Imyanitov, N. S., Kuvaev, B. E., Rudkovskii, D. M., *Zh. Prikl. Khim. (Leningrad)* (1967) **40**, 2821.
26. Fenton, D. M., U.S. Patents **3,490,872** and **3,539,298** (1970).
27. Kang, H., Mauldin, C. H., Cole, T., Slegeir, W., Cann, K., Pettit, R., *J. Am. Chem. Soc.* (1977) **99**, 8323.

RECEIVED February 22, 1978.

9

Homogeneous Catalysis of the Water Gas Shift Reaction: Pentacarbonyliron and the Metal Hexacarbonyls as Active Catalyst Precursors

C. C. FRAZIER,[1] R. M. HANES, A. D. KING, JR., and R. B. KING

Department of Chemistry, University of Georgia, Athens, GA 30602

Methanol or 1-butanol solutions of the mononuclear metal carbonyls $M(CO)_6$ (M = Cr, Mo, and W) and $Fe(CO)_5$ in the presence of aqueous sodium or potassium hydroxide are active homogeneous catalysts for the water gas shift reaction ($CO + H_2O \rightleftarrows CO_2 + H_2$). The effects of temperature, pressure, and base concentration on the rate of hydrogen production from CO and H_2O in the presence of $Fe(CO)_5$ and NaOH have been investigated. The observation by IR spectroscopy that $HFe(CO)_4^-$ reacts with CO under pressure in 1-butanol or THF to give $Fe(CO)_5$ suggests the following catalytic cycle for the water gas shift reaction catalyzed by basic solutions of $Fe(CO)_5$: (1) $HFe(CO)_4^- + CO \rightarrow Fe(CO)_5 + H^-$; (2) $H^- + H_2O \rightarrow OH^- + H_2$; (3) $Fe(CO)_5 + OH^- \rightarrow Fe(CO)_4C(O)OH^-$; (4) $Fe(CO)_4C(O)OH^- \rightarrow HFe(CO)_4^- + CO_2$.

Increased recent interest in the homogeneous catalysis of the water gas shift reaction (Reaction 1) by ruthenium (1) and rhodium (2) carbonyl derivatives has prompted us to reexamine Reppe's observation during World War II (3) that $Fe(CO)_5$ in the presence of a base can catalyze this reaction. The aqueous sodium hydroxide used as a base by

$$CO + H_2O \rightleftarrows H_2 + CO_2 \qquad (1)$$

[1] Current address: Department of Chemistry, University of Minnesota, Deluth, MN.

Reppe led to systems that were catalytic in metal but not in base for the water gas shift reaction. We have found recently that by using aqueous 1-butanol rather than pure water as the solvent in the $NaOH/Fe(CO)_5$ system, a catalyst can be generated for the water gas shift reaction which is not only catalytic in iron but also in base above 120°C. To facilitate mechanistic study of the $NaOH/Fe(CO)_5$-catalyzed water gas shift reaction, the rate of hydrogen evolution now has been determined as a function of initial CO pressure, reaction temperature, and base concentration. In addition, a similar investigation of the group VI metal carbonyls $M(CO)_6$ (M = Cr, Mo, and W) has identified these complexes as very active water gas shift catalysts in the presence of alcoholic sodium or potassium hydroxide. This chapter discusses the kinetic results and the information obtained by an IR spectroscopic examination of the catalytic solutions obtained from the mononuclear metal carbonyls $Fe(CO)_5$ and $M(CO)_6$ (M = Cr, Mo, and W) and hydroxide ion using a specially designed high pressure IR cell (4).

Experimental Procedures

All reactions were carried out in 700-mL stainless steel, high pressure reaction vessels. The reaction solution was added, along with a Teflon-coated stirring bar, to a vessel that was flushed and loaded with CO to the desired pressure. The vessel was heated in an insulated oven, which rests on a magnetic stirring motor. Temperature control (± 1°C after the desired reaction temperature was reached) was maintained using a proportional temperature controller with a thermocouple inserted in a thermowell, which extended below the solution level of the reaction vessel as a sensor. Heating the reaction vessel from room temperature to 160°C typically required from 40 to 45 minutes.

Gas samples were periodically removed through a valve-controlled port at the top of the reaction vessel. A portion of each sample was injected into a Varian Aerograph Model 920 gas chromatograph with either a 5A molecular sieve column for measuring H_2 and CO or a silica column for measuring CO_2 and CO. Known H_2/CO mixtures were used for calibration of the molecular sieve column.

The $Fe(CO)_5$ solutions used in these experiments were prepared by dissolving NaOH in 30 mL (1.67 mol) of distilled water and combining this base solution with 170 mL of 1-butanol previously added to the reaction vessel. This solution was bubbled with N_2 for 20–30 min before addition of 0.3 mL (0.00223 mol) of $Fe(CO)_5$. The reaction vessel was closed under N_2 and connected to a high pressure manifold for flushing and loading with CO.

The $M(CO)_6$ solutions (M = Cr, Mo, and W) were prepared by dissolving a weighed sample of the metal hexacarbonyl in 100 mL of solvent (usually methanol) and adding the appropriate amount of base as 10M aqueous KOH. After completion of an experimental run, the aqueous layer was separated from butanol, evaporated, and a portion of the dried residue incorporated into a KBr pellet for examination by IR spectroscopy. In some experiments solid material precipitated on cooling the reaction vessel to room temperature. These solids were also analyzed by IR spectroscopy. Formate (1600 and 1360 cm^{-1}), carbonate (1440 cm^{-1}), and bicarbonate (1650, 1605, and 1310 cm^{-1}) were identified by their characteristic IR frequencies.

To exclude the possibility of heterogeneous rather than homogeneous catalysis, precipitated solids were filtered from the supernatant liquid and added to a fresh solution of solvent and base for further reaction. The observed catalytic activity was insignificant. The supernatant liquid, in contrast, demonstrated activity similar to that measured in the initial run.

The stainless steel high pressure IR cell with Irtran-1 windows and associated high pressure equipment and spectrometer has been described elsewhere (4). Air-sensitive iron carbonyl solutions, which were to be examined by IR spectroscopy, were loaded into the high pressure cell under N$_2$ and were then quickly placed under an atmosphere of CO to insure their stability.

The Catalytic System Derived from Fe(CO)$_5$

Hydrogen production turnover numbers have been measured as temperature, pressure, and base concentration were varied (see Tables I and II) in an effort to determine the mechanism of the catalytic system derived from 1-butanol solutions of Fe(CO)$_5$ and base. Turnover numbers are given as moles of hydrogen per mole of metal per six hours to allow all of the experiments to be compared on a meaningful basis. Under certain conditions, some experimental runs use all of the added CO in less than one day. To insure that the turnover numbers represent kinetically useful information, the turnover obtained at an early stage in the reaction are presented.

Interpretation of the results of these experiments has unfortunately been hampered by competing side reactions of both CO and CO$_2$ with

Table I. Effect of Base Concentration on the Reactivity of the Fe(CO)$_5$-Catalyzed Water Gas Shift Reaction[a]

Run Number	Initial CO Pressure (atm)	Base:Metal Mol Ratio	Temperature (°C)	Mol H$_2$ per Mol Metal per 6 Hr
1	23.1	0	150	~0.002
2	28.2	14	160	78
3	28.2	28	163	54
4	28.2	224	160	24
5	28.2	448	160	3
6	14.6	7.5	161	109
7	14.6	28	160	83

[a] All runs used 170 mL of 1-butanol, 30 mL H$_2$O, and 0.30 mL of Fe(CO)$_5$.

Table II. Effect of Pressure and Temperature on the Reactivity the Fe(CO)$_5$-Catalyzed Water Gas Shift Reaction[a]

Run Number	Initial CO Pressure (atm)	Base:Metal Mol Ratio	Temperature (°C)	Mol H$_2$ per Mol Metal per 6 Hr
1	28.2	28	137	21
2	28.2	28	145	23
3	28.2	28	164	54
4	28.2	28	181	57
5	28.2	28	183	60
6	7.8	28	162	19
7	14.6	28	160	83
8	21.4	28	160	70
9	28.2	28	163	54

[a] All runs used 170 mL of 1-butanol, 30 mL of H$_2$O, and 0.30 mL of Fe(CO)$_5$.

base to produce formate and bicarbonate, respectively. Within the CO pressure range that has been used to date, the rate of formate production at 160°C in the absence of Fe(CO)$_5$ according to Reaction 2 has been determined to be significant at the base concentrations (0.31M) used most in these studies. Formate production also has been observed in

$$CO + OH^- \rightarrow HCO_2^- \qquad (2)$$

hydroformylation reactions of olefins in basic aqueous solutions of Fe(CO)$_5$ (5). However, by observing the total system pressure as well as hydrogen production during catalytic runs at 7.8 and 28.2 atm CO at 160°C, we have determined that while there is an initial rapid rate of formate production, the rate of formate production diminishes appreciably as the catalytic production of hydrogen proceeds. Hydrogen production measured at low pressures of CO$_2$ with excess H$_2$O follows first-order kinetics as shown in Figures 1 and 2. The nonzero intercept of these graphs is an artifact resulting from the loss of CO caused by formate production. While these observations could indicate that the H$_2$ and CO$_2$ products result from catalyzed formate decomposition, the more likely interpretation is that both the rate of formate formation and of hydrogen production are rapid in the presence of an initially high concentration of base and fall off as OH$^-$ concentration drops as the ion combines with CO to produce formate.

When formate ion is added to the bomb along with the basic butanol/H$_2$O solution of Fe(CO)$_5$, hydrogen production under the usual temperature and pressure reaction conditions is essentially indistinguishable from the observed rate of runs without added formate. Control experiments with formate ion added to the charge and under N$_2$ pressure, not

CO, produce negligible quantities of hydrogen. These results confirm the supposition that formate ion does not have a direct role in accounting for the water gas shift products.

The data in Table I show that as the concentration of base is increased in the reaction mixture hydrogen production accelerates, levels off, and then decreases when large quantities of base are added. At the highest base:metal ratios given, copious amounts of formate are produced, severely depleting the CO reservoir, which may in part explain the observed decrease in the rate of hydrogen production at the highest base:metal ratios.

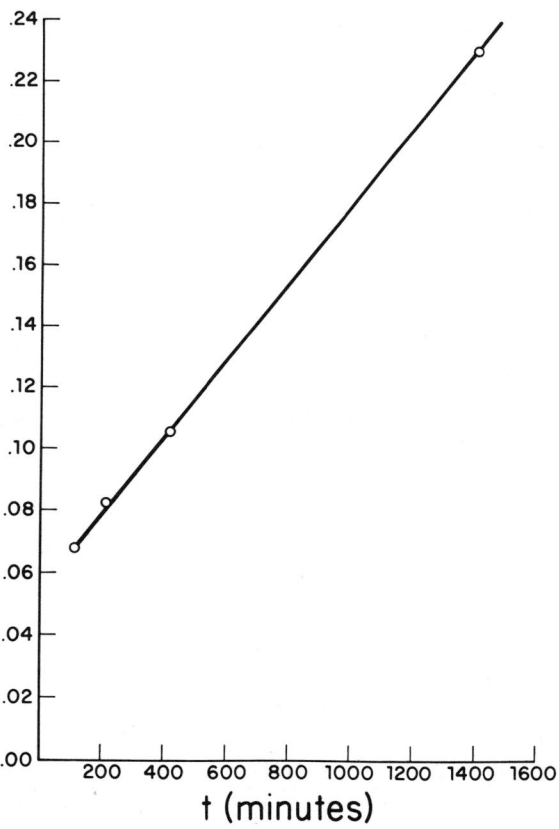

Figure 1. Plot of $-\log\left(1 - \frac{[H_2]}{[H_2] + [CO]}\right)$ vs. time for the reaction of CO at 7.8 atm and 162°C with a solution of $Fe(CO)_5$ in aqueous butanol containing 0.31M NaOH

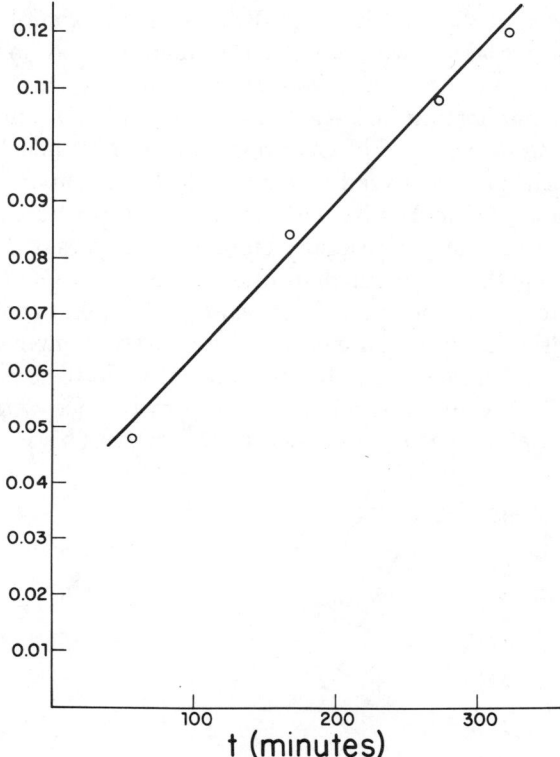

Figure 2. Plot of $-\log\left(1 - \frac{[H_2]}{[H_2] + [CO]}\right)$ vs. time for the reaction of CO at 28.2 atm and 163°C with a solution of $Fe(CO)_5$ in aqueous butanol containing 0.31M NaOH

As CO pressure is increased from 7.8 to 28.2 atm the rate of hydrogen production rises and then falls, as can be seen from the data in Table II. Since a mixture of solid sodium bicarbonate is observed in the bomb upon cooling at the conclusion of a run with 7.8 atm of initial CO pressure, it can be inferred that Reaction 3 also plays a part in controlling the pH of the reaction mixture. When experiments using 14.6 and 28.2 atm of CO are terminated, formate is the predominant inorganic anion found. Thus the composition of inorganic solids, carbonate, bicarbonate, and formate which can be found in the bomb is determined by the initial concentration of base, initial CO pressure, and the amount of CO_2 that

$$CO_2 + 2OH^- \rightarrow CO_3^{2-} + H_2O \tag{3}$$

has accumulated at the end of a run. Because of the complex interaction between these variables, we are currently quantitatively measuring CO, H_2, and CO_2 vs. time under a variety of experimental conditions in order to define the complicating side reactions of CO and CO_2 and to establish the true relation of base and of CO pressure to the key mechanistic steps of the metal carbonyl-catalyzed water gas shift reaction.

The data in Table II also demonstrate that the rate of hydrogen production increases as the reaction temperature is raised. The rate constants governing the apparent first-order uptake of CO for a series of comparable reactions performed at various temperatures are shown plotted logarithmically as a function of reciprocal temperature (K) in Figure 3. An activation energy of 20 kcal/mol is derived from the slope of this plot. This value is surprisingly close to the activation energy of the iron oxide heterogeneous water gas shift reaction (6).

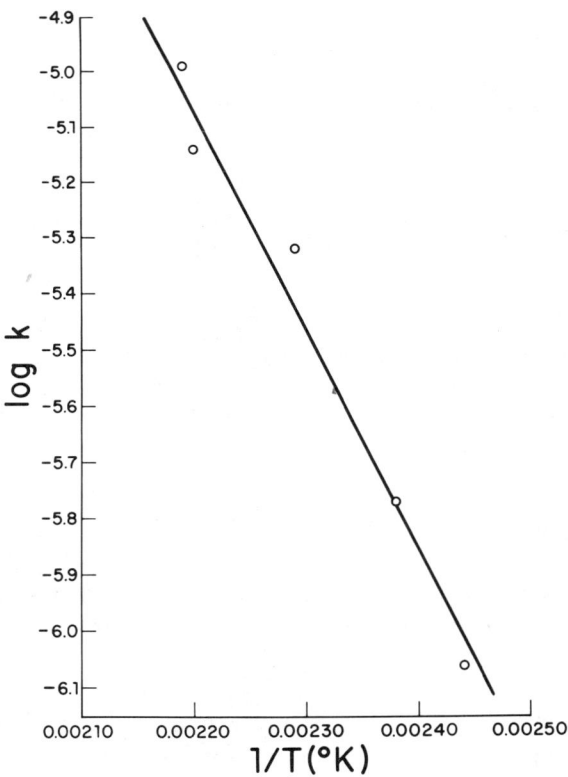

Figure 3. Plot of log k vs. 1/T (K) leading to an activation energy of 20 kcal/mol for the reaction of CO at 28.2 atm with a solution of $Fe(CO)_5$ in aqueous butanol containing 0.31M NaOH

Table III. Systems Derived from the Group VI Metal Carbonyls $M(CO)_6$ (M = Cr, Mo, W) as Catalysts for the Water Gas Shift Reaction[a]

Run No.	Carbonyl	Base	Solvent	Initial CO P (atm)	Metal Conc. (M)	Base:Metal (Mol)	T	H_2:M Per Day (Mol)
1	$Cr(CO)_6$	KOH	MeOH	7.8	0.00182	1100	140	280
2	$Mo(CO)_6$	KOH	MeOH	4.3	0.00207	87	120	5
							140	24
3	$Mo(CO)_6$	KOH	MeOH	4.3	0.0022	760	115	9
							135	10
							160	40
4	$Mo(CO)_6$	KOH	MeOH	11	0.00208	800	120	0
							130	3
							145	130
5	$Mo(CO)_6$	none	MeOH	4.3	0.00255	0	160	0
6	$Mo(CO)_6$	KBH_4	MeOH	4.3	0.0036	3	140	30
7	$W(CO)_6$	KOH	MeOH	7.7	0.00132	1265	95	0
							110	13
							130	140
8	$W(CO)_6$	KOH	MeOH	7.7	0.00105	1600	170	~900

[a] All runs used 100 mL of MeOH. When KOH is indicated, this was added as a 10M aqueous solution.

The Catalytic Systems Derived from $M(CO)_6$ ($M = Cr$, Mo, and W)

Table III presents turnover numbers for the production of hydrogen by the water gas shift reaction (Reaction 1) using catalysts derived from the metal hexacarbonyls $M(CO)_6$ ($M = Cr$, Mo, and W). Runs 2 through 4 clearly show that as the reaction temperature increases in the experiments with $Mo(CO)_6$, the rate of hydrogen production accelerates. Run 7 indicates that the same effect occurs for runs using $W(CO)_6$ as the catalyst precursor. Run 5 demonstrates that a base is required for catalysis. Run 6 illustrates that bases other than hydroxide can act as catalysts. Since KBH_4 is known to react with $M(CO)_6$ ($M = Cr$, Mo, and W) in donor solvents at elevated temperatures to produce the corresponding $HM_2(CO)_{10}^-$ anions (7), the results of run 6 suggest that these anions may be involved in the catalytic cycle.

High Pressure Spectroscopic Studies

The IR spectra of 1-butanol solutions of $Fe(CO)_5$ and $M(CO)_6$ ($M = Mo$ and W) containing aqueous $NaOH$ or KOH were examined under pressure using a stainless steel high pressure IR cell with Irtran 1 windows (4) in attempts to identify the various metal carbonyl species present in the reaction solutions under catalytic conditions. In the cases of $M(CO)_6$ ($M = Mo$ and W), the only metal carbonyl $\nu(CO)$ frequencies observed at temperatures and pressures comparable with those used for catalysis were those that correspond to the respective metal hexacarbonyl. In addition, a band at 2300 cm^{-1} gradually appeared in the IR spectra of such metal hexacarbonyl solutions above 110°C under CO pressure. This band can be assigned to the CO_2 produced in the water gas shift reaction (Reaction 1) proceeding under these conditions.

Similar IR spectroscopic studies of basic solutions of $Fe(CO)_5$ under CO pressures led to somewhat more complex results. Addition at room temperature of $Fe(CO)_5$ (0.10 mL, 0.76 mmol) to 49.4 mL of a N_2-saturated solution of 1:20 water–butanol containing 0.12 g (3 mmol) of $NaOH$ resulted in the rapid formation of $HFe(CO)_4^-$. No further changes were observed in the IR spectrum of such an $HFe(CO)_4^-$ solution when it was kept at 25°C for 4 hr under 330 atm CO. However, upon gradual heating under CO pressure, the characteristic strong 1885 cm^{-1} band of $HFe(CO)_4^-$ gradually disappeared with the concurrent appearance of a 1995 cm^{-1} band indicative of the presence of regenerated $Fe(CO)_5$. These spectral changes are depicted in Figure 4. The formation of $Fe(CO)_5$ from $HFe(CO)_4^-$ and 330 atm CO was complete at 93°–98°C. Cooling the solution to room temperature while still under

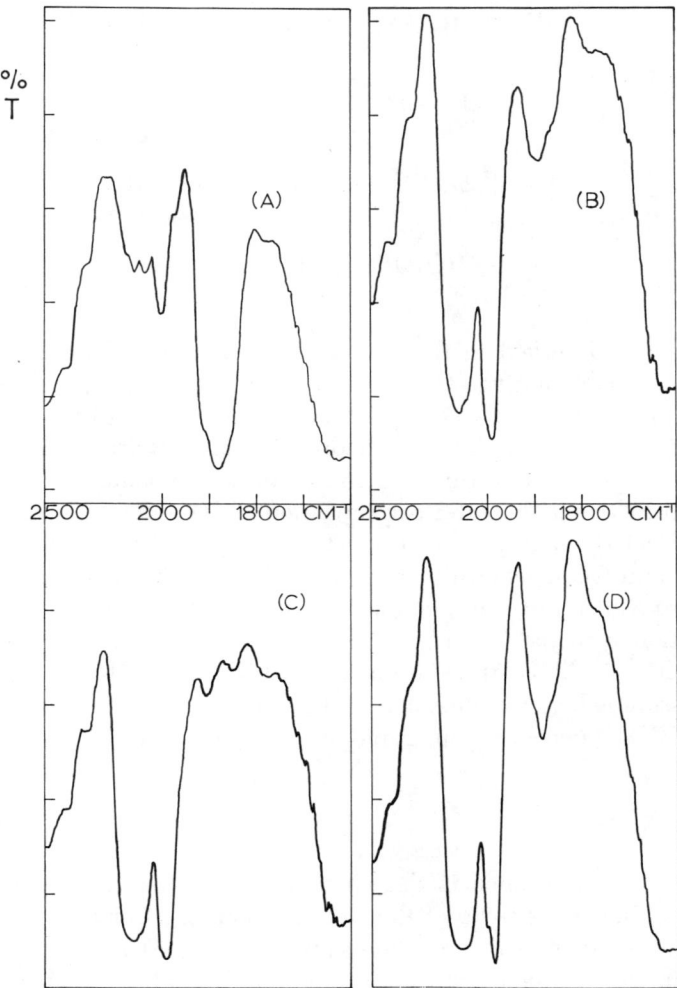

Figure 4. IR spectra of a solution of 0.10 mL of $Fe(CO)_5$, 0.12 g of NaOH, 1.4 mL of H_2O, and 48 mL of 1-butanol under 330 atm CO. (A) At 28°C; (B) after heating to 82°–91°C; (C) after heating to 93°–98°C; (D) after cooling the heated solution back to 38°C. Weak bands from the 1-butanol solvent are observable at 2015 cm^{-1} in Figure 1A and at 1906, 1851, and 1775 cm^{-1} in Figure 1C. As CO dissolves in the solvent a broad band at 2130 cm^{-1} progressively intensifies.

330 atm CO in the high pressure cell resulted in partial regeneration of $HFe(CO)_4^-$ from reaction of $Fe(CO)_5$ with residual base.

The above observations suggest that a complete cycle for the water gas shift reaction can be described by the following reactions:

$$HFe(CO)_4^- + CO \rightleftarrows Fe(CO)_5 + H^- \quad (4a)$$

$$H^- + H_2O \rightarrow OH^- + H_2 \quad (4b)$$

$$OH^- + Fe(CO)_5 \rightarrow Fe(CO)_4C(O)OH^- \quad (4c)$$

$$Fe(CO)_4C(O)OH^- \rightarrow HFe(CO)_4^- + CO_2 \quad (4d)$$

The changes observed in the IR spectrum in the $\nu(CO)$ region of an anhydrous THF solution of $[(C_6H_5)_3P]_2N^+HFe(CO)_4^-$ upon heating the solution under 330 atm CO imply that Reaction 4a can be written as an equilibrium when anhydrous nonhydroxylic solvents are used so that the hydride ion produced in Reaction 4a cannot be protonated to form hydrogen as in Reaction 4b. After carbonylation of $[(C_6H_5)_3P]_2N^+HFe(CO)_4^-$ to give $Fe(CO)_5$ and presumably $[(C_6H_5)_3P]_2N^+H^-$ is complete at 60°C and 330 atm CO, cooling the system while maintaining the CO pressure leads to reversion of some of the $Fe(CO)_5$ to $HFe(CO)_4^-$, presumably by reaction with the $[(C_6H_5)_3P]_2N^+H^-$. Additional $Fe(CO)_5$ reverts to $HFe(CO)_4^-$ as the CO pressure is lowered in stages. However, after the CO pressure is below approximately 140 atm, the conversion of $Fe(CO)_5$ to $HFe(CO)_4^-$ becomes quite rapid and is complete within minutes.

Conclusions

The studies outlined in this chapter suggest that a variety of metal carbonyls, including some of the simplest mononuclear metal carbonyls, can generate active catalysts for the water gas shift reaction (Reaction 1) by simple treatment with hydroxide ion. Investigation of the mechanisms of such catalytic reactions is complicated by side reactions of the CO reactant and the CO_2 product with the strongly basic system. However, now that these side reactions, involving production of formate and carbonate, are recognized, it should be possible to design experiments that permit identification of the key steps in these metal carbonyl-catalyzed water gas shift reactions.

Acknowledgment

We are indebted to the Division of Basic Energy Sciences of the U.S. Department of Energy for support of this work under Contract EY-76-S-09-0933.

Literature Cited

1. Laine, R. M., Rinker, R. G., Ford, P. C., *J. Am. Chem. Soc.* (1977) **99**, 252.
2. Cheng, C. H., Hendricksen, D. E., Eisenberg, R., *J .Am .Chem. Soc.* (1977) **99**, 2791.
3. Reppe, W., Reindl, E., *Liebigs Ann.* (1953) **582**, 116.
4. King, R. B., King, A. D., Jr., Iqbal, M. Z., Frazier, C. C., *J. Am. Chem Soc.* (1978) **100**, 1687.
5. Kang, H., Mauldin, C. H., Cole, T., Slegeir, W., Cann, K., Pettit, R., *J. Am. Chem. Soc.* (1977) **99**, 8323.
6. Laudien, K., Witzmann, W., *Chem. Tech.* (1967) **19**(4), 232.
7. Hayter, R. G., *J. Am. Chem. Soc.* (1966) **88**, 4376.

RECEIVED March 3, 1978.

10

Oxygen-Exchange and Ligand Substitution Reactions in $Cr(CO)_6$ and $\mu\text{-}H[Cr(CO)_5]_2^-$, and the Water Gas Shift Reaction

DONALD J. DARENSBOURG, MARCETTA Y. DARENSBOURG, ROBERT R. BURCH, JR., JOSEPH A. FROELICH, and MICHAEL J. INCORVIA

Department of Chemistry, Tulane University, New Orleans, LA 70118

> *An essential step in the homogeneous catalyzed water gas shift reaction involves nucleophilic attack of hydroxide ion at the carbon atom of a metal-bound CO, with a subsequent process leading to the extrusion of CO_2. These two processes have been studied for the reaction of oxygen-18 enriched NaOH with $Cr(CO)_6$ in a biphasic medium in the presence of a phase-transfer catalyst. Oxygen exchange was observed to occur at a faster rate than metal hydride, $\mu\text{-}H[Cr(CO)_5]_2^-$, formation with concomitant production of CO_2. The kinetic parameters for the reaction of $\mu\text{-}H[Cr(CO)_5]_2^-$ with CO in alcoholic solvent to afford $Cr(CO)_6$ and H_2 have been determined.*

The catalytic production of methane from abundant coal resources is expected to become increasingly more important as a means for meeting the exponentially growing demands for clean energy fuel (1, 2). This is particularly so since the annual gas reserve production ratio has been steadily decreasing in recent years. In the formation of methane, a mixture of gases (synthesis gas) containing hydrogen, carbon monoxide, carbon dioxide, and methane is first produced by the reaction of coal with oxygen and steam; i.e., the coal gasification step. When the H_2:CO in synthesis gas is ≥ 3, the formation of methane can be described by Reaction 1. Synthesis gas, high in hydrogen content, is also essential as a feedstock in the Fischer–Tropsch process (3, 4).

$$3H_2 + CO \rightarrow CH_4 + H_2O \tag{1}$$

Unfortunately, the H_2:CO ratio is generally much less than 3:1, and therefore it has to be adjusted by the shift reaction in which carbon monoxide is oxidized to carbon dioxide with concomitant production of hydrogen (Reaction 2). After removal of CO_2 and sulfur compounds, the

$$CO + H_2O \rightarrow CO_2 + H_2 \tag{2}$$

shifted gas mixture can be subjected to catalytic methanation to afford a product gas that contains some excess hydrogen ($< 10\%$), but must not contain $> 0.1\%$ carbon monoxide. The sulfur compounds, generally considered to readily deactivate the metal-based catalysts, include hydrogen sulfide and volatile sulfur compounds such as thiophene, mercaptans, carbon disulfide, carbonyl sulfide, and thioethers. Nickel is the most widely used commercial heterogeneous catalyst for methanation although ruthenium, cobalt, and iron metals have also been used. Indeed, ruthenium (0.5% on alumina) has been recognized as a very active catalyst, its only drawback being its high cost (5, 6).

Supported ruthenium has also been shown to be an extremely effective water gas shift catalyst at operating temperatures above 350°C (7). An essential step in this catalytic process involves the binding of CO to the metal. In this regard, on a fully reduced, supported ruthenium sample, CO absorption affords a band in the IR at 2040 cm^{-1}, which is ascribed to a linearly bound carbon monoxide group (8).

Much of the justification for the extensive study of transition metal cluster chemistry is embedded in the assumption that reactions of metal clusters are realistic structural models for reactions at metal surfaces in such processes as heterogeneous catalysis (9, 10, 11). For example, the metal carbonyl clusters, $Ir_4(CO)_{12}$ and $Os_3(CO)_{12}$, were demonstrated to be effective homogeneous catalysts for methanation (12). Additionally, Demitras and Muetterties (13) have found $Ir_4(CO)_{12}$ to be a homogeneous catalyst in the Fischer–Tropsch synthesis of aliphatic hydrocarbons. Homogeneous catalysis of the water gas shift reaction by metal carbonyl clusters (e.g., $Ru_3(CO)_{12}$) in alkaline solution has been reported by Laine, Rinker, and Ford (14), and more recently by Pettit's group (15). Nevertheless, mononuclear metal carbonyls (e.g., $Fe(CO)_5$ and the group VIb metal hexacarbonyls) have been demonstrated to have considerable activity above 120°C as soluble catalysts for Reaction 2 (16).

The initial step in the homogeneous-catalyzed water gas shift reaction undoubtedly involves nucleophilic attack of hydroxide ion at the carbon atom of a metal-bound carbon monoxide. Muetterties has observed that $[Re(CO)_6]^+$ exchanges oxygen atoms with atoms in water, probably through the intermediacy of $Re(CO)_5COOH$ (*17*). Rather exhaustive investigations of this oxygen-exchange process, both from a mechanistic as well as a synthetic viewpoint, have recently been undertaken. These have included reports on the incorporation of oxygen-18 into cationic metal carbonyl derivatives of manganese and rhenium (*18, 19, 20, 21*). It was shown in reactions involving substituted group VIIb metal carbonyl cationic species, where there are two electronically different CO groups, that the oxygen atoms on the more electron-poor carbonyls (i.e., carbonyl ligands with the larger CO stretching force constant (*22*)) were exchanged at a faster rate. Thus it was possible to prepare stereospecifically labelled metal carbonyl derivatives using the water-exchange process. In addition, the corollary observation that the relative rates of oxygen exchange decrease with increasing substitution at the metal center with electron donating ligands was noted, $M(CO)_6^+ > M(CO)_5L^+ >> M(CO)_4L_2^+$. The $L_n(CO)_{5-n}M(COOH)$ intermediates were also found to undergo CO_2 elimination with metal hydride formation (the skeletal sequence shown in Reaction 3), a process common to the energy-important metal-catalyzed water gas shift reaction.

$$M-CO^+ + {}^-OH \rightleftharpoons [M-\overset{\overset{\displaystyle O}{\|}}{C}-OH] \rightarrow M-H + CO_2 \qquad (3)$$

The carbonyl groups in the neutral iso-electronic group VIb metal hexacarbonyls and their Lewis-base substituted derivatives do not undergo oxygen exchange with water under the conditions where the more highly activated cationic group VIIb metal carbonyls' oxygens readily exchange. Basic solutions however will effect this oxygen exchange process in these derivatives. This was accomplished using a biphasic reaction medium with the organic phase consisting of the neutral carbonyl derivative in benzene and an aqueous phase, consisting of sodium hydroxide with a small quantity of tetra-*n*-butylammonium iodide as the phase-transfer catalyst (*23*). Indeed, in basic alcoholic solutions under an atmosphere of carbon monoxide, we and others (*16*) have found that the group VIb metal hexacarbonyls can serve as catalysts for the water gas shift reaction.

Although mononuclear metal carbonyls are purportedly less effective as catalysts for this process when compared with metal carbonyl clusters (*14, 15*), investigations of these systems will provide for a better understanding of the fundamental steps in the homogeneous metal-catalyzed water gas shift reaction. Therefore, the primary objective of this work was to examine: (i) the reversible nature of the reaction of hydroxide ion with $Cr(CO)_6$, along with the concomitant formation of μ-H[Cr(CO)$_5$]$_2^-$ and CO_2; and (ii) the ligand substitution reactions of μ-H[Cr(CO)$_5$]$_2^-$ with CO, both thermally and photochemically (Scheme 1).

Scheme 1

$$Cr(CO)_6 + {}^-OH \rightleftharpoons [Cr(CO)_5COOH]^- \rightarrow \mu\text{-}H[Cr(CO)_5]_2^- + CO_2$$

$$\overset{2CO}{\underset{H_2O}{\nearrow}}$$

$$2Cr(CO)_6 + H_2 + {}^-OH \qquad (4)$$

Oxygen-Exchange Reactions of $Cr(CO)_6$

Figure 1 illustrates the $\nu(CO)$ IR spectral traces obtained during the course of oxygen-18 incorporation into chromium hexacarbonyl using the phase-transfer catalyzed technique (23). These spectra were determined at various times on the hexane-soluble portion of the residues which resulted after the removal of the solvent mixture. The observed and calculated $\nu(CO)$ values for the ten possible isotopically substituted chromium hexacarbonyl species, $[Cr(C^{16}O)_{6-n}(C^{18}O)_n]$ ($n = 0$–6), are listed in Table I. Band assignments were made by noting the rates of appearance and decay of bands simultaneously aided by iterative calculations involving a restricted CO force field (24, 25). The carbonyl stretching parameters obtained from the approximate force field are the following: $k_1 = 16.49$, $k_c = 0.27_4$, and $k_t = 0.53_8$. The principal features of these spectra other than the parent absorption at 1987.1 cm^{-1} (antisymmetric stretching of trans-$C^{16}O$ groups) are bands attributed to the presence of $C^{18}O$ trans to $C^{16}O$ (1960–1950 cm^{-1}) or a band attributed to $C^{18}O$ trans to $C^{18}O$ (1940.9 cm^{-1}). Similar CO stretching parameters have previously been reported by Perutz and Turner using ^{13}CO frequency data (26). However, it should be pointed out again here that the CO-factored force field, useful in assigning isotopic frequency shifts for ^{13}C isotopes, is not simultaneously useful in accurately assigning frequency shifts for ^{18}O isotopes (27, 28).

The reaction of $Cr(CO)_6$ with ^-OH to afford $\mu\text{-}H[Cr(CO)_5]_2^-$ has previously been reported (see review by F. Calderazzo on metal carbonyls in Ref. 29; see also Ref. 30). Under the conditions of our oxygen exchange reaction with $Cr(CO)_6$, highly oxygen-18 enriched $\mu\text{-}H[Cr(CO)_5]_2^-$ derivatives are formed. However, oxygen exchange occurs much more rapidly than does CO_2 elimination in the proposed $[Cr(CO)_5COOH]^-$ intermediate (Reaction 4). It should be noted here that the relative rate of oxygen exchange vs. CO_2 elimination is pH dependent (20). Presumably this is the result of the following skeletal sequence:

$$[CrCOOH]^- + OH^- \rightarrow [CrCOO]^= + H_2O \rightarrow$$
$$[Cr]^= + CO_2 \xrightarrow{H_2O} [CrH]^- + {}^-OH.$$

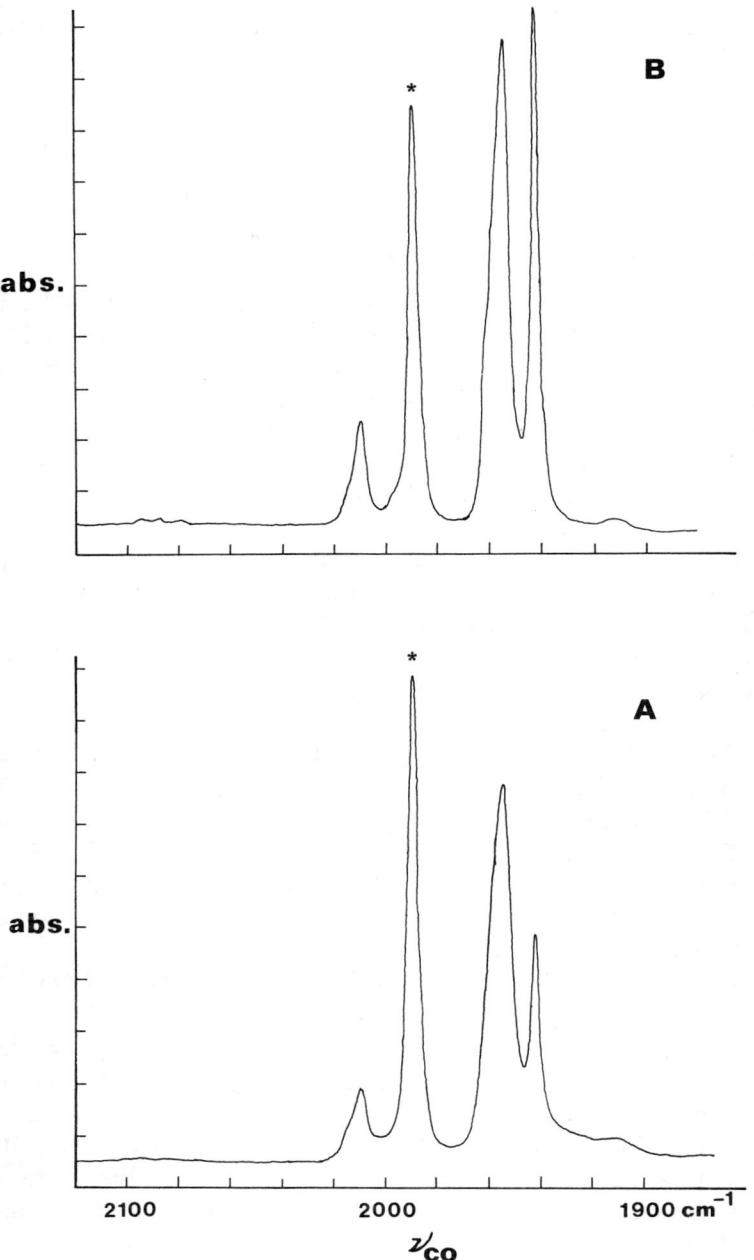

Figure 1. Time-dependent $\nu(CO)$ spectra in hexane solution for the oxygen-exchange reaction of $Cr(CO)_6$ with hydroxide ion: (A) 3 hr; (B) 10 hr. Parent $\nu(CO)$ absorption indicated by an asterisk.

Table I. Observed and Calculated ν(CO) Frequencies for the Ten Possible Isotopic Species of $Cr(CO)_6$ with $C^{18}O$ [a]

Isotope	Geometry	$\nu(CO)$ Frequencies (cm^{-1})					
1		2118.7	2020.3	2020.3	1987.7 (1987.1)	1987.7	1987.7
2	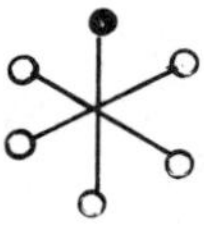	2112.5	2011.8	2020.3	1953.7 (1953.1)	1987.7	1987.7
3	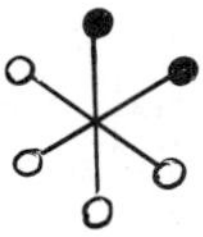	2105.6	2015.5	2008.9 (2008.9)	1956.5	1950.8	1987.7
4		2106.3	1983.1	2020.3	1939.8 (1940.9)	1987.7	1987.7
5		2097.7	2008.9	2008.9	1959.2	1950.8	1950.8
6		2098.7	2013.8	1983.5	1952.7	1939.8	1987.7
7		2089.7	2008.9	1983.9	1939.8	1954.5	1950.8
8		2091.0	1997.7	1971.5	1939.8	1939.8	1987.7
9		2080.1	1971.5	1995.2	1939.8	1939.8	1952.3
10	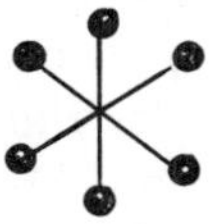	2067.6	1971.5	1971.5	1939.8	1939.8	1939.8

[a] Observed frequencies are listed in parentheses directly below the calculated values.

Experiments are currently underway to measure relative rates as well as activation parameters for oxygen exchange vs. metal hydride formation and CO_2 elimination for a variety of metal carbonyl species in homogeneous media as a function of pH.

Although the mechanistic details of bridging hydride formation as produced from $Cr(CO)_6$ and ^-OH (Reaction 4) are presently not completely understood, a process involving CO_2 elimination from the hydroxycarbonyl intermediate with concomitant production of $[HCr(CO)_5]^-$ parallels our earlier work on the reaction of the iso-electronic cationic species, $[Mn(CO)_6]^+$, with H_2O (19). There is limited spectral and chemical evidence for the presence of the elusive $[HCr(CO)_5]^-$ in the reaction of $NaBH_4$ with $Cr(CO)_6$ to produce μ-$H[Cr(CO)_5]_2^-$ (31). The $\nu(CO)$ IR monitor of this reaction is consistent with initial formation of the formyl, $[(CO)_5CrC(O)H]^-$, and subsequent decarbonylation to yield as yet not isolated $[HCr(CO)_5]^-$. The conversion of the species tentatively ascribed to mononuclear hydride into bridging hydride is promoted by small amounts of O_2 or Ph_3C^+. The latter reactions are both consistent with production of $[Cr(CO)_5]$ which is rapidly scavenged by remaining $[HCr(CO)_5]^-$. Additionally, the $[HCr(CO)_5]^-$ species reacts with $Cr(CO)_6$, presumably either by nucleophilic attack on the neutral carbonyl by the hydride or, at higher temperatures, by addition to the carbon monoxide dissociated $[Cr(CO)_5]$ moiety.

Throughout our discussion above we have implied that the hydroxycarbonyl intermediate is the only intermediate in both the oxygen-exchange process and the CO_2 elimination with a metal hydride formation step. However, the distinct possibility for the presence of an additional intermediate involving a metal-complexed formate species of the type

$$M\overset{O}{\underset{O}{\diagup\!\!\!\diagdown}}CH$$

in the latter process exists in these metal carbonyl catalyzed reactions. A formato complex, $HCOOCo[PPh_3]_3$, has been described which results from the insertion of CO_2 in $HCo(N_2)[PPh_3]_3$ or from the reaction of formic acid with $HCo(N_2)[PPh_3]_3$ (32). The formate ion has been reported as one of the products of the reaction of hydroxide ion with metal carbonyls, e.g., $Fe(CO)_5$ (15) and $Ir_4(CO)_{12}$ (33) with ^-OH. Additionally, the catalyst solution of $Ru_3(CO)_{12}$ used in the water gas shift reaction was found to be a very active catalyst for the decomposition of formate to H_2 plus CO_2 (14). In this connection, it is particularly noteworthy that direct IR spectral evidence, taken in conjunction with rate studies, indicates that the surface formate ion is the reaction intermediate of the water gas shift reaction on ZnO and MgO (34). There is, nevertheless, reason to suspect some possible differences in the heterogeneous and homogeneous reaction systems, perhaps because of the influence of surface oxygen atoms in the heterogeneous processes.

Ligand Substitution Reactions of $\mu\text{-}H[Cr(CO)_5]_2^-$

During the preparation of the $\mu\text{-}H[Cr(CO)_5]_2^-$ derivative from $Cr(CO)_6$ (30) and hydroxide ion, H_2 has also been observed as a product. This most likely arises from the reaction of the intermediate, $[Cr(CO)_5H^-]$, with H_2O to give $Cr(CO)_5H_2$, followed by rapid production of $[Cr(CO)_5]$ and H_2 (35, 36), a process analogous to that of $Fe(CO)_4H^-$ with water to form the well-known $H_2Fe(CO)_4$ species (15). The $[Cr(CO)_5H]^-$ species can also be scavenged by $[Cr(CO)_5]$, leading to the production of $\mu\text{-}H[Cr(CO)_5]_2^-$, although this process should be retarded in solutions of high CO concentrations. Therefore, to have $Cr(CO)_6$ behave as a catalyst for the water gas shift reaction, it is necessary to provide enough energy for the reaction of $\mu\text{-}H[Cr(CO)_5]_2^-$ with CO and H_2O to yield $Cr(CO)_6$, H_2, and ^-OH, thus making the overall reaction catalytic in both $Cr(CO)_6$ and hydroxide ion (Scheme 2).

Scheme 2

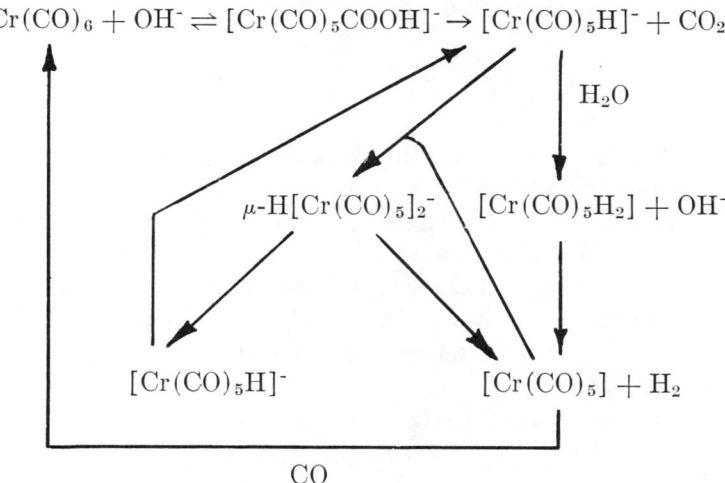

Ligand substitution reactions of $\mu\text{-}H[M(CO)_5]_2^-$ derivatives have been studied in detail with phosphine and phosphite ligands to afford $M(CO)_4L_2$ complexes (37, 38). These studies clearly demonstrate that CO loss from the bridging hydride species is a more facile process than dimer disruption. For example, $\mu\text{-}H[Cr(CO)_5]_2^-$ was found to undergo ligand exchange with ^{13}CO in refluxing THF solution (37). At this time we wish to report on an extension of these studies where the emphasis is on determining the energetics of the reaction of $\mu\text{-}H[Cr(CO)_5]_2^-$ with CO to yield $Cr(CO)_6$.

The rate of substitution of $\mu\text{-}H[Cr(CO)_5]_2^-$ with CO in CO-saturated 2-pentanol to afford $Cr(CO)_6$ was observed to follow the first-order rate

$$\text{rate} = k[\mu\text{-H}[Cr(CO)_5]_2^-] \tag{5}$$

law (5). Values of the rate constants k, calculated from the first-order rate expression, are given in Table II for several temperatures. Figure 2 illustrates the typically observed IR traces of the disappearance of starting material $\nu(CO)$ band (E_u mode) with the simultaneous appearance of the product $\nu(CO)$ band (T_{1u} mode), along with the linear first-order rate plot that was obtained over the entire reaction. The activation parameters for the dissociative process (Reaction 6) of the tetraethylammonium derivative were found to be $\Delta H^* = 22.3(1.1)$ kcal/mol and $\Delta S^* = -17.8(3.5)$ eu, where the error limits represent 95% confidence limits. It is important to note that the yield of $Cr(CO)_6$ from Reaction 6 was demonstrated to be quantitative. Hydrogen, the other product

$$\mu\text{-H}[Cr(CO)_5]_2^- \xrightarrow[\text{slow}]{k} [HCr(CO)_5]^- + [Cr(CO)_5] \tag{6}$$

$$\downarrow \begin{array}{c} \text{2-pentanol} \\ \text{CO} \end{array} \text{fast}$$

$$2Cr(CO)_6 + H_2$$

obtained from Reaction 6, was measured by gas chromatography (using a 5A molecular sieve, 60/80 mesh column).

We have begun a systematic study of the effect of the cationic counter-ion on the kinetics of Reaction 6. Preliminary results indicate a pronounced dependence of the reaction rate on the nature of the gegenion. For example, the rate constant (k) at 97°C was observed to be ~ 4.5 times larger when the counter-ion was the potassium ion compared with that obtained for the tetraethylammonium salt ($t_{1/2} \approx 1$ hr). This observation points out a consideration of much practical importance when designing a catalytic system of the type discussed in this communication,

Table II. First-Order Rate Constants for the Reaction of $Et_4N^+\mu\text{-H}[Cr(CO)_5]_2^-$ with CO in 2-Pentanol[a]

Temp (°C)	$10^5 k$, sec^{-1} [b]
72.0	1.88 ± 0.09
85.6	6.44 ± 0.32
85.6	6.47 ± 0.13
85.6[c]	6.42 ± 0.43
97.0	18.30 ± 0.4

[a] Solution continuously saturated with carbon monoxide.
[b] Error limits for rate constant data represent 95% confidence limits.
[c] Measurement carried out in the dark, other carried out in room light.

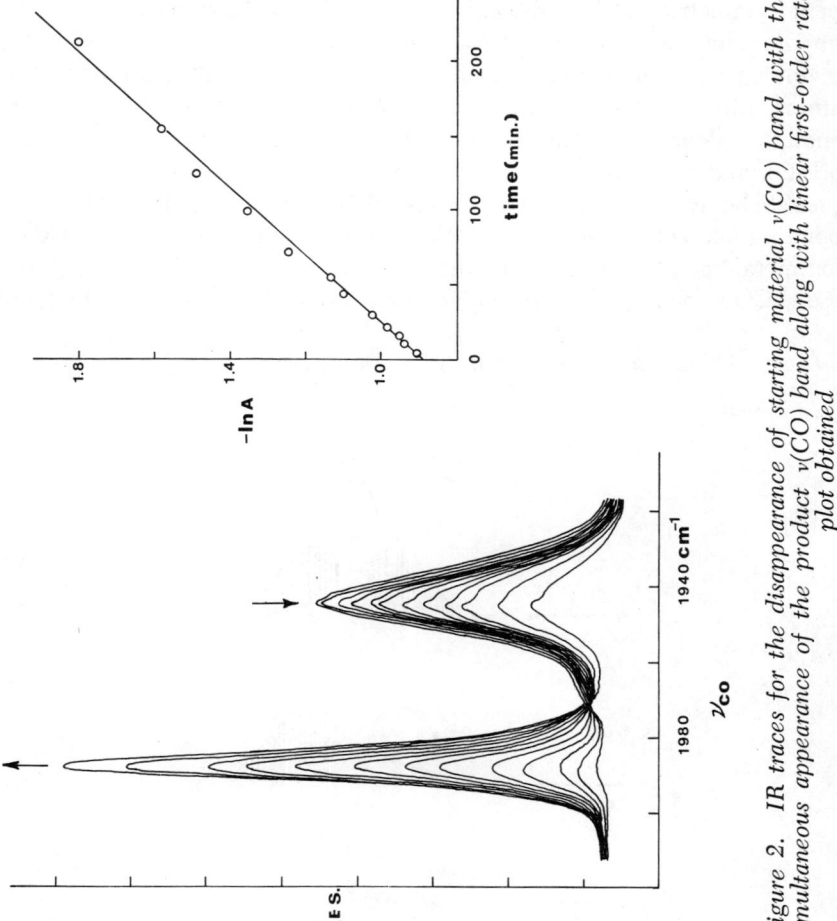

Figure 2. IR traces for the disappearance of starting material $\nu(CO)$ band with the simultaneous appearance of the product $\nu(CO)$ band along with linear first-order rate plot obtained

i.e., the nature of the counter-ion (M^+) in the MOH co-catalyst can have a significant influence on the energetics of subsequent reactions in the catalytic cycle.

A single crystal neutron diffraction study of $Et_4N^+\mu\text{-}H[Cr(CO)_5]_2^-$ has been carried out by Dahl and co-workers which indicates the anion to be of D_{4h} symmetry (i.e., eclipsed equatorial CO groups on the two metal centers with a linear CO_{ax}–Cr · · · Cr–CO_{ax} arrangement) with a bent symmetric Cr–H–Cr bond (Cr · · · Cr = 3.386(6)Å) (*39*). The bonding has been described as a three-center bond of a "closed" type involving the simultaneous symmetrical overlap of orbitals on the three atoms, directed towards the center of a triangle (*40*). There is just enough valence electrons in the moiety to fill the bonding molecular orbital, and none to occupy any of the other possible molecular orbitals (*see* Scheme 3). Again this description indicates partial Cr · · · Cr bonding character (*40, 41, 42*). We have indeed obtained some evidence for metal–metal interaction in the molecular ion $\mu\text{-}H[Mo(CO)_5]_2^-$ from the $\nu(CO)$ spectral changes which result upon stereospecific substitution

Scheme 3. *MO Energy Levels*

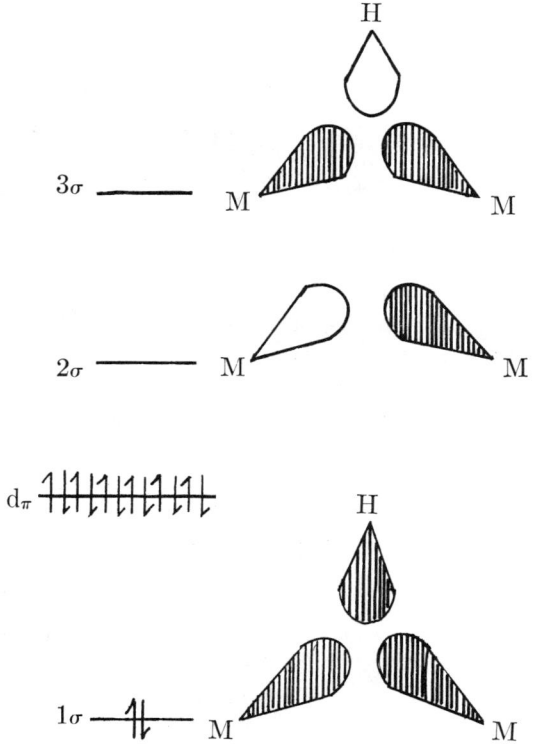

of an equatorial site with ^{13}CO on one $Mo(CO)_5$ unit only (43). Scheme 3 places the filled $d\pi$ orbitals of the two $M(CO)_5$ moieties higher in energy than the filled three-center bonding MO, as proposed by Harris and Gray (44).

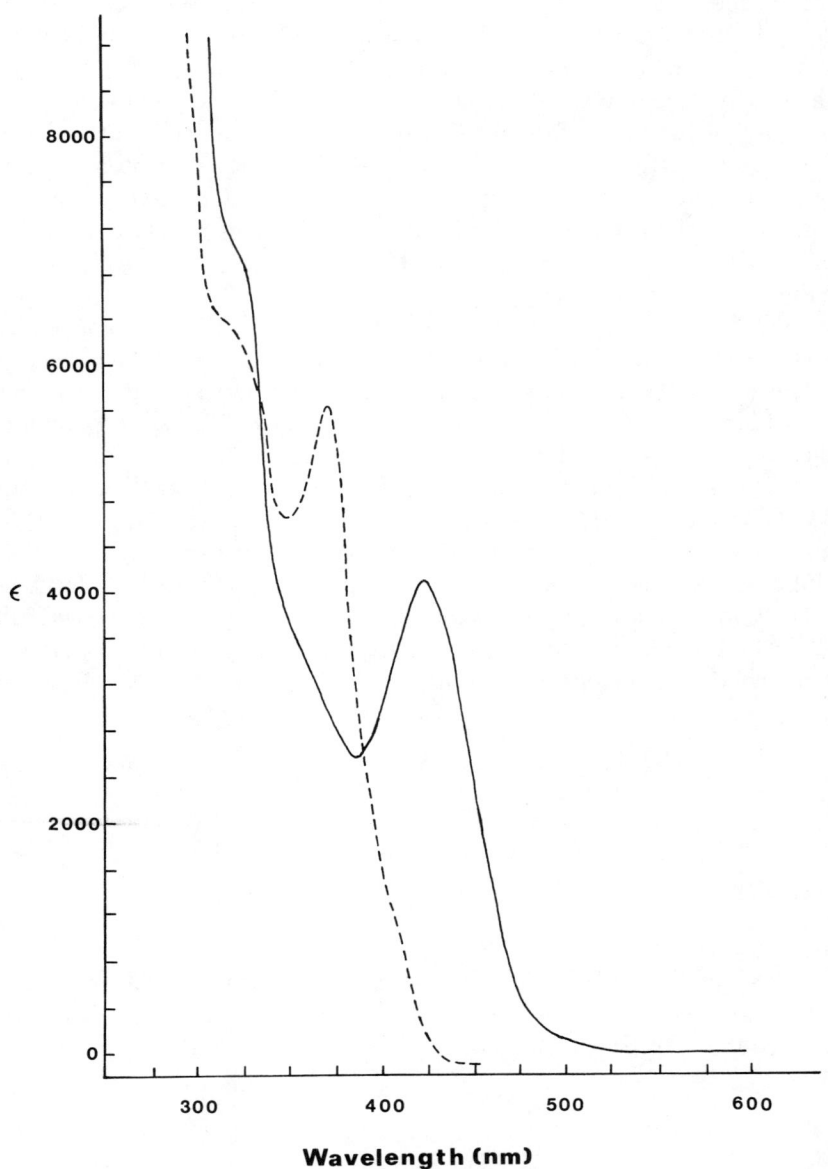

Figure 3. Electronic spectra at room temperature in THF of $Et_4N^+\mu$-H[M-$(CO)_5]_2^-$ species: (———) M = Cr; (- - -) M = W

Figure 3 contains the low energy absorption bands in the room-temperature electronic spectra of $Et_4N^+\mu\text{-}H[M(CO)_5]_2^-$ (M = Cr or W). As noted in Figure 3, the lowest energy absorption is significantly red-shifted for the chromium derivative. This strong, lowest energy absorption band in the tungsten analog has been assigned to the $d\pi \rightarrow 2\sigma$ transition. Similarly, an assignment of the band at 423 nm (ϵ = 4100) in the chromium derivative seems appropriate. Population of the 2σ level, which is antibonding with respect to the Cr–H–Cr linkage (the extent of antibonding character of this level will nevertheless depend on the degree of direct orbital interaction between the two chromium atoms (42)), should facilitate dimer disruption, thereby providing a photoassisted pathway for returning to the hexacarbonyl. This is particularly attractive since the low energy absorption of $\mu\text{-}H[Cr(CO)_5]_2^-$ is in the visible region of the spectrum.

We have therefore carried out some preliminary photochemical experiments on $Et_4N^+\mu\text{-}H[Cr(CO)_5]_2^-$ in the presence of CO. A filter system which has its band pass maximum at 410 nm was used to irradiate into the low energy band of $\mu\text{-}H[Cr(CO)_5]_2^-$ at 423 nm. The quantum efficiency (Φ) for the disappearance of the dimer was found to be 0.036 ($\pm$ 0.008) in THF solution. However, only slightly greater than one mole of $Cr(CO)_6$ (1.1 $\pm$ 0.1) was observed for every mole of $\mu\text{-}H[Cr(CO)_5]_2^-$ consumed. Similar photochemical reaction modes were noted in alcoholic solutions. No other carbonyl-containing product(s) have thus far been detected in the reaction; presumably the $[HCr(CO)_5^-]$ component resulting from the heterolytic dimer cleavage undergoes further secondary photochemical reactions (Reaction 7). It is therefore quite clear that

$$\mu\text{-}H[Cr(CO)_5]_2^- \xrightarrow{h\nu} [HCr(CO)_5^-] + [Cr(CO)_5] \quad (7)$$

$$\downarrow \qquad \searrow^{CO} \qquad \downarrow^{CO}$$

$$? \qquad\qquad Cr(CO)_6$$

photolysis of the $\mu\text{-}H[Cr(CO)_5]_2^-$ ion in the presence of CO with long wavelength light will not provide for a quantitative pathway for the production of $Cr(CO)_6$ and H_2.

Conclusions

Although detailed relative catalytic activity data for the use of mononuclear group VIb metal hexacarbonyls vs. metal carbonyl clusters (e.g., $Cr(CO)_6$ vs. $Ru_3(CO)_{12}$) as homogeneous catalysts for the water gas shift reaction has not been provided in this communication, some comparisons between the two systems are in order. In alkaline aqueous ethoxyethanol solutions ($T = 100°C$, $P_{CO} \approx 1$ atm), $Cr(CO)_6$ has been observed to be more active ($\sim$ five times) than $Ru_3(CO)_{12}$ for short reaction periods (45). Unfortunately, the $Cr(CO)_6$ catalyst has a shorter lifetime, becoming inactive after 20–25 turnovers. The initial step in both processes involves nucleophilic attack by hydroxide ion at the carbon atom of a neutral metal carbonyl derivative, with a subsequent reaction leading to formation of anionic metal carbonyl hydride species and CO_2. The anionic metal hydrides via reaction with the protic solvents eventually afford hydrogen gas, and the catalytic cycle is completed by reaction with CO. A striking difference noted in the two reaction systems is that in the $Ru_3(CO)_{12}$ reaction mixture, metal hydride formation is much more rapid than in the $Cr(CO)_6$ catalyst mixture and occurs with little (if any) oxygen exchange (45). Thus, whereas it has been suggested that reductive elimination of H_2 from $H_4Ru_4(CO)_{12}$ is the rate-determining step in the $Ru_3(CO)_{12}$ catalyzed process (46), metal hydride formation ($[Cr(CO)_5H]^-$) can be rate limiting in the production of H_2 from $Cr(CO)_6$ in the absence of substantial formation of μ-$H[Cr(CO)_5]_2^-$.

Acknowledgment

Financial support from the National Science Foundation through Grant CHE 76-04494 and from Tulane University for this project is greatly appreciated.

Literature Cited

1. Greyson, M., "Catalysis," P. H. Emmett, Ed., Vol. 4, Chap. 6, Reinhold, New York, 1956.
2. Mills, G. A., Steffgen, F. W., *Catal. Rev.* (1973) **8**, 159.
3. Pichler, H., *Adv. Catal.* (1952) **4**, 271.
4. Vannice, M. A., *Catal. Rev.* (1976) **14**, 153.
5. Karn, F. S., Shultz, J. F., Anderson, R. B., *Ind. Eng. Chem., Prod. Res. Develop.* (1965) **4**, 265.
6. McKee, D. W., *J. Catal.* (1967) **8**, 240.
7. Shelef, M., Gandhi, H. S., *Ind. Eng. Chem., Prod. Res. Develop.* (1974) **13**, 80.
8. Davydov, A. A., Bell, A. T., *J. Catal.* (1977) **49**, 332.
9. Muetterties, E. L., *Bull. Soc. Chim. Belg.* (1975) **84**, 959.
10. Muetterties, E. L., *Bull. Soc. Chim. Belg.* (1976) **85**, 451.
11. Muetterties, E. L., *Science* (1977) **196**, 839.
12. Thomas, M. G., Beier, B. F., Muetterties, E. L., *J. Am. Chem. Soc.* (1976) **98**, 1296.
13. Demitras, G. C., Muetterties, E. L., *J. Am. Chem. Soc.* (1977) **99**, 2796.
14. Laine, R. M., Rinker, R. G., Ford, P. C., *J. Am. Chem. Soc.* (1977) **99**, 253.
15. Kang, H. C., Mauldin, C. H., Cole, T., Slegeir, W., Cann, K., Pettit, R., *J. Am. Chem. Soc.* (1977) **99**, 8323.

16. Frazier, C. C., Hanes, R. M., King, R. B., King, A. D., ADV. CHEM. SER. (1978) **173**, 94.
17. Muetterties, E. L., *Inorg. Chem.* (1965) **4**, 1841.
18. Darensbourg, D. J., Drew, D., *J. Am. Chem. Soc.* (1976) **98**, 275.
19. Darensbourg, D. J., Froelich, J. A., *J. Am. Chem. Soc.* (1977) **99**, 4726.
20. Ibid (1977) **99**, 5940.
21. Darensbourg, D. J., *Isr. J. Chem.* (1977) **15**, 247.
22. Darensbourg, D. J., Darensbourg, M. Y., *Inorg. Chem.* (1970) **9**, 1691.
23. Darensbourg, D. J., Froelich, J. A., *J. Am. Chem. Soc.* (1978) **100**, 338.
24. Cotton, F. A., Kraihanzel, C. S., *J. Am. Chem. Soc.* (1964) **84**, 4432.
25. Schachtschneider, J. H., Snyder, R. G., *Spectrochim. Acta* (1963) **19**, 85, 117.
26. Perutz, R. N., Turner, J. J., *Inorg. Chem.* (1975) **14**, 262.
27. Jones, L. H., *Inorg. Chem.* (1976) **15**, 1244.
28. Burdett, J. K., Perutz, R. N., Poliakoff, M., Turner, J. J., *Inorg. Chem.* (1976) **15**, 1245.
29. Wender, I., Pino, P., "Metal Carbonyls in Organic Synthesis," Interscience, New York, 1968.
30. Grillone, M. D., Kedzia, B. B., *J. Organomet. Chem.* (1977) **140**, 161.
31. Darensbourg, M. Y., Burch, R. R., Jr., unpublished data.
32. Pu, L. S., Yamamoto, A., Ikeda, S., *J. Am. Chem. Soc.* (1968) **90**, 3896.
33. Angaletta, M., Malatesta, L., Caglio, G., *J. Organomet. Chem.* (1975) **94**, 99.
34. Ueno, A., Onishi, T., Tamaru, K., *Trans. Faraday Soc.* (1970) **66**, 756.
35. Behrens, H., Weber, R., *Z. Anorg. Allgem. Chem.* (1957) **291**, 122.
36. Rhomberg, M. G., Owen, B. B., *J. Am. Chem. Soc.* (1951) **73**, 5904.
37. Darensbourg, M. Y., Walker, N., *J. Organomet. Chem.* (1976) **117**, C68.
38. Darensbourg, M. Y., Walker, N., Burch, R. R., Jr., *Inorg. Chem.* (1978) **17**, 52.
39. Roziere, J., Williams, J. M., Stewart, R. P., Jr., Petersen, J. L., Dahl, L. F., *J. Am. Chem. Soc.* (1977) **99**, 4497.
40. Olsen, J. P., Koetzle, T. F., Kirtley, S. W., Andrews, M., Tipson, D. L., Bau, R., *J. Am. Chem. Soc.* (1974) **96**, 662.
41. Love, R. A., Chin, H. B., Koetzle, T. F., Kirtley, S. W., Whittesey, B. R., Bau, R., *J. Am. Chem. Soc.* (1976) **98**, 4491.
42. Handy, L. B., Ruff, J. K., Dahl, L. F., *J. Am. Chem. Soc.* (1970) **92**, 7312.
43. Darensbourg, D. J., Burch, R. R., Jr., Darensbourg, M. Y., *Inorg. Chem.* (1978) **17**.
44. Harris, D. C., Gray, H. B., *J. Am. Chem. Soc.* (1975) **97**, 3073.
45. Darensbourg, D. J., Froelich, J. A., unpublished data.
46. Ford, P. C., "Abstracts of Papers," 175th National Meeting, ACS, March, 1978, INOR 77.

RECEIVED February 22, 1978.

11

Catalytic Reductions Using Carbon Monoxide and Water in Place of Hydrogen

R. PETTIT, K. CANN, T. COLE, C. H. MAULDIN, and W. SLEGEIR

University of Texas at Austin, Austin, TX 78712

> *Carbonyl complexes of rhodium, ruthenium, osmium, iridium, and platinum, in the presence of H_2O and a weak base (e.g., trimethylamine), act as catalysts for the conversion of propene to a mixture of butanal and methylpropanal; with the exception of the platinum system, these catalysts are considerably more active than $Fe(CO)_5$ as reported by Reppe. Under the same conditions, but in the absence of olefin, the carbonyls act as catalysts for the conversion of CO and H_2O to CO_2 and H_2. The metal carbonyls, together with $Fe(CO)_5$, in the presence of H_2O, CO, and a weak base such as Me_3N, serve as catalysts for the conversion of nitrobenzene, dinitrobenzene, and 2,4- and 2,6-dinitrotoluene to the corresponding aminobenzene derivatives.*

Carbon monoxide is readily available from low-grade carbonaceous material such as coal, lignite, etc., and it is readily conceivable that this material could play an increasingly important role in the energy and "petrochemical" fields as the sources of petroleum continue to diminish. In several cases, such as the Fischer–Tropsch synthesis of hydrocarbons (*1*) and the Koch synthesis of carboxylic acids (*2*), the potential significance of CO is well recognized, and it is very likely that other new uses will emerge following further research work. In this chapter we shall discuss our results to date concerning the use of CO and water in place of hydrogen as a reducing medium.

In principle, any reduction of a substrate S by hydrogen to yield SH_2 (Reaction 1) can be conducted with CO + H_2O as indicated in Reaction 2, and for each hydrogen molecule involved the latter process is always approximately 7 kcal/mol more thermodynamically favorable.

$$S + H_2 \rightarrow SH_2; F = -x \text{ kcal/mol} \qquad (1)$$

$$S + CO + H_2O \rightarrow SH_2 + CO_2; F = -(x+7) \text{ kcal/mol} \qquad (2)$$

The key problem in the potential use of CO as in Reaction 2 is of course the design of catalysts which allow the reaction to proceed at an acceptable rate. Just as hydrogen does not readily add to most substrates (the addition of hydrogen to ethylene, for example, does not occur, and the reaction is in fact forbidden by the Woodward–Hoffmann rules), and in most cases requires intervention of a catalyst, such will also be true when $CO + H_2O$ is used.

Reppe Modification of the Hydroformylation Reaction

Our initial efforts in the area of reductions with $CO + H_2O$ involved a study of the mechanism of what we shall term the Reppe modification of the hydroformylation reaction (3). In the normal hydroformylation process, the elements of hydrogen and CO are added to an olefin to generate an aldehyde (Reaction 3). In the Reppe modification of this process, the same reaction is achieved using $CO + H_2O$ in place of hydrogen (Reaction 4). In the normal process, the catalyst used is usually a

$$R - CH = CH_2 + CO + H_2 \rightarrow RCH_2CH_2CHO \qquad (3)$$

$$R - CH = CH_2 + 2CO + H_2O \rightarrow RCH_2CH_2CHO + CO_2 \qquad (4)$$

carbonyl derivative of cobalt or rhodium, whereas in the Reppe modification, iron carbonyl in conjunction with a Lewis or Brönsted base is used. A comparison of the two processes reveals several intriguing points of interest. Thus, in the normal process with cobalt carbonyl as the catalyst, temperatures of 150°C or higher and pressures of 3000 psi are used, whereas in the Reppe modification the conditions are much less severe; 100°C temperature and 500 psi of CO pressure are sufficient. Furthermore, when $Fe(CO)_5$ is used as a catalyst in the normal process with hydrogen, it performs very poorly (4); likewise, we find that when cobalt carbonyl is used under Reppe's conditions it too performs very poorly (zero activity).

To understand the manner in which the $Fe(CO)_5$ plus base catalyst system uses $CO + H_2O$ in place of hydrogen, speculating that such an understanding would be of value to the design of other catalysts capable of effecting reductions with $CO + H_2O$, we investigated the mechanism of the Reppe reaction. The pertinent results are summarized below.

We found that the hydroformylation reaction strongly depended on the pH of the medium in the reactor (5). When an autoclave was charged with ethylene, water, $Fe(CO)_5$, and Na_2CO_3 under 500 psi of CO, the $Fe(CO)_5$ dissolved to form $NaHFe(CO)_4$ according to Reaction 5. The formation of the $HFe(CO)_4^-$ anion from $Fe(CO)_5$ and base

$$Na_2CO_3 + Fe(CO)_5 + H_2O \rightarrow NaHFe(CO)_4 + CO_2 + NaHCO_3 \quad (5)$$

is the old and well-established reaction, and the anion is a well-characterized species (6, 7). Upon complete dissolution of the $Fe(CO)_5$ in this manner, the pH of the medium is around 12.0, but only when the pH drops to the vicinity of 10.7 (attributable to the reaction of CO + $OH^- \rightarrow HCOO^-$) does the formation of propanol begin to occur. (Propanal is the initial product, but under the conditions used it is readily reduced to propanol).

In a similar experiment in which 1,5-cyclooctadiene is used in place of ethylene, again, only when the pH drops to about 10.7 is there observed isomerization of 1,5- to 1,3-cyclooctadiene. When the same experiment is run without any olefin present, then hydrogen begins to appear in the reactor when the pH again reaches approximately 10.7.

In a similar experiment to that described with ethylene, but with addition of acetaldehyde, it is found that the acetaldehyde is immediately reduced to ethanol at a pH of 12.0, but again only when the pH is lowered to 10.7 does the hydroformylation of ethylene begin to take place.

A plausible explanation of this pH dependence is that a significant concentration of $H_2Fe(CO)_4$ begins to form at a pH of about 10.7 and that this is the species which initiates the hydroformylation reaction. $H_2Fe(CO)_4$ is known to be a powerful catalyst for the isomerization of olefins (8), and it also would be a sensible candidate for the formation of molecular hydrogen through simple thermal decomposition (9). The $HFe(CO)_4^-$ anion is not the catalyst for these three reactions; however, it is capable of reducing aldehydes to alcohols.

We consider then that the mechanism of the Reppe modification of the hydroformylation reaction closely parallels that of the normal process as described by Heck and Breslow (10). The principal steps are given in the following scheme:

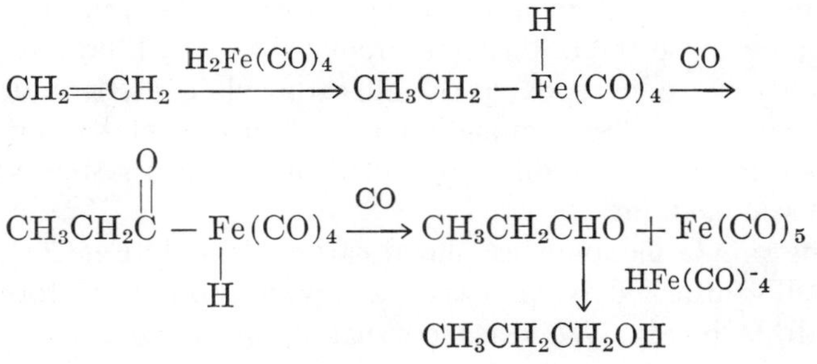

$$CH_2{=}CH_2 \xrightarrow{H_2Fe(CO)_4} CH_3CH_2-\underset{}{\overset{H}{Fe(CO)_4}} \xrightarrow{CO}$$

$$CH_3CH_2\overset{O}{\overset{\|}{C}}-\underset{H}{\overset{}{Fe(CO)_4}} \xrightarrow{CO} CH_3CH_2CHO + Fe(CO)_5$$

$$\downarrow HFe(CO)_4^-$$

$$CH_3CH_2CH_2OH$$

The normal hydroformylation reaction and the Reppe modification are mechanistically closely related; the key point which emerges is that in the latter process it is easier to form the species $H_2Fe(CO)_4$ from the reaction of $Fe(CO)_5$ and aqueous base than it is from $Fe(CO)_5$ and molecular hydrogen. For this reason, the combination of $CO + H_2O$ provides a superior reducing system for the reductive addition of CO to an olefin than does molecular hydrogen.

Other Catalysts for the Hydroformylation Reaction with $CO + H_2O$

The formation of the $H_2Fe(CO)_4$ catalyst described above presumably occurs via nucleophilic attack of hydroxyl ion on a CO ligand of $Fe(CO)_5$ with generation of the anionic metallocarboxylic acid followed by decarboxylation and protonation (Reaction 6).

$$(CO)_4Fe(CO) \xrightarrow{OH^-} (CO)_4Fe\overset{O}{\overset{\|}{-C}}-OH \xrightarrow{-CO_2} (CO)_4FeH^{\ominus} \xrightarrow{H^+} H_2Fe(CO)_4 \quad (6)$$

1

When aqueous amines are used as the base instead of Na_2CO_3, presumably the amine rather than hydroxyl ion acts as the attacking nucleophile; hydrolysis of the resulting metallocarboxyamide derivative then affords the metallocarboxylic acid **1** (*11*).

These considerations now provide a guideline for the development of other potential catalysts for the use of $CO + H_2O$ in the hydroformylation of olefins. If the catalyst is to function in the same manner as just described for $Fe(CO)_5$, then a minimum requirement is that the system form a metal carbonyl which will be readily attacked by a weak base to form an anion analogous to **1**. A weak base is essential because CO_2 is an inevitable by-product, and only the carbonate salts of weak bases regenerate the base and CO_2 upon heating. Thus, if the system is to be catalytic in base as well, then clearly only a weak base can be used. This would appear to be the critical requirement, for the literature indicates that metallocarboxylic acids readily decarboxylate (*12*), and the final step in Reaction 6, the protonation of a hydridometalcarbonyl anion, would seem to offer no problem provided the catalyst system was not in a highly basic medium.

Of the simple mononuclear metal carbonyls, only $Fe(CO)_5$ appears to be readily attacked by a weak base with subsequent formation of metal hydride bonds. We reasoned that if the metal carbonyl system contained the structural feature shown in formula 2, i.e., a metal–metal bond system containing a bridging carbonyl group and a terminal carbonyl ligand, then nucleophilic attack of OH⁻ might be facilitated. This would be attributable to the possible delocalization of negative charge from the metal atom to an oxygen atom through the resonance interaction indicated in 3.

$$\underset{2}{\text{M}-\text{M}=\text{C}=\text{O} \atop \underset{\text{O}}{\overset{\text{C}}{\underset{\|}{\diagdown\diagup}}}} \xrightarrow{\text{OH}^-} \underset{}{\text{M}-\text{M}-\text{COOH} \atop \underset{\text{O}}{\overset{\text{C}}{\underset{\|}{\diagdown\diagup}}}} \longleftrightarrow \underset{3}{\text{M}-\text{M}-\text{COOH} \atop \underset{\text{O}-}{\overset{\text{C}}{\underset{\|}{\diagdown\diagup\!\diagup}}}}^{-}$$

These considerations thus pointed to a large number of metal cluster carbonyl compounds which have the structural features indicated in 2. As mentioned earlier, the decarboxylation of compounds of type 3 to give metal hydrides was expected to proceed easily.

Several metal cluster systems have now been tried in the hydroformylation reaction of propylene and CO + H₂O (*13*), and as is seen from the data in Table I, some of them are found to be much superior to Fe(CO)₅ as catalysts. This is especially true of Rh₆(CO)₁₆ and Ir₄(CO)₁₂.

Since each of these catalyst systems was expected to proceed via intermediate formation of a metal hydride anion (from 3) and protonation to a neutral dihydrogen metal species, then in the absence of any olefin, each could be expected to produce molecular hydrogen upon thermal decomposition. That is, each also should be a catalyst for the water gas shift reaction (Reaction 7). Such was found to be the case; in the

$$\text{CO} + \text{H}_2\text{O} \rightarrow \text{CO} + \text{H}_2 \qquad (7)$$

last column of Table I there is listed the moles of hydrogen produced per mole of catalyst used under identical reaction conditions as with the hydroformylation reaction except for the omission of olefin. It is seen that each of the catalyst systems used are catalysts for the formation of hydrogen from CO and H₂O.

It must be pointed out that the exact nature of the catalytic species has not been established in the case of the metal cluster compounds listed in Table I. Rearrangement of the added cluster compounds to some other cluster species under the reaction conditions is readily conceivable, and in several cases, very probable.

Table 1. Hydroformylation of Propane

Experiment[a]		T(°C)
1	$Fe(CO)_5$	110
2	$Rh_6(CO)_{16}$	125
3	$Ru_3(CO)_{12}$	100
4	$H_4Ru_4(CO)_{12}$	100
5	$Os_3(CO)_{12}$	180
6	$H_2Os_3(CO)_{10}$	180
7	$H_4Os_4(CO)_{12}$	180
8	$Ir_4(CO)_{12}$	125
9	$(Bu_4N)_2[Pt_3(CO)_6]_5$	125

[a] All experiments run for 10 hr in a 300 mL stirred autoclave containing 0.05 mmol of catalyst, 22 mL of 25% aqueous trimethylamine, 78 mL of THF, 350 psi of CO, and 150 psi of propylene. Small amounts of propane were formed in experiments 2, 5, 6, 7, and 8.

Reduction of Aromatic Nitro Compounds to Amines

In 1925, in a German patent (*14*), it was revealed that alkaline solutions of $Fe(CO)_5$ would reduce aromatic nitro compounds to amines in a stoichiometric manner. More recently, Ladensberg and co-workers (*15*) have shown that salts of the trinuclear anion $HFe_3(CO)_{11}^-$ will also effect the reduction, and Watanabe and co-workers (*16*) have obtained similar results with the mononuclear anion $HFe(CO)_4^-$. Each of these species could have been formed under the conditions used in the first report, and presumably one or both would be the reactive reducing species. (H. Alper has recently shown that $Fe(CO)_5$, in the presence of $NaOD/D_2O$, will convert nitrobenzene into ϕND_2; however, the nature of the reducing species has not been established (*17*).)

With $HFe(CO)_4^-$ as the reducing agent, Watanabe and co-workers report that the reaction remains stoichiometric rather than catalytic when conducted under a pressure of CO. The stoichiometry, however, is most remarkable in that 1 mol of $HFe(CO)_4^-$ salt will reduce 1.8 mol of nitrobenzene to aniline. Since, in terms of electron-transfer reduction, the conversion of nitrobenzene to aniline involves six electrons per molecule, then the $HFe(CO)_4$ salt is acting as an 11 (10.8) electron-transfer agent. Irrespective of the exact mechanism of the electron-transfer process, clearly most of these electrons must be provided by the CO ligands, which in turn become oxidized to CO_2.

A rational explanation of the process would be a succession of two electron-transfer steps as indicated in Reactions 8 and 9. With complete depletion of the CO ligands and final oxidation of Fe^0 to Fe^{++} (Reaction 10), then a total of 12 electrons could be provided by $Fe(CO)_5$.

and Water Gas Shift with CO + H$_2$O

Moles of C$_4$ Aldehyde/Moles of Catalyst	n-/Isoaldehyde Ratio	C$_4$ Aldehyde/ C$_4$ Alcohol	Water Gas Shift Reaction (Moles of H$_2$/Moles of Catalyst)[b]
5.2	1.0	4.5	5
300	1.4	40	1700
47	11.5	43	3300
79	11.0	37	3400
13	1.9	6.6	270
6	1.2	~300	270
9	1.4	~300	400
250	1.8	~300	300
0.5	1.9	—	700

[b] Identical conditions as in (a) expect that the propylene is omitted; a temperature of 150°C was used for experiments 2 and 8. Optimum conditions for each catalyst system have not been determined.

Simple inspection of Reactions 8, 9, and 10 leads to the conclusion that the reaction should be made to be catalytic in Fe(CO)$_5$ if it were conducted under a CO pressure in the presence of excess base. The Fe(CO)$_4$ generated as in Reaction 8 should revert to Fe(CO)$_5$ upon treatment with CO, and provided inevitable loss of Fe^0 to Fe^{++} is avoided (CO does not reduce the ferrous ion under mild conditions), then only catalytic quantities of Fe(CO)$_5$ would be required.

This prediction has been realized. Various aromatic nitro compounds have been reduced to the corresponding amines by treatment at 25°C in aqueous solution of glyme containing catalytic quantities of Fe(CO)$_5$ and large amounts of triethylamine under a pressure of 1700 psi of CO (18). To observe catalysis, however, it was necessary to maintain the concentration of the oxidant, i.e., the nitrobenzene, low at all times. If this were not done, then rapid loss of the CO ligands occurred (Reaction 9) and irreversible formation of iron oxides (Reaction 10) resulted. The concentration of the oxidant was maintained low inside the reactor vessel simply by pumping in the nitrobenzene over a

$$\text{Fe(CO)}_5 \xrightarrow{+ \text{OH}^-} \text{HFe(CO)}_4^- \rightarrow \text{Fe(CO)}_4 + \text{H}^+ + 2\text{e}^- \quad (8)$$

$$\text{Fe(CO)}_{m(m \leq 4)} \xrightarrow{+ \text{OH}^-} \text{HFe(CO)}^-_{m-1} \rightarrow \text{Fe(CO)}_{m-1} + \text{H}^+ + 2\text{e} \quad (9)$$

$$\text{Fe}^\circ \rightarrow \text{Fe}^{++} + 2\text{e}^- \quad (10)$$

Table II. Catalytic Reductions with Fe(CO)$_5$[a]

Compound	Product	% Nitroarene Reduced	% Fe(CO)$_5$ Remaining[b]
Nitrobenzene (11.0 g)	aniline	100	95
m-Dinitrobenzene (8.2 g)	m-phenylene diamine	100	12
2,4-Dinitrotoluene (16.8 g)	2,4-diaminotoluene	100	32
2,6-Dinitrotoluene (9.3 g)	2,6-diaminotoluene	100	56

[a] Reaction conditions: the nitro compound dissolved in 50 mL of glyme was pumped over a 10–12 hr period into a stirred 300 mL reaction vessel containing 120 mL of glyme, 6g H$_2$O, 30 g of Et$_3$N, and 1 g Fe(CO)$_5$. The reaction was run at room temperature under 1700 psi of CO.
[b] Determined by IR absorption intensities.

relatively long period of time (approximately 12 hr). Even so, some of the Fe(CO)$_5$ was destroyed and appeared as oxides of iron. The pertinent data are shown in Table II.

In a separate, noncatalytic type of experiment, we demonstrated that nitrobenzene could be rapidly reduced to aniline in alkaline solutions of KHFe(CO)$_4$ maintained at a pH of 12.0; this indicated that the HFe(CO)$_4$ anion rather than H$_2$Fe(CO)$_4$ is capable of effecting the reduction. In our earlier studies dealing with hydroformylation (13), we have shown that several metal cluster species readily generate metal hydride anions upon treatment with mild bases, and these were also tested as candidates for the catalytic reduction of nitrobenzene to aniline with CO + H$_2$O as the reducing agent. The results obtained are summarized in Table III.

Table III. Reduction of Nitrobenzene with CO + H$_2$O[a]

Catalyst	T	Time (hr)	Reduction (%)	H$_2$[b] (%)
Rh$_6$(CO)$_{16}$	125	1	100	2–6
Ru$_3$(CO)$_{12}$	100	2	71	52
H$_4$Ru$_4$(CO)$_{12}$	100	2	73	56
Os$_3$(CO)$_{12}$	180	1	100	< 3
H$_2$Os$_3$(CO)$_{10}$	180	1	100	< 3
H$_4$Os$_3$(CO)$_{12}$	180	1	100	< 3
Ir$_4$(CO)$_{12}$	150	10	47	2–6
[Bu$_4$N]$_2$[Pt$_3$(CO)$_6$]$_5$	125	10	42	< 1
Re$_2$(CO)$_{10}$	180	2	10	< 2

[a] Reaction condition: 50.0 mmol of catalyst, 35 g of 25% aqueous trimethylamine, and 65 mL of THF added to a 300-mL pressure reactor. CO (500 psi) was introduced and the reactor heated to a temperature indicated in the table.
[b] The percent of hydrogen in the gas phase at the end of the reaction.

Each of the catalysts listed earlier in the hydroformylation reaction are found to be catalysts for the reduction of nitrobenzene to aniline; the rhodium, osmium, and ruthenium species are particularly effective although with osmium higher temperatures must be used. (A. F. M. Igbal previously has reported that several derivatives of rhodium, including $Rh_6(CO)_{16}$ can act as catalysts for the reduction of nitrobenzene to aniline using $CO + H_2O$ (19).) The catalysts listed are also much sturdier than was found in the case of iron carbonyl, and in the experiments listed in Table III the total amount of nitrobenzene and catalyst in a 1000:1 molar ratio was added to the reaction at the outset. In no case was there observed any precipitate of metal oxides or carbonates, and presumably much higher catalytic turnover numbers could be realized if the reaction were run in a continuous-type reactor.

In addition to providing electrons for the homogeneous thermal reduction of nitroarenes, the $HFe(CO)_4^-$ anion is also capable of providing electrons for the reduction of MnO_2, Ce^{+IV}, and other oxidizing agents in the mode of a primary cell battery. For example, the cell indicated in Figure 1, in which MnO_2 paste taken from a flashlight battery and

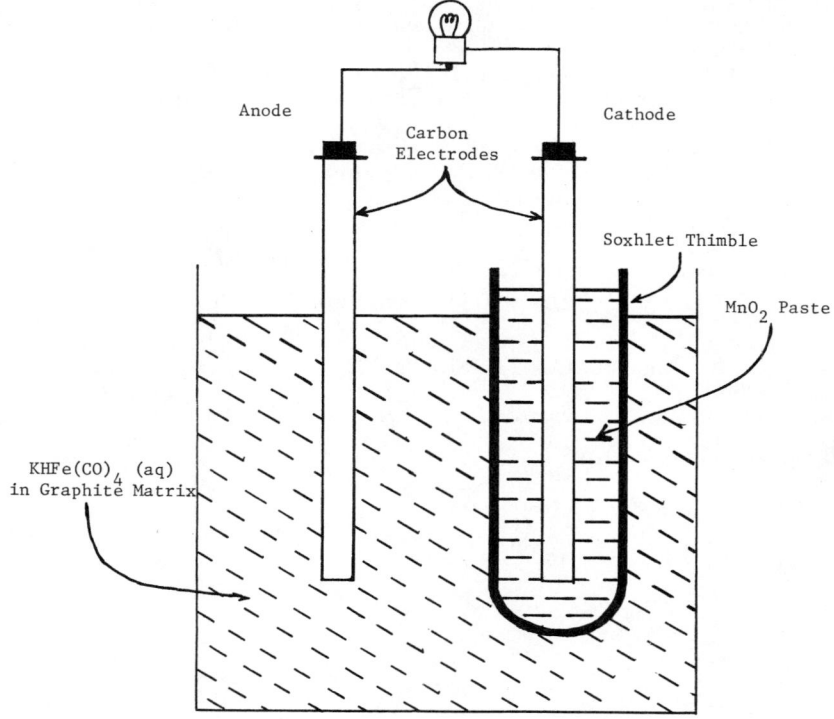

Figure 1. A simple primary cell with $HFe(CO)_4^-$ acting as the source of electrons

contained in a Soxhlet thimble is immersed in a solution of $KHFe(CO)_4$ in water, provides a voltage of 1.4. Presumably, the reaction occurring at the anode is that indicated by Reaction 8, but we have not as yet confirmed this point.

Acknowledgment

The authors thank the National Science Foundation and the Robert A. Welch Foundation for financial assistance.

Literature Cited

1. Storch, H. H., Golumbic, N., Anderson, R. B., "The Fischer-Tropsch and Related Syntheses," Wiley, New York, 1951.
2. Koch, H., *Brennst.—Chem.* (1955) **36**, 321.
3. Reppe, W., Vetter, H., *Justus Liebigs Ann. Chem.* (1953) **582**, 133.
4. Bird, C. W., "Transition Metal Intermediates in Organic Chemistry," p. 117, Academic, New York, 1967.
5. Pettit, R., Mauldin, C., Cole, T., Kang, H., *Ann. N. Y. Acad. Sci.* (1977) **295**, 151.
6. Smith, M. B., Bau, R., *J. Am. Chem. Soc.* (1973) **95**, 2388.
7. Dahl, L. F., Blount, J. F., *Inorg. Chem.* (1965) **4**, 1373.
8. Steinberg, H. W., Markby, R., Wender, I., *J. Am. Chem. Soc.* (1975) **79**, 6116.
9. Hieber, W., Vetter, H., *Z. Anorg. Allg. Chem.* (1933) **212**, 145.
10. Heck, R. F., Breslow, D. S., *J. Am. Chem. Soc.* (1961) **83**, 4023.
11. Edgell, W. F., Yang, M. T., Builkin, B. J., Bayer, R., Koizumi, N., *J. Am. Chem. Soc.* (1965) **87**, 3080.
12. Kurck, T., Höfler, M., Noack, M., *Chem. Ber.* (1966) **99**, 1153.
13. Kang, Hi Chun, Mauldin, C. H., Cole, T., Slegeir, W., Cann, K., Pettit, R., *J. Am. Chem. Soc.* (1977) **99**, 8323.
14. I. G. Farbenindustrie, A.G., German Patent No. **441,179** (Jan. 18, 1925).
15. Landesberg, J. M., Katz, L., Olsen, Carol, *J. Org. Chem.* (1972) **37**, 930.
16. Watanabe, Y., Mitsudo, T., Yamashita, M., Takegami, Y., *Bull. Chem. Soc. Jpn.* (1975) **48**, 1478.
17. Abbayes, H., Alper, H., *J. Am. Chem. Soc.* (1977) **99**, 98.
18. Pettit, R., Cann, K., Cole, T., Slegeir, W., *J. Am. Chem. Soc.* (1978) **100**, 3969.
19. Igbal, A. F. M., *Tetrahedron Lett.* (1971) 3385.

RECEIVED March 3, 1978.

12

Mechanistic Studies Related to the Metal-Catalyzed Reduction of Carbon Monoxide to Hydrocarbons

CHARLES P. CASEY and STEPHEN M. NEUMANN

Department of Chemistry, University of Wisconsin, Madison, WI 53706

> *Anionic metal formyl complexes are produced upon reaction of K^+ $HB[OCH(CH_3)_2]_3^-$ with metal carbonyls. The formyl compounds can be observed in solution by ^1H NMR (14–16 δ) and have been isolated as stable solids in several cases: $(CH_3CH_2)_4N^+\{[3,5\text{-}(CH_3)_2\text{-}C_6H_3O]_3P\}(CO)_3FeCHO^-$ is kinetically quite stable but decomposes upon heating to 70°C to $(CO)_4FeH^-$ ($E_a = 29.7 \pm 2$ kcal mol^{-1}). No metal formyl compound can be observed in equilibrium with the corresponding metal hydride. In contrast, the equilibrium between $L(CO)_3FeCOCH_3^-$ and $(CO)_4FeCH_3^-$ lies entirely on the side of the acetyl iron compound. Metal formyl compounds can act as hydride donors to ketones, alkyl halides, and metal carbonyls. Metal carbene complexes react with molecular hydrogen, leading to reductive cleavage of the carbene ligand.*

Metal Formyl Complexes

Metal formyl complexes have been proposed as important intermediates in the metal-catalyzed reduction of CO by H_2 (*1, 2, 3, 4*). While the insertion of CO into alkyl and aryl carbon–metal bonds is well known (*5*), the insertion of CO into a metal–hydrogen bond to give a metal formyl complex has not been observed. (The intermediacy of metal formyl compounds in the substitution reactions of metal hydrides has been considered.) To ascertain the reasons for the failure to observe metal formyl complexes in the reactions of metal hydrides with CO, we have developed a new synthesis of metal formyl complexes and have studied their properties.

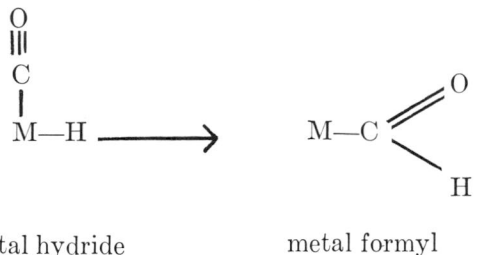

metal hydride   metal formyl

Synthesis and NMR Observation of Metal Formyl Complexes. At the outset of our research, the only reported metal formyl complex was $[(C_6H_5)_3P]_2N^+(CO)_4FeCHO^-$, which had been prepared by Collman and Winter in 1973 (9) from acetic formic anhydride and $Na_2Fe(CO)_4$. Attempts to extend this synthetic route to the preparation of metal formyl complexes from $(C_5H_5)Fe(CO)_2^-$, $Cr(CO)_5^{-2}$, $(C_6H_5)_3PMn(CO)_4^-$, and $[(C_6H_5)_3P]_2Mn(CO)_3^-$ uniformly met with failure (10).

We have developed a new synthesis of metal formyl compounds from the addition of metal trialkoxyborohydrides to metal carbonyls (10, 11). The formyl proton characteristically appears at very low field, 14–16 δ, in the ^1H NMR spectrum of metal formyl complexes. This low field resonance has allowed us to rapidly survey the reactions of trialkoxyborohydrides with a series of metal carbonyls. Initially, $Na^+HB(OCH_3)_3^-$ was used as the borohydride reducing agent, but we have subsequently found that $K^+HB(O\text{-}i\text{-}Pr)_3^-$ is a more rapid and effective hydride donor (10). We have obtained NMR evidence for the formation of metal formyl complexes in the reactions of $K^+HB(O\text{-}i\text{-}Pr)_3^-$ with $Fe(CO)_5$ (14.9 δ); $(C_6H_5O)_3PFe(CO)_4$ (14.8 δ, d, $J = 44$); $(C_6H_5)_3PFe(CO)_4$ (15.5 δ, d, $J = 24$); $Cr(CO)_6$ (15.2 δ); $W(CO)_6$ (15.9 δ); and $Re_2(CO)_{10}$ (16.0 δ). In some cases we have isolated the metal formyl complexes. In other cases, such a $Cr(CO)_6$, the maximum observed conversion to $(CO)_5Cr\text{-}CHO^-$ was 76% after 25 min at room temperature, and the formyl complex underwent subsequent decomposition with a half-life of 40 min at room temperature.

Isolation of Stable Metal Formyl Complexes. To date we have isolated and characterized the following stable metal formyl complexes: $[(C_6H_5)_3P]_2N^+(CO)_4FeCHO^-$, $(CH_3CH_2)_4N^+[(C_6H_5O)_3P](CO)_3FeCHO^-$, $(CH_3CH_2)_4N^+[(3,5(CH_3)_2C_6H_3O)_3P](CO)_3FeCHO^-$, and $(CH_3CH_2)_4N^+(CO)_9Re_2CHO^-$.

Kinetic Stability of Metal Formyl Complexes. Metal formyl complexes have approximately the same kinetic stability as the corresponding metal acetyl complexes. Thermal decomposition of $(CH_3CH_2)_4N^+[(C_6H_5O)_3P](CO)_3FeCHO^-$ in THF at 65°C gives a mixture of two metal hydrides in a 4:1 ratio: $(CO)_4FeH^-$, formed by loss of phosphite and hydride migration, and $[(C_6H_5O)_3P](CO)_3FeH^-$, formed by loss of CO and hydride migration. The decarbonylation of the related tri-(3,5-dimethylphenyl)phosphite complex $(CH_3CH_2)_4N^+[(ArO)_3P](CO)_3FeCHO^-$ proceeded by exclusive loss of phosphite to give $(CO)_4FeCHO^-$ and $(ArO)_3P$ as the only observed products.

$$[(C_6H_5O)_3P](CO)_3FeCHO^- \rightarrow (CO)_4FeH^- + [(C_6H_5O)_3P](CO)_3FeH^- \quad (1)$$

$$[(ArO_3P](CO)_3FeCHO^- \rightarrow (CO)_4FeH^- + (ArO)_3P \quad (2)$$

A detailed kinetic study of Reaction 2 was carried out. The rate of formation of metal hydride from metal formyl complex was followed by ^1H NMR. First-order kinetics were observed for Reaction 2 to more than two half-lives, indicating that the rate of reaction was independent of the concentration of phosphite. In related experiments we have found that the initial rate of Reaction 2 is independent of added phosphite. Only the phosphorus-containing species shown in Reaction 2 were observed by ^{31}P NMR. The half-life for decomposition of $(CH_3CH_2)_4N^+[(ArO)_3P]$-$(CO)_3FeCHO^-$ in THF at 67.3°C was found to be 1.1 hr. Measurement of the rate of decomposition of the metal formyl complex over the temperature range 47°–79°C gave an activation energy for the process of 29.7 ± 2 kcal/mol. ($\Delta H^{\ddagger} = 29.0 \pm 1.5$ kcal, $\Delta S^{\ddagger} = 7.9 \pm 6.1$ eu at 63°C).

The mechanism of Reaction 2 is thought to proceed by rate-determining phosphite dissociation followed by rapid hydride migration from the formyl carbon to iron. Analogous mechanisms for the decarbonylation of metal acyl complexes have been demonstrated (5). The fact that the rate of Reaction 2 is independent of phosphite concentration indicates either that migration of hydride from the formyl group to the metal in the coordinatively unsaturated intermediate is either much faster than capture of the intermediate by phosphite or that migration of hydride from the formyl group to the metal is concerted with the loss of phosphite from starting material.

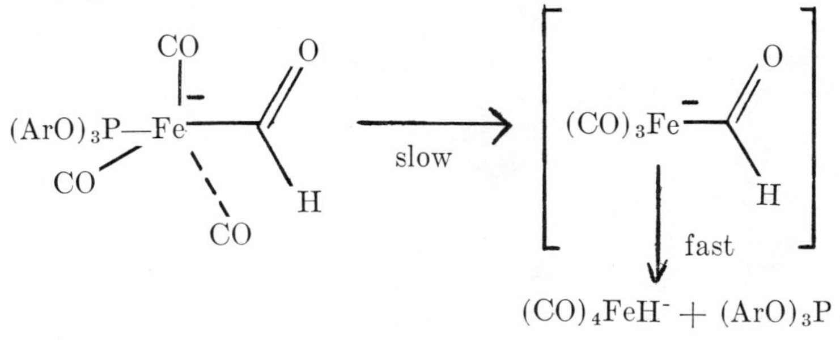

The corresponding acetyl iron complex $[(ArO)_3P](CO)_3FeCOCH_3^-$ does not decarbonylate under the reaction conditions since the acetyl compounds are greatly favored thermodynamically relative to the methyl iron compound $(CO)_4FeCH_3^-$. However, the phosphite exchange of the acetyl complex proceeds via a coordinatively unsaturated intermediate very similar to that proposed in the rate-determining step of the decomposition of the metal formyl complex. The rate of exchange of $(C_6H_5O)_3P$ exchange with $[(ArO)_3P](CO)_3FeCOCH_3^-$ (Reaction 3) was measured

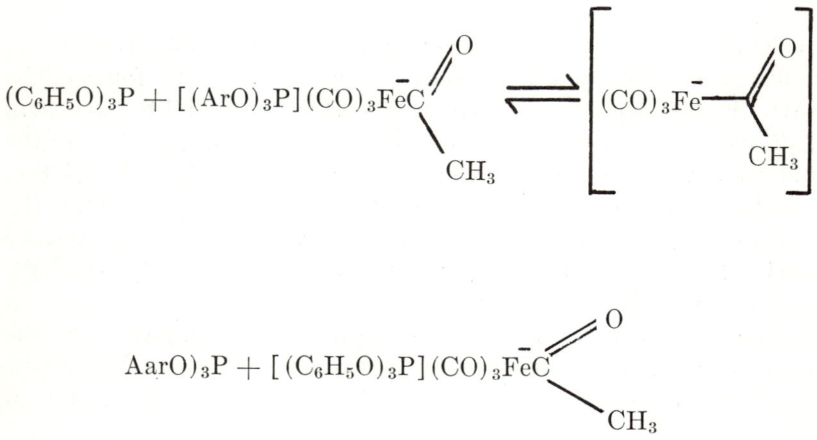

$$(C_6H_5O)_3P + [(ArO)_3P](CO)_3\overset{-}{Fe}C\overset{\displaystyle O}{\underset{\displaystyle CH_3}{\diagdown}} \rightleftarrows \left[(CO)_3\overset{-}{Fe}-C\overset{\displaystyle O}{\underset{\displaystyle CH_3}{\diagdown}}\right] \quad (3)$$

$$AarO)_3P + [(C_6H_5O)_3P](CO)_3\overset{-}{Fe}C\overset{\displaystyle O}{\underset{\displaystyle CH_3}{\diagdown}}$$

by ^{31}P NMR. Preliminary kinetic data indicate that the rate of Reaction 3 is about 20 times slower than the rate of decomposition of metal formyl complex in Reaction 2 (kinetics of the ligand exchange reaction of $L(CO)_3FeCOC_6H_5^-$ have been measured (*12*).) Thus, the formyl group and the acetyl group affect the lability of phosphite ligands in these iron complexes in very similar ways.

Equilibrium between Metal Formyl Complexes and Metal Hydrides. Now that we have an efficient route to metal formyl complexes, we can study the thermodynamic stability of these complexes. We have found that metal formyls are much less thermodynamically stable than the corresponding metal hydrides. Thus, metal formyl species have never been observed in the reactions of metal hydrides with CO because of the thermodynamic instability of the formyl complexes and not because of their kinetic instability.

The equilibrium between $(CH_3CH_2)_4N^+[(ArO)_3P](CO)_3FeCHO^-$ and $(CH_3CH_2)_4N^+(CO)_4FeH^-$ was studied by 270 MHz ^1H NMR. At equilibrium, even in the presence of 1.5M $(ArO)_3P$, there was no observable metal formyl complex. Since 1% of the metal formyl compound could have been detected, a limit on the equilibrium constant for Reaction 4 can be estimated: $K_{eq} = \{[(ArO)_3P](CO)_3FeCHO^-\}/[(CO)_4FeH^-]\cdot[(ArO)_3P] \leq 1.7 \times 10^{-2} M^{-1}$. This corresponds to a free energy difference of at least 3.1 kcal mol^{-1} in favor of the metal hydride. In contrast, the equilibrium between $[(ArO)_3P](CO)_3FeCOCH_3^-$ and $(CO)_4FeCH_3^-$ + $(ArO)_3P$ strongly favors the acyl iron compound.

$$(CO)_4FeH^- + (ArO)_3P \rightleftharpoons [(ArO)_3P](CO)_3FeCHO^- \qquad (4)$$

The greatly different thermodynamic stabilities of metal acyl and metal formyl compounds is probably attributable to the greater strength of the M–H bond (estimated 50–60 kcal mol^{-1}) compared with the M–C bond (estimated 30–40 kcal mol^{-1}). We are now attempting to measure the heat of reaction for the conversion of a metal-formyl compound to a metal hydride. This will allow a much better estimate of the energy difference between a metal formyl complex and a metal hydride since the energy difference was too large to measure by equilibration of the species.

Hydride Transfer Reactions of Metal Formyl Complexes. We have found that metal formyl complexes can act as hydride donors to electrophiles such as ketones, alkyl halides, and metal carbonyls. Et_4N^+trans-$[(C_6H_5O)_3P](CO)_3FeCHO^-$ reacts with 2-butanone overnight at ambient temperature to give a 95% yield of 2-butanol. The possibility that 2-butanone is reduced by $(CO)_4FeH^-$ formed in situ from decomposition of the metal formyl complex is excluded since the metal formyl complex reacts with 2-butanone much faster than it decomposes to $(CO)_4FeH^-$ and since no reaction between $(CO)_4FeH^-$ and 2-butanone was observed by IR spectroscopy.

Reaction of a THF solution of $Et_4N^+[(C_6H_5O)_3P](CO)_3FeCHO^-$ with CF_3CO_2H (10 equivalents) led to the formation of 27% CH_3OH and no observable formaldehyde (< 5%). Methanol might arise from acid cleavage of the metal formyl complex to give formaldehyde, which is subsequently reduced to methanol by hydride donation from a second equivalent of metal formyl complex. Alternatively, methanol could arise via O-protonation to give a hydroxycarbene complex, which is subsequently reduced to a hydroxymethyl complex by a second equivalent of metal formyl complex. Cleavage of the hydroxymethyl complex by acid would give methanol. Either of the two routes leads to a maximum 50% yield of methanol.

Collman and Winter have reported that $[(C_6H_5)_3P]_2N^+(CO)_4FeCHO^-$ reacts with 1-iodo-octane to give octane (75%) and a trace of nonanal; octane formation was attributed to decarbonylation of $[(C_8H_{17})(CHO)$-

$$\text{Et}_4\text{N}^+ [(\text{C}_6\text{H}_5\text{O})_3\text{P}](\text{CO})_3\text{Fe}-\overset{-}{\text{C}}\overset{\displaystyle =\text{O}}{\underset{\text{H}}{}} \quad \xrightarrow{\text{H}^+} \quad \underset{\text{H}\quad\text{H}}{\overset{\displaystyle\overset{\text{O}}{\|}}{\text{C}}} \xrightarrow{\mathbf{1}} \xrightarrow{\text{H}^+} \text{CH}_3\text{OH}$$

(with $[(\text{C}_6\text{H}_5\text{O})_3\text{P}](\text{CO})_3\text{Fe}=\text{C}\begin{smallmatrix}\text{O-H}\\\text{H}\end{smallmatrix} \xrightarrow{\mathbf{1}} [(\text{C}_6\text{H}_5\text{O})_3\text{P}](\text{CO})_3\text{Fe}-\text{CH}_2\begin{smallmatrix}\text{OH}^-\\\end{smallmatrix}$ below)

Fe(CO)$_4$], followed by hydride migration and reductive elimination of octane (9). We have found that Et$_4$N$^+$*trans*-[(ArO)$_3$P](CO)$_3$FeCHO$^-$ (Ar = 3,5-dimethylphenyl) reacts with *n*-C$_7$H$_{15}$I when stirred overnight at ambient temperature in THF to give *n*-C$_7$H$_{16}$ (71%) and (CO)$_4$FeP-(OAr)$_3$; less than 0.2% *n*-C$_7$H$_{15}$CHO was observed. Since formyl complexes appear to function as hydride donors, we propose that heptane is formed by nucleophilic displacement of iodide by the formyl hydrogen atom.

Transformylation Reactions. Metal formyl complexes can transfer hydride to metal carbonyl compounds to produce new metal formyl complexes. These "transformylation" reactions can be used to determine the relative stability of a series of metal formyl complexes. The reaction of Et$_4$N$^+$[(C$_6$H$_5$O)$_3$P](CO)$_3$FeCHO$^-$ with Re$_2$(CO)$_{10}$ in THF-d_8 was followed by ^{1}H NMR. The characteristic doublet at 14.9 δ caused by the iron formyl complex is rapidly replaced by a singlet at 16.04 δ attributed to Et$_4$N$^+$*cis*-(CO)$_9$Re$_2$CHO$^-$, formed in 82% yield as determined by NMR integration. (CO)$_4$Fe[P(OC$_6$H$_5$)$_3$] was detected by IR as a co-product of the reaction.

The rhenium formyl complex was independently synthesized by reaction of K$^+$HB(O-*i*-Pr)$_3^-$ with Re$_2$(CO)$_{10}$ in THF at 0°C, followed by aqueous basic workup and cation exchange with Et$_4$N$^+$Br$^-$. It was recrystallized from THF-hexane, and isolated in 32% yield as a yellow, air-stable solid.

Transformylation reactions between various metal formyl and metal carbonyl compounds indicate the following order of stability of formyl complexes relative to their metal carbonyl precursors: $[(C_6H_5)_3P]_2N^+$-$(CO)_4FeCHO^- > Et_4N^+(CO)_9Re_2CHO^- > Et_4N^+ trans\text{-}[(C_6H_5O)_3P](CO)_3$-$FeCHO^-$. The possibility that these transformylations proceed via decarbonylation to give a metal hydride, which then transfers hydride to a metal carbonyl, is excluded since the reaction of $Et_4N^+[(C_6H_5O)_3P]$-$(CO)_3FeCHO^-$ with $Re_2(CO)_{10}$ or $Fe(CO)_5$ is faster than its decomposition to $(CO)_4FeH^-$ and since $Et_4N^+(CO)_4FeH^-$ does not react with $Re_2(CO)_{10}$ or $(CO)_4FeP(OC_6H_5)_3$ under the reaction conditions.

$$[(C_6H_5O)_3P](CO)_3FeCHO^- + Re_2(CO)_{10}$$
$$\Updownarrow$$
$$(C_6H_5O)_3PFe(CO)_4 + (CO)_9Re_2CHO^- + Fe(CO)_5$$
$$\Updownarrow$$
$$Re_2(CO)_{10} + (CO)_4FeCHO^-$$

Hydrogenation of Metal Carbene Complexes

We have studied the reaction of molecular hydrogen with transition-metal carbene complexes as a model for the termination step in the synthesis of hydrocarbons and alcohols from CO and hydrogen (13).

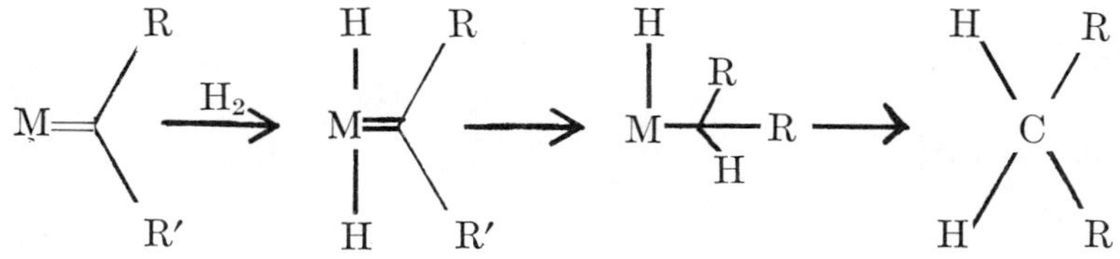

We have found that reaction of $(CO)_5WC(C_6H_5)_2$ with H_2 (69 atm) at 100°C gives $CH_2(C_6H_5)_2$ in 40% yield. Under these conditions, $(CO)_5WC(C_6H_5)_2$ undergoes rapid CO dissociation as shown by ^{13}CO exchange studies. Similarly $(CO)_5WC(OCH_3)C_6H_5$ reacts with H_2 (1.8 atm) at 140°C to give methyl benzyl ether in 92% yield. Alkenes and ketones present as side-chain groups in the metal-carbene complexes are not reduced under these reaction conditions.

Apparently, the reaction does not involve a metal hydrogenation catalyst but involves a stoichiometric reaction of molecular hydrogen with

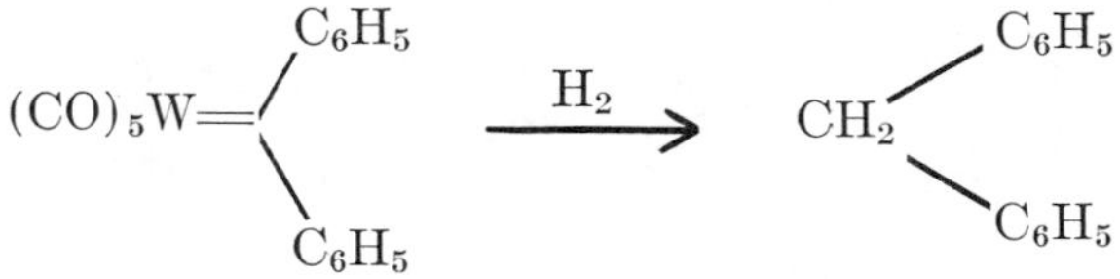

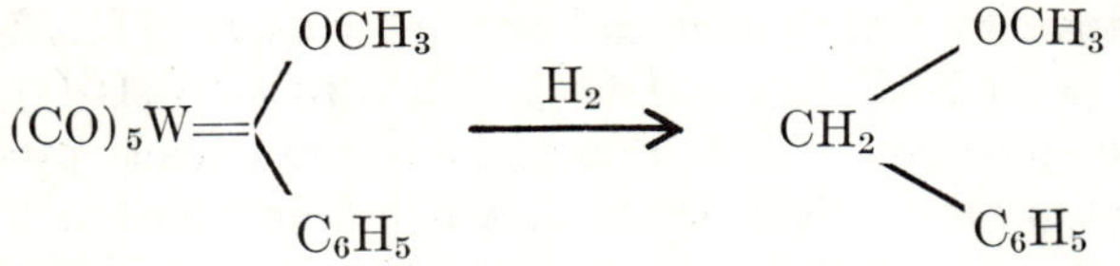

$(CO)_5W=C(OCH_3)(C_6H_5) \xrightarrow{H_2} CH_2(OCH_3)(C_6H_5)$

$(CO)_5Cr= \text{[2-acetyl-2-methyl-tetrahydrofuran-ylidene]} \xrightarrow{H_2} \text{[2-acetyl-2-methyl-tetrahydrofuran]}$

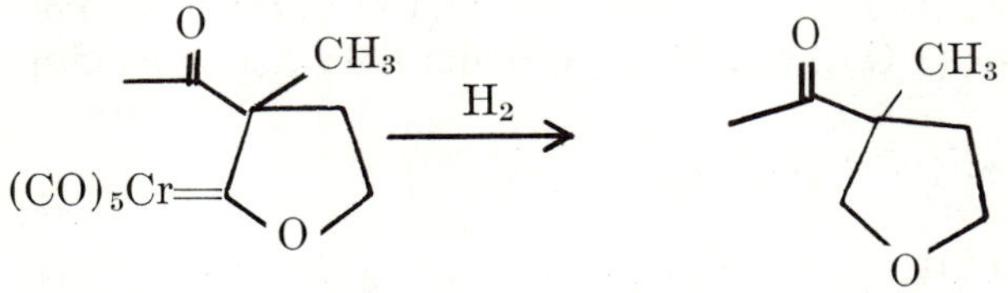

$(CO)_5Cr= \text{[3,3-bis(3-methyl-2-butenyl)-tetrahydrofuran-ylidene]} \xrightarrow{H_2} \text{[3,3-bis(3-methyl-2-butenyl)-tetrahydrofuran]}$

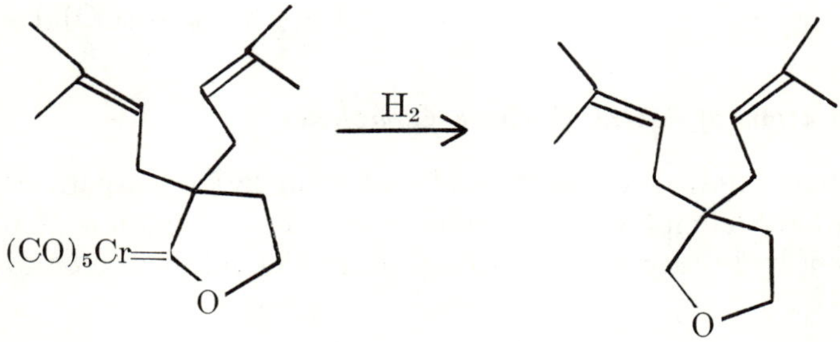

$(CO)_5Cr=\text{[tetrahydrofuran-ylidene]} \xrightarrow{\Delta \text{ or } h\nu} (CO)_4Cr=\text{[tetrahydrofuran-ylidene]} + CO$

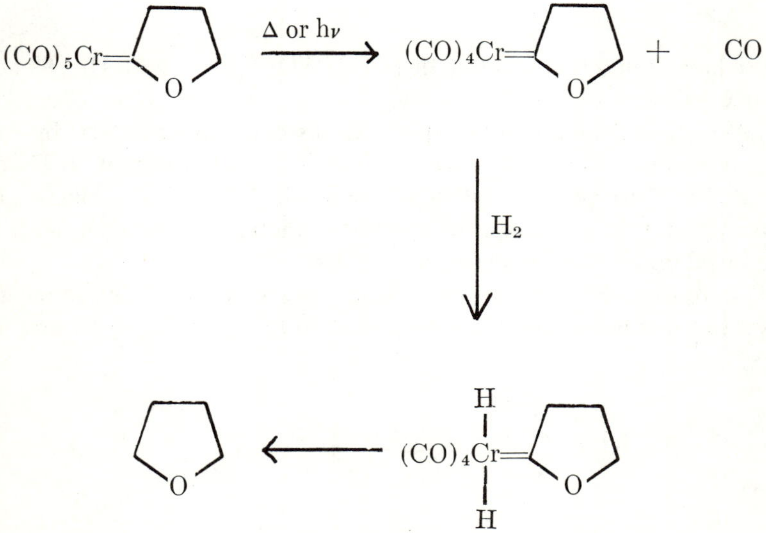

$\downarrow H_2$

$\text{[tetrahydrofuran]} \longleftarrow (CO)_4Cr(H)_2=\text{[tetrahydrofuran-ylidene]}$

a coordinatively unsaturated metal carbene complex. This implies that photochemical hydrogenation of carbene complexes can be accomplished under mild conditions since photolysis of the metal carbene complex can lead to the same coordinatively unsaturated intermediate. Preliminary experiments indicate that photochemical hydrogenation can occur below room temperature.

Acknowledgment

Support from the Division of Basic Energy Sciences of the Department of Energy is gratefully acknowledged.

Literature Cited

1. Henrici-Olivé, G., Olivé, S., *Angew. Chem. Int. Ed. Engl.* (1976) **15**, 136.
2. Vannice, M. A., *J. Catal.* (1975) **37**, 449, 462.
3. Storch, H. H., Golumbic, N., Anderson, R. B., "The Fischer-Tropsch and Related Synthesis," Wiley, New York, 1951.
4. P. H. Emmett, Ed., "Catalysis," Vol. 4, Reinhold, New York, 1956.
5. Calderazzo, F., *Angew. Chem., Int. Ed. Engl.* (1977) **16**, 299.
6. Byers, B. H., Brown, T. L., *J. Organomet. Chem.* (1977) **127**, 181.
7. Berry, A., Brown, T. L., *J. Organomet. Chem.* (1971) **33**, C67.
8. Basolo, F., Pearson, R. G., "Mechanisms of Inorganic Reactions," 2nd ed., p. 555, Wiley, New York, 1967.
9. Collman, J. P., Winter, S. R., *J. Am. Chem. Soc.* (1973) **95**, 4089.
10. Casey, C. P., Neumann, S. M., *J. Am. Chem. Soc.* (1976) **98**, 5395.
11. Winter, S. R., Cornett, G. W., Thompson, E. A., *J. Organomet. Chem.* (1977) **133**, 339.
12. Conder, H. L., Darensbourg, M. Y., *Inorg. Chem.* (1974) **13**, 506.
13. Casey, C. P., Neumann, S. M., *J. Am. Chem. Soc.* (1977) **99**, 1651.

RECEIVED February 22, 1978.

13

Carbon Monoxide–Metal Oxide Interactions

Surface Site Requirements for Electron Transfer Processes

KENNETH J. KLABUNDE, RICHARD A. KABA, and RUSSELL M. MORRIS

Department of Chemistry, University of North Dakota, Grand Forks, ND 58202

Electron transfer from metal oxide surfaces to CO can be quite facile, occurring at room temperature. This process can be important as an initial CO activation step in metal oxide catalyzed reduction schemes. We have attempted to clarify what types of metal oxides interact (MO + CO → MO$^+$ ... CO$^-$·) with CO in this way, and what surface features these active metal oxides possess. Only MgO, CaO, SrO, BaO, and ThO$_2$ were electron transfer active. These oxides have in common the possession of both Lewis basic sites and one electron reducing site. It appears that CO is first adsorbed on Lewis base sites followed by slow migration to electron transfer reducing sites. The studies leading to this conclusion are discussed.

Active Sites on Metal Oxides for Electron Transfer Processes. Numerous studies have appeared throughout the literature dealing with ESR studies of paramagnetic centers generated on metal oxide surfaces by a variety of methods. Scheme 1 summarizes some of this work but is limited to MgO, which serves as an illustrative example. High energy radiation (neutrons or γ-rays) cause many crystalline defects where electrons can be trapped to form paramagnetic F or F' centers (bulk defects) and S or S' centers (surface defects) (1). The F and S centers are "metal excess" sites caused by the presence of an anion vacancy and then electron trapping. There are corresponding V centers, which are cation vacancies that trap positive holes. It is probable that

Scheme 1. *Paramagnetic Centers Generated on MgO*

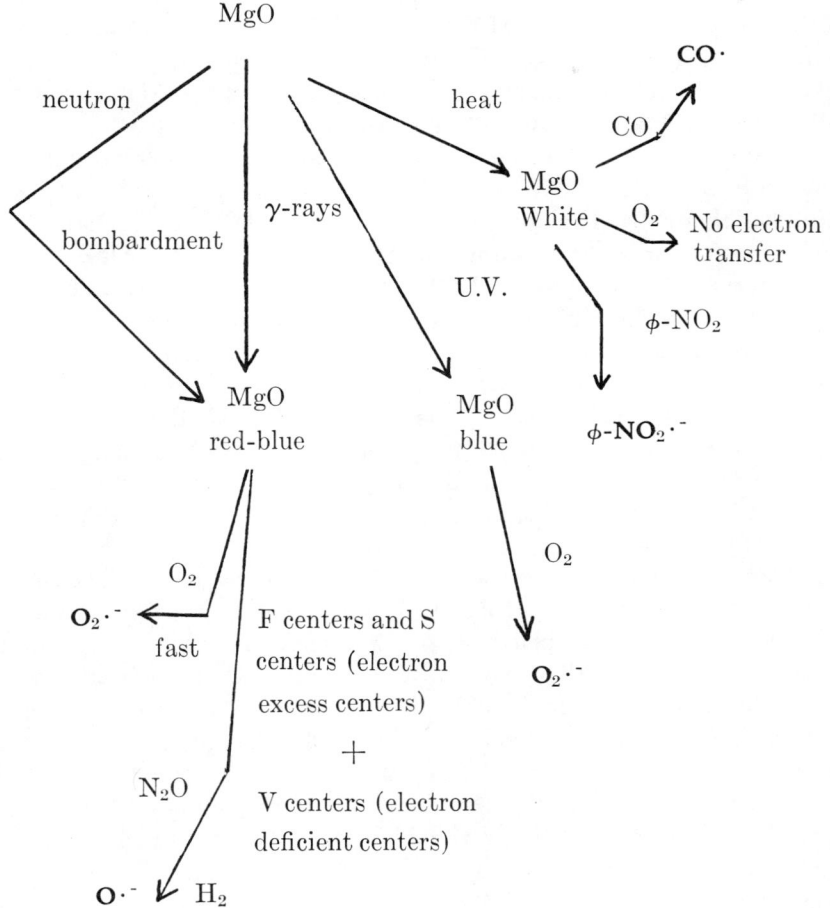

less energy is required to form a cation vacancy (leading to V centers) than anion vacancies (leading to S and F centers) (1).

The chemistry of the trapped electron centers (S and F) has been studied extensively with added small molecules. In particular, oxygen addition has been studied in some detail (2, 3). Many metal oxides other than MgO have been studied similarly (4–10). Extensive work has shown that $O^-\cdot$, $O_2^-\cdot$, and $O_3^-\cdot$ can be obtained by interacting the F and S centers with N_2O (3), O_2, and N_2O (with $O_2^-\cdot$ present) respectively (11, 12, 13). The $O^-\cdot$ species is prepared particularly well by UV irradiation of MgO in the presence of H_2 (3), followed by N_2O addition. Furthermore, several groups have begun investigations of the chemistry of the

$O^-\cdot$ species from the aspect of surface anion–molecule reactions (14, 15). In particular, alkane dehydrogenation by $O^-\cdot$ on MgO (14) and ethylene and ethylene oxide to form $H_2C=\overset{.}{C}^-\cdot$ and $H_2C-\overset{.}{C}H_2O^-$ radicals have been studied (15, 16). O_2, CO, and ethylene form $O_3^-\cdot$, $CO_2^-\cdot$, and $C_2H_4^-\cdot$ according to Naccache (16). There is much fascinating chemistry still to be investigated regarding these kinds of interactions.

A number of other molecules have been allowed to interact with the F and S centers. Even CO_2 can be reduced by the centers (17, 18, 19). Similarly, thermally activated and UV-activated MgO exposed to SO_2 causes the formation of two different $SO_2^-\cdot$ radicals (20). Also, $H_2S_2^-\cdot$ can be formed by addition of H_2S to UV-activated MgO (21). Thus, it appears that the chemistry of $O^-\cdot$ on MgO and F and S centers on MgO or other oxides are rich areas for investigations, and ESR is a powerful tool to use in these studies (22).

Generally, it appears that UV-activated MgO is similar in chemistry to γ-ray or neutron-bombarded MgO (or other oxides), although there is still room for more detailed studies regarding UV wavelength dependence and "quantum yield" studies. However, there are gross differences between the active sites on MgO generated thermally (nonparamagnetic) and those generated by irradiation techniques (paramagnetic). Thermally generated sites are still "reducing sites" in the sense that electron transfer to added molecules takes place. At the same time, however, backbonding or back donation can allow the reduced molecule to drain back charge density to the oxide surface but still maintain a paramagnetic bonding picture. These active sites, whatever they look like, are selective. Thus, O_2 does not react with thermally generated active sites on MgO to yield paramagnetic species (1, 23), nor does CO_2 form paramagnetic species when exposed to thermally activated MgO, although CO_2 is adsorbed (nonparamagnetically) at room temperature to the extent of near monolayer coverage (1).

In a fascinating paper, Tench and Nelson discuss thermally generated active sites on MgO, and their electron transfer to adsorbed nitro compounds to form paramagnetic anion radicals (24). According to proton splitting anisotropy, it was concluded that nitrobenzene lies flat on the MgO surface after accepting the electron. These authors discuss three possibilities for the mechanism of the electron transfer, or in particular the exact surface site required. They reported that adsorption of H_2O or CO_2 (nonparamagnetically) completely inhibits the electron transfer to nitro compounds. They consider three site possibilities: (1) transition metal impurity ions; (2) electrons trapped at intrinsic defects (F or S centers); and (3) lattice oxygen ions on the surface. They reject (1) because of the very low concentrations of transition metal ions vs. spins observed and (2) because F and S centers are ESR observable and essentially none are present on thermally activated MgO. They conclude that (3) is the mechanism, which means that an $O^=$ ion must give up one electron to the nitro compound. Now the reaction $O^= \rightarrow O^- + e^-$ is exothermic in the gas phase by ~ 6.5 eV (24, 25, 26), but in the MgO lattice the $O^=$ ion should be greatly stabilized by the coulombic field. Thus, the ease with which $O^=$ gives up an electron will depend greatly on its degree of coordination. Tench and Nelson conclude that if the coordination of the $O^=$ ion is correct, then compounds of electron affinity of 0.7 or greater can accept an electron (such as nitro compounds), but molecules of < 0.5 eV (such as O_2) cannot (24).

Thus, in thermally activated MgO, and presumably in other oxides thermally activated that do not show F and S centers, we have a completely different type of active center than in irradiated samples (F and S unpaired electron centers). What does this thermally generated active site look like? And what does the "hole" left behind look like? Very little has been established concerning these questions, except that it is very improbable that transition metal impurities play a role (23, 24, 27, 28, 29).

Electron Transfer to CO from Thermally Activated MgO and ThO_2. The classic work of Lunsford and Jayne in 1966 first showed that CO adsorption on thermally activated MgO yielded a paramagnetic species (30). This paper reported that CO interacts in such a way as to produce a radical with uniaxial anisotropy of the g factor, and with bonding similar to that believed to exist in metal carbonyls (strong π-acceptor characteristic of CO). They observed values of 2.0021 for g_{11} and 2.0055 for $g_\perp$. It was proposed that the radical formed was actually neutral or even slightly positive owing to the strong donation of an electron pair to the MgO, and receiving in return an electron in an antibonding π orbital. Further, Lunsford and Jayne (30) believed that iron ion impurities were necessary and were related to the electron transfer process (currently this is thought not to be true (27).

Brey and co-workers extended these studies to thermally activated ThO_2, and observed several paramagnetic species after CO addition (29). These authors outline several models for the adsorption process, as described in Scheme 2. They believe that (A) is formed by CO adsorption on a site with a preexisting surface defect that has immediately available an unpaired electron (similar to the Lunsford model (30)). They believe (B) and (C) are formed by slow diffusion of defects in ThO_2 which finally reach the surface and react with adsorbed CO to probably form bridged CO radical species. A fourth ESR signal is probably attributed to a $CO^{+}\cdot$ species (D), although overall the species would probably be nearly neutral. And with $CO^{+}\cdot$, (D) could interact with $O^=$ to form $CO_2^-\cdot$ (E) (29).

Scheme 2. Assignment of CO Surface Structures to Observed ESR Signal (29)

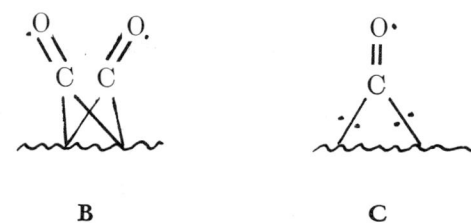

Meriaudeau, Breysse, and Claudel (31) also studied CO on ThO_2 with particular emphasis on the possible formation of CO^+ since they had proposed this species earlier as an intermediate in the catalytic oxidation of CO (32). ^{13}CO as well as ^{12}CO experiments were carried out with the conclusion that the only species they can ascribe to a CO species has an axial symmetry signal similar to (C) in Scheme 2. Using the spin densities obtained from their ^{13}CO work, Meriaudeau and co-workers (31) were able to calculate that the CO radical species has its unpaired electron in an orbital of high p character and spends 15 or 33% of its time around the carbon atom (two different paramagnetic CO absorbates). Their CO-surface bonding interpretation depends on the fact that a marked change in electrical conductivity of the ThO_2 during CO chemisorption was found. They believe that the CO–ThO_2 electronic exchange results in the formation of slightly positive CO adsorbates and a surface becoming negative as a whole to a depth of a few atomic layers. This creates energy-band curvature near the surface and accounts for the change in electrical conductivity (31).

Thermally Activated Alkaline Earth Oxides (MgO, CaO, SrO, BaO) as Catalysts. It is curious that thermal activation of alkaline earth oxides, besides promoting electron transfer processes, also leads to more active hydrogenation and isomerization catalysts. In an extremely significant paper, Hattori, Tanaka, and Tanabe (33) describe MgO as a hydrogenation catalyst and show that it apparently operates in an ionic mode ($H_2 \rightarrow H^+ + H^-$ on the surface) that allows D_2 uptake into 1,3-butadiene without H–D scrambling to form cis-2-butene-1,4-d_2. Alkaline earth oxides showed an activity order of CaO > SrO > MgO > BaO >> BeO for this reaction when activated between 800°–1000°C (*34*). Alkenes can also be hydrogenated (*35*) by these oxides by using higher reaction temperatures. Other processes have also been studied with the alkaline earth oxides as catalysts: isomerization of alkenes (*36, 37, 38*), polymerization of styrene (*39, 40*), esterification of benzaldehyde (*41*), and H_2–D_2 exchange (*28*).

Results and Discussion

What types of surface sites are involved in these electron transfer and catalytic processes on thermally activated alkaline earth oxides? Tanabe and his co-workers believe that Lewis base sites and one-electron "reducing sites" are both important and have shown for CaO that one type of site can be favored over the other by different thermal activation procedures (*39, 42*).

We have used Tanabe's idea in principle for the quantitative study of the number of reducing sites on MgO. The measurements were made by determining the number of spins of $C_6H_5NO_2^{-}\cdot$ formed on MgO samples after each had been heat treated at a different temperature (400°–1000°C) and then cooled to 25°C. The samples were aged for several weeks to insure complete radical anion formation. These experiments were carried out to: (1) determine if the number of reducing sites generated by heat treatment of MgO in vacuo changed with change in temperature, as Tanabe had found for heat treatment of CaO in air; and (2) to determine if the $C_6H_5NO_2^{-}\cdot$ formation paralleled $CO^{-}\cdot$ formation on identically treated MgO samples. Figure 1 is a plot of relative $CO^{-}\cdot$ formation vs. absolute $C_6H_5NO_2^{-}\cdot$ formation, both vs. temperature of heat treatment. Note the excellent correlation of $C_6H_5NO_2^{-}\cdot$ production with $CO^{-}\cdot$ production. We believe this is strong evidence that the same type of reducing sites are involved in both systems—CO and $C_6H_5NO_2$. It remains to be seen if other alkaline earth oxides will behave similarly and if other molecules will behave similarly on these oxides. These data also support our contention that transition metal impurities are not responsible for formation of the radical species observed.

We have determined surface areas of the heat-treated MgO samples by BET methods. The values ranged from 213 m^2/g to 83 m^2/g (83 m^2/g at 300°C heat treatment, 213 at 400°, 139 at 500°, 140 at 600°, 130 at 700°, 134 at 800°, 129 at 900°, and 122 at 1000°C). During the experiment it appears that at 400°–500°C the main portions of H_2O, O_2, CO_2, and other gases are desorbed in vacuo. The surface area becomes lower with higher heat treatment and then stabilizes at about 500°C, and this is where maximum radical formation activity was found.

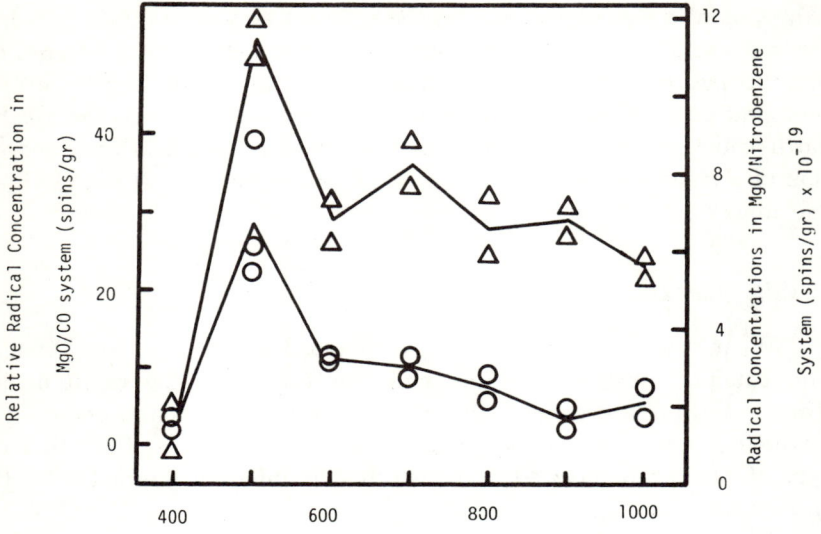

Figure 1. Relative radical concentration (spins/g) in the MgO/CO (circles) system and absolute radical concentration (spins/g) of the MgO/nitrobenzene (triangles) system as a function of the heat treatment temperature of MgO. NB. The scales for the two systems are different; also some samples have not reached their maximum radical growth value, especially in the MgO/CO system.

Knowing the surface areas, we were able to calculate crystallite sizes and thereby determine the ratio of MgO surface to MgO internal as 0.12. Also knowing the number of spins of $C_6H_5NO_2^-\cdot$ formed, we were able to calculate the number of MgO surface molecules to $C_6H_5NO_2^-\cdot$ at approximately 15:1 (for the 500°C heat treated sample). Also, although more data is still needed, we estimate the MgO surface molecule to $CO^-\cdot$ formation as approximately 400:1 for the same MgO sample. These results demonstrate that these electron transfer processes are major surface processes of significant importance and not simply caused by minor effects and/or small impurities.

Is MgO unique in these electron transfer processes? Are alkaline earth oxides unique? We carried out studies on a wide range of oxides, carbonates, hydroxides, and sulfides looking for activity for formation of paramagnetic species with CO. These were all heat treated at 600°C or as high as their melting point or decomposition point would allow. Table I lists the compounds studied. It can be seen that only the alkaline earth oxides and ThO_2 were found to be active. What do these have in common that the other materials studied do not? Much previous work indicates that these all possess fairly strong Lewis base surface sites as well as "reducing sites" (24, 42–46). In addition to these, ThO_2 also has Lewis acid sites on its surface. None of the other materials possess basic and reducing sites together (47). Furthermore, since the CO radical formation is slow compared with CO adsorption and since Lewis acids such as CO_2 poison the surface towards any CO radical formation, we believe (27) that CO absorption first takes place on Lewis base sites ($O^=$ sites) in approximate monolayer coverage, much like Tench and Nelson (24) describe for H_2O or CO_2. We also believe that once the CO is adsorbed on basic sites, slow migration of CO to reducing sites takes place (23, 27). The order of activity for the CO radical formation is MgO > CaO > SrO > BaO (very weak) (27).

As previously mentioned, it has been suggested by earlier workers (30) that iron impurities play an important part in the formation of CO radicals on MgO. The data in Table I suggests that this is not the case for our samples (i.e., it seems unlikely that iron impurities would only lead to radical formation in the alkaline earth oxides and thorium oxide). Furthermore, our determination of the iron content of the various samples (MgO, 0.0018% Fe; CaO, 0.0005% Fe; and SrO, 0.0019% Fe) certainly did not reflect the observed order of activity of the alkaline earth oxides towards CO radical formation discussed earlier. Thus, we believe it is unlikely that iron impurities play a role in the present work.

The EPR signal of the adsorbed CO radical on MgO, Figure 2, shows an unsymmetric g value. The signal is not destroyed until the sample is heated to > 150°C. CaO behaves quite similarly. Water addition at room temperature to the CO radical MgO system immediately causes a change in the EPR signal (see Figure 2). The available data suggest

Table I. Oxides, Hydroxides, Carbonates, and Other Materials Tested for Activity for Formation of Paramagnetic Species[a]

Active		Non-Active
MgO	BeO	Fe_2O_3 (0.05%) : Al_2O_3
CaO	Na_2O (220°C)	GeO_2
SrO	NaOH(5%) : SiO_2	ZnO
BaO	NaOH (250°C)	ZrO_2
ThO_2	NaOOCH (190°C)	La_2O_3
	Na_2CO_3	CeO_2
	Al_2O_3	Nd_2O_3
	Al_2O_3 (45%) : SiO_2	Eu_2O_3
	SiO_2	Dy_2O_3
	TiO_2	UO_3/U_2O_5
	FeS(7%) : SiO_2	PbO
		SnO_2

[a] All materials were heat treated at 600°C unless otherwise indicated in vacuo (10^{-6} Torr) for 12–18 hr before exposing to 150 Torr CO.

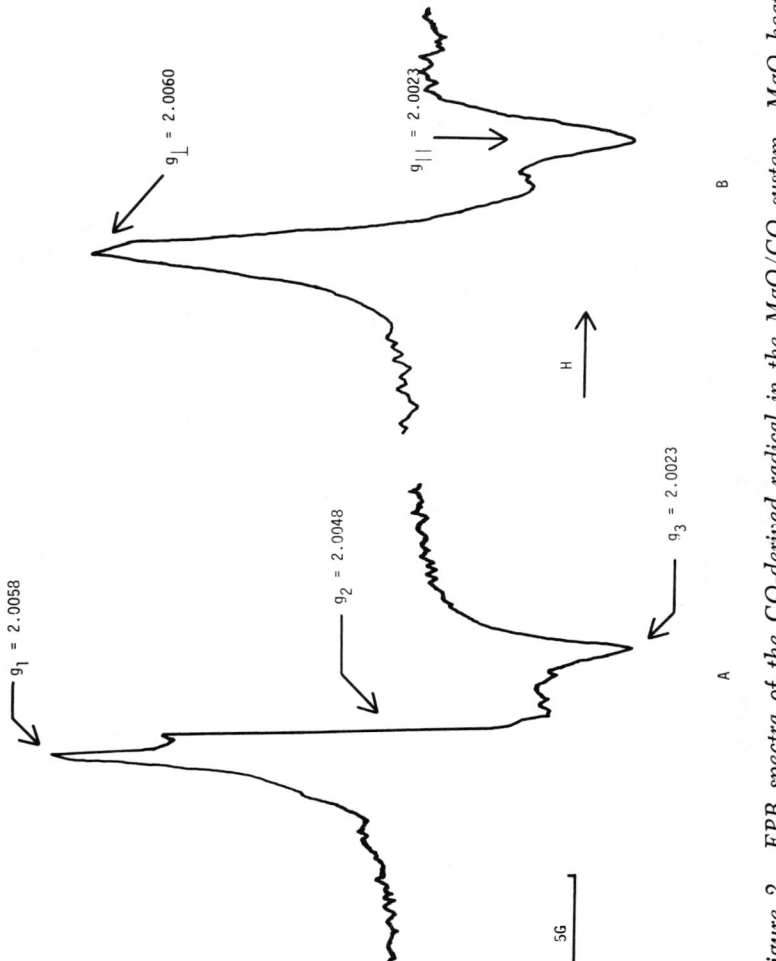

Figure 2. EPR spectra of the CO-derived radical in the MgO/CO system. MgO heat treated at 800°C. (A) After ca. two weeks exposure to 160 Torr CO; (B) MgO/CO system, spectra (A), upon exposure to ca. 25 Torr D_2O.

that the H_2O (or D_2O) is adsorbed in the vicinity of the adsorbed radical, changing its environment and thus the EPR signal rather than forming a new radical.

In later publications we will describe some CO radical chemistry on alkaline earth oxides with reagents such as O_2, H_2, H_2O, D_2O, and organics. Also, we are testing MgO and CaO in model compound reduction sequences to determine if radical processes such as those described here are actually important in the CO–H_2O reducing medium.

Relationship to Catalysis, Coal Liquefaction, and Electron Movement in Metal Oxides

Understanding in detail the chemistry of CO and CO–H_2O mixtures over metal oxide surfaces has considerable importance for a number of reasons: (1) CO–H_2O is a more effective reducing medium relative to pure H_2 for reduction and depolymerization of low rank coals where metal oxides and other minerals can act as in situ catalysts (48, 49, 50, 51, 52); (2) CO–H_2O is better than H_2 for the reduction of some organic functional groups (53, 54); (3) some metal oxides, especially basic oxides, are used as promotors in Fischer–Tropsch catalytic chemistry and their role is not understood mechanistically (55, 56, 57); and (4) metal oxides are often used as catalysts for the CO + H_2O ⇌ CO_2 + H_2 water–gas shift reaction (e.g., see Ref. 58; see also Ref. 59).

Our work and that of Tench (24) has shown that the electron transfer sites on these metal oxides (e.g., MgO and CaO) are selective and yet present in high surface concentration. Preliminary activation by thermal or photochemical oxide means is required, however. This means that these processes can be related to photon-induced electron transfer (60, 61, 62, 63, 64) in semiconductor metal oxides in that similar surface sites may be required (i.e., similar coordination geometries).

Acknowledgment

Extremely helpful discussions with Kozo Tanabe are greatly appreciated. Also, we acknowledge with gratitude the generous support of the Energy Research and Development Administration (ERDA-FE, E(49-18)-2211) of the U.S. Department of Energy and the National Science Foundation (traineeship for R.M.M.).

Literature Cited

1. Nelson, R. L., Tench, A. J., Harmsworth, B. J., *Trans. Faraday Soc.* (1967) **63**, 1427.
2. Che, M., Tench, A. J., *Chem. Phys. Lett.* (1973) **18**, 199.
3. Lunsford, J. H., *Cat. Rev.* (1973) **8**, 135.

4. Che, M., Tench, A. J., Coluccia, S., Zecchina, A., *J. Chem. Soc., Faraday Trans. 1* (1976) **72,** 1553.
5. Meriaudeau, P., Vedrine, J. C., *J. Chem. Soc., Faraday Trans. 2* (1976) **72,** 472.
6. Fujita, Y., Kwan, T., *Bull. Chem. Soc. Jpn.* (1958) **31,** 379.
7. Barry, T. I., Stone, F. S., *Proc. R. Soc. A* (1960) **255,** 124.
8. Kwan, T., Sancier, K. M., Fujita, Y., Setaka, M., Fukuzawa, S., Kirino, Y., *J. Res. Inst. Catal., Hokkaido Univ.* (1968) **16,** 53.
9. Kennedy, D. R., Ritchie, M., Mackenzie, J., *Trans. Faraday Soc.* (1958) **54,** 119.
10. Gravelle, P. C., Juillet, F., Meriaudeau, P., Teichner, S. J., *Discuss. Faraday Soc.* (1971) **52,** 140.
11. Tench, A. J., Holroyd, P., *J. Chem. Soc., Chem. Comm.* (1968) 471.
12. Lunsford, J. H., Jayne, J. P., *J. Chem. Phys.* (1966) **44,** 1487.
13. Tench, A. J., Lawson, T., *Chem. Phys. Lett.* (1970) **7,** 459.
14. Aika, K. I., Lunsford, J. H., *J. Phys. Chem.* (1977) **81,** 1393.
15. Taarit, Y. B., Symons, M. C. R., Tench, A. J., *J. Chem. Soc., Faraday Trans. 1* (1977) **73,** 1149.
16. Naccache, C., *Chem. Phys. Lett.* (1971) **11,** 323.
17. Lunsford, J. H., Jayne, J. P., *J. Phys. Chem.* (1965) **69,** 2182.
18. Meriaudeau, P., Vedrine, J. C., Taarit, Y. B., Naccache, C., *J. Chem. Soc., Faraday Trans. 2* (1975) **71,** 736.
19. Meriaudeau, P., Taarit, Y. B., Vedrine, J. C., Naccache, C., *J. Chem. Soc., Faraday Trans. 2* (1977) **73,** 76.
20. Schoonheydt, R. A., Lunsford, J. H., *J. Phys. Chem.* (1972) **76,** 323.
21. Lin, M. J., Lunsford, J. H., *J. Phys. Chem.* (1976) **80,** 2015.
22. Gardner, C. L., Casey, E.-J., *Catal. Rev.* (1974) **9,** 1.
23. Kaba, R. A., Klabunde, K. J., unpublished work.
24. Tench, A. J., Nelson, R. L., *Trans. Faraday Soc.* (1967) **63,** 2254.
25. Massey, H. S. W., "Negative Ions," p. 17, University Press, Cambridge, 1950.
26. Winters, E. R. S., *Adv. Catal.* (1958) **10,** 196.
27. Klabunde, K. J., Kaba, R. A., Morris, R. M., *Inorg. Chem.*, in press.
28. Boudart, M., Delbouille, A., Derouane, E. G., Indovina, V., Walters, A. B., *J. Am. Chem. Soc.* (1972) **94,** 6622.
29. Brey, W. S., Jr., Gammage, R. B., Virmani, Y. P., *J. Phys. Chem.* (1971) **75,** 895.
30. Lunsford, J. H., Jayne, J. P., *J. Chem. Phys.* (1966) **44,** 1492.
31. Meriaudeau, P., Breysse, M., Claudel, B., *J. Catal.* (1974) **35,** 184.
32. Claudel, B., Juillet, F., Trambouze, Y., Veron, J., "Proceedings of the Third International Congress of Catalysis," (1965) **1,** 214.
33. Hattori, H., Tanaka, Y., Tanabe, K., *J. Am. Chem. Soc.* (1976) **98,** 4652.
34. Tanaka, Y., Hattori, H., Tanabe, K., *Chem. Lett.* (1976) 37.
35. Hattori, H., Tanaka, Y., Tanabe, K., *Chem. Lett.* (1975) 659.
36. Schachter, Y., Pines, H., *J. Catal.* (1968) **11,** 147.
37. Tani, N., Misono, M., Yoneda, Y., *Chem. Lett.* (1973) 591.
38. Baird, M. J., Lunsford, J. H., *J. Catal.* (1972) **26,** 440.
39. Iizuka, T., Hattori, H., Ohno, Y., Sohma, J., Tanabe, K., *J. Catal.* (1971) **22,** 130.
40. Hattori, H., Yoshii, N., Tanabe, K., *Catal. Proc. Int. Congr., 5th, 1972 (1973)* paper 10, 233.
41. Tanabe, K., Saito, K., *J. Catal.* (1974) **35,** 247.
42. Hattori, H., Satoh, A., *J. Catal.* (1976) **45,** 32.
43. Take, J., Kikuchi, N., Yoneda, Y., *J. Catal.* (1971) **21,** 164.
44. Flockhart, B. D., Leith, I. R., Pink, R. C., *Trans. Faraday Soc.* (1969) **65,** 542.
45. Krylov, O. V., Morkova, Z. A., Tret'yakov, I. I., Fokina, E. A., *Kinet. Katal.* (1965) **6,** 128.

46. Mohri, M., Tanabe, K., Hattori, H., *J. Catal.* (1974) **32**, 144.
47. Tanabe, K., private communications.
48. Fisher, F., Schrader, H., *Brennst. Chem.* (1921) **2**, 257.
49. Appell, H. R., Wender, I., Miller, R. D., *Chem. Ind., Engl.* (1969) **47**, 1703.
50. Appell, H. R., Wender, I., Miller, R. D., *Am. Chem. Soc., Div. Fuel Chem., Prepr.* (1968) **12**, 220.
51. Severson, D., Souby, M., Kube, W., *Bur. Mines Information Circular 8650* (1974) 236.
52. Souby, M., Severson, D., Kube, W., *Proc. N. D. Acad. Sci.* (1976) **28**, 50.
53. Jones, D., Baltisberger, R. J., Klabunde, K. J., Woolsey, N. F., Stenberg, V. I., *J. Org. Chem.* (1978) **43**, 175.
54. Pettit, R., "Conference on the Place of Transition Metals in Organic Synthesis," *Ann. N.Y. Acad. Sci.* (1976) **295**, 151.
55. Deluzarche, A., Hindermann, J. P., Kieffer, R., Muth, A., Papadopoulos, M., Tanielian, C., *Tetrahedron Lett.* (1977) 797.
56. Shah, Y. T., Perrotta, A. J., *Ind. Eng. Chem., Prod. Res. Dev.* (1976) **15**, 123.
57. Hanrici-Olivé, G., Olivé, S., *Angew. Chem., Int. Ed. Engl.* (1976) **15**, 136.
58. Oliver, R. B., British Patent **1,099,802** (1968); *Chem. Abstr.* (1968) **68**, 53839r.
59. Atroshchenko, V. I., Bibr, B., Zhidkov, B. A., Zasorin, A. P., Ivanova, L. N., *Nauchn. Osn. Podbora Proizvod. Katal., Acad. Nauk SSSR, Sibirsk. Otd.* (1964) 177; *Chem. Abstr.* (1965) **63**, 4991f.
60. Kraeutler, B., Bard, A. J., *J. Am. Chem. Soc.* (1977) **99**, 7729.
61. Fujishima, A., Honda, K., *Nature (London)* (1972) **238**, 37.
62. Freund, T., Gomes, W. P., *Catal. Rev.* (1969) **3**, 1.
63. Schrauzer, G. N., Guth, T. D., *J. Am. Chem. Soc.* (1977) **99**, 7189.
64. Wrighton, M. S., Ginley, D. S., Wolczanski, P. T., Ellis, A. B., Morse, D. L., Linz, A., *Proc. Nat. Acad. Sci., U.S.A.* (1975) **72**, 1518.

RECEIVED February 22, 1978.

14

Structure and Reactivity of Ni(II)-d^8 Complexes with Monodentate Tertiary Phosphine: CO Fixation

C. SAINT-JOLY, M. DARTIGUENAVE, and Y. DARTIGUENAVE

Laboratoire de Chimie de Coordination du CNRS et Université P. Sabatier, 205 Rte de Narbonne, 31030 Toulouse, France

Nickel(II) phosphine complexes have been reported to be efficient catalysts in carbonylation reactions. To investigate this reaction mechanism, we have studied the reaction of CO on the related Ni(II) complexes: $NiX_2(PMe_3)_n$ (n = 2, 3) and $[NiX(PMe_3)_m]BF_4$ (m = 3, 4). Pentacoordinate carbonyl nickel(II) species (without reduction of Ni(II) to Ni(O)) were isolated (1) by direct substitution of PMe_3 by CO in the pentacoordinate complex and (2) by addition of CO on the trans square-planar tetracoordinate complex. These compounds are trigonal–bipyramidal complexes with CO in equatorial position. The Ni–CO distance (1.73 Å) is the shortest reported Ni–CO distance. Since these carbonylation reactions can be viewed as substitution of an equatorial PMe_3 by CO in a d^8 TBP, they can be related to the substitution reactions in square-planar d^8 metal complexes.

Numerous carbonylation reactions are known that are catalyzed by nickel, palladium, and platinum complexes. The reaction products can be a variety of organic compounds such as esters, aldehydes, etc. (1, 2). Since very few detailed studies on the possible mechanism of these reactions have been reported, it is commonly assumed that acylmetal complexes are the basic intermediates in the reactions. For example, Scheme 1 outlines the mechanism proposed by Garrou and Heck for the palladium-catalyzed conversion of arylhalides into esters (1, 2).

Scheme 1

$$PdX_2(PPh_3)_2 + CO + 2\,n\text{-}BuOH \rightarrow Pd(CO)(PPh_3)_2 + 2HX + (n\text{-}BuO)_2CO$$
$$Pd(CO)(PPh_3)_2 + RX \rightarrow PdRX(PPh_3)_2 + CO$$
$$PdRX(PPh_3)_2 + CO \rightarrow Pd(RCO)X(PPh_3)_2$$
$$Pd(RCO)X(PPh_3)_2 + n\text{-}BuOH \rightarrow PdHX(PPh_3)_2 + n\text{-}BuCOOR$$
$$PdHX(PPh_3)_2 + CO \rightarrow Pd(CO)(PPh_3)_2 + \underline{HX}$$

neutralized by NR_3

In the nickel chemistry, attempts to isolate acylnickel(II) compounds have failed for a long time (1, 2, 3, 4). Moreover, the resulting organic compounds were carbonyl free, indicating an apparent instability of the supposed acylnickel(II) intermediate (1, 2, 5, 6, 7). However, in 1972 Pankowski and Bigorgne (8) succeeded in isolating the pentacoordinate $NiI_2(CO)(PMe_3)_2$ complex, and in 1973 Klein (5, 6, 7) prepared the stable acylnickel(II) complexes $NiX(CH_3CO)(PMe_3)_2$ in normal conditions of temperature and pressure.

$$NiX(CH_3)(PMe_3)_2 + CO \xrightarrow[\text{fast}]{\text{pentane}} NiX(CH_3CO)(PMe_3)_2$$

$$d^8 \quad \nu_{CO} = 1635\text{--}1650 \text{ cm}^{-1}$$
$$X = Cl, Br, I$$

The crystal structure determination (9) shows the presence of a trans square-planar Ni(II) complex with normal bond lengths and angles. The acetyl plane is perpendicular to the coordination plane of the Ni(II). The Ni–C distance of 1.84(1) Å can be compared directly with the values of 1.82(3) and 1.80 Å obtained in the Ni(0):$Ni(CO)_4$ and $Ni_2(CO)_6$-(P_2Ph_4) complexes where π back bonding is operative and to the mean value of 1.84 Å obtained for Ni–C in the Ni(II) complexes $[Ni(CN)_5]^{3-}$ (10) and 1.82(2) Å in $NiI_2(CO)(Fdma)$ (11) (Fdma = ferrocene-1,1'-bis(dimethylarsine)). It is tempting to relate the stabilization of the acylcomplex $NiX(CH_3CO)(PMe_3)_2$ to the presence in the molecule of the halide ligand since Klein did not succeed, until now, to isolate stable acylcomplexes by reaction of CO on the $Ni(CH_3)_2(PMe_3)_3$ and $[Ni(CH_3)(PMe_3)_4]^+$ complexes.

Since the insertion of CO in a metal–carbon bond is a well known reaction, its mechanism is still not well understood and the choice between CO insertion in a M–C bond or alkyl migration to the CO bonded on M is difficult (12). However, in the only reported example—the CH_3-Mn-$(CO)_5$/CO system—the authors favor the second hypothesis. Thus, if the acyl complex $NiX(CH_3CO)(PMe_3)_2$ is considered the result of migration of the CH_3 ligand to the CO group bonded on the nickel center, a possible mechanism for this reaction is reported in Figure 1.

Table I. Analytical Data and Physical Properties for

	Color	MP(°C)[a]	C(%) Calc.	C(%) Found
$NiI_2(CO)(PMe_3)_2$	brown	160	17.07	17.03
$NiBr_2(CO)(PMe_3)_2$	red brown	172	21.09	20.78
$NiCl_2(CO)(PMe_3)_2$	red brown	—	27.14	26.92
$[NiBr(CO)(PMe_3)_3]BF_4$	orange	—	24.99	24.40
$[NiCl(CO)(PMe_3)_3]BF_4$	orange	—	unstable	

[a] All melting points are uncorrected.

This mechanism is related to the ligand replacement reaction in square-planar d^8 complexes (13) where the first step is addition on a square-planar complex giving two trigonal–bipyramidal isomers. Isomer 1 is the more probable since migration of the CH_3 group from an equatorial site to another equatorial site produces the square-planar ligand arrangement that is observed experimentally. On the other hand, isomer 2 would produce an approximately tetrahedral ligand arrangement, which is the excited state of the stable square-planar species (14).

Although, to our knowledge, no pentacoordinate $NiXR(CO)(PR_3)_2$ complexes have been reported (but some are with Pd(II) or Pt(II)), but following Heck their existence can be deduced from the chemical reactions. Thus, we have investigated the related reaction of CO with the Ni(II) complexes: $NiX_2(PMe_3)_n$ (n = 1,2; X = halide) and [NiX-

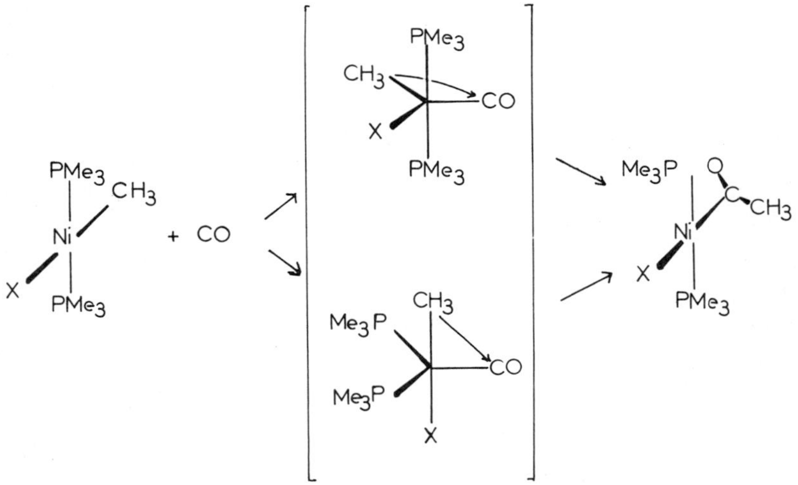

Figure 1. Possible mechanism for formation of acyl complex. Application of substitution mechanism on square planar d^8 complexes.

the Molecular and Cationic Carbonyl Ni(II) Complexes

H(%)		P(%)		X(%)	
Calc.	*Found*	*Calc.*	*Found*	*Calc.*	*Found*
3.69	3.56	12.57	12.57	51.52	51.78
4.55	4.65	15.54	15.23	40.09	39.88
5.86	5.82			22.89	23.62
5.65	5.69				

$[(PMe_3)_4]BF_4$, keeping in mind the previous isolation of the $NiI_2(CO)(PMe_3)_2$ compound by Pankowski and Bigorgne.

Besides $NiI_2(CO)(PMe_3)_2$, few Ni(II) complexes of CO have been reported in the literature: $NiX(CH_3CO)(PMe_3)_2$ (5, 6, 7, 9); $NiPh(CO)(PR_3)_2$ (15); $NiX(CO)(\eta^3\text{-allyl})$ (3); $NiX(CO)(\eta^5\text{-}C_5H_5)$ (4); $NiI_2(CO)(Fdma)$ (11). They are usually prepared by oxidative addition on Ni(0) complexes. Usually, the reported attempts to isolate carbonyl complexes of Ni(II) have led to reduction either in the valence state of Ni(II) (Ni(II) → Ni(0)) or in the ligand (16, 17). These authors are used to relate this apparent reluctance of Ni(II) to coordinate CO to two factors: (1) the relative stable and contracted 3d orbitals of Ni(II) ion that render back donation into the carbonyl η^* function less effective; (2) the kinetic instability of the carbonyl complexes (15, 16, 17). We will, in this study, put the emphasis on the second factor.

All of the molecular carbonyl Ni(II) complexes have been prepared by dissolving the molecular $NiX_2(PMe_3)_n$ (X = Cl,Br,I; n = 2,3) complexes in benzene (X = Br,I) and/or in ethanol (X = Cl,Br,I) at room temperature. The solutions rapidly react with CO at room temperature and atmospheric pressure, with a color change from green-blue to green and brown. Brown crystals of $NiX_2(CO)(PMe_3)_2$, insoluble in ethanol, are isolated by filtration and washed with ethanol saturated with CO. When prepared in benzene, addition of pentane saturated with CO is necessary to precipitate the brown complexes (Table I).

Reaction of the cationic $[NiX(PMe_3)_4]BF_4$ complexes (X = Cl,Br,I) with CO was carried out in a mixture of benzene–dichloromethane at room temperature and atmospheric pressure. CO is immediately absorbed and an impure brown solid can be precipitated by addition of pentane. However, recrystallization of this brown complex in dichloromethane saturated with CO gives yellow-to-gold microcrystals of $[NiX(CO)(PMe_3)_3]BF_4$, which are stable enough to be isolated in the presence of CO for X = Br. When X = I, because of the dissociation of $[NiI(PMe_3)_4]BF_4$ following the reaction:

$$2[\text{NiI}(\text{PMe}_3)_4]\text{BF}_4 \xrightarrow{\text{CH}_2\text{Cl}_2} \underset{\text{insoluble}}{\text{NiI}_2(\text{PMe}_3)_3} + [\text{Ni}(\text{PMe}_3)_4](\text{BF}_4)_2 + \text{PMe}_3,$$

only the molecular $\text{NiI}_2(\text{CO})(\text{PMe}_3)_2$ was isolated (19).

All of these complexes have been characterized as truly pentacoordinate low spin Ni(II) complexes (Table II). The molecular complexes are reasonably stable in the solid state and can be handled in air for a short time. The cationic complexes must be kept under CO to prevent dissociation. All are unstable in solution, if CO is not present in excess.

$$\text{NiX}_2(\text{CO})(\text{PMe}_3)_2 \rightleftharpoons \text{NiX}_2(\text{PMe}_3)_2 + \text{CO}$$

PMe₃ is a better ligand than CO as shown by the reaction:

$$\text{NiX}_2(\text{CO})(\text{PMe}_3)_2 + \text{PMe}_3 \rightleftharpoons \text{NiX}_2(\text{PMe}_3)_3 + \text{CO}$$

To understand the CO fixation mechanism at the Ni(II) center, it is essential to determine the stereochemistry of the complexes. If, as it is more likely, the trigonal–bipyramidal geometry of $\text{NiX}_2(\text{PMe}_3)_3$ (18) and $[\text{NiX}(\text{PMe}_3)_4](\text{BF}_4)$ (19) is preserved for the carbonyl complexes, the probable isomers will be those with the halogens in the equatorial positions of the trigonal bipyramid, in absence of special ligand requirements (19, 20).

Table II. Characterization Data

	$\mu_{eff}(BM)$ $(25°C)^a$	$\nu_{CO}(cm^{-1}\ a)^b$	Dipole Moment[c]
$\text{NiI}_2(\text{CO})(\text{PMe}_3)_2$	0.64	2015 s	3.90
$\text{NiBr}_2(\text{CO})(\text{PMe}_3)_2$	0.49	1965 w 2010 s	3.95
$\text{NiCl}_2(\text{CO})(\text{PMe}_3)_2$	0.34	1960 w 2005 s 1955 w	
$[\text{NiBr}(\text{CO})(\text{PMe}_3)_3]\text{BF}_4$	1.02	2030 s	
$[\text{NiCl}(\text{CO})(\text{PMe}_3)_3]\text{BF}_4$			

[a] In the solid state, corrected from the diamagnetism of the ligands (15).
[b] In the solid state, as nujol mull.
[c] In C₆H₆ saturated with CO.
[d] In CH₂Cl₂ saturated with CO; positive shift downfield from 62.5% H₃PO₄. (*) The slow exchange limit is not reached.

Scheme 2

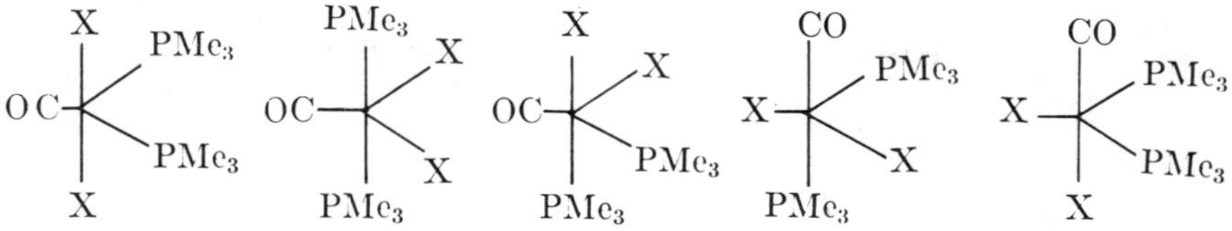

[Structures 1-5 showing square pyramidal/trigonal bipyramidal Ni complexes with X, CO, and PMe$_3$ ligands]

1   2   3   4   5

For the molecular $NiX_2(CO)(PMe_3)_2$ complexes (Scheme 2), only isomer 2 is consistent with the presence of one ν_{co} vibration (2015 cm^{-1}, X = I) and a single $^{31}P\{^1H\}$NMR line at −80°C (6.55 ppm, X = I). This NMR line is in the ppm range of axial PMe$_3$ ligand in the Ni(II) pentacoordinate complexes $[NiX_n(PR_3)_{5-n}]$ (*19*) as shown in Figure 2. Thus the presence of a still fast phosphine exchange process at −80°C (either inter- or intramolecular) is very unlikely since the $^{31}P\{^1H\}$NMR signal should be in between the P_{ax} and P_{eq} $^{31}P\{^1H\}$ range. An identical result—stabilization of isomer 2—has been obtained in the exchange study of PMe$_3$ by P(OMe)$_3$ in the $NiX_2(PMe_3)_3$ complexes, where the variable-temperature $^{31}P\{^1H\}$NMR has showed substitution of the equatorial PMe$_3$ by P(OMe)$_3$ (*21*).

for the Carbonyl Ni(II) Complexes

$^{31}P\{^1H\}$ FT NMR (−80°C)^d	Molecular Weight (g)^e	Electronic Transitions (25°C, 10^{-3} cm^{-1} (ϵ))^f
6.55	408 (398)	15.87(420); 19.03(1300); 21.8(2400) 28.7
16.6	436 (492)	17.9(590); 21.9(1500); 24.8(3600) 29.6
21.6	decomp	19.2(270); 22.7(500); 26.7
+2.4 (−80°C) { +15.7 (−120°C) { −26.5	decomp	18.18 25.5
+6.1 (−80°C)	decomp	

^{e} In C$_6$H$_6$ saturated with CO; the calculated values are in parentheses.
^{f} In CH$_2$Cl$_2$ saturated with CO (C$_{Ni}$ = 10^{-2} M/L).

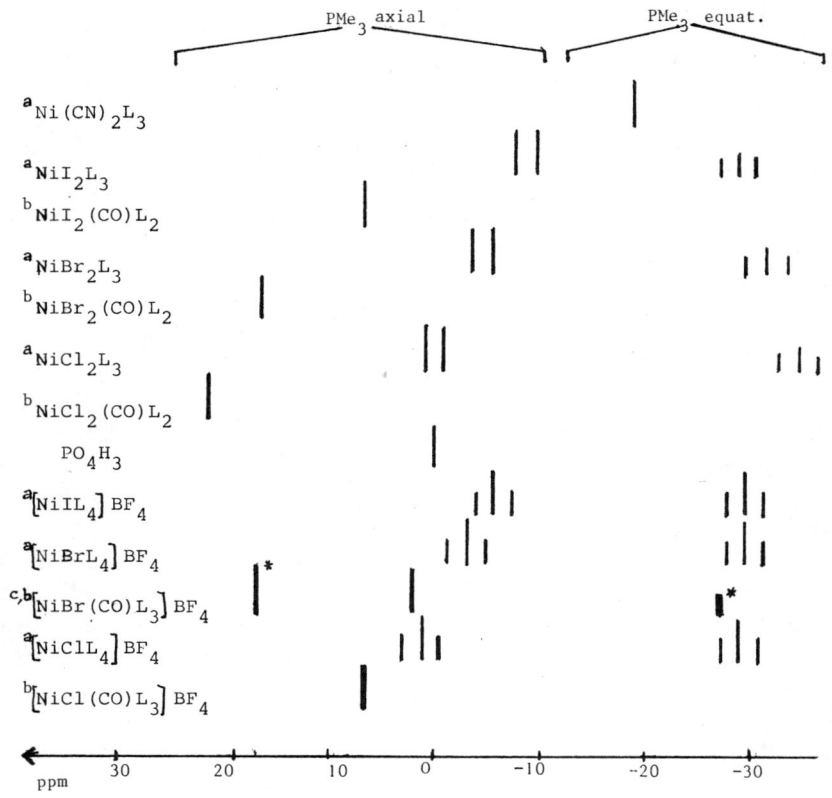

Figure 2. $^{31}P\{^1H\}$FT NMR data at: (a) $-90°C$; (b) $-80°C$; (c) $-120°C$; (*) (CH_2Cl–$CHClF_2$) for the trigonal bipyramidal Ni(II)–PMe_3 complexes in CH_2Cl_2 solutions ($P(OMe)_3$ as reference)

For the cationic complexes [NiX(CO)(PMe$_3$)$_3$]BF$_4$ (Scheme 3), the presence of only one ν_{CO} vibration in the solid state (ν_{CO} = 2030 cm^{-1}, X = Br) and in solution shows the existence of only one isomer. The more probable is isomer 6 or 8, because of the halide preference for the equatorial site. The $^{31}P\{^1H\}$NMR spectrum at $-80°C$ consists of one broad line at $\delta = 2.9$ ppm ($X = Br$), indicating that the slow limit exchange still has not been reached. The spectrum is still not completely resolved at $-120°C$ (in 1/1 $CH_2Cl_2/CHClF_2$ solution) but presents the shape expected for an AB_2 spectrum (two lines at $\delta = 17.5$ ppm and -26.5 ppm in the ratio 2:1). Thus the lines are in the ppm range expected for axial and equatorial phosphine ligands and the probable isomer is isomer 6, a result that agrees also with the $P(OMe)_3$ substitution process on the cationic $[NiX(PMe_3)_4]BF_4$ complexes (21).

These carbonylation reactions can be viewed as substitution of an equatorial ligand PMe_3 by CO in a d^8 trigonal bipyramid and thus can be related to the well-known substitution reaction in planar d^8 transition metal complexes (13), the trigonal bipyramid being the ground state and the square planar the transition state or intermediate.

The experimental data thus can be explained by the Schemes 4 and 5:

Scheme 4

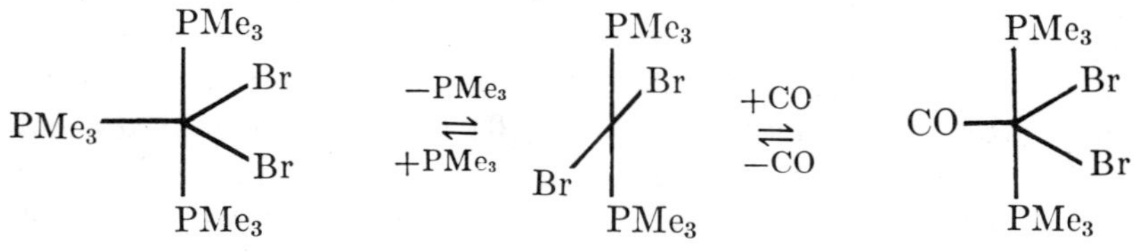

Scheme 5

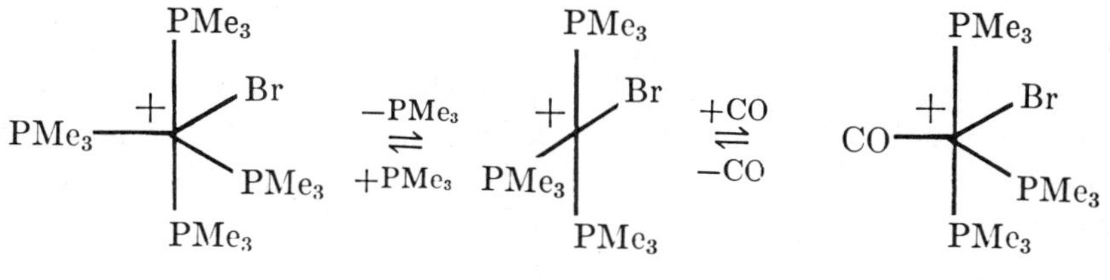

The S_{N2} mechanism, with addition of CO to the square-planar $NiX_2(PMe_3)_2$ or $[NiX(PMe_3)_3]BF_4$ as the intermediate, is reasonable since: (1) the same $NiX_2(CO)(PMe_3)_2$ and $[NiX(CO)(PMe_3)_3]BF_4$ complexes have been obtained by action of CO on the square-planar $NiX_2(PMe_3)_2$ and $[NiX(PMe_3)_3]BF_4$ species which can be isolated; and (2) addition of excess PMe_3 prevents any CO fixation. However, because of the importance of solvent assistance, this mechanism may not be unique, and the reactions are still under investigation.

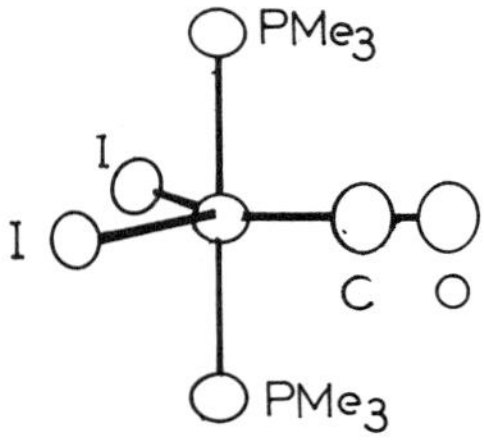

Figure 3. X-ray structure of $NiI_2(CO)(PMe_3)_2$. Pnma—regular TBP Ni–C = 1.73 Å, the shortest reported Ni–C distance (Ni–C = 1.84 Å in $Ni(CO)_4$) (18).

The structural determination of only one pentacoordinate carbonyl Ni(II) complex $NiI_2(CO)(Fdma)$ has been reported. It is a trigonal bipyramid with the two iodine atoms in equatorial positions. But CO occupies an axial site, a result that does not agree with our conclusions on $NiX_2(CO)(PMe_3)_2$, but which can be attributable to the Fdma requirements. Therefore, it was of interest to examine the crystal structure of the related $NiI_2(CO)(PMe_3)_2$ (22); it crystallizes from ethanol in brown microcrystals, in the Pnma system (Z = 8) (Figure 3). This molecule presents the characteristic of having a mirror plane with the five atoms Ni, P(1), P(2), C, and O, which is a mirror plane m of the space group. Two considerations are important: (1) the nickel atom possesses a nearly regular trigonal–bipyramidal coordination with the CO and two iodine ligands occupying the equatorial positions; (2) a very short Ni–CO bond length (1.73 Å), which is to our knowledge the shortest Ni–CO distance ever reported, even shorter than in Ni(0) and Ni(I) complexes and in $NiI_2(CO)(Fdma)$ (1.84 Å). This can be related to appreciable π back bonding in the Ni–C bond, which is expected to be more important in $NiI_2(CO)(PMe_3)_2$ where CO is equatorial than in $NiI_2(CO)(Fdma)$ where CO is axial. This can be explained since there is only one CO group with a strong π-bonding capacity which shares with the two halide ligands (with low π-bonding requirements) the same d_{xy}, $d_{x^2-y^2}$ metal orbitals and thus has the exclusive use of the electrons of these orbitals (to prevent the accumulation of negative charge on nickel). This result is in agreement with the structural determination of $NiX_2(PMe_3)_3$ complexes where smaller, or at least equal, Ni–P_{eq} bond lengths (compared with the Ni–P_{ax} distances) are observed (18), while in $[Ni(P(OR)_3)_5]^{++}$, the Ni–P_{eq} distances are longer than those in the Ni–P_{ax}, as expected from theoretical grounds (23).

Conclusion

(1) The synthesis of new pentacoordinate carbonyl Ni(II) complexes, molecular $NiX_2(CO)(PMe_3)_2$ (X = Cl,Br,I) and cationic $[NiX(CO)(PMe_3)_3]^+$ (X = Cl,Br), shows that Ni(II) can react with CO at room temperature and atmospheric pressure without reduction of Ni(II) to Ni(O).

(2) The mechanism is of S_{N2} type, that is, addition of CO on the square-planar species $NiX_2(PMe_3)_2$ or $[NiX(PMe_3)_3]^+$ gives the penta-coordinate carbonyl complex. The solvent plays a part which is still under study.

(3) The Ni–CO bond is short, as can be shown by x-ray determination, but decarbonylation of the complex remains easy, which is in agreement with the kinetic instability already reported for these complexes.

(4) These experimental results agree well with the first part of the proposed mechanism of formation of acyl Ni(II) complexes.

Literature Cited

1. Heck, R. F., "Organotransition Metal Chemistry," pp. 201–261, Academic, New York, 1974.
2. Garrou, P. E., Heck, R. F., *J. Am. Chem. Soc.* (1976) **98**, 4115.
3. Jolly, P. W., Wilke, G., "The Organic Chemistry of Nickel," Vol. 1, p. 364, Academic, New York, 1974.
4. Ibid., p. 450.
5. Klein, H. F., *Angew. Chem.* (1973) **85**, 403.
6. Klein, H. F., Karsch, H. H., *Chem. Ber.* (1976) **109**, 2524.
7. Ibid., 2515.
8. Pankowski, M., Bigorgne, M., *J. Organomet. Chem.* (1972) **35**, 397.
9. Huttner, G., Orama, O., Bejenke, V., *Chem. Ber.* (1976) **109**, 2533.
10. Raymond, K. N., Corfield, P. W. N., Ibers, J. A., *Inorg. Chem.* (1968) **7**, 1362.
11. Pierpont, C. G., Eisenberg, R., *Inorg. Chem.* (1972) **11**, 828.
12. Calderazzo, F., *Angew. Chem., Int. Ed. Engl.* (1977) **16**, 299.
13. Langford, C. H., Gray, H. B., "Ligand Substitution Processes," W. A. Benjamin, New York, 1966.
14. Shapley, J. R., Osborn, J. A., *Acc. Chem. Res.* (1973) **6**, 313.
15. Wada, M., Oguro, K., *Inorg. Chem.* (1976) **15**, 2346.
16. Booth, G., Chatt, J., *J. Chem. Soc. A* (1962) 2099.
17. Corain, B., Favero, G., *J. Chem. Soc., Dalton Trans.* (1975) 283.
18. Dawson, J. W., McLennan, T. J., Robinson, W., Merle, A., Dartiguenave, M., Dartiguenave, Y., Gray, H. B., *J. Am. Chem. Soc.* (1974) **96**, 4428.
19. Dartiguenave, M., Dartiguenave, Y., Saint-Joly, C., Gleizes, A., Galy, J., Meier, P., Merbach, A. E., *Inorg. Chem.*, in press.
20. Stalick, J. K., Ibers, J. A., *Inorg. Chem.* (1970) **9**, 453.
21. Meier, P., Merbach, A. E., Dartiguenave, M., Dartiguenave, Y., *J. Am. Chem. Soc.* (1976) **98**, 6402 and to be published.
22. Saint-Joly, C., Gleizes, A., Dartiguenave, M., Dartiguenave, Y., Galy, J., unpublished data.
23. Rossi, A. R., Hoffman, R., *Inorg. Chem.* (1975) **14**, 369.

RECEIVED February 22, 1978.

15

Binuclear Metal Complexes of Cofacial Diporphyrins

C. K. CHANG

Michigan State University, East Lansing, MI 48823

A group of novel binuclear metal ligands composed of two alkyl porphyrins covalently linked in a cofacial configuration has been synthesized. The interplanar distance of the diporphyrins can be varied from 6.4 to 4.2 Å by changing the length of the linkage. The presence of exciton interaction in these dimers was evidenced by a substantial blue shift (10–30 nm) of the Soret peak. Both homo- and heterodimetalloporphyrins have been prepared, e.g., Fe–Fe, Cu–Cu, Mg–Mg, Co–Co, and Fe–Cu. Dioxygen was found to form both 1:1 and 2:1 complexes with Co(II) and Fe(II) diporphyrins. The ring separation played a major role in determining the metal-to-oxygen ratio. Implications of the dimer studies on biological oxygen reduction and the "special-pair" chlorophylls are discussed.

We recently have been developing the chemistry of a group of novel binuclear metal ligands composed of two porphyrins covalently linked in a true parallel configuration. These cofacial diporphyrins have great significance in many branches of chemistry. As organic molecules, in addition to being challenging synthetic targets, they can present a multitude of properties by the mere token of their size and by the resulting interaction of the two 18 π-electron porphyrin rings. As inorganic compounds, they have the unusual capability of constraining two metal ions at selected distances and thus can display interesting properties arising from metal–metal interactions. Furthermore, from the point of view of biochemistry, they represent a class of elaborately designed bioinorganic models for many essential biological systems; among these we cite: (1) "special pair" chlorophyll model in photosynthetic unit; (2) chlorophyll aggregates model for studying excitation energy transfer processes; (3) cytochrome oxidase model capable of multielectron reduction of oxygen; (4) monooxygenase model by which molecular oxygen can be "activated" via two-electron transfer; and (5) polynuclear complexes with certain catalytic activity. It is no doubt that the study of cofacial diporphyrins and their metal complexes will lead to some very interesting chemistry. In this chapter our very recent work in this area is summarized, and some initial findings concerning their biological implications are presented.

Synthesis

The compounds to be discussed include the following:

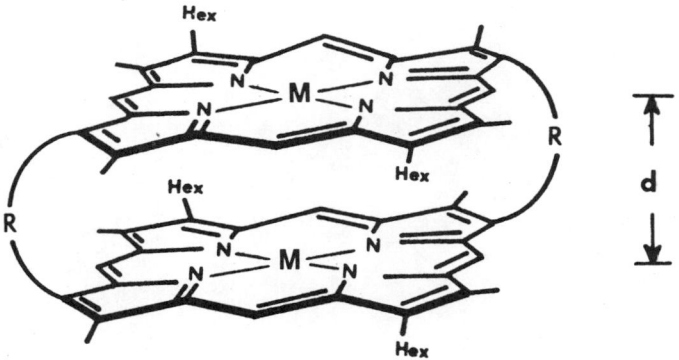

1 $R = -CH_2CH_2CON(n\text{-Bu})CH_2CH_2CH_2-$ $d = 6.4$ Å
2 $R = -CH_2CON(n\text{-Bu})CH_2CH_2CH_2-$ $d = 5.4$ Å
3 $R = -CH_2CON(n\text{-Bu})CH_2CH_2-$ $d = 4.2$ Å

where the two Ms can be identical or different. A rational synthesis of these stacked macrocyclic compounds would be based on a number of premises: (1) a porphyrin monomer carrying diagonally functionalized side chains (C_{2h} symmetry) should be prepared in large quantity; (2) the coupling of the two porphyrins should be in such a manner that the resultant dimer can consist of dissimilar porphyrins and/or hetero metal ions; and (3) the composite diporphyrins should have relatively high solubility and good chemical stability so that a wide range of experiments can be performed in solution. Since porphyrins with C_{2h} symmetry cannot be obtained by modifying the naturally occurring type IX porphyrins, our objective at the outset of this work was to find a practical synthesis for porphyrins having the desired functional groups. It will be noticed that tetraarylporphyrin (TPP) derivatives have not been selected because of the possible steric hindrance associated with the vertical benzene rings which can prevent the formation of a tightly spaced dimer.

We have discovered that the carboxydipyrromethene 6, which is more conveniently prepared than the unsubstituted analog 7, can be brominatively decarboxylated and cyclized without isolation to porphyrins with very good yields (1, 2). This method is especially attractive in that the cyclization can be performed in large scale (0.2 mol) and suffers no decrease in yield. This feature combined with the fact that most dipyrromethene precursors also can be prepared in large quantities from readily available materials have made it possible for the first time for porphyrins with elaborated substitution patterns to be realistically prepared in a large scale (Scheme 1).

Scheme 1

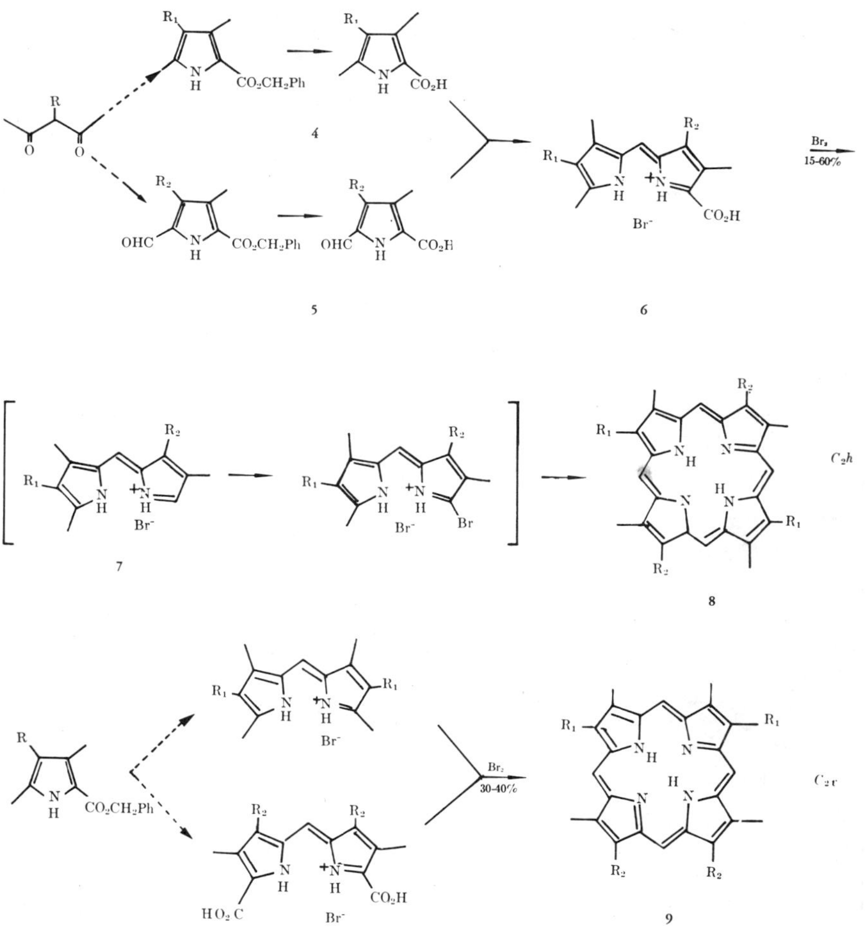

15. CHANG   Cofacial Diporphyrin Complexes

$$8\ R_1 = CH_2CH_2CO_2Me \xrightarrow{LiAlH_4} R_1 = CH_2CH_2CH_2OH \xrightarrow{CH_3SO_2Cl} R_1 = CH_2CH_2CH_2OSO_2CH_3$$
$$\text{or } CH_2CO_2Me \qquad\qquad \text{or } CH_2CH_2OH \qquad\qquad \text{or } CH_2CH_2OSO_2CH_3$$
$$R_2 = \text{hexyl} \qquad\qquad\qquad R_2 = \text{hexyl} \qquad\qquad\qquad R_2 = \text{hexyl}$$

$$\xrightarrow{n\text{-BuNH}_2}$$

$$10\ R_1 = CH_2CH_2CH_2NH(CH_2)_3CH_3 \qquad (1)$$
$$11\ R_1 = CH_2CH_2NH(CH_2)_3CH_3$$

$$8 \longrightarrow ClOC-\bigcirc-COCl$$
$$RHN\sim\bigcirc\sim NHR \longrightarrow 1 \text{ or } 2 \text{ or } 3 \qquad (2)$$
$$\text{10 or 11} \qquad\qquad 40\text{–}60\%$$

The porphyrin ester **8** was then reduced and converted to the secondary amine by Reaction 1 (3). The coupling of two porphyrins via amide formation (Reaction 2) was effected by a high dilution, slow mixing procedure that used a syringe pump (1, 3). The resulting diporphyrin can consist of two diastereoisomers, **12** and **13**; each, in turn, can exist as a pair of enantiomers. Attempts to separate them by HPLC have not been successful.

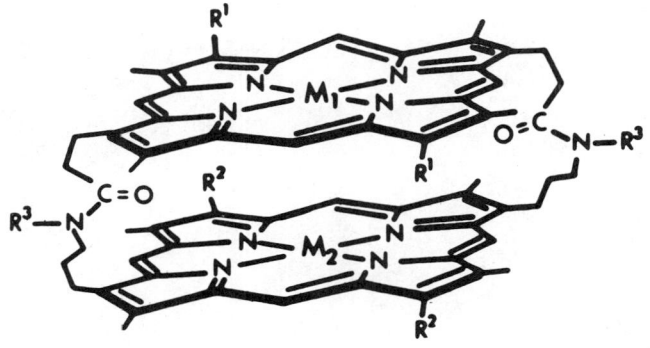

12

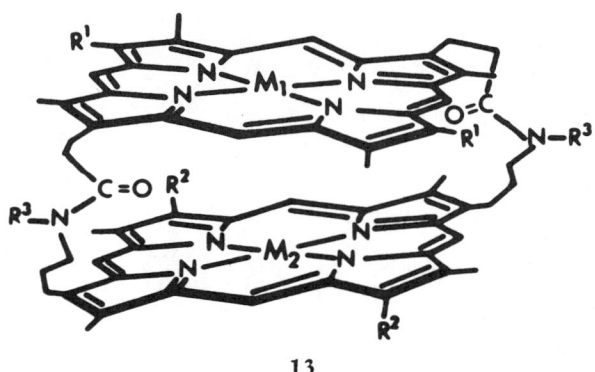

13

Metal ions can be inserted to the diporphyrin ligands by standard procedures. If a metal complex of the diamine 10 or 11 was coupled with a free base porphyrin, the resultant dimer would have only one metal ion while subsequent metal insertion could lead to a mixed dimetal system. Using this approach, we have successfully prepared $Cu-H_2$, Fe–Cu, $Mg-H_2$, and Fe–Mg diporphyrins. The stability of most of the dimetal complexes appears rather similar to the monomers. The Mg–Mg system, however, is more sensitive to acid. Upon washing with dilute acids, one Mg is expelled to give a more stable $Mg-H_2$ dimer.

The NMR spectra of the free base diporphyrins revealed very small shifts for the peripheral substituents. This is expected since the β substituents of one porphyrin fell in the blank region of the anisotropic shielding of the second ring. The inner nitrogen protons, however, were shifted drastically to high field, presumably because of enhanced ring

Table I. Characteristic Data

Compound	Soret Band[a]	NMR N–H ($CDCl_3$)	Fluorescence[b] Q(0,0)
8 Me ester	398 nm (169.9 ᵉmM)	−3.8 δ	619 nm
1 Dimer-7	383 nm (191)	−6.2 δ	628 nm
2 Dimer-6	381 nm (209)	−6.6 δ	630 nm
3 Dimer-5	373 nm (201)	−8.5 δ	630 nm

[a] Absorption spectra recorded in CH_2Cl_2 at 23°C.
[b] 77°K in toluene glass.
[c] Fluorescence quantum yields assuming yield for etioporphyrin I = 0.09.

current (3, 4). The magnitude of the shift depends on the ring separation. This interplanar distance can be estimated by studying the dipolar interaction of two paramagnetic metal ions (5) such as Cu^{2+} ($I = 3/2$). The EPR spectrum of the Cu–Cu diporphyrin 3 is shown in Figure 1. The two unpaired copper electrons are localized largely in the porphyrin plan and cannot pair; one can clearly see the triplet spectrum with the seven-line hyperfine splittings as well as the intense signals in the normally forbidden half-field region ($\Delta M_s = \pm 2$) (6, 7, 8). From the apparent zero-field splitting (D) obtained from the full-field lines, one can estimate the separation between the coppers to be 4.2 Å. EPR parameters for other Cu–Cu diporphyrins are listed in Table I.

Exciton Interaction

The absorption maxima of the diporphyrins (Table I and Figure 2) differed from those of monomeric porphyrins in that: (1) the Soret band of diporphyrins shifted to the blue; (2) the visible bands shift slightly to the red; and (3) the Soret band has a prominent red tail extending out to the 500 nm region. These spectral abnormalities can be understood in terms of exciton coupling (4). If a pair of degenerate dipolar states (X, Y) on porphyrin A is to interact with a similar pair (X′, Y′) on porphyrin B, the nature of the exciton coupling will depend on the dimer geometry.

The energy level diagram in Figure 3 shows the relationship between the energy of a particular transition for an isolated molecule, ΔE_o, and the energy of the allowed component of the same transition in the

of Cofacial Diporphyrins

ΦF^c	Cu–Cu Zero-Field Splitting[d]	Interplanar[e] Separation
0.094	—	—
0.035	0.011 cm^{-1}	6.4 Å
0.021	0.0205 cm^{-1}	5.4 Å
0.007	0.0415 cm^{-1}	4.2 Å

[d] EPR spectra were recorded on a Varian E-4 spectrometer, 77°K in CH_2Cl_2/toluene.
[e] Calculated according to Ref. 7.

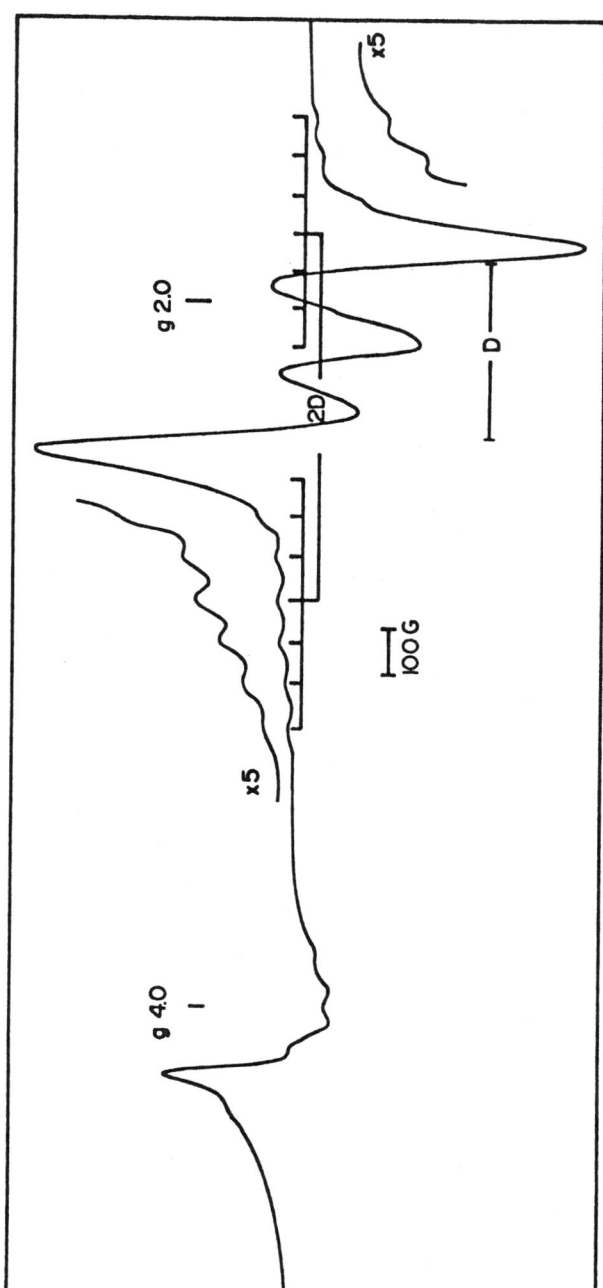

Figure 1. EPR spectrum of the di-copper complex of diporphyrin 3 in CH_2Cl_2/toluene at 77°K (4)

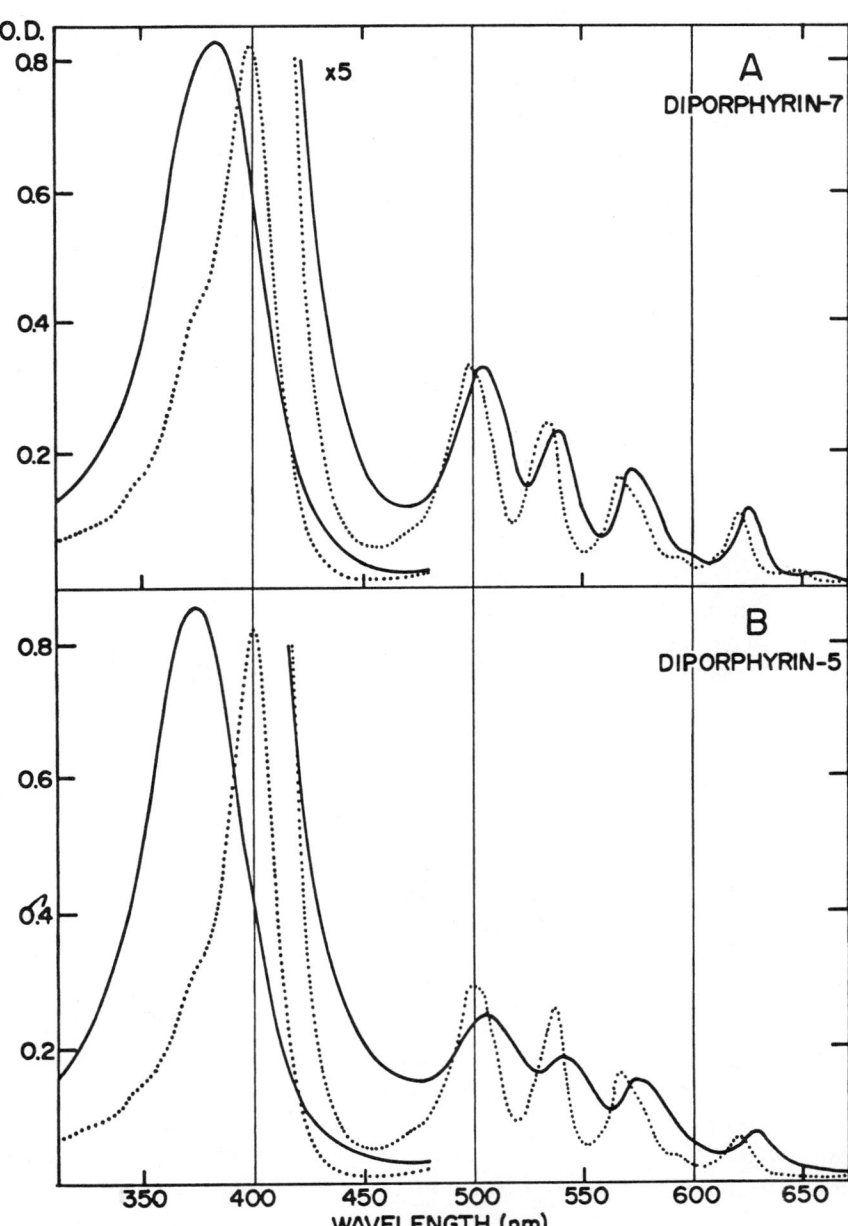

Figure 2. (A) Diporphyrin 1 (solid line) vs. monomer 8 dipropionic acid ester (dotted line). (B) Diporphyrin 3 (solid line) vs. monomer 8 diacetic acid ester (dotted line) (4).

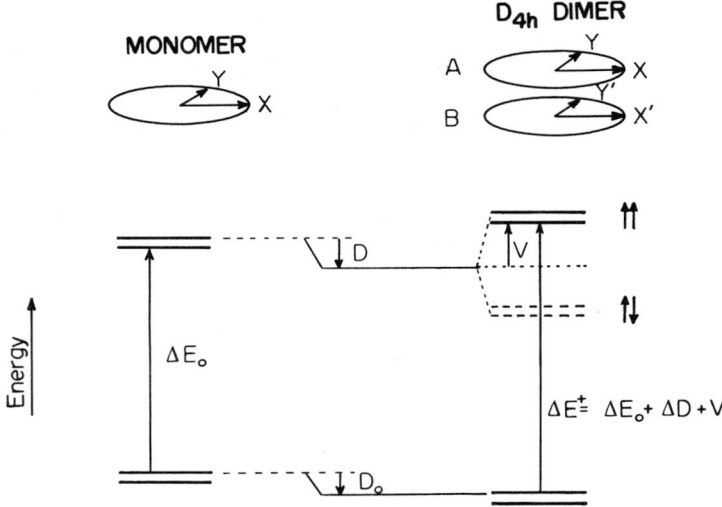

Figure 3. Exciton splitting in a dimer

dimer. The ΔD term is the "solvent parameter" which represents the difference in the effect of solvation on the energy of the ground and excited states. The exciton coupling between two parallel transition dipoles is basically set by Equation 3, where μ and R are the dipole moment and

$$V = \frac{\mu^2}{R^3} G \qquad (3)$$

the perpendicular distance between the two dipoles, respectively. G is a geometric factor related to the orientation of the two porphyrin planes.

In the case of the idealized geometry with D_{4h} symmetry (no tilting and sliding of the two rings), there is a higher energy, degenerate, allowed pair of exciton states ($X + X'$ and $Y + Y'$) and a lower energy, forbidden pair of states ($X - X'$ and $Y - Y'$), whose net transition dipole moments are equal to zero. Consequently only one line will be observed in the absorption spectrum, and it is shifted to the blue with respect to the absorption of the monomer. In such case G equals unity, and from available geometrical data one can estimate a range of magnitudes for the exciton coupling term V in the diporphyrins. However, if the dimer geometry deviates from the D_{4h} symmetry, the tilting and sliding of the porphyrin planes would cause fluctuation of the higher exciton levels as well as developing intensity in the lower exciton states (9). Although the intensity of the lower exciton transition (ΔE^-) cannot compare with that of the ΔE^+, ΔE^- can be responsible for the appearance of the "Soret tail" in the 450-nm region.

$$2000 \text{ cm}^{-1} < V < 6000 \text{ cm}^{-1}$$
$$6.4 \text{ Å} > R > 4.2 \text{ Å} \qquad (4)$$

The above estimation is supported further by spectral information shown in Figure 2. Let us first consider the blue shift of the Soret band (B state). To first approximation, the B^+ band should occur at:

$$\Delta E_B^+ = \Delta E_B^\circ + \Delta D + V \qquad (5)$$

where ΔE_B° is the Soret absorption of monomer. ΔD usually causes a red shift but its value is difficult to estimate. In diporphyrin 1, from the ΔE_B^+ at 384 nm and the red tail where the center of ΔE_B^- band probably occurred around 480 nm, one can set an upper limit for the exciton coupling as half of this energy gap or $V = 2500$ cm^{-1}, comparable with the V obtained by Equation 3. In diporphyrin 3, with a shortened R, V is enhanced more than threefold, and this was reflected by the wider separation of the blue maximum and the red tail. In the visible region (Q state) since the transition dipole strength is much weaker than the B band, the exciton coupling is much smaller. Here the solvent red shift term ΔD can be comparable with V in magnitude; therefore, the slight red shift and the more diffused band shape possibly is a result of both the manifestation of ΔD and inhomogeneous solvent broadening (10).

The Soret blue shift appears to be a common feature of most cofacial metal diporphyrins as well, although the magnitude of the shift varies from metal to metal. Collman et al. (11) have synthesized a TPP type, face-to-face binary porphyrin with 6.5 Å separation and reported a 15-nm blue shift for the Soret peak. The characteristic spectral shifts also have been observed in zinc porphyrin aggregates (12), μ-oxo scandium dimers (10), and sometimes even in nonrigidly held "clam shell" dimers (13). In the past year, however, two other groups (14, 15) have described briefly the synthesis of similar type diporphyrins, presumably with a ring separation in the 6 Å range and have reported the absence of the Soret blue shift. In view of the large body of evidence developed from our work, meaningful discussion of the discrepancy must await unambiguous proof of true cofacial geometry in the latter two diporphyrins.

Fluorescence emission data of the diporphyrins also are given in Table I: the Q(0,0) band is red shifted and fluorescence yield is decreased. Although some reduction in Φ_F is predicted by the exciton model (9), the extent of quenching found in the diporphyrins is certainly impressive. It is likely that the enormous red tail of the Soret band further enhances the self quenching of the dimer fluorescence yield.

Cation Radicals of the Mg–Mg Diporphyrin

Electrolysis of Mg–Mg diporphyrins in CH_2Cl_2 [$(C_4H_9)_4NClO_4$ or $(C_4H_9)_4NPF_6$ as electrolyte] yielded various oxidation products (16). Cyclic voltammetry showed that the first oxidation peak for (Mg–Mg) $\rightleftarrows$ (Mg–Mg)$^{++}$ took place at about +0.6 V (vs. Ag/AgCl electrode) and that the second oxidation for (Mg–Mg)$^{++}$ $\rightleftarrows$ (Mg–Mg)$^{4+}$ occurred at about + 1.0 V. Partial electrolysis of the Mg–Mg diporphyrin 1 in degassed CH_2Cl_2 at +0.53 V (stopped when coulometer indicated that about 0.2 molar equivalents of the dimer had been oxidized) yielded a violet solution with an absorption peak at 670 nm, believed to contain (Mg–Mg)$^+$ monocation radicals. EPR measurements on this solution showed a signal, $g = 2.003$, which has a peak-to-peak separation of only 1.05 G (Figure 4). Under identical conditions, MgOEP$^+$ radicals were found to have a linewidth of 2.5 G (17). The extremely narrow linewidth of the dimer is surprising, and we speculate that oligomerization of the monocation radicals may have taken place in solution. Nevertheless, the narrowing in EPR linewidth clearly indicates that there is extensive electron exchange between the two rings, similar to the properties exhibited by in vivo special pair (bacterio)chlorophylls P700 (18) and P870 (19). Further studies on the properties of the ion radical species as a function of the ring separation are being carried out. We also have prepared several Mg–H$_2$ diporphyrins; it is expected that photolysis would produce charge-separated Mg$^+$–H$_2^-$ species similar to the (PA $\rightleftarrows$ P$^+$A$^-$) donor-acceptor bacteria photosynthetic units (20).

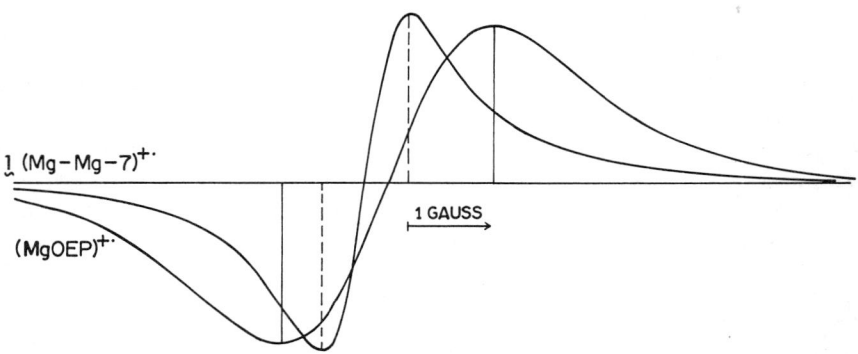

Figure 4. EPR spectra of cation radicals of Mg–Mg-7 (1) and MgOEP. Electrolyses were carried out at + 0.53 V (vs. Ag/AgCl) in CH_2Cl_2 with Bu_4NPF_6 as electrolyte. Spectra were recorded on a Varian E-4 instrument at 23°C.

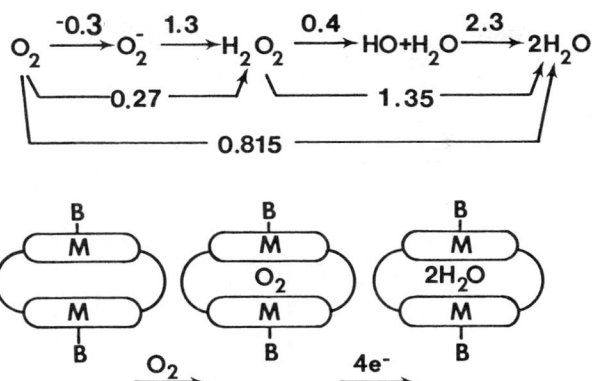

Figure 5. Midpoint reduction potentials of oxygen (according to Wang (25)) and multielectron transfer

Multielectron Reduction of Molecular Oyxgen

Reduction of molecular oxygen is difficult from the chemical point of view because one not only must come up with a mechanism to react with the triplet ground state oxygen but also has to avoid the formation of toxic, high energy intermediates such as superoxide (Figure 5) (21). Nevertheless, nature has solved this problem marvellously by using multinuclear metal catalysts. The best example is provided by cytochrome oxidase which contains two hemes and two copper ions and carries out a four-electron reduction of oxygen to hydrogen (22). The idea of multielectron reduction of oxygen can be realized in metal diporphyrins if oxygen can form a sandwiched complex with metals and receive electrons from them. Indeed oxygen was found to form either 1:1 or 2:1 complexes with Co(II) and Fe(II) diporphyrins, depending on the metal separation.

When a bulky ligand, 1-triphenylmethylimidazole, was mixed with Co(II)–Co(II) diporphyrin 1 (6.5 Å gap) and exposed to oxygen, both visible and EPR spectra documented the formation of a double (1:1) Co–O_2 complex. The oxygenation can be shown reversible: evacuation resulted in eliminating the superoxo complex and restoring the Co(II) signal (Figure 6A). Co(II)–Co(II) diporphyrin 3 (4.2 Å gap), on the other hand, reacted completely in a different way. Addition of oxygen to the [ϕ_3CIm-Co(II)]$_2$ complex at room temperature instantaneously produced a species consistent with the formulation of 2Co/O_2 (gasometry Co:O_2 = 1:0.55). This complex, written as μ-peroxo [Co–O_2–Co], is diamagnetic and has no EPR signals. When a trace of I_2 was added to this solution, a well-defined isotropic spectrum was obtained (Figure 6B) consisting of 15 lines (g_{iso} = 2.024 $|A_{Co}|$ = 10 G). Such a spectrum

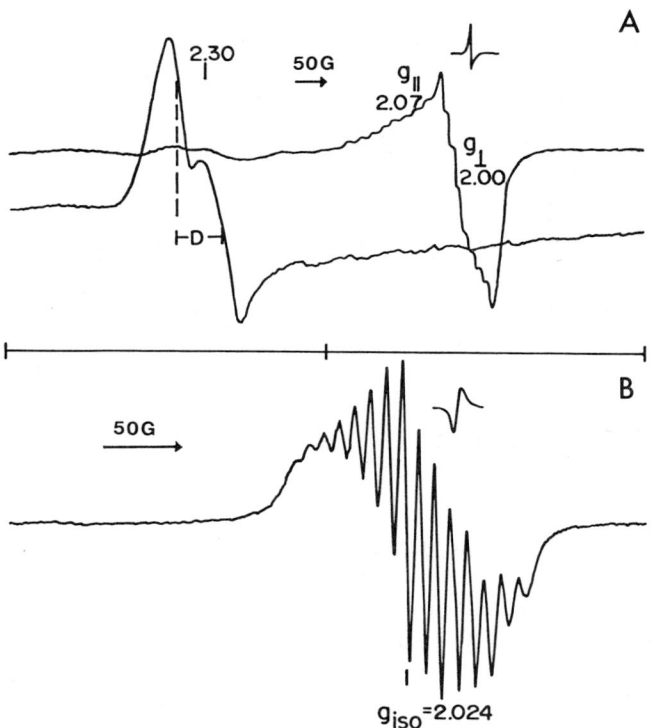

Figure 6. EPR spectra of oxygen-containing samples of $[\phi_3 CImCo(II)]_2$ diporphyrins. (A) Upper trace: complex of 1 with 1 atm of O_2 at 77°K; bottom trace: after sample was evacuated at −20°C and recorded at 77°K. (B) Room temperature spectrum of the dioxygen adduct of cobalt complex of 3 after addition of small amount of I_2. This is a typical binuclear μ-superoxo dicobalt spectrum. All experiments were carried out in CH_2Cl_2/toluene mixtures.

would be expected if the μ-peroxo dicobalt complex became oxidized to a μ-superoxo dicobalt complex in which the two equivalent ^{59}Co nuclei would give a total of $(2 \times 2 \times 7/2) + 1 = 15$ lines (23). The μ-superoxo formulation has further been substantiated by ^{17}O experiments (24).

Behavior of the iron complex of diporphyrins in general parallels that of the cobalt system. Oxygenation of five-coordinate $[\phi_3 CIm\text{-}Fe(II)]_2$ complexes of 1 and 2 at −45°C in DMF or CH_2Cl_2 resulted in a visible spectrum ($\alpha = 562$ nm, $\beta = 529$ nm) similar to that of the oxygen adduct of the "myoglobin active site model" (26, 27, 28) or the "crowned heme" (1). Addition of oxygen to the same complex of 3 nevertheless resulted in instantaneous oxidation of the heme, even at −45°C. Preliminary kinetic measurements indicated that the rate of autoxidation of this bis-heme is at least 10^3 times faster than the monomeric myoglobin site

(29). This is consistent with the scheme that dioxygen adds to the bis-heme to give a μ-peroxo dimer [Fe–O$_2$–Fe] which then decomposes to yield other Fe(III) species (30). With monomeric hemes, formation of the μ-peroxo complex is usually the rate limiting step (31, 32). In Fe–Fe 3, since the two iron ions are positioned at favorable distances, the energy barrier leading to the sandwiched oxygen complex is lowered, thereby resulting in a faster oxidation rate.

The exceedingly fast autoxidation rate exhibited by some of the metal diporphyrins suggests that these complexes can be used as an efficient catalyst for the electrochemical reduction of oxygen. Several studies (33, 34, 35) have suggested the use of metal phthalocyanines or porphyrins strongly absorbed on various graphite surfaces as catalysts. Although a reduction in the oxygen overpotential generally has been observed using these absorbed substances, problems in reproducibility, stability, and activity over a wide range of pHs have been acknowledged. We have examined the redox behavior of a number of metal, diporphyrin-coated graphite electrodes. Preliminary results indicated that the current–potential (i–E) curves generated by linear sweep voltammetry in oxygen-saturated solutions were rather similar to those of monomeric metalloporphyrins (Figure 7) (36). No explanation can be offered at present

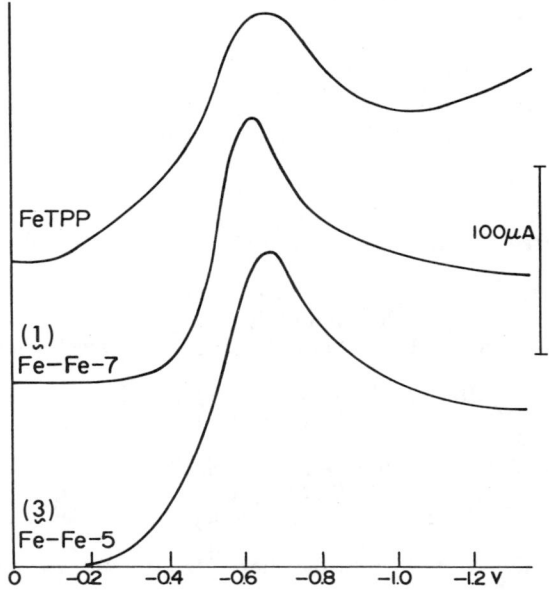

Figure 7. Linear sweep voltammograms of oxygen reduction by iron porphyrins deposited on carbon electrode in 0.1N KOH, oxygen-saturated solution; 25°C, 20 mV/sec

because of the complexity in analyzing the surface-coupled oxygen reduction mechanism. However, experiments are underway to attach the catalyst covalently to electrode surface so as to minimize the heterogenity of the surface.

Prospect

Various lines for future development of the chemistry of cofacial diporphyrins can be recognized. Apart from the few areas we have discussed in this chapter, several other examples of investigation can be selected: (1) the coordination and reduction of molecular nitrogen and unsaturated organic compounds such as Ru–N_2–Ru and Rh–CH=CH–Rh; (2) the electrochemical and photochemical reactions of Mn–Mn diporphyrins with water, since binuclear manganese complexes are believed to play important role in the oxidation of water to give oxygen in green plant photosynthesis; (3) the synthesis of metal porphyrins of novel geometry. Many 2:1 porphyrin–metal compounds can be prepared using diporphyrin as ligand. We believe that the study of diporphyrins can lead to many challenging research areas which will have significant impact on both fundamental as well as applied research.

Acknowledgment

I would like to thank Research Corporation and the donors of the Petroleum Research Fund, administered by the American Chemical Society, for partial support of this research and to C. B. Wang for synthetic effort.

Literature Cited

1. Chang, C. K., *J. Am. Chem. Soc.* (1977) **99**, 2819.
2. Wang, C. B., Chang, C. K., unpublished data.
3. Chang, C. K., Kuo, M. S., Wang, C. B., *J. Heterocycl. Chem.* (1977) **14**, 943.
4. Chang, C. K., *J. Heterocycl. Chem.* (1977) **14**, 1285.
5. Boyd, P. D. W., Smith, T. D., Price, J. H., Pilbrow, J. R., *J. Chem. Phys.* (1972) **56**, 1253.
6. Blumberg, W. E., Peisach, J., *J. Biol. Chem.* (1965) **240**, 870.
7. Chasteen, N. D., Belford, R. L., *Inorg. Chem.* (1970) **9**, 169.
8. Mengersen, C., Subramanian, J., Fuhrhop, J.-H., *Mol. Phys.* (1976) **32**, 893.
9. Kasha, M., Rawls, H. R., El-Bayoumi, M. A., *Pure Appl. Chem.* (1965) **11**, 371.
10. Gouterman, M., Holten, D., Lieberman, E., *J. Chem. Phys.* (1978).
11. Collman, J. P., Elliott, C. M., Halbert, T. R., Tovrog, B. S., *Proc. Natl. Acad. Sci. USA* (1977) **74**, 18.
12. Zachariasse, K. A., Whitten, D. G., *Chem. Phys. Lett.* (1973) **22**, 527.
13. Ichimura, K., *Chem. Lett.* (1977) 641.

14. Ogoshi, H., Sugimoto, S., Yoshida, Z., *Tetrahedron Lett.* (1977) 169.
15. Kagan, N. E., Mauzerall, D., Merrifield, R. B., *J. Am. Chem. Soc.* (1977) **99**, 5484.
16. Fuhrhop, J.-H., Kadish, K. M., Davis, D. G., *J. Am. Chem. Soc.* (1973) **95**, 5140.
17. Felton, R. H., Dolphin, D., Borg, D. C., Fajer, J., *J. Am. Chem. Soc.* (1969) **91**, 196.
18. Warden, J. T., Bolton, J. R., *Acc. Chem. Res.* (1974) **7**, 189.
19. Corker, G. A., *Photochem. Photobiol.* (1976) **24**, 617.
20. Clayton, R. K., *Proc. Natl. Acad. Sci. USA* (1972) **69**, 44.
21. Hamilton, G. A., *Prog. Bioorg. Chem.* (1971) **1**, 83.
22. Caughey, W. S., Wallace, W. J., Volpe, J. A., Yoshikawa, S., "The Enzymes," P. D. Boyer, Ed., 3rd ed., Vol. XIII, p. 299, Academic, New York, 1976.
23. Weil, J. A., Kinnaird, J. K., *J. Phys. Chem.* (1967) **71**, 3341.
24. Chang, C. K., *J. Chem. Soc., Chem. Comm.* (1977) 800.
25. Wang, J. H., *Acc. Chem. Res.* (1970) **3**, 90.
26. Chang, C. K., Traylor, T. G., *Proc. Natl. Acad. Sci. USA* (1973) **70**, 2647.
27. Chang, C. K., Traylor, T. G., *J. Am. Chem. Soc.* (1973) **95**, 5819.
28. Brinigar, W. S., Chang, C. K., *J. Am. Chem. Soc.* (1974) **96**, 5595.
29. Chang, C. K., Powell, D., Traylor, T. G., *Croat. Chem. Acta* (1977) **49**, 295.
30. Chin, D.-H., Gaudio, J. D., La Mar, G. N., Balch, A. L., *J. Am. Chem. Soc.* (1977) **99**, 5486.
31. Cohen, I. A., Caughey, W. S., *Biochemistry* (1968) **7**, 636.
32. Ochiai, E.-I., *Inorg. Nucl. Chem. Lett.* (1974) **10**, 453.
33. Kozawa, A., Zilionis, V. E., Brodd, R. J., *J. Electrochem. Soc.* (1970) **117**, 1470.
34. Zagal, J., Sen, R. K., Yeager, E., *Electroanal. Chem. Interfacial Electrochem.* (1977) **83**, 207.
35. Kuwana, T., Fujihara, M., Sunakwa, K., Osa, T., *Electroanal. Chem. Interfacial Electrochem.* (1978) **88**, 299.
36. Deborski, G., Armstrong, N., Chang, C. K., unpublished data.

RECEIVED February 22, 1978.

16

Homogeneous Oxidative Coupling Catalysts

Products of the Oxidation of Copper(I) Chloride by Oxygen in Polar, Aprotic Media

GEOFFREY DAVIES[1], MOHAMED F. EL-SHAZLY,
DEBORAH R. KOZLOWSKI, CHARLES E. KRAMER,
MARTIN W. RUPICH, and ROBERT W. SLAVEN

Department of Chemistry, Northeastern University, Boston, MA 02115

> *The products of oxidation of Cu(I) chloride by oxygen in aprotic ligand/solvent systems are useful catalysts for a variety of oxidative coupling reactions of molecular oxygen. The ligand/solvent environment determines the nature of the products. The copper-reduced oxygen products obtained in pyridine are growing polymers with a pyridine-stabilized CuO core, which can be separated from the py_2-$CuCl_2$ co-product, and are the initiators for the oxidative coupling of phenols. Saturated, methylated amines are ineffective stabilizers of the copper-reduced oxygen interaction while amide and lactam ligands stabilize distinct clustered primary oxidation products. Cryoscopic measurements and a determination of the structure of a catalyst derivative with coordinated N-methyl-2-pyrrolidinone ligands are valuable in furthering our understanding of the structure of the products formed in other ligand/solvent systems.*

Zuberbühler (1) has pointed to a need for systematic studies of the oxidation of Cu(I) species by oxygen in non-aqueous (aprotic) media. Such investigations would complement what is already known about the corresponding reactions in aqueous solution (1, 2) and also might provide useful clues concerning the Cu^I–oxygen interaction at the hydrophobic sites (3, 4) believed to be present in copper oxidases. It is

---
[1] Author to whom reprint requests should be sent.

reasonable to expect at least a formal resemblance between the products of oxidation of Cu(I) in such model systems and the resting state of such multicentered copper oxidases as ascorbic acid oxidase and the laccases (4). An impetus for research on model systems is that the products of oxidation of Cu(I) chloride in solvents like pyridine will catalyze the oxidative coupling of phenols by molecular oxygen, as do in vivo copper oxidases (5).

Our studies of the products of oxidation of Cu(I) chloride by oxygen in aprotic solvent systems were prompted by frustrated attempts (6) to extend their ability (7) to catalyze Reaction 1 to the corresponding oxidative coupling of substituted p-phenylenediamines (8), Reaction 2.

$$2\,\langle O \rangle\text{-}NH_2 + O_2 \xrightarrow{py} \langle O \rangle\text{-}N{=}N\text{-}\langle O \rangle + 2H_2O \quad (1)$$

$$n\,H_2N\text{-}\langle\underset{R}{O}\rangle\text{-}NH_2 + nO_2 \xrightarrow{py} {\Large[}{=}N\text{-}\langle\underset{R}{O}\rangle\text{-}N{=}{\Large]}_n + 2nH_2O \quad (2)$$

Thus, when the products of oxidation of Cu(I) chloride by oxygen in pyridine were used as catalysts for Reaction 2, the molecular weights of the azo-polymer products obtained were too low to allow their fabrication into useful new materials. However, the high-temperature stability and other desirable properties of these polymers (6, 8) encouraged us to investigate the nature of such catalyst systems in detail with the goal of improving their efficiency and specificity in oxidative coupling processes.

Our work was first concentrated on catalyst identification in pyridine since the oxidation of Cu(I) chloride by oxygen in this solvent has a reproducible stoichiometry over a wide range of experimental conditions (6), and the reaction mixture also catalyzes the oxidative coupling of acetylenes (9) and phenols (10–16). As a result of this work in pyridine, we have extended our studies to other solvent/ligand systems to better understand the interesting catalytic products. One major goal of our research is to discover ligand systems that stabilize highly active, isolatable oxidative coupling catalysts for structural and mechanistic studies. As a result, greater substrate specificity in homogeneous coupling processes than that found for the pyridine catalyst system might be achieved. The results then might have some bearing on analogous copper-catalyzed reactions in vivo.

This chapter is divided into four main sections. First we summarize the catalytic and other properties of the products of oxidation of Cu(I) chloride by oxygen in pyridine. The results in this system suggest four main synthetic routes to oxidized Cu(I) chloride oxidative coupling catalysts, which have been applied in a study of pyridine derivatives, amines, amides, and lactams as potential initiator-stabilizing ligands. Requirements for effective initiator stabilization are indicated, and the likely implications of the first structural information on the catalytic species formed in these systems (with a lactam ligand) are discussed.

Reaction of Cu(I) Chloride with Oxygen in Pyridine

The stoichiometry of the reaction of Cu(I) chloride with oxygen in pyridine at 5°–80°C with molar pyridine:copper ratios > 0.5 is given by Reaction 3 (*6, 17, 18, 19*). The reaction products can be separated by

$$4CuCl + O_2 \xrightarrow{py} 2CuCl_2 + 2\,\text{``CuO''} \tag{3}$$

GPC (pyridine eluant (*17, 18*)); the brown component, "CuO," which initiates the oxidative coupling of 2,6-dimethylphenol (*10–16*), Reaction 4, has the cryoscopic properties of polymeric species with CuO units that,

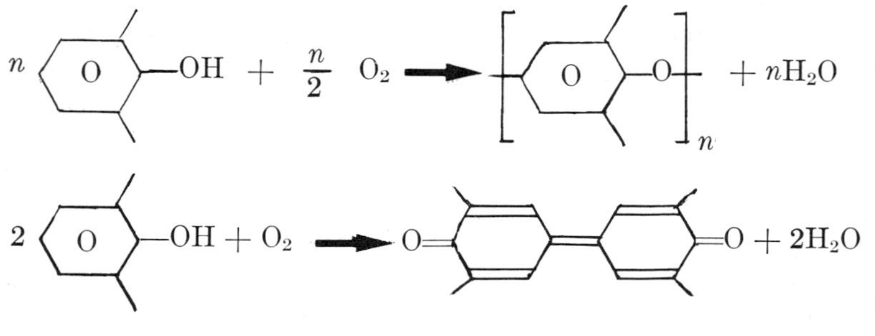

$$n\;\text{Ar-OH} + \frac{n}{2}\,O_2 \longrightarrow [\text{Ar-O}]_n + nH_2O \tag{4a}$$

$$2\;\text{Ar-OH} + O_2 \longrightarrow O{=}\text{Ar}{=}\text{Ar}{=}O + 2H_2O \tag{4b}$$

for their existence in solution, depend on a large excess of pyridine. PMR measurements on concentrated initiator solutions (*17*) provide evidence for stabilizing coordinated pyridine, which is also obvious from attempts to isolate solid initiator species for structural characterization. Thus, evaporation of separated initiator solutions gives insoluble, inactive Cu(II) oxide, which also is produced when pyridine-saturated air or oxygen is passed over solid Cu(I) chloride. Cu(II) oxide formation is obviously a major driving force in Reaction 3, with pyridine providing temporary initiator stabilization through coordination at the CuO centers. Reaction 3 is essentially irreversible. Attempts at reversing it with Cu(I)-specific ligands such as 2,9-dimethyl-1,10-phenanthroline lead to oxidation of the substituting ligand rather than to the release of molecular oxygen (*17*).

The unusual, featureless spectra of dilute initiator solutions in pyridine (which are typical of analogous products in other amine systems) are little affected by long standing at room temperature or by heating (Figure 1); however, a steady decrease in reactivity towards pyridine solutions of water, HCl, or $HClO_4$ is observed on standing and leads to the development of long induction periods (Figure 2) in the oxidative coupling of 2,6-dimethylphenol (Reaction 4) (18). This undesirable property is attributed to continuing initiator polymerization that leads, on prolonged heating, to species which, like Cu(II) oxide, are insoluble in pyridine.

Dissolution of Cu(I) chloride in pyridine before exposure to oxygen leads to a reaction with the stoichiometry of Reaction 3 but at much lower rates than are observed with solid CuCl (*see* above) or with slurries of CuCl in mixtures of pyridine and methylene chloride or *o*-dichlorobenzene (in which CuCl is virtually insoluble) (6, 10, 17). In addition, the products from predissolved CuCl are inferior as catalysts for

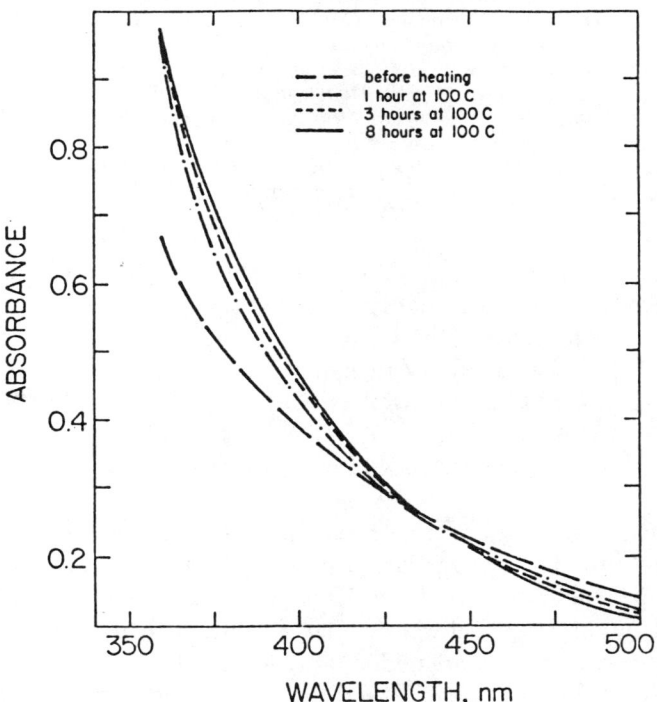

Figure 1. Spectral changes which occur on heating chromatographically separated, pyridine-stabilized "CuO" initiator species (4.2 × 10^{-4} g-atom Cu/L) in pyridine at 100°C. Data from Ref. 18.

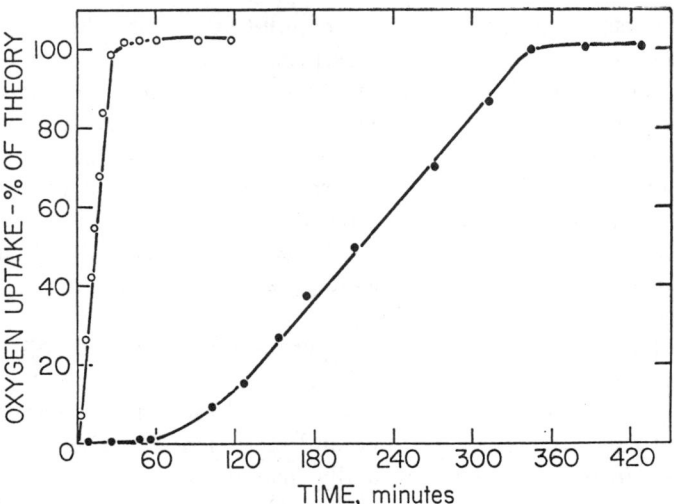

Figure 2. Development of induction periods in the oxidative coupling of 2,6-dimethylphenol as catalyzed by "CuO" species in pyridine. Percentage theoretical oxygen uptake vs. time in the presence of freshly prepared, unseparated products from Reaction 3 in pyridine (○) and the same product mixture heated at 60°C for 13 days (●). Reaction conditions: 3.5 mmol 2,6-dimethylphenol and 0.4 mmol total copper in 50 mL pyridine under 1 atm O_2 at 25°C. Data from Ref. 18.

phenolic coupling compared with those obtained from slurries. This strongly suggests that coordination of molecular oxygen and one pyridine ligand per Cu(I) center [from our titration experiments in o-dichlorobenzene (17)] is necessary for Reaction 3 to proceed and that coordination of oxygen is blocked in dissolved Cu(I) species, which have a very high affinity for pyridine (20). The spectra of oxidized slurry products remain virtually unchanged from the earliest stages of Reaction 3 in pyridine.

The usual tools for characterization of Cu(II) complexes are not applicable to the initiator species in pyridine. Thus, aside from the facts that active initiators cannot be isolated as solids and that their spectra consist of an intense, featureless charge transfer band which extends through the visible spectral region (Figure 1), the species are neutral (17), ESR-nondetectable (17), and cannot be reduced at a dropping mercury cathode in the range 0–(−1.65 V) (vs. SCE, pyridine solvent, and tetraethylammonium perchlorate as electrolyte) (18).

Three important requirements are not satisfied by the products of Reaction 3 in pyridine. First, although the chromatographically separated

"CuO" consistituent is a good oxidative coupling initiator and catalyst, it cannot be isolated intact from solution. Second, the catalyst solutions undergo deactivating polymerization on standing or heating. Finally, the catalytic products are nonspecific in oxidative processes. We have extended our work to other ligand/solvent systems in an attempt to satisfy these goals: to date we have investigated close to 40 other systems chosen from groups of pyridine derivatives, amines, amides, and lactams (Tables I, II, and III).

The following four general synthetic approaches have been used, based on our experience with the parent pyridine system.

Direct Synthesis (DS). Like pyridine, liquid ligands, such as **2, 3, 4, 10, 27, 32, 35,** and **36** can be used as solvent media for CuCl oxidation. In all of these systems, the product solutions contain species that catalyze oxidative coupling reactions of oxygen. Experimental problems with these ligands include the high affinity of amides and lactams for water, which is particularly troublesome in GPC of product solutions (ligand used as eluant) and leads to product decomposition (17, 18). Under stringent experimental conditions, lack of success in chromatographic separations can be equivocal in those cases where only one oxidation product is actually formed or where slow equilibria exist between different product species. In some cases (e.g. with **32** ligand/eluant), very small yields of chloride-free initiator solution fractions can be obtained from the leading sample edge eluting from long (16 feet) GPC columns. However, as with pyridine, such fractions from monodentate ligands invariably produce either insoluble Cu(II) oxide (e.g., with **35**) or non-stoichiometric, highly water-sensitive products (e.g. with **27** or **32**).

Although experimentally difficult and restricted, this approach does allow elimination of those ligands that are oxidatively unstable in the presence of Cu(I) chloride oxidation products and excess oxygen.

Ligand Exchange (LE). The products of Reaction 3 in pyridine can be cleanly separated by GPC with pyridine as the eluant; the products are potential precursors. However, despite the fact that PMR evidence (pyridine proton broadening) might suggest the existence of labile coordinated pyridine at initiator centers, we have been unsuccessful in preparing other initiator species through substitution of pyridine by other ligands in pyridine. With some ligands (e.g. **9** and **11**) there is little evidence for exchange even on long standing of initiator/ligand mixtures in pyridine, despite the fact that very stable copper complexes with the substituting ligand are to be expected in some cases.

This LE approach, while attractive in that it avoids the complicated product mixture which might arise if unseparated products from Reaction 3 are used as precursors, suffers from the disadvantage that the precursors require a large excess of pyridine for solution stability. In addition, unless the substituting ligand induces initiator depolymerization, the products will be polymeric (*18*).

In Situ Generation (ISG). If Reactions 5 and 3 could be combined in a suitable aprotic solvent, then the use of catalytic amounts of Cu(II)

$$2Cu^0 + 2Cu^{II} \rightarrow 4Cu^{I} \qquad (5)$$

$$4Cu^{I} + O_2 \rightarrow 2Cu^{II} + 2 \text{ "CuO"} \qquad (3)$$

$$2Cu^0 + O_2 \rightarrow 2 \text{ "CuO"} \qquad (6)$$

in Reaction 6 would eliminate the need for chromatographic separation of the products of Reaction 3. Experiments in pyridine, **27, 32,** and **35** with copper foil or sponge show that this combination of processes actually occurs, but the use of catalytic (< 1 atom %) amounts of $CuCl_2$ gives rates of oxygen uptake which are very much less than those for CuCl oxidation under similar experimental conditions. The result of these low rates of oxygen uptake is clearly seen in pyridine, where the rate of Reaction 5 is the highest: long induction periods in the oxidative coupling of phenols are observed. We have shown previously that this effect is attributed to initiator polymerization (*18*). Also worth noting is the fact that theoretical oxygen uptake in this mode takes nearly three weeks in **32** under typical (millimolar Cu^0) conditions and gives a brown solution which is inactive for the oxidative coupling of 2,6-dimethylphenol. We suspect that initiators produced by ISG in **32** also readily polymerize in solution since $CuCl_2$ cluster incorporation protects initiators from this deactivating process (*see* below).

Ligand/Solvent Combinations (LS). Extension of our search for effective stabilizing ligands for copper–oxygen species demands the use of solid ligands and thus creates the need for suitable solvent media. Although we have only recently begun to fully appreciate the potential role of the solvent in the course of Cu(I) chloride oxidation, we have found that *o*-dichlorobenzene and methylene chloride serve as viable solvent media for many LS systems. The latter solvent has particular advantages in that it has a low affinity for water, is easily evaporated to afford solid products, and causes sufficient swelling of styrene/di-vinylbenzene gel-permeation resins to give efficient chromatographic separations. However, the following observations indicate that the range of solvents may have to be extended as more ligand systems are explored.

(1) Initiator species in pyridine precipitate and decompose to Cu(II) oxide on addition of *o*-dichlorobenzene but are soluble in methylene chloride.

(2) Insoluble Cu(II) oxide is the only observed copper–oxygen product of Reaction 3 in neat **19** or in mixtures of this ligand with methylene chloride, but the use of o-dichlorobenzene instead of methylene chloride gives homogeneous product solutions which gel on standing (consistent with initiator polymerization) and are too unstable for chromatographic separation.

The rates of oxidation of CuCl in pyridine/methylene chloride mixtures are higher than those in neat pyridine, perhaps because of less extensive py coordination of Cu(I) sites, which would interfere with oxygen coordination prior to oxidation (*see* above).

Substituted Pyridine Ligands (Table I)

Following our partial success with pyridine as the ligand/solvent medium, we extended our work to the formally aprotic pyridine derivatives in Table I, using the DS, LE, or LS (solid ligands only) approaches. These ligands were chosen to effect steric and electronic variations at the initiator centers; pyridine ligands with N-group pendants were used as potential capping ligands in an attempt to block initiator polymerization.

In room-temperature DS syntheses with these ligands, reaction of Cu(I) chloride with oxygen often proceeds in at least two stages, the first of which conforms to Reaction 3, and the remainder of which are associated with ligand oxidation. In some cases ,e.g. with **3**, Cu(I) and ligand oxidation proceed at comparable rates. Ligand **4** produces a nearly insoluble product mixture through Reaction 3; although the supernatant solution catalyzes the oxidative coupling of 2,6-methylphenol, it is too dilute to make initiator separation a practical proposition.

In both DS and LE approaches, ligand oxidation is observed for all ligands except **9, 11,** and **12** (LE), for which there is no evidence for ligand exchange even under reflux conditions. LE with **5** (which is expected to be a stronger π-acid than py) under nitrogen does lead to some substitution (spectral changes), but this ligand dissociates from the initiator products in acetone, benzene, and methylene chloride, leaving insoluble Cu(II) oxide. The complicated chromatographic behavior of LE products obtained with **5** is consistent with slow exchange of py and concurrent polymerization of products.

LE with ligand **6** consumes 1.5 mol O_2 per mol of ligand and gives a green solution which is an inactive oxidative coupling catalyst. LE with ligands **7** and **8** gives red product solutions after the consumption of 2.5 and 3.5 mol O_2 per mol of ligand, respectively. For ligands **6, 7,** and **8**, oxygen uptake rates are highest for **8**, suggesting initial oxidation at the benzylic CH_2 group. GPC (pyridine eluant) provides no evidence

Table I. Pyridine Derivatives

2: 2-methylpyridine
3: 2-ethylpyridine (pyridine with CH₂CH₃)
4: 2,6-dimethylpyridine
5: 4-benzoylpyridine (Ph–C(=O)– on pyridine)
6: 2,6-diacetylpyridine
7: 2,6-bis(N-phenyliminomethyl)pyridine
8: 2,6-bis(N-benzyliminomethyl)pyridine
9: tri(2-pyridyl)amine
10: quinoline
11: 2,2'-biquinoline
12: 1,10-phenanthroline (R=H)
13: 2,9-dimethyl-1,10-phenanthroline (R=Me)

for azobenzene formation (*see* Reaction 1) and gives brown, copper-containing solids on addition of concentrated HCl to brown column fractions (HCl rapidly converts initiator species in pyridine to soluble $CuCl_2$).

Our investigations with **11, 12,** and **13** stem from our earlier belief (*17*) that the catalytic product from Reaction 3 in pyridine was a pyridine-coordinated Cu(I) peroxide. LE with ligands **11** and **13** might reverse Reaction 3 because the Cu(I) complexes with these ligands are very stable (*21*). However, initiator species in pyridine show little affinity for **11** (*see* above) (although rapid exchange is observed for Cu(I) chloride oxidation products in **32** and **35**). We are currently investigating LS synthesis of initiator derivatives with **11, 32, 34, 35, 36,** *N,N*-dimethylnicotinamide, 1-methyl-2-pyridone and *N,N*-diethyl-3-toluamide. Ligand **12** gives a large number of nonstoichiometric products in LE and **13** is oxidized slowly during ligand exchange. The chromatographically separated products from LE with **13**, while spectrally similar to the Cu(dmp)$_2^+$ moiety (broad spectral maxima between 400 and 450 nm in pyridine), invariably have high oxygen contents consistent with ligand oxidation.

Simple substituted pyridine derivatives are too easily oxidized in the DS and LE modes to be useful as initiator stabilizers. We are currently investigating **9** and **11** in the LS mode. Preliminary results indicate that **9** acts as a stabilizing ligand for initiator species, which can be isolated by gel permeation chromatography.

Amine Ligands (Table II)

DS with ligands **14, 15, 16,** and **17** gives green or blue product mixtures that are very weak oxidative coupling catalysts after isolation as solids. Oxygen uptake is rapid with these ligands. Ligands **24** and **25** are unsuitable for use because they are too easily air oxidized. Consumption of more than the theoretical 0.25 mol O$_2$ per mol of Cu(I) with ligands **18, 20,** and **22** in DS or LS modes demonstrates the necessity for removal of protons from all N–H groups for successful initiator stabilization. For example, while ligand **22** consumes 4 mol O$_2$ per mol of ligand in methylene chloride, ligand **23** conforms to the expected Cu(I) chloride oxidation stoichiometry (*see* Reaction 3).

Ligands **19** and **21** give the highest rates of solution-phase Cu(I) chloride oxidation so far observed in the absence of ligand oxidation. The products in DS or LS (methylene chloride solvent) with these ligands are insoluble solids and the respective Cu(II) complexes, whereas LS (*o*-dichlorobenzene) gives homogeneous product solutions that contain highly active oxidative coupling catalysts. However, these latter solutions gel on standing or concentration (precluding GPC separation) and ultimately produce the same solids as obtained in DS or LS (methylene chloride). These solids are very poor oxidative catalysts and show essentially no affinity for the amine ligands.

Table II. Amine Ligands

NH$_2$—CH$_2$—CH$_2$—NH$_2$ (Me)$_2$N—CH$_2$—CH$_2$—N(Me)$_2$

14   **15**

NH$_2$—CH$_2$—CH$_2$—CH$_2$—NH$_2$ CH$_3$—CH(NH$_2$)—CH$_2$—NH$_2$

16   **17**

(pyrrolidine ring with N—R)

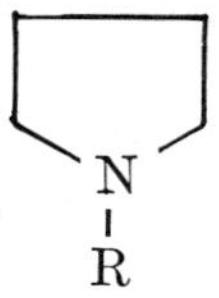

18: R=H
19: R=Me

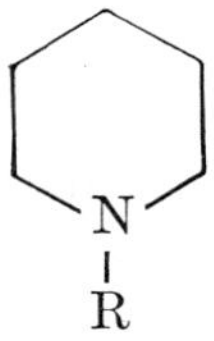

(piperidine ring with N—R)

20: R=H
21: R=Me

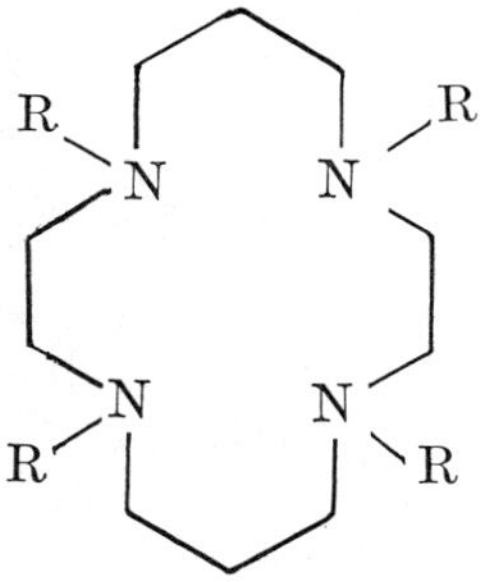

(cyclam macrocycle with four N—R groups)

22: R=H
23: R=Me

(pyrrolidine ring with N—R)

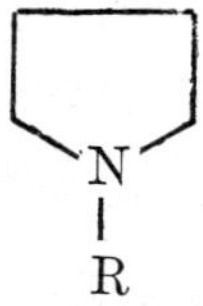

24: R=H
25: R=Me

Cyclam, **22**, and tetramethylcyclam, **23**, were chosen for study because they combine the desirable properties of **19** and **21** (rapid Cu(I) chloride oxidation reduces the time available for initiator polymerization) with the potential stabilizing influence of a macrocycle. Ligand **23** is interesting in that the four methyl groups are on the same side of the near-planar N$_4$ core in metal complexes (22, 23) and thus might prevent initiator polymerization through a "picket fence" effect (24). The products of the rapid oxidation of Cu(I) chloride in the presence of an equimolar quantity of 23 in methylene chloride or o-dichlorobenzene are an insoluble form of Cu(II) oxide (IR max at 460 cm^{-1}), the Cu(II) complex of 23, and free ligand.

Methyl-substituted amines like 19, 21, and 23 can be expected to give rise to relatively rapid Cu(I) chloride oxidation since, being harder ligands than pyridines and amides, they preferentially stabilize the Cu(II) oxidation state of the metal (25). What is evident here is that the planar geometry and strong σ character of 23 are not compatible with the requirements of the copper-reduced oxygen interaction despite an extremely strong tendency for metal chelation by the ligand. The lattice energy of Cu(II) oxide is sufficient to enable its ready production rather than the formation of a macrocyclic "CuO" center. Similar effects and rates of CuCl oxidation are observed on oxidation of preformed CuLCl or of slurries of CuCl in the LS mixture (in contrast with the results for pyridine noted above). The initiator polymerization found for pyridine, 19, and 21 as ligands indicates that monodentate nitrogen ligands are ineffective in blocking this process.

Amides and Lactams (Table III)

It has been known for some time that ligands 27, 35, and 36 can be used for direct synthesis of oxidized Cu(I) chloride solutions which are useful catalysts for oxidative coupling processes (26, 27). Amide and lactam ligands contain the nitrogen atoms of amines (which generally promote Cu(I) oxidation, see above) and also serve as crude models for the peptide link of proteins, which is the general environment of copper found in oxidases (28).

The need for methylation of all N–H groups and the necessary absence of acidic protons is again demonstrated by the ready oxidation of 26, 31, and 33 ligands in the DS, LE, and LS modes. Attempted chromatography of the products obtained in 27 (commercial GPC resins, 27 as eluant) gave gave very poor separations, the best being obtained with Biorad Biobeads SX-8 resins (nominal exclusion limit 1000 daltons). Small amounts of chloride-free material could be obtained from the leading edge of eluting samples, but the solids obtained on evaporation were extremely water sensitive and gave irreproducible analytical results, usually with high copper contents approaching those of Cu(II) oxide.

None of the oxidation products formed in the presence of 27, 29, 32, 34, 35, and 36 can be cleanly separated by GPC. Spectral and polarographic analyses of product solutions give no evidence for discrete Cu(II) chloride formation (29) (in sharp contrast to the results obtained in 19 and to pyridine (17, 18)). This evidence indicates that different

Table III. Amide and Lactam Ligands

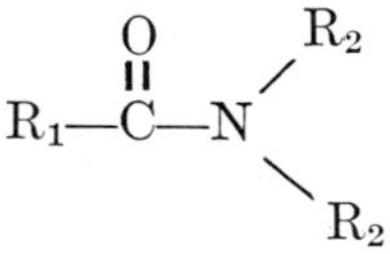

$$R_1-\underset{\underset{O}{\|}}{C}-N\begin{matrix}R_2\\R_2\end{matrix}$$

26: $R_1=H; R_2=Me$
27: $R_1=R_2=Me$
28: $R_1=Me; R_2=Ph$
29: $R_1=Ph; R_2=Me$
30: $R_1=R_2=Ph$

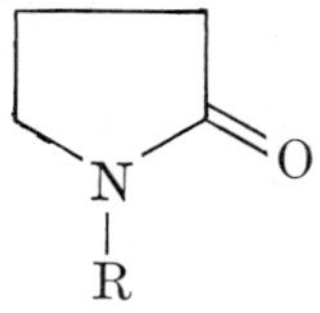

31: R=H
32: R=Me

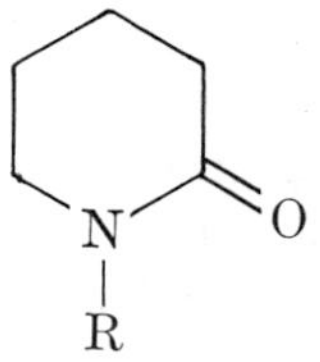

33: R=H
34: R=Me

Me$_2$SO
35

((Me)$_2$N))$_3$PO
36

oxidation products are ultimately formed in the presence of amide ligands on the one hand and amines on the other. Two other observations support this conclusion. First, the properties of the products obtained with amides seem far less time-dependent and second, Reaction 4b and/or the production of low molecular weight polymers from 2,6-dimethylphenol seems favored over polyphenylene oxide formation (Reaction 4a).

Ligands 28 and 30 do not promote Cu(I) chloride oxidation in the LS mode (methylene chloride or o-dichlorobenzene). Ligand 32 provides oxidation products which are far less sensitive to water than any obtained with the other ligands listed in Table III. The oxidation of Cu(I) chloride in 32 (DS) produces a brown, catalytically active solution which resists all attempts at separation. Precipitation of the products by addition of glyme results in a yellow-green solid with an ESR spectrum identical to that of the product solution and different from that of CuCl$_2$ in 32 (Figure 3). If this solid is precipitated, washed with glyme, and allowed to stand as a slurry for 7–20 days, remarkable transformations are observed. The yellow solid splits into two sections, the lower of

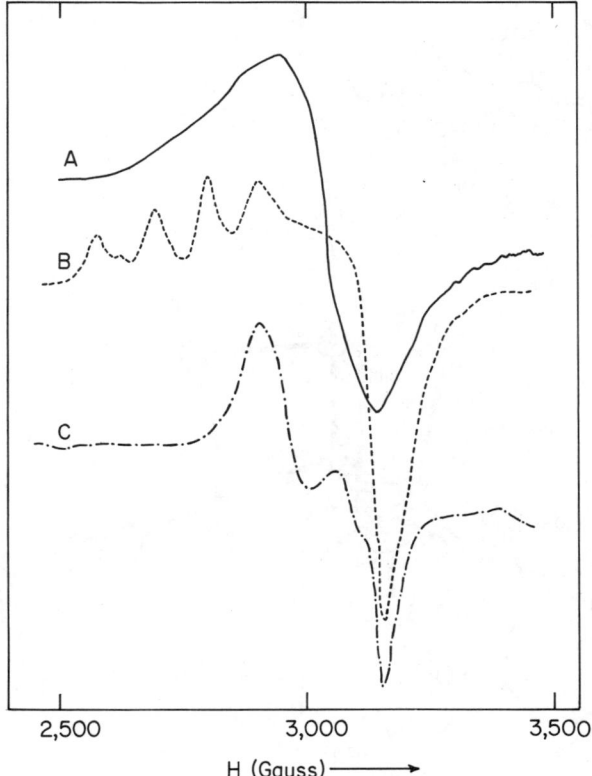

Figure 3. ESR spectra of the following: (A) $CuCl_2$ in frozen $L = 32$; (B) product of oxidation of $Cu(I)$ chloride by oxygen in 32 (precipitated solid or solution in 32); (C) solid $Cu_4Cl_6L_4(OH_2)$. Spectra obtained with a Varian E-9 spectrometer at $-100°C$.

which is brown and the upper of which is a pale-green gel. Brown-orange catalytic crystals slowly appear in the lower section. Analysis of these crystals gives the empirical formula $Cu_4Cl_6L_4O(OH_2)$; their ESR spectrum is included in Figure 3. The stoichiometry for formation of this product is given by Reaction 7 (L=32).

$$12\ CuLCl + 3O_2 + 6H_2O \rightarrow 2Cu_4Cl_6L_4O(OH_2) + 4CuL(OH)_2 \quad (7)$$

As indicated in this equation, the gelatinous green phase that accompanies crystal growth is a Cu(II) hydroxide complex. The structure (30) of the crystalline product is shown in Figure 4.

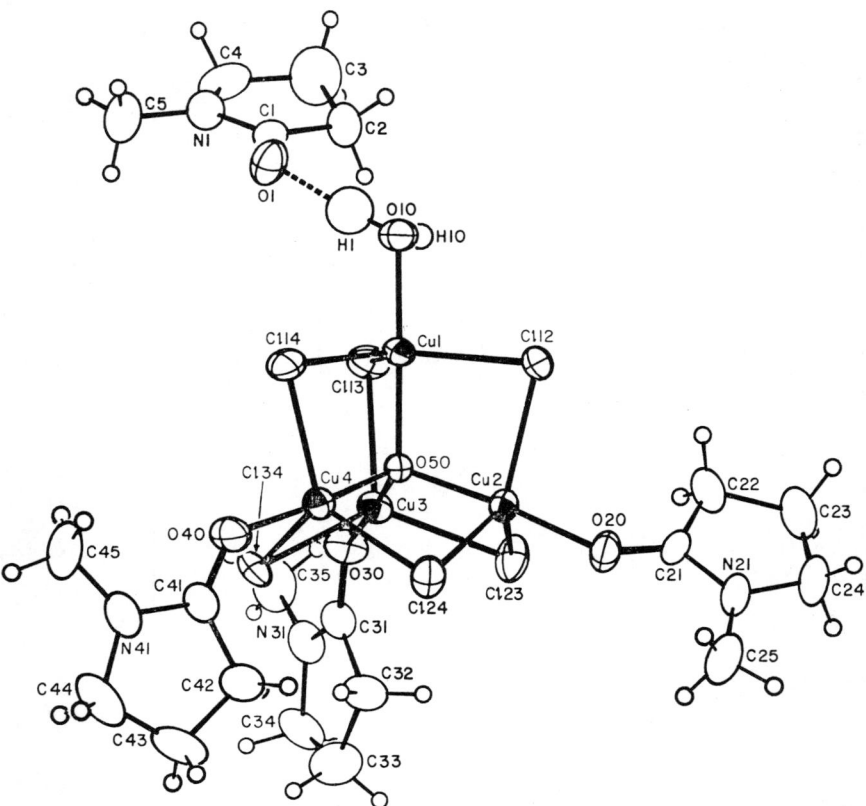

Figure 4. An ORTEP-II view of the geometry of the catalytically active cluster $L_3Cu_4OCl_6(OH_2) \cdot L$ (L = N-methyl-2-pyrrolidinone) (30)

The room-temperature magnetic moment of a ground sample of the crystals is 2.2 μ_B (1.92 μ_B at 77°K, Faraday balance), consistent with the five-coordinate Cu(II) centers of the structure (*31, 32, 33, 34*). Low-temperature magnetic measurements are being undertaken to assess the extent of exchange coupling between the metal centers. Small crystals that have appeared in homogeneous product solutions in **32** also are being investigated, and we are currently using the above technique in an attempt to obtain crystalline products with **27, 29, 34, 35,** and **36** as ligands.

If $L_3Cu_4Cl_6(O)OH_2 \cdot L$ is allowed to stand in excess **32**, catalytically inactive $L_4Cu_4Cl_6O$ is produced. The latter also can be synthesized by a literature procedure (*33*). Evidently, a terminal oxygen atom is a requirement for catalytic activity in these clusters (*30*).

It is of interest to consider the mechanism of oxidation of slurries of solid Cu(I) chloride in the presence of monodentate ligands such as pyridine and **32** and to explore the relationship between the crystalline derivative and the primary product of oxidation.

The structure of solid Cu(I) chloride (zinc blende, $a_o = 5.406$ Å (*35*)) is compatible with the formation of a clustered product. We have shown cryoscopically in **32** that the primary product of oxidation is a stable tetrameric cluster, Reaction 8 (ligands omitted) (*30*). Based on

$$(\text{CuCl})_n + \frac{n}{4} \text{O}_2 \rightarrow \frac{n}{4} \text{Cu}_4\text{Cl}_4\text{O}_2 \qquad (8)$$

our observations and on results obtained in other systems (*1*), we favor an inner-sphere mechanism for Reaction 8. Assuming similar primary clustered products in all cases, it is evident that pyridine does not stabilize the primary product cluster, which undergoes fracture to give two discrete products, *see* Reaction 3. However, as noted earlier, coordinated pyridine does provide limited stability to the resulting Cu–O–Cu–O core. Polymerization of initiator species, Reactions 9 and 10, presumably is

$$2 \text{ Cu-O-Cu-O} \rightarrow (\text{Cu-O-Cu-O}_2)_2 \qquad (9)$$

$$(\text{-Cu-O-Cu-O-})_m + (\text{-Cu-O-Cu-O-})_n \rightarrow (\text{-Cu-O-Cu-O-})_{m+n} \qquad (10)$$

allowed by the existence of labile-coordinated pyridine (*see* above) and strongly nucleophilic-exposed oxygen atoms [we have found that even highly polymerized initiators are significantly stronger protic bases than is pyridine (*18*)].

Saturated amines such as **19** stabilize neither the primary cluster nor the Cu–O–Cu–O core produced on cluster fracture. However, we have isolated a ligand class (the amides) which stabilizes clustered products that can serve as useful models for the active sites of copper oxidases.

Acknowledgment

This work was supported by National Science Foundation Grant CHE7522453, which is gratefully acknowledged. Our understanding of these catalytic copper species has been greatly improved by a structural determination of the catalyst derivative with N-methyl-2-pyrrolidinone by Melvin R. Churchill and Frank J. Rotella, who graciously contributed their results prior to publication. The experimental assistance of Jack Treger, Richard Himmelwright, Raymond Hoover, William Reiff, Edward Solomon, and Edward Witten is also greatly appreciated.

Literature Cited

1. Zuberbühler, A. D., "Metal Ions in Biological Systems," H. Sigel, Ed., Vol. 5, Ch. 7, Dekker, New York, 1975.
2. Hemmerich, P., "The Biochemistry of Copper," p. 15, Academic, New York, 1966.
3. Osterberg, R., *Coord. Chem. Rev.* (1974) **12**, 309.
4. Gray, H. B., Coyle, C. L., Dooley, D. M., Grunthaner, P. J., Hare, J. W., Holwerda, R. A., McArdle, J. V., McMillin, D. R., Rawlings, J., Rosenberg, R. C., Sailasutá, N., Solomon, E. I., Stephens, P. J., Wherland, S., Wurzbach, J. A., ADV. CHEM. SER. (1977) **162**, 145.
5. Beinert, H., *Coord. Chem. Rev.* (1977) **23**, 119.
6. Kramer, C. E., Doctoral thesis, Northeastern University (1974).
7. Terentev, A. O., Mogliansky, Y. D., *J. Gen. Chem. USSR (Engl. Transl.)* (1958) **28**, 2002.
8. Bach, H. C., Black, W. B., *J. Polym. Sci. C* (1969) **22**, 799.
9. Hay, A. S., *J. Org. Chem.* (1960) **25**, 1275.
10. Endres, G. F., Hay, A. S., Eustance, J. W., *J. Org. Chem.* (1963) **28**, 1300.
11. Finkbeiner, H., Hay, A. S., Blanchard, H. S., Endres, G. F., *J. Org. Chem.* (1966) **31**, 549.
12. Stamatoff, G. S., U.S. Patent **3,229,910** (1966).
13. Hay, A. S., U.S. Patent **3,432,466** (1969).
14. Ogata, Y., Morimoto, T., *Tetrahedron* (1965) **21**, 2791.
15. Tsuruya, S., Shirai, T., Kawamura, T., Yonezawa, T., *Makromol. Chem.* (1970) **B2**, 57.
16. Kametani, T., Ihara, M., Takemura, M., Satoh, Y., Teresawa, H., Ohta, Y., Fukumoto, K., Takahashi, K., *J. Am. Chem. Soc.* (1977) **99**, 3805.
17. Kramer, C. E., Davies, G., Davis, R. B., Slaven, R. W., *J. Chem. Soc., Chem. Commun.* (1975) 606.
18. Bodek, I., Davies, G., *Inorg. Chem.* (1978) **17**.
19. Bodek, I., Davies, G., *Inorg. Chim. Acta* (1978) **27**, 213.
20. Nigh, W. G., "Oxidation in Organic Chemistry," Vol. 5B, p. 1, Academic, New York, 1973.
21. James, B. R., Williams, R. J. P., *J. Chem. Soc.* (1961) 2007.
22. Wagner, F., Mocella, M. T., D'Aniello, Jr., M. J., Wang, A. H-J., Barefield, E. K., *J. Am. Chem. Soc.* (1974) **96**, 2625.
23. Wagner, F., Barefield, E. K., *Inorg. Chem.* (1973) **12**, 2435.
24. Collman, J. P., *Acc. Chem. Res.* (1977) **10**, 342.
25. Patterson, G. S., Holm, R. H., *Bioinorg. Chem.* (1975) **4**, 257.
26. Tsuruya, S., Kawamura, T., Yonezawa, T., *J. Polym. Sci.* (1971) **9**, 1659.
27. Bach, H. C., *Polym. Prepr., Am. Chem. Soc., Div. Polym. Chem.* (1967) **8**, 610.
28. Frieden, E., *Chem. Eng. News.* (1974) March 25, p. 42.
29. Tsuruya, S., Yonezawa, T., Kato, H., *J. Phys. Chem.* (1974) **78**, 811.
30. Davies, G., El-Shazly, M. F., Rupich, R. W., Churchill, M. R., Rotella, F. J., unpublished data.
31. Bertrand, J. A., Kelley, J. A., *J. Am. Chem. Soc.* (1966) **88**, 4746.
32. Bock, H., Dieck, H. T., Pyttlik, H., Schnoeller, Z., *Anorg. Allg. Chem.* (1968) **357**, 54.
33. Dieck, H. T., *Inorg. Chim. Acta* (1973) **7**, 397.
34. Churchill, M. R., DeBoer, B. G., Mendak, S. J., *Inorg. Chem.* (1975) **15**, 2496 and references therein.
35. Wyckoff, R. W. G., "Crystal Structures," Second Ed., Vol. 1, p. 110, Wiley, New York, 1963.

RECEIVED March 7, 1978.

17

Novel Cleavage and Oligomerization Reactions of Nickel(0) Complexes

Applications to Homogeneous Deoxygenation and Desulfurization

JOHN J. EISCH and KYOUNG R. IM

Department of Chemistry, State University of New York at Binghamton, Binghamton, NY 13901

> *The ease of interaction of Ni(0) complexes with organic substrates has been shown to depend upon both the ligands on nickel and the solvent. The presence of α,α'-bipyridyl with the Ni(0) complex and the alkyne led to the isolation of a nickelacyclopropene, an observation in accord with the recently proposed metallocyclic pathway for the Ni(0)-catalyzed trimerization of alkynes. Allylic and benzylic ethers and epoxides have been observed to undergo oxidative insertion of Ni(0) into their C–O bonds with solvent (TMEDA > THF > Et_2O > C_6H_6) and ligand (Et_3P > Ph_3P; α,α'-bipy > COD) effects consistent with an electron-transfer attack by Ni(0). With such sulfur heterocycles as dibenzothiophene, phenoxathiin, phenothiazine, and thianthrene, a 1:1 admixture of $(COD)_2Ni$ with α,α'-bipyridyl gave as the principal product the desulfurized, ring-contracted cyclic product.*

By replacing certain heterogeneous transition metal catalysts by homogeneous or polymer-bonded catalysts, many important reactions, such as hydrogenation, unsaturated hydrocarbon oligomerization, and hydroformylation, show significant gains in stereo-, regio-, and loco-selectivity (1, 2) (Loco- is suggested as a prefix to denote the molecular site (functional group) at which a given reaction occurs selectively (Latin, *locus*: place or site). The term, chemospecific, which has already been suggested for this situation (2) does not seem to call to mind the intended meaning.) As part of a long-range, physical organic investigation of transition metal oxidative additions, the present study has examined the insertion reactions of Ni(0) complexes into various relatively weak π and σ bonds. Chosen for particular attention were the alkyne π bond and the carbon–heteroatom σ bonds (C–E, where E = O, S, X, or a metalloid). By understanding the influence of ligands and solvents on the rate and extent of such stoichiometric insertions, we hope to develop such reactions into rationally founded catalytic processes. For example, the catalytic scission of C–S bonds, presently effected with hydrogen and heterogeneous catalysts at elevated temperatures (3, 4, 5, 6, 7), might be achievable at a much lower temperature with an appropriate soluble transition metal complex. Such mild desulfurizing conditions might mean savings of time and energy for industrial processes used in desulfurizing coal-derived fuels and chemicals. In addition, agents capable of cleaving C–S bonds at low temperatures could be valuable in selectively degrading the macrostructures of various coals themselves into fragments manageable for structural studies. The increasing reliance of the industrial nations on coal for energy and for chemicals raises the problem of desulfurization to new prominence. Indeed, since the days of Sodom and Gomorrah, the presence of sulfur and its oxides in the environment has been recognized as a serious problem for human communities.

Interaction of Ni(0) Complexes with Alkynes

Recent studies on the allylation of alkynes with bis(π-allyl)nickel have revealed that the Ni(0) generated in this process causes the trimerization and, more importantly, the reductive dimerization of a portion of the alkyne (8). A deuterolytic work-up led to the terminally di-deuterated diene (5), supporting the presence of a nickelole precursor (4) (Scheme 1). The further interaction of 4 with 1, either in a Diels–Alder fashion (6) or by alkyne insertion in a C–Ni bond (7), could lead to the cyclic trimer 8 after extrusion of Ni(0), thereby accounting for the trimerizing action of Ni(0) on alkynes. This detection of dimer 5 then provided impetus for the synthesis of the unknown nickelole system to learn if its properties would accord with this proposed reaction scheme. Therefore, E,E-1,4-dilithio-1,2,3,4-tetraphenyl-1,3-butadiene (9) was treated with bis(triphenylphosphine)nickel(II) chloride or 1,2-bis(diphenylphosphino ethane)nickel(II) chloride to form the nickelole 10 (9) (Scheme 2). The nickelole reacted with dimethyl acetylenedicarboxylate to yield 11 and with CO to produce 12. Finally, in keeping with the hypothesis offered in Scheme 1, 10a did act as a trimerizing catalyst toward diphenylacetylene (13) to yield 14.

Scheme 1

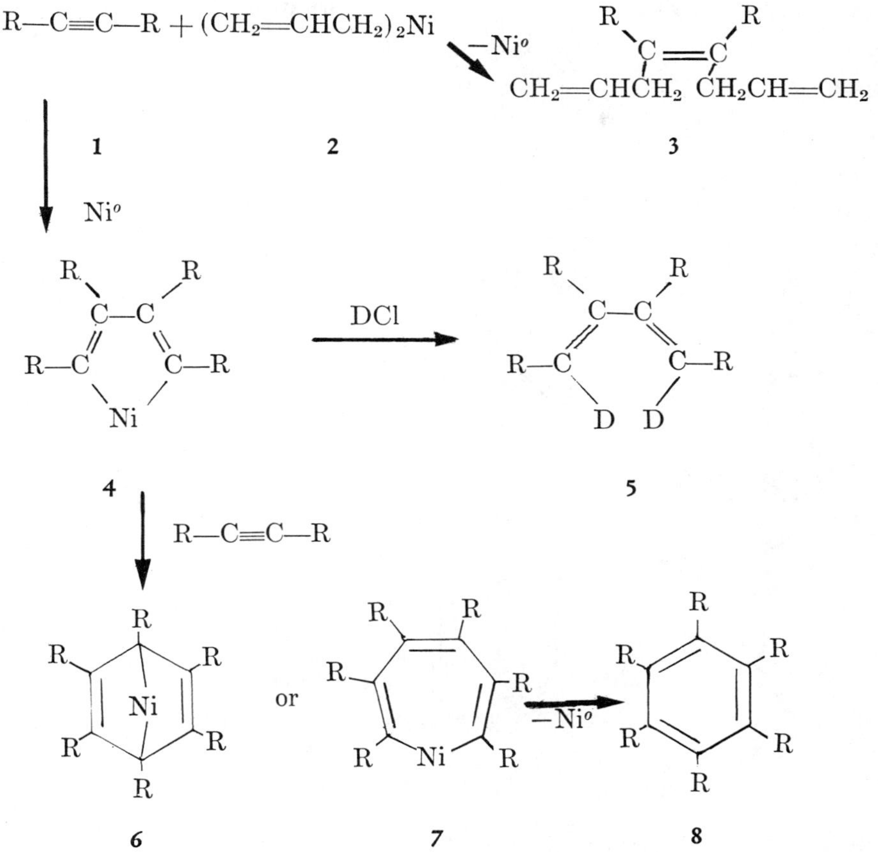

$$R-C\equiv C-R + (CH_2=CHCH_2)_2Ni \xrightarrow{-Ni^0}$$

1, 2, 3 (with structure 3 showing C=C with R groups and CH$_2$=CHCH$_2$ allyl groups)

↓ Ni^0

4 (nickelacyclopentadiene) → DCl → 5 (diene with D substituents)

↓ R—C≡C—R

6 (nickel-bridged benzene) or 7 (nickelacycloheptatriene) → $-Ni^0$ → 8 (hexasubstituted benzene)

The profound effect of ligands on the course of alkyne–Ni(0) interactions was observed with bis(1,5-cyclooctadiene)nickel(0) (15). Although heating 13 in hexane with 15 brought about reductive dimerization and trimerization, a combination of 15 and 1 equiv of α,α'-bipyridyl (cf. infra) gave a black solid (16, IR band at 1770 cm^{-1}) whose protolysis under an argon atmosphere with 85% H$_3$PO$_4$ yielded only cis-stilbene (17). Simple exposure of a slurry of 16 to air regenerated 13. On the other hand, treatment of 16 with 6N HCl under argon produced a brick-colored solid (18) whose protolysis with glacial acetic acid in air produced only trans-stilbene (19). When the black solid 16 was treated instead with a sequence of 6N-DCl and glacial o-deuterioacetic acid, α,α'-dideuterio-trans-stilbene was obtained. The formulation of 16 as a nickelacyclopropene is supported by the 1770 cm^{-1} band, which resembles that characteristic of other metallocyclopropenes (10). Thus, as

Scheme 2

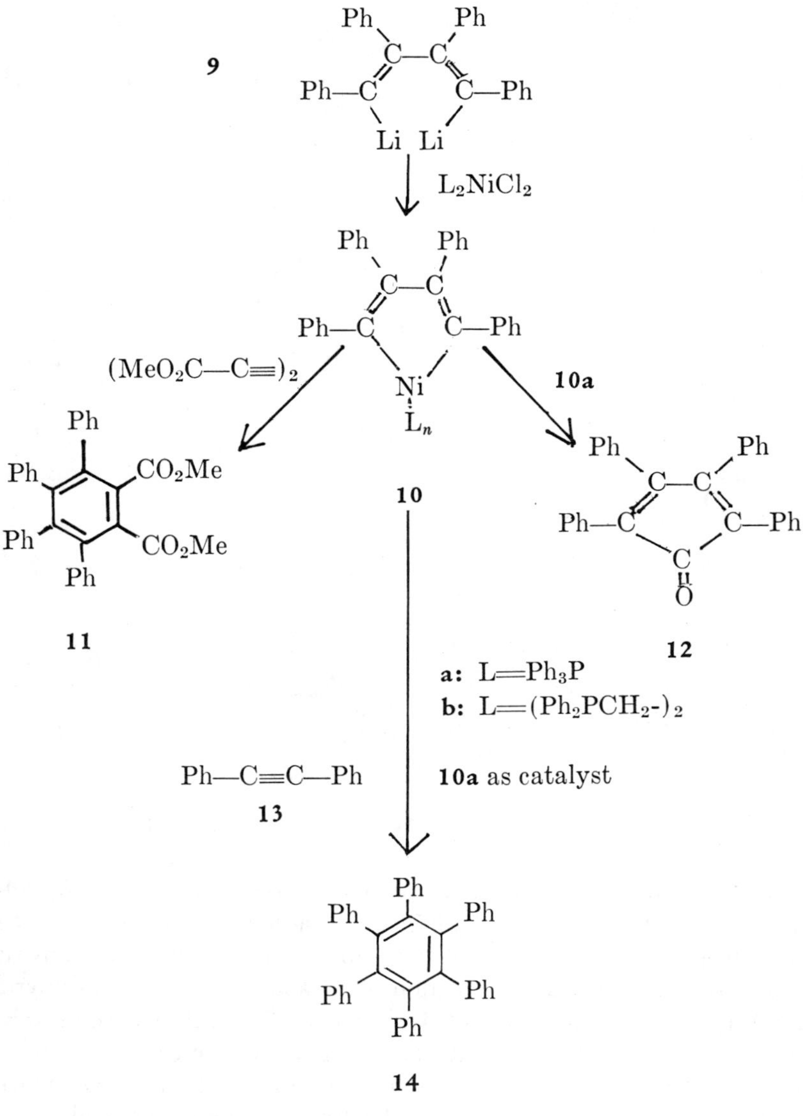

a: L=Ph$_3$P
b: L=(Ph$_2$PCH$_2$-)$_2$

portrayed in Scheme 3, the presence of bipyridyl seems to arrest the alkyne–Ni(0) interaction at the 1:1 stage (**16**). When **16** was heated in THF with two or more equivalents of dimethyl acetylenedicarboxylate (**21**), **20** and the cyclotrimer of **21** were isolated. For the catalytic cyclotrimerization of alkynes, accordingly, this finding supports a stepwise passage from a 1:1 (e.g., **16**) to a 2:1 or nickelole adduct (in this

Scheme 3

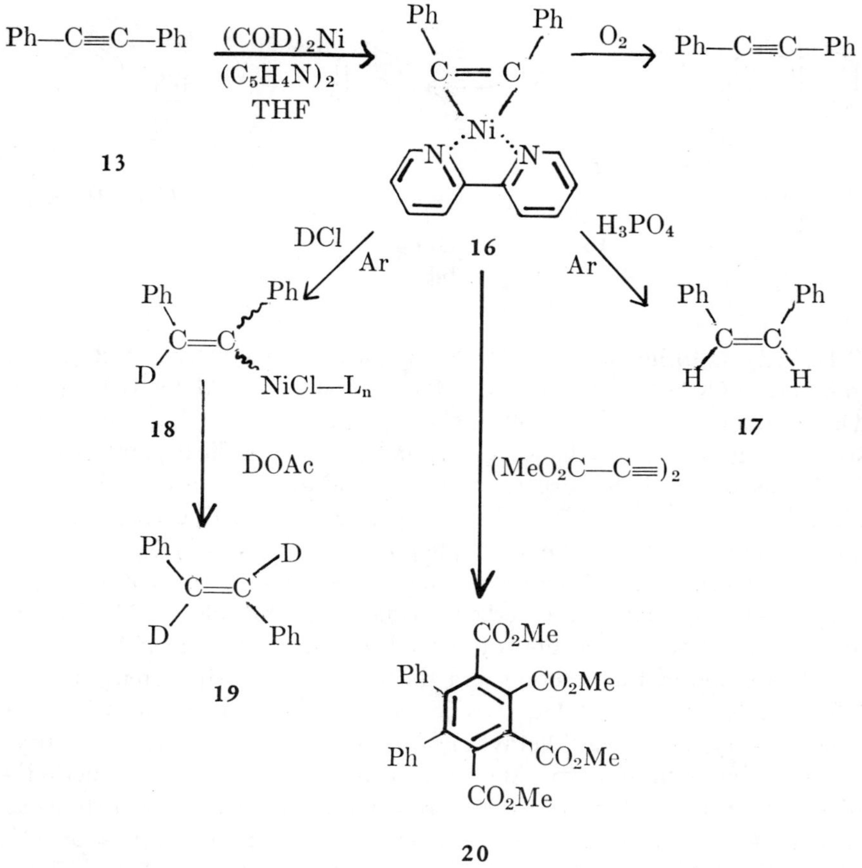

case, **4**, where R_1 and $R_2 =$ Ph and R_3 and $R_4 = CO_2Me$) to a 3:1 adduct (**8** and **20**, via **6** or **7**). More importantly, these results can mean that bipyridyl fosters oxidative addition onto the alkyne π bond and stabilizes the Ni(II)-like nickelacyclopropene intermediate (**16**), both by its back-bonding and chelating properties.

Insertions of Ni(0) Complexes into Carbon–Oxygen Bonds

Further evidence for the fostering effect of donor ligands on the oxidative additions of Ni(0) was gained from a cleavage study of ethers and epoxides. Allylic ethers, such as allyl phenyl ether (**22**), underwent smooth insertion of Ni(0) with either **15** in THF or tetrakis(triethylphosphine)nickel(0) (**23**) in benzene (Scheme 4). The following reactivity trends are noteworthy: (1) as to solvent, TMEDA > THF >>

Scheme 4

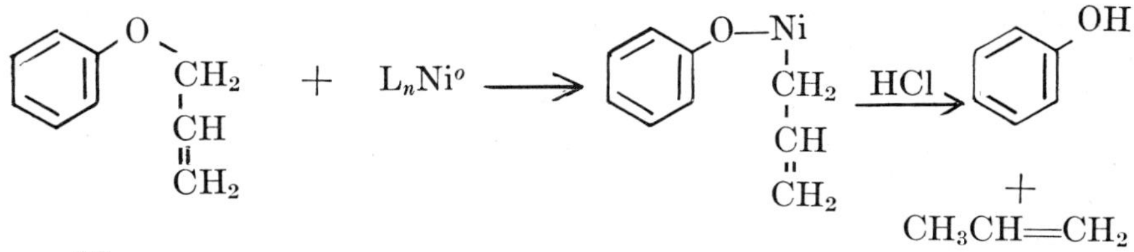

[Structure: phenyl-O-CH₂-CH=CH₂ (22)] + L$_n$Ni⁰ → [phenyl-O—Ni with CH₂-CH=CH₂] $\xrightarrow{HCl}$ [phenol OH] + CH$_3$CH=CH$_2$

22

15: L$_n$ = 2COD
23: L$_n$ = 4Et$_3$P

C_6H_6; (2) as to ligands, $(COD)_2Ni$ + bipy > $(COD)_2Ni$; and (3) as to ether, PhOCH$_2$CH = CH$_2$ >> Ph-O-CH$_2$Ph >> MeOCH$_2$Ph (*11*). The formation of small amounts (< 5%) of *o*- and *p*-allylphenols is consistent with the intermediacy of allyl and phenoxy radical pairs in these Ni(*0*) insertions. [The formation of ca. equal amounts of *o*- and *p*-allylphenols is not reconcilable with a competing thermal Claisen rearrangement, where exclusively the *o*-allylphenol is produced (*11*)].

Corroborative evidence for this mechanistic view was acquired from a study of a special case of ethers, namely the epoxides (*11*). In this instance, C–O bond cleavage is followed by elimination of NiO (Scheme 5). The scope of the reaction and the influence of Ni(*0*) reagents are informative: (1) substituted styrene oxides and other epoxides bearing adjacent π systems (e.g., diethyl *cis*-2,3-epoxysuccinate) undergo deoxygenation, but saturated epoxides (e.g., *cis*-1,2-epoxyoct-1-yl (trimethyl)-silane) seem to be inert; (2) the deoxygenation appears to be a stepwise process, for either *cis*- or *trans*-β-trimethylsilylstyrene oxide (**24** or **25**) eventually yield the same composition of olefins **26** and **27** (~ 95:5); however, the ratio of olefinic isomers from **24** at small conversion was 60:40 (**26:27**); (3) although (COD)$_2$Ni in THF or (Ph$_3$P)$_4$Ni in benzene do not readily react with **24**, either a 1:1 mixture of (COD)$_2$Ni and bipyridyl in THF or (Et$_3$P)$_4$Ni in benzene quantitatively form **26** and **27** (95:5). Now, the nonstereoselectivity of the kinetically controlled product composition (range of **26:27** = 60:40 to 95:5) indicates a stepwise rupture of the C–O epoxide bonds, so that intermediates like **28** and **29** can interconvert before NiO elimination. Thus, both the smooth deoxygenation of **24** and **25** and the slower conversion of **27** into **26** seem most likely to occur via single electron transfer from Ni(*0*) to the epoxides or to the olefin **27**, respectively. The other observations, such as the accelerating effect of electron-attracting π systems on the epoxide and of strong donor ligands on nickel (bipyridyl or Et$_3$P), are also consistent with such an electron-transfer mechanism.

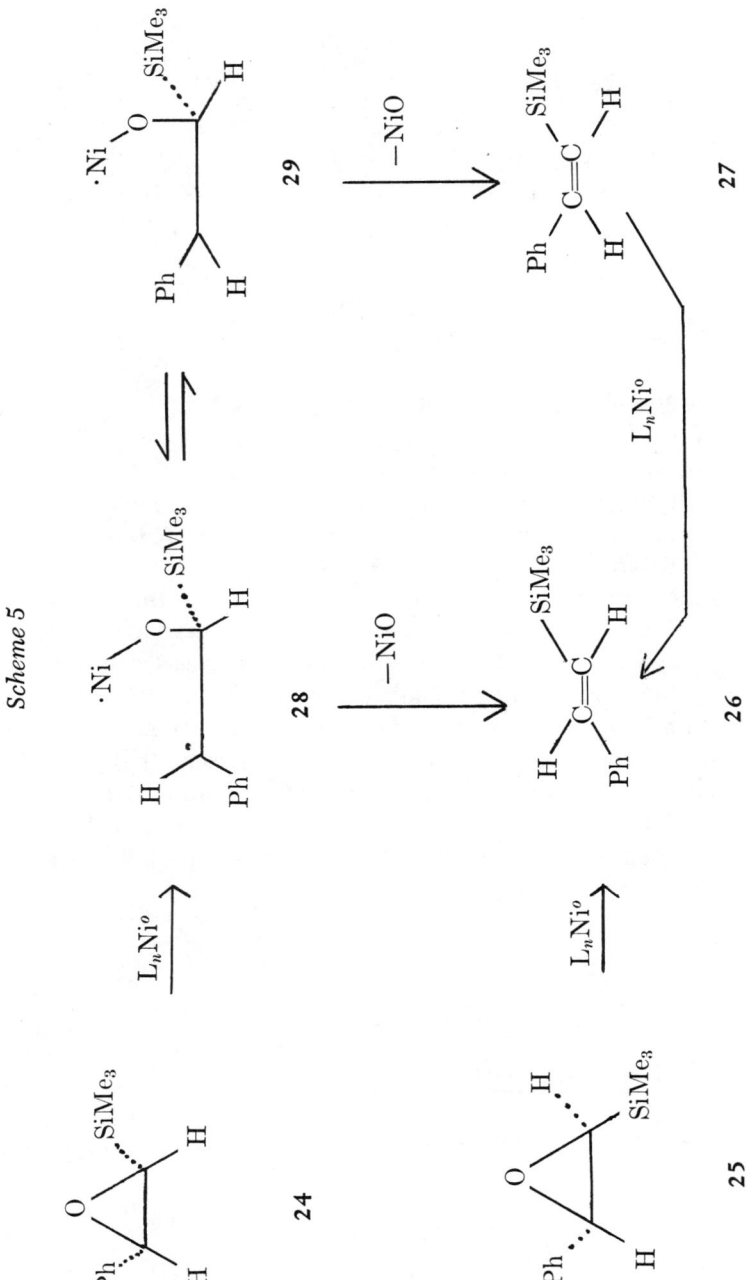

Scheme 5

Scheme 6

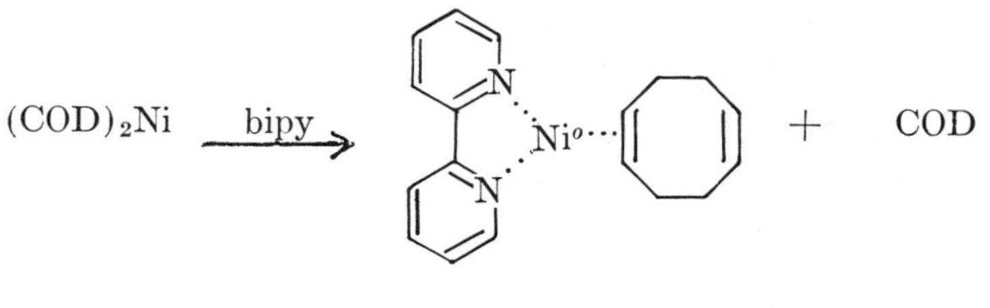

$(COD)_2Ni \xrightarrow{bipy}$ [α,α'-bipyridyl Ni(COD)] + COD

28

The efficacy of a 1:1 combination of $(COD)_2Ni$ (yellow) and α,α'-bipyridyl can be ascribed to the formation of α,α'-bipyridyl (η^2-1,5-cyclooctadiene)Ni(0) (**28**) (purple in THF), which NMR studies have shown to be formed under these conditions (*12*) (Scheme 6). Reagent **28** is evidently a much more reactive electron donor. A Japanese research group has demonstrated the effectiveness of this Ni(0) source in forming cycloalkanes from α,ω-dihaloalkanes (*13*). Studies in this laboratory also have adduced evidence for radical processes in these alkyl halide reactions. Thus, when β,β,β-triphenylethyl bromide was treated with **28** in THF and the reaction mixture subsequently worked up with DCl, only undeuterated **29** and **30** were formed. Since any organonickel precursor would have yielded deuterated **29** or **30**, it is clear that radicals were generated (**31** and **32**) that acquired hydrogen from the THF before or after a 1,1-phenyl shift (*14*) (Scheme 7). Taken together with the absence of deuterium in **30**, the phenyl shift is especially diagnostic of the intermediacy of radical **31** (*15*).

Scheme 7

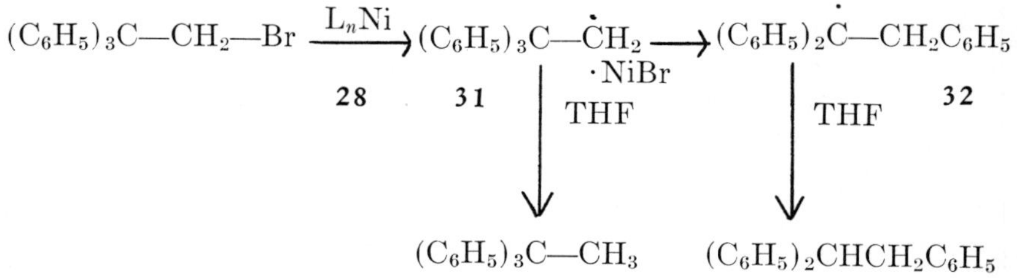

$(C_6H_5)_3C—CH_2—Br \xrightarrow{L_nNi}_{28} (C_6H_5)_3C—\overset{\cdot}{C}H_2 \xrightarrow{} (C_6H_5)_2\overset{\cdot}{C}—CH_2C_6H_5$
$\qquad\qquad\qquad\qquad\qquad\quad$ **31** $\quad\cdot NiBr \qquad\qquad$ **32**
$\qquad\qquad\qquad\qquad\qquad\qquad\quad\downarrow THF \qquad\qquad \downarrow THF$

$\qquad\qquad\qquad\qquad\qquad (C_6H_5)_3C—CH_3 \qquad (C_6H_5)_2CHCH_2C_6H_5$
$\qquad\qquad\qquad\qquad\qquad\qquad$ **29** $\qquad\qquad\qquad\qquad$ **30**

Desulfurization and Hydrodesulfurization of Sulfur Heterocycles

The foregoing insight into the mechanism of Ni(0) insertion reactions proved most helpful in developing soluble desulfurizing agents for aromatic sulfides (*16*). As test substrates for Ni(0) desulfurizations, heterocycles 33, 34, 35, and 36 were chosen since these are typical of the

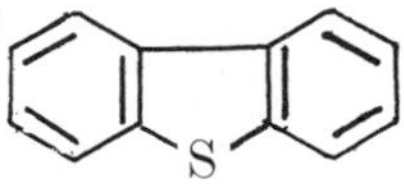

33

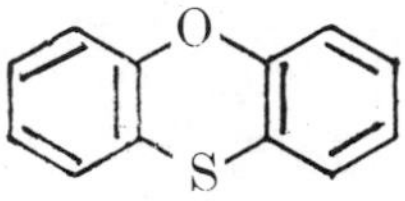

34

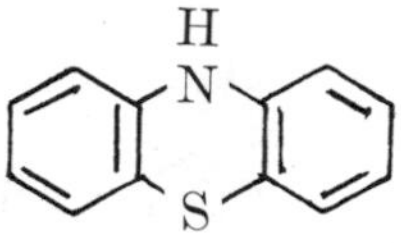

35

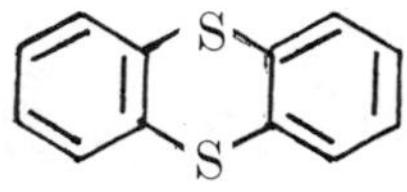

36

stable organosulfur compounds found as contaminants in liquid fuels and organic chemicals derived from coal. Quickly it became evident that the bipy(COD)Ni reagent was again superior to (COD)$_2$Ni for inserting into a C–S bond; but a net single C–S insertion leading to 37 (and upon protolysis to 38) turned out to be a minor pathway (< 5%) (Scheme 8). With two equiv of 28 in THF at 50°C, the major product formed by treatment with 6N HCl or with LiAlH$_4$ was biphenyl (39) in a 55–75% yield. When the reaction was conducted for shorter periods at lower conversion, both 39 and biphenylene 40 were detected. The nature of intermediate(s) 41 will be discussed below when additional results permit some generalization.

The behavior of heterocycles 34, 35, and 36 toward bipy(COD)Ni and was instructive (Scheme 9): here complete desulfurization was the exclusive course, with ring contraction (42) being the preponderant outcome for 34 and 35. Only traces of the hydrodesulfurized product 43 were formed. Thianthrene was converted into dibenzothiophene (33), whose further reactions followed those depicted in Scheme 8.

In the work-up of such ring contractions, it was found that a brief (10 min) treatment of the reaction mixture with LiAlH$_4$, followed by hydrolysis, was a more rapid, higher-yielding procedure than treatment with 6N HCl. When, on the other hand, 1 equiv of LiAlH$_4$ was added

Scheme 8

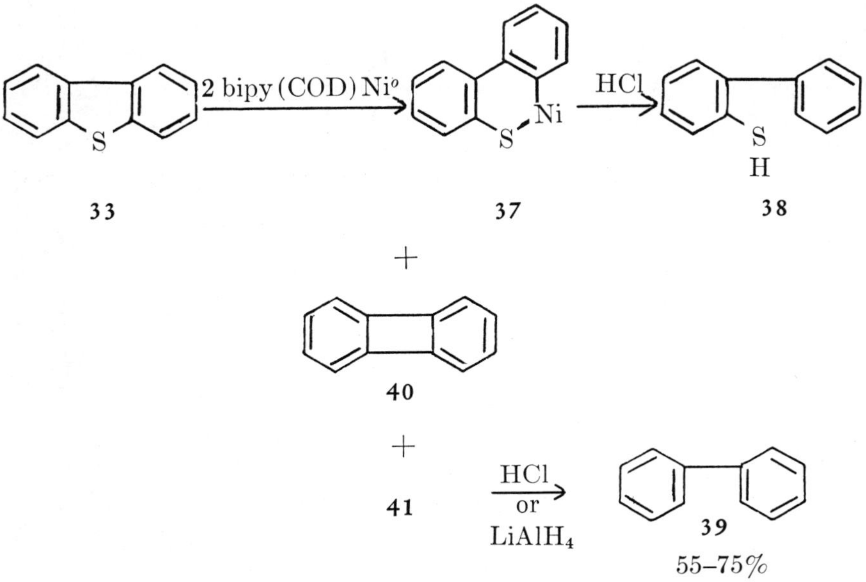

33 → 2 bipy(COD)Ni⁰ → **37** → HCl → **38**

+

40

+

41 → HCl or LiAlH₄ → **39** 55–75%

Scheme 9

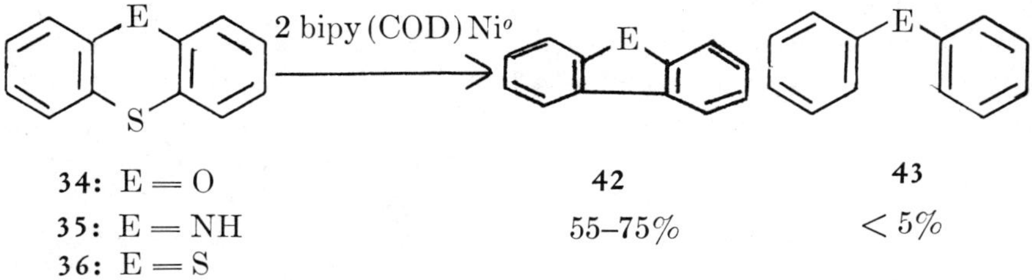

34: E = O
35: E = NH
36: E = S

→ 2 bipy(COD)Ni⁰ → **42** 55–75% + **43** < 5%

Scheme 10

33: E = σ bond
34: E = O
35: E = NH

→ 2 bipy(COD)Ni⁰ / LiAlH₄ → **43**

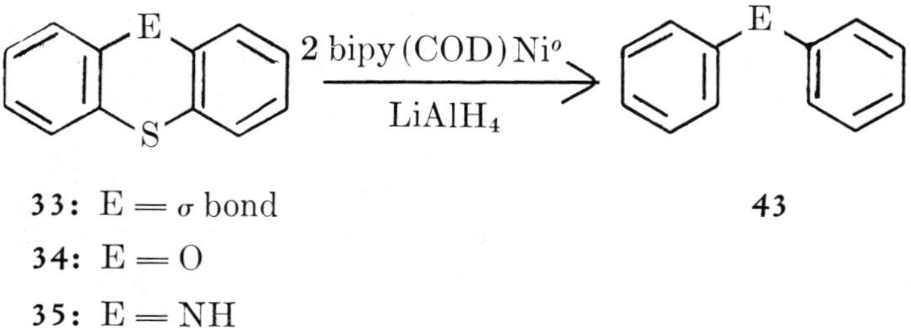

to 2 equiv of **28** before introducing the heterocycle substrate, subsequent heating led exclusively to the hydrodesulfurized product for **33, 34,** and **35**. Thianthrene (**36**) gave both **33** and **39** (Scheme 10).

The ring contraction observed with all of these heterocycles (**40** and **42**), taken together with the inability of (COD)$_2$Ni alone to effect desulfurization, indicates that bipy(COD)Ni operates principally by an electron-transfer mechanism rather than by insertion in a C–S bond (cf. low yield of **38**). Since 2 equiv of **28** are more effective, prior coordination of one Ni(0) at sulfur seems to be important (Scheme 11). That biphenyl-

Scheme 11

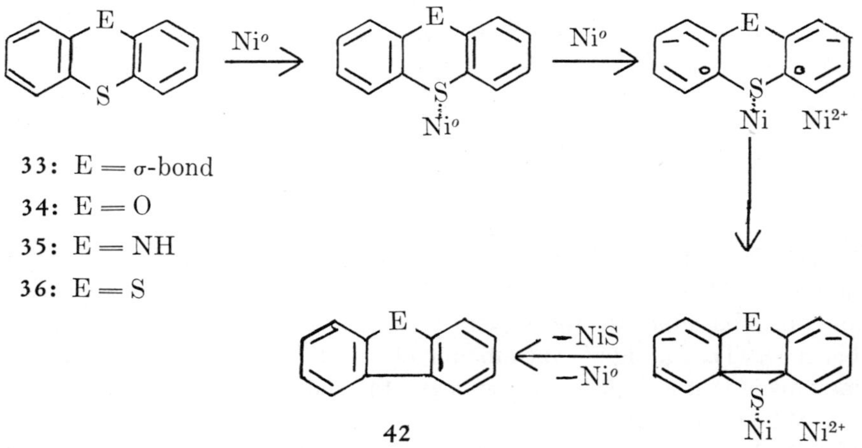

33: E = σ-bond
34: E = O
35: E = NH
36: E = S

ene is a kinetically controlled product in the desulfurization of dibenzothiophene (*see* Scheme 8) suggested that it must be subsequently recleaved by the further action of Ni(0) complexes. This possibility was verified in a separate experiment by treating samples of biphenylene, in turn, with (COD)$_2$Ni, (Et$_3$P)$_4$Ni, bipy(COD)Ni, and (COD)$_2$Ni + LiAlH$_4$. (LiAlH$_4$ alone did not attack **40**.) All of these reagents in warm THF cleaved biphenylene and yielded, upon hydrolysis, biphenyl (*14*).

More informative about the manner in which these nickel reagents cleave biphenylene (**40**) were reactions in which deuterated hydride or proton sources were used. Thus, treatment of **40** with (Et$_3$P)$_4$Ni in benzene, followed by work-up with 12N DCl under argon, gave a 70:30 mixture of 2,2′-dideuteriobiphenyl (**43**) and 2-deuteriobiphenyl (**44**). A similar run conducted in THF gave a 30:70 ratio of **43** and **44**. The deuterated biphenyls presuppose organonickel precursors **45** and **46** (Scheme 12), which must also be the precursors **41** of the biphenyl

Scheme 12

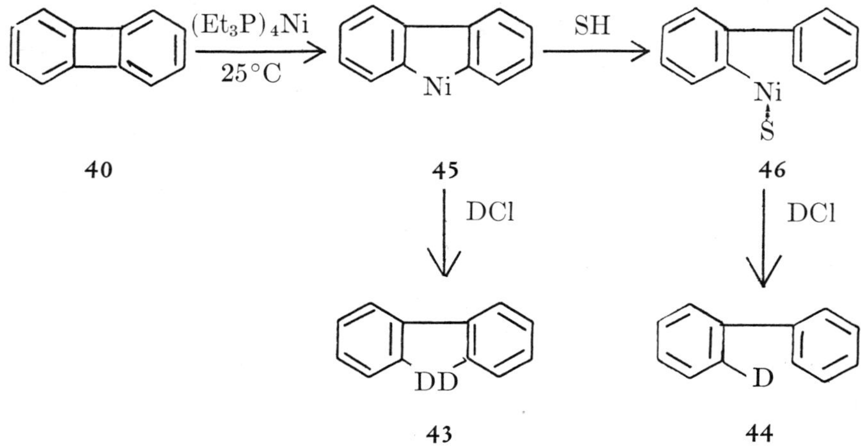

	C_6H_6	43 70%	44 30%
SH =	THF	30%	70%

formed from dibenzothiophene (*see* Scheme 8). The larger amount of monodeuterated biphenyl (**44**) in THF suggests that radical intermediates may be attacking the solvent, a better hydrogen atom donor than benzene. That a significant amount of **44** is formed, even in benzene, may mean that a Et_3P ligand can serve the source of the hydrogen (*14*).

When the cleavage of biphenylene was conducted with a 2:1 ratio of $LiAlD_4$ and $(COD)_2Ni$ in THF at 25°C, treatment with H_2O gave a 65:35 mixture of biphenyl (**39**) and 2,2′-dideuteriobiphenyl (**43**) (Scheme 13). This result showed that at least one-third of the hydrogenolysis of **40** had occurred before hydrolysis. A similar run that was worked up with D_2O gave a 35:65 mixture of **39** and **43**. Accordingly, another one-third of the biphenyl originates from a metallole (**47**, M = Ni or Al) and the last one-third of **39** must arise by hydrogen abstraction from the solvent (Scheme 13) (*14*).

The exact nature of the reagents generated by admixing varying ratios of $(COD)_2Ni$ or bipy(COD)Ni with $LiAlH_4$ remains to be ascertained. For a 2:1 ratio of $LiAlH_4$ and $(COD)_2Ni$, some unpublished work suggests the formation of a composition, $Li_2(AlH_3 \cdot Ni \cdot AlH_3)$ (*17*). However, the most effective ratio of bipy(COD)Ni and $LiAlH_4$ for hydrodesulfurizing **33**, **34**, **35**, and **36** appears to be 2:1. For the present, it seems reasonable to suggest that such combinations can involve anionic nickel or Al–Ni bonds.

Scheme 13

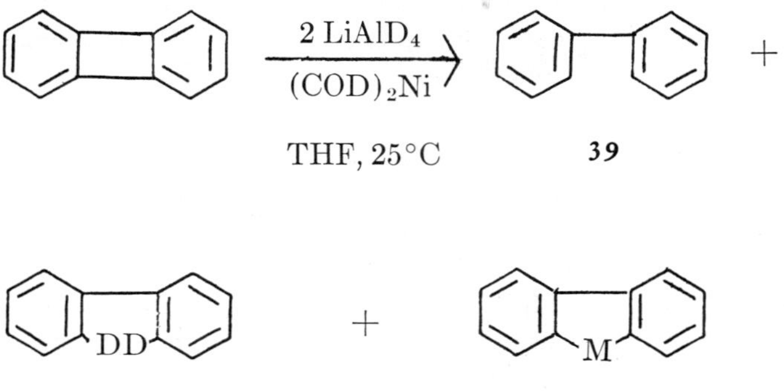

[Biphenylene] → (2 LiAlD₄ / (COD)₂Ni, THF, 25°C) → [biphenyl] **39** +

[biphenyl-DD] **43** + [dibenzofuran-M structure] **47**

Although the foregoing results unambiguously prove that some of the desulfurization of dibenzothiophene proceeds via biphenylene as an intermediate, it remains to be established whether this is the principal pathway to **39** (with **45** and **46** as precursors). Ongoing studies hope to settle this question by the desulfurization of 2,7-dimethyldibenzothiophene (**48**); if the corresponding biphenylene (**49**) were the principal precursor of the bitolyl, one might expect to obtain approximately equal amounts of the *m,m'*- and *p-p'*-bitolyls (**50** and **51**). If desulfurization would lead directly to nickeloles (cf. **45**) and biphenylylnickel derivatives (cf. **46**), then *p,p'*-bitolyl should be the principal outcome (Scheme 14).

Scheme 14

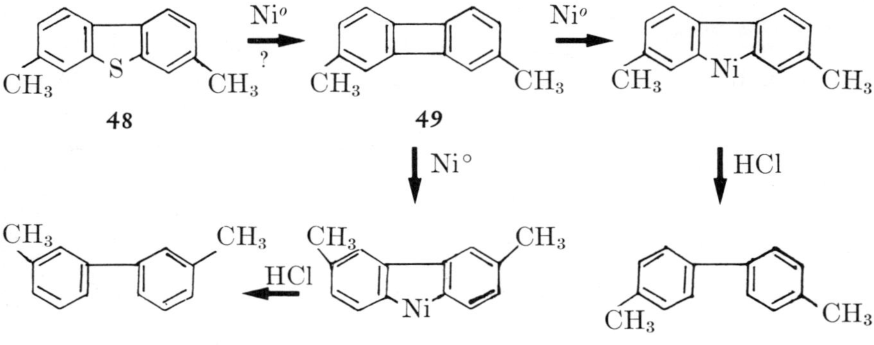

48 →(Ni°, ?) **49** →(Ni°) [methyl-nickel biphenylyl]

↓ Ni°   ↓ HCl

50 ←(HCl) [nickelole with CH₃ groups]   **51**

Prospects for Further Developments

The achievement of desulfurizing relatively stable aromatic sulfides with Ni(0) complexes under stoichiometric conditions now sets the stage for a study of catalytic processes under an atmosphere of hydrogen. An understanding of the nature of such homogeneous reagents, which are effective under mild conditions (25°–50°C under 1 atm), can guide the design of similar homogeneous catalysts. Accordingly, not only nickel complexes but also those soluble complexes of iron, cobalt, molybdenum, titanium, and tungsten merit close attention in the search for new, homogeneous desulfurizing catalysts. These metals are especially promising since most of them have been found effective as heterogeneous hydrodesulfurizing catalysts at elevated temperatures (3, 4, 5, 6, 7). A key problem to be solved in passing from the stoichiometric to the catalytic system will be the regeneration of the metal complex from the metal sulfide under mild conditions.

Even as stoichiometric reagents, such nickel complexes as bipy-(COD)Ni have already demonstrated their value in desulfurizing and ring-contracting processes, in hydrodesulfurization, in selective deoxygenation of epoxides, and in the controlled cleavage of allyl ethers. Their ability to form metallocyclopentadienes from biphenylenes or from thiophenes can develop into a valuable alternative route to such reactive metal heterocycles.

Finally, the chemistry uncovered in studying the mode of action of these Ni(0) reagents lends strong support to an electron-transfer view of their reactions. In addition, the finding that dibenzothiophene divests itself of sulfur, at least partially, through the relatively strained biphenylene and the reactive dibenzonickelole intermediates, is unusual. It demonstrates that nature is often more tortuous in its reaction paths than the human mind can readily imagine.

Acknowledgment

The authors wish to acknowledge support of various aspects of this research by the National Science Foundation (CHE-76-10119) and the Energy Research and Development Administration of the Department of Energy (EF-77-G-01-2739).

Literature Cited

1. Forster, D., Roth, J. F., Eds., "Homogeneous Catalysis—II," ADV. CHEM. SER. (1974) **132**.
2. Trost, B. M., Salzmann, T. N., *J. Am. Chem. Soc.* (1973) **95**, 6840.
3. Wu, W. R. K., Storch, H. H., *Bur. Mines Bull.* (1968) **633**.

4. Ivanovskii, F. P., Beskova, G. S., Dontsova, V. A., *Khim. Prom.* (1966) **42**, 845 (*Chem. Abstr.* (1967) **66**, 20788).
5. Novak, V., Cirova, A., *Ropa Uhlie* (1971) 299 (*Chem. Abstr.* (1971) **75**, 91563).
6. Ishiguro, T., French Pat. **1,466,905** (*Chem. Abstr.* (1976) **67**, 75131).
7. vanKlinken, J., Kouwenhoven, H. W., Van Weeren, P. A., German Pat. **1,951,007** (*Chem. Abstr.* (1971) **73**, 27303).
8. Eisch, J. J., Damasevitz, G. A., *J. Organomet. Chem.* (1975) **96**, c19.
9. Eisch, J. J., Galle, J. E., *J. Organomet. Chem.* (1975) **96**, c23.
10. Collman, J. P., *Acc. Chem. Res.* (1968) **1**, 141.
11. Eisch, J. J., Im, K. R., *J. Organomet. Chem.* (1977) **139**, c45.
12. Dinjus, E., Gorski, I., Uhlig, E., Walther, H., *Z. Anorg. Allg. Chem.* (1976) **422**, 75.
13. Takahaski, S., Suzuki, Y., Sonogashira, K., Hagihara, N., *J. Chem. Soc., Chem. Commun.* (1976) 839.
14. Eisch, J. J., Im, K. R., unpublished studies (1977).
15. Curtin, D. Y., Hurwitz, M. J., *J. Am. Chem. Soc.* (1952) **74**, 5381.
16. Eisch, J. J., Im, K. R., *J. Organomet. Chem.* (1977) **139**, c51.
17. House, G., Wilke, G., unpublished studies, cited in Jolly, P. W., Wilke, G., "The Organic Chemistry of Nickel," Vol. I, pp. 261–262, Academic, New York, 1974.

RECEIVED March 3, 1978.

18

Gaseous Evolution of Molecular Hydrogen and Oxygen in Photochemical Splitting of Water by Platinized Chlorophyll a Dihydrate Polycrystals

Laboratory Simulation of the Primary Light Reaction in Plant Photosynthesis

L. GALLOWAY, D. R. FRUGE, and F. K. FONG

Department of Chemistry, Purdue University, West Lafayette IN 47907

> The proposal that $(Chl\ a \cdot 2H_2O)_2$ is the photoreaction center Chl a aggregate for the water splitting reaction in plant photosynthesis led us to the belief that chlorophyll dihydrate polycrystals can be used in efficient water photolysis in vitro. In this chapter, the observation of gaseous evolution under white light illumination of platinized chlorophyll dihydrate polycrystals is described. The photoelectrolytic products were determined to be molecular hydrogen and oxygen by diffusion pyrolysis and mass spectrometry. The photochemical activity of $(Chl\ a \cdot 2H_2O)_n$ is attributed to the presence of water in Chl a aggregation via the C9 keto C = O $\cdots$ H(H)O $\cdots$ Mg bonding interaction. The demonstration of the decomposition of water by $(Chl\ a \cdot 2H_2O)_n$ lends support to the suggestion that a single photosystem in plant photosynthesis may be capable of splitting water in vivo.

Recently we reported Chl a photogalvanic effects attributable to water splitting reactions that result from illumination of the chlorophyll a dihydrate aggregate $(Chl\ a \cdot 2H_2O)_n$ *(1)*. The photooxidation of $(Chl\ a \cdot H_2O)_{n \geq 2}$ by water was subsequently observed in ESR experiments

(2). However, the quantum efficiency of the observed effects was low, and we were unable to detect the discharge of hydrogen by direct analytical means. In this chapter, we describe the conditions under which we were successful in bringing about gaseous hydrogen evolution attributable to water splitting in a photoelectrolytic cell containing platinized chlorophyll polycrystals (Chl a · $2H_2O)_n$. The demonstration of water photoelectrolysis by the chlorophyll is of interest because of the role of Chl a in plant photosynthesis. It is also topical in view of the current search for a direct process for harvesting solar energy to produce gaseous hydrogen for fuel. Considerable attention has been focused on n-type semiconducting photoanodes such as TiO_2 and $SrTiO_3$ (3, 4, 5). However, these materials operate in the near UV wavelength region where the solar radiant energy density is low. In contrast, the action spectrum of the photoreactivity of (Chl a · $2H_2O)_n$ with water spans the visible and far red wavelength regions (1). The polycrystals of chlorophyll a dihydrate thus became our prime target for investigation.

In our earlier photochemical conversion experiments, the chlorophyll was plated on a shiny platinum electrode (1). The photolytic reactions were detected by measuring the electron flow in an external circuit of a liquid junction photovoltaic cell consisting of the Pt–Chl a half cell and a Chl a-free half cell. It occurred to us that the quantum efficiency of this assembly can be limited by the poor contact between the smooth metal surface and the chlorophyll. It appears that only those Chl a molecules in direct contact with platinum are effectively engaged in the photochemical process. In this context, Jacki Roettger of this laboratory has demonstrated recently that an increase in the Chl a thickness of the Pt–Chl a electrode has no systematic effect on the cell performance. We accordingly sought to overcome the problem of low quantum efficiency by filling in the crevices that separate the polycrystalline Chl a aggregates from each other and from the smooth metal electrode surface by finely divided platinum particles.

Preparation of Platinized (Chl a · $2H_2O)_n$ Electrode. The Effect of Oxygen on the Photovoltaic Activity

A shiny Pt foil was platinized by passing a 30 mA current for 10 min through a $7 \times 10^{-2}M$ chloroplatinic acid solution containing $6 \times 10^{-4}M$ lead acetate. A layer of polycrystalline chlorophyll, containing 1.5×10^{17} Chl a molecules, was deposited on the platinized electrode surface, using the procedure described by Tang and Albrecht (6). The Chl a-plated electrode then was platinized again in the same chloroplatinic acid solution, except that the 30 mA current was passed for only 15 seconds.

The action spectra of the photovoltaic response (*1*) of the platinized Chl a electrode at pH = 7, measured in a cell (*see* Figure 1) using as the second half-cell a platinized electrode not covered with Chl a, are given in Figure 2. The 740-nm maximum of the spectral response shown in Figure 2 confirms that $(Chl\ a \cdot 2H_2O)_n$ (*7, 8*) is primarily responsible for the observed photovoltaic effects. Under an argon atmosphere the observed response of the Chl a cell is photocathodic. A remarkable change was observed when the electrolyte solution was saturated with oxygen. The photogalvanic current reversed in sign. On purging the oxygen-saturated solution with argon, the original photocathodic response was restored.

To enhance the photovoltaic response, the pH values (*1*) of the Chl a and Chl a-free half cells were maintained at 3 and 11, respectively, in another experiment. After the half cells were degassed by the passage

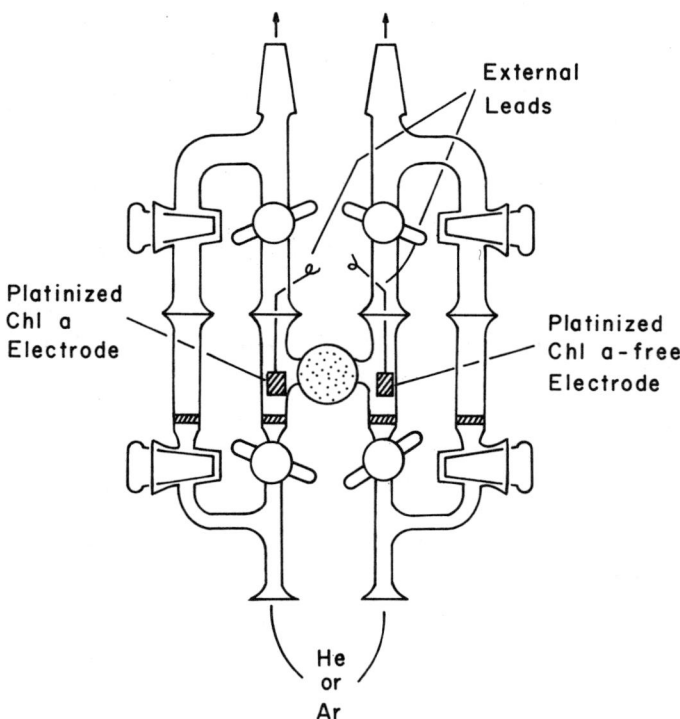

Figure 1. *The platinized Chl a cell. The Chl a-free electrode is used as a half cell in a liquid-junction photovoltaic cell. In photolytic reaction, only the platinized Chl a electrode is used in the production of molecular hydrogen and oxygen from water.*

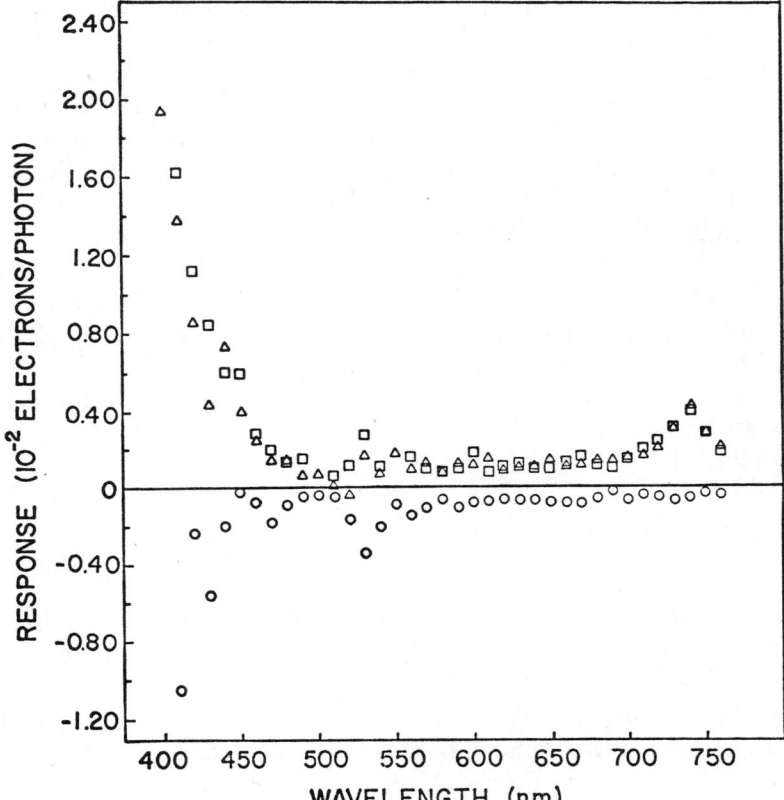

Figure 2. Effect of oxygen in the photovoltaic response of the platinized Chl a electrode. (a) △, Electrolyte purged with argon gas, measurements made under a positive pressure of Ar; (b) ○, electrolyte saturated with oxygen; (c) □, oxygen removed by passage of argon gas through the electrolyte solution for 30 min. The ordinate units are given in 10^{-2} electrons per incident photon. The efficiency of the liquid junction cell was calculated by dividing the photovoltaic response (electrons sec^{-1}) by the incident photoresponse (photons sec^{-1}).

of argon gas for about 30 minutes, the photovoltaic response of the cell was monitored with the entire output from the 1000 W tungsten-halogen lamp focused on the platinized Chl a electrode. An initial photocurrent of about 1.5 μA was obtained. After two hours of continuous illumination, a reversal in sign of the photocurrent was observed, possibly indicative of a buildup in the oxygen content of the electrolyte solution.

The power of the radiation incident on the sample was determined, using a Laser Precision radiometer, to be 1.7 W. Approximately one half of this power occurs beyond 800 nm, out of the reach of Chl a photochemistry. The observed photocurrent, 1.5 μA, amounts to a conversion efficiency of about 10^{-6}, assuming that the average energy of the photochemically active photons is 2 V. The peak monochromatic quantum efficiency at 740 nm is 4×10^{-3} (see Figure 2). In the "green gap" where the Chl a absorption is weak, the average monochromatic conversion efficiency is about 8×10^{-4}. The average quantum efficiency using monochromatic light is thus three orders of magnitude greater than that obtained using white light. It was found (10) that at an incident flux corresponding to the 1.7 W incident power, the photoresponse of the Pt–Chl a cell observes a semilinear flux dependence. The monochromatic incident fluxes used in the experiment represented in Figure 2 were $\lesssim 10^{14}$ photons $\sec^{-1} cm^{-2}$, at which the photoresponse varies linearly with flux. It is evident from Figure 2 of Ref. 10 that a linear extension of the low-flux photoresponse data at the maximum available power of the source would extrapolate to a quantum efficiency about 10^3 times higher than that observed.

Gaseous Evolution of Molecular Hydrogen and Oxygen. Mass Spectrometric and Pyrolytic Analyses

Qualitative Observations. Continued illumination of the cell with an open external circuit led to the observation of gaseous evolution from the Pt–Chl a electrode, notably the formation of gas bubbles in the illuminated area followed instantaneously the repositioning of the illuminated area. To eliminate the possibility that the gas bubbles may have resulted from the degassing of argon because of heating by the light source, the cell was purged with helium and the experiment was repeated under a positive pressure of helium. The solubility of helium in water increases with increasing temperature, being 0.94 and 1.21 mL/100 mL water at 25° and 75°C, respectively. Gaseous evolution was again observed under identical illumination conditions. When the Chl a-free platinized platinum electrode was illuminated under these conditions, no signs of bubbling were detected.

Mass Spectrometric Analysis. After the platinized Chl a electrode in the cell shown in Figure 1 was illuminated for 30 minutes, the gaseous content over the electrolyte solution was evacuated directly into the sample chamber of a Consolidated Electrodynamics Corp. 21-110-B mass spectrometer. The resulting mass spectrum is shown in Figure 3. In addition to the expected He^+ line at mass 4, a strong peak at mass 2 with an attendant trace peak at mass 3 was observed. The latter peaks are respectively attributed to H_2^+ and to the triatomic ion H_3^+ (11, 12). These identifications were confirmed by using pure hydrogen and helium as source. The gaseous content above the electrolyte solution in the Chl a-free cell also was analyzed by mass spectrometry after a similar light

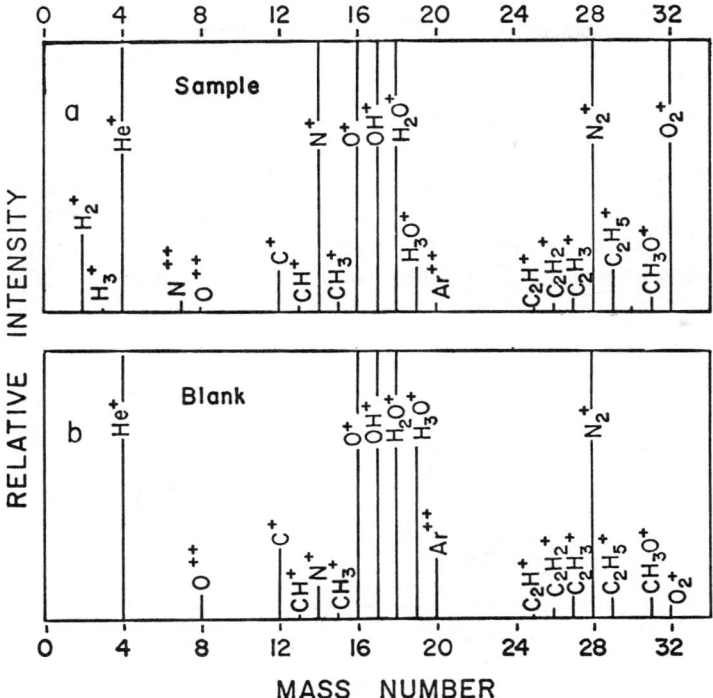

Figure 3. Mass spectrometric determination of the gaseous products of the photoelectrolytic process. (a) Gaseous sample collected from illuminated platinized Chl a cell; (b) gaseous content over a similarly irradiated, platinized electrode in the absence of Chl a. Molecular hydrogen and oxygen, the two main products of the water splitting reaction, are readily manifested by the lines observed at masses 2 and 32 in (a). Helium gas was used to purge the cell prior to the light reaction. See text for a more detailed analysis.

treatment of the Chl a-free platinized electrode. No lines at masses 2 and 3 were detected (see comparison in Figure 3). Small quantities of H_2^+ are known to accompany hydrocarbon fragments at masses 13, 15, 25, 26, 27, and 29. These fragments are found to occur in similar intensity ratios in both the sample and blank determinations shown in Figure 3. The possibility that the observed H_2^+ line may have originated from hydrocarbon fragmentation is thus ruled out. The intense mass 32 line in the sample spectrum is attributed to air leakage through the external leads admitted into the sample cell via epoxied metal–glass joints for the measurement of the photovoltaic response. By eliminating these joints we were able to reduce the mass 32 line to a level at which a quantitative evaluation of the oxygen evolved can be made in terms of a statistical distribution of isotopic oxygen (13).

Mass spectrometric analyses of the photolytic products from various mixtures of $H_2^{16}O$, $D_2^{16}O$, and $H_2^{18}O$ are compared with the corresponding mass spectra of electrolyzed samples containing identical isotopic mixtures in Figures 4, 5, and 6. The results of an experiment using 1:1 D_2O-H_2O are shown in Figure 4. In this experiment argon gas was used to purge the cell prior to the light reaction. If we assume that the observation of the mass 4 (D_2^+) line has resulted from water splitting and use the mass 20 (D_2O^+) line as an internal reference for calibration, we estimate, from a comparison of the relative intensity ratios of lines at masses 20 (D_2O^+)

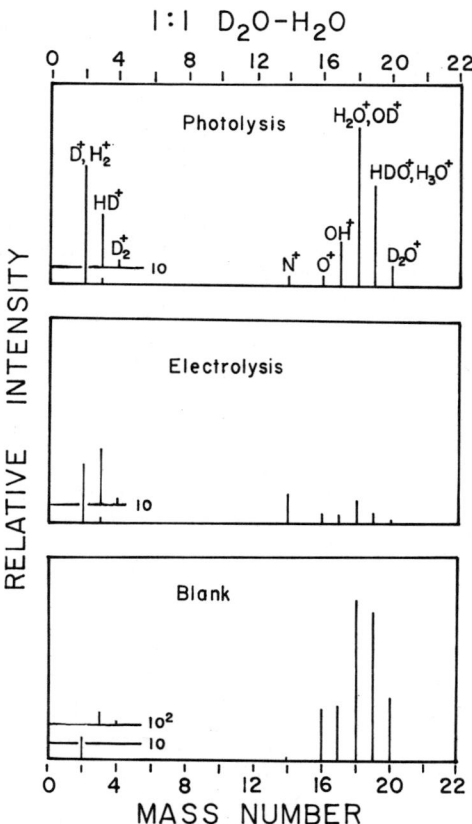

Figure 4. Comparison of the mass spectrometric determination of the photolytic products of 1:1 D_2O–H_2O with that of conventional electrolysis. Under comparable instrumental settings, the lines at masses 3 and 4 are not observed in the blank. Argon gas was used to purge the cell prior to the light reaction. Note changes in scale in the low-mass region.

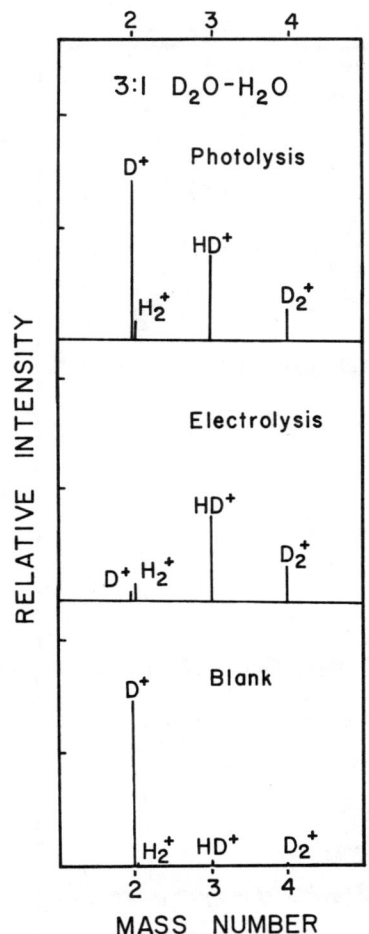

Figure 5. Comparison of the mass spectra of the products of the photolysis and electrolysis of 3:1 $D_2O–H_2O$. Argon gas was used to purge the cell prior to the light reaction.

and 4 (D_2^+) observed in the photolytic and electrolytic runs, that water photolysis occurs at a rate of 9×10^{-6} mol hr^{-1}, corresponding to a gaseous (hydrogen and oxygen) evolution rate of 0.3 mL hr^{-1}. A significant line at mass 2 is present in the blank. We attribute this to the fragmentation of D_2O to D^+ and OD^+, which is consistent with the observation of water fragments at masses 17, 18, and 19 in the photolytic, electrolytic, and blank runs. In the blank, the lines at masses 3 and 4 are not observed under settings similar to those of the photolytic and electrolytic experiments. We were unable to detect the H^+ signal at mass 1 on account of instrumental limitations.

The above considerations suggested the use of higher instrumental resolution so that the interference at mass 2 from D_2O fragmentation could be differentiated from the H_2^+ line. The results of a higher resolu-

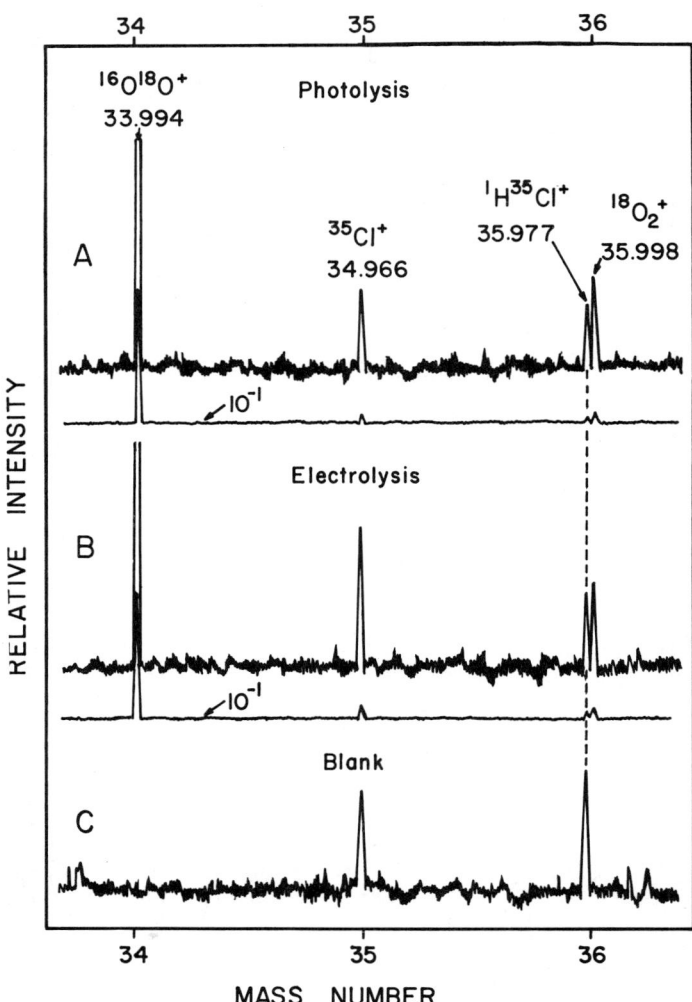

Figure 6. Comparison of mass spectrometric analyses of the photolytic and electrolytic products using 5:1 $H_2^{16}O-H_2^{18}O$. The occurrence of the $^{35}Cl^+$ and $^1H^{35}Cl^+$ lines is attributable to the presence of KCl in the aqueous electrolyte. Helium gas was used to purge the cell prior to the light reaction.

tion determination of the molecular species H_2^+, HD^+, and D_2^+ obtained in the photolysis and electrolysis 3:1 D_2O-H_2O are shown in Figure 5. The D^+ line was significantly reduced on freezing the sample prior to the evacuation of the gaseous contents of the cell into the mass spectrometer. This observation corroborates the supposition that the D^+ line was observed as a result of water fragmentation. In a separate experiment, in which helium gas was used to purge the cell prior to the light reaction, the presence of molecular oxygen in the photochemical splitting of water is ascertained by using 5:1 $H_2^{16}O$–$H_2^{18}O$ (*see* Figure 6). The occurrence of the lines at masses 34 ($^{16}O^{18}O^+$) and 35.998 ($^{18}O_2^+$) compares well with the corresponding lines observed for the electrolytic sample. Both lines are absent in a blank run in which the Chl a-free electrode in Figure 1 was treated under conditions identical to those of the photolytic experiment.

Quantitative Determination by Diffusion Pyrolysis. In an attempt to quantify the rate of the photochemical water-splitting process, we designed a pyrolytic apparatus, shown schematically in Figure 7, that allowed us to determine the quantity of the hydrogen gas produced by photolysis by burning it in a stream of oxygen gas carried by gaseous helium. The level of background oxygen in the helium flow was registered as a current using the Hersch oxygen indicator (*15*). Gaseous helium was admitted through the entire apparatus at a constant flow rate until the oxygen current reached a steady-state value. At this point the valves, denoted by open circles in Figure 7, were adjusted so that the helium flow was directed along a pathway (marked by arrows in Figure 7) that bypassed the sample cell.

With both valves above and below the platinized Chl a electrode closed, the photolytic reaction was allowed to carry on for a known duration of time. After a brief period (~ 1 min) of cooling, the valve at (a) was opened. The distance between (a) and (b) (at which point the gas line reached the hydrogen and oxygen reactor) was 45 cm. The hydrogen and oxygen produced by water photolysis diffused along the

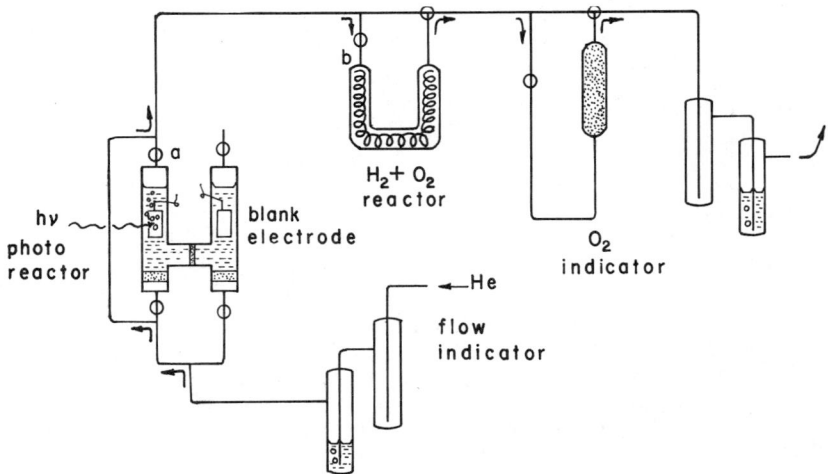

Figure 7. The diffusion pyrolysis apparatus. The arrows indicate the direction of helium flow during operation.

helium flow. Molecular hydrogen, being 16 times lighter than oxygen, arrived at (b) approximately four times as quickly as the photolytic oxygen, causing a reduction in the background oxygen current level as it was burned by the platinum heating coil in the hydrogen and oxygen reactor (*see* Figure 8). On depletion of hydrogen through combustion, the oxygen current level was restored to the steady-state background level. Subsequently, the slower diffusing oxygen from the platinized Chl a cell reached the oxygen indicator, resulting in an increase in the oxygen current.

In Figure 8, the pyrolytic analyses of two photolysis experiments in which the photochemical reaction was carried out for five and 15 minutes (respectively shown in Figures 8a and 8b) are compared with a corre-

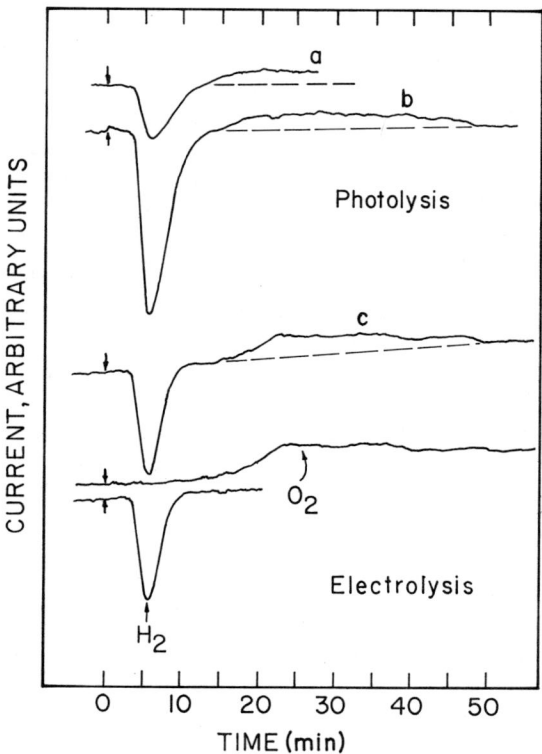

Figure 8. Comparison of pyrolytic analyses of the photolysis and electrolysis of water at pH = 3. (a,b) Illumination of the Pt–Chl a sample with the entire output of a 1000 W tungsten-halogen lamp for 5 and 15 min, respectively. (c) Lower, hydrogen generated by passing 10 mA for 20 sec; middle, oxygen generated by passing a current of 10 mA for 20 sec; upper, sum of middle and lower curves.

sponding analysis of hydrogen and oxygen generated by electrolysis (Figure 8c) in which the Chl a-free (blank) half cell in Figure 1 acted as a counter electrode. First, the blank electrode was used as the anode in the electrolysis, so that hydrogen was produced in the absence of light in the enclosed sample cell with the Pt–Chl a electrode operating as the cathode. After a 10 mA current was passed for 20 seconds, the sample cell was opened so that the hydrogen gas was admitted into the helium flow toward the hydrogen and oxygen reactor. The expected reduction in the oxygen current was observed after a time lag similar to that observed in the photolytic experiments.

On completion of the electrolytic hydrogen measurement, the sample cell was flushed with helium to get rid of any hydrogen that remained in the electrolyte. With the blank electrode operating as a cathode and the Pt–Chl a electrode as the anode, the entire procedure described in the preceding paragraph was repeated for analyzing the oxygen produced by electrolysis in the sample cell. The summation of these two analyses resulted in the current–time plot given in Figure 8c, which is compared with the corresponding plots obtained in the photolytic experiments. The results shown in Figure 8 were obtained at a helium flow rate of 28 bubbles per minute through 0.25 in. i.d. glass tubings in the flow indicator. The observed $H_2:O_2$ ratio appears to be greater in the product from photolysis (Figure 8a, 8b) than from electrolysis (Figure 8c). In extended work completed after the oral presentation of this paper, we obtained evidence that platinum itself can be responsible for the description of water in a thermal reaction of the platinum with water, in which the products are hydrogen and some oxidation compound of platinum (14).

Discussion

Our model for the Chl a–H_2O photoelectrolytic cell was based (16) on the p-type semiconductor properties (17, 18) of (Chl a · $2H_2O)_n$ (8, 9). The experimental behavior described above can be explained by considering the polycrystalline chlorophyll as a semiconducting cathode and the finely dispersed platinum particles as a metallic anode immersed in an acidic aqueous electrolyte according to the description of Wrighton (4), Mavroides (5), Ohnishi (19), and their co-workers. In dark equilibration, there is an energy band bending at the Chl a surface, producing a Schottky barrier at the Chl a–electrolyte interface. Electron-hole pairs generated at the semiconductor surface by light absorption across the band gap leads to a separation by the electric field of the barrier, eliminating wasteful recombination back reactions. Electrons are transferred from the cathode surface to the H^+/H_2 free energy level of the aqueous phase, liberating hydrogen in the reaction:

$$2\text{Chl a} + 2H_2O \rightarrow 2\text{Chl a}^+ + H_2 + 2OH^-$$

The holes move into the bulk of $(\text{Chl a} \cdot 2H_2O)_2$ and are then transmitted through the platinum anode, to the H_2O/O_2 level of the electrolyte, giving off oxygen in the reaction:

$$2\text{Chl a}^+ + H_2O \rightarrow 2\text{Chl a} + \frac{1}{2}O_2 + 2H^+$$

The photooxidation of $(\text{Chl a} \cdot 2H_2O)_{n>2}$ by water recently has been established by ESR observations (2).

It has been noted recently (2, 7, 16) that among known aggregates of chlorophyll (8, 9, 20–28) $(\text{Chl a} \cdot H_2O)_2$ (A) and $(\text{Chl a} \cdot 2H_2O)_{n \geq 2}$ (B) are distinguished by their ability to be photooxidized in the presence of water, giving rise to well-characterized photovoltaic behavior (1, 7, 16). Significantly, it was suggested that the photooxidized dihydrate aggregate, $(\text{Chl a} \cdot 2H_2O)_n^+$, not $(\text{Chl a} \cdot H_2O)_2^+$, is sufficiently strong an oxidant to be readily reduced by water (2). The photochemical activity of Chl a–H_2O aggregates has been attributed to photoactivated proton shifts (31, 32) between the magnesium-bound water molecule and the carbonyl group to which it is hydrogen bonded, resulting in the acquisition by the magnesium atom of a negative charge (31, 32). The magnesium atom being conjugated to ring V, the negative charge it acquires during photoactivation is expected to be readily transmitted via π-electron resonance to the C9 keto group, which is presumably the site at which the electron leaves both aggregates (A) and (B) during the primary light reaction. We can thus rationalize the apparently symbiotic roles of the magnesium atom and the ring V cyclopentanone ring in the photochemical activity of (A) and (B). We note that the charge transfer mechanism described here appears to be feasible only in (A) and (B). In C_2 Chl a dimers in which the $C=O \cdots H(H)O \cdots Mg$ interactions involve the C9 keto group (22, 23, 26, 27, 28) as in the dimer of Chl a polyhydrate (26) or in covalently linked dimers (22, 23) in which the C10 $C=O \cdots H(H)-O \cdots Mg$ interactions are sterically forbidden (26, 27), delocalization of the negative charge acquired by the magnesium atom in photoactivation to the C9 keto carbonyl is expected to be stabilized by the water proton to which the carbonyl group is hydrogen bonded. It has been reported that, unlike $(\text{Chl a} \cdot H_2O)_2$ and $(\text{Chl a} \cdot 2H_2O)_n$, the C9-linked C_2 symmetrical dimer can only be photooxidized in the presence of an added electron acceptor, such as tetranitromethane (28).

The present demonstration of the photochemical splitting of water by $(Chl\ a \cdot 2H_2O)_n$ lends support to the recently proposed photosynthesis model in which a single photosystem may be capable of the in vivo water splitting reaction via a two-photon activation mechanism (*1*). It was at one time commonly believed that two Chl a photosystems are required in bringing about water splitting according to the so-called "series scheme" or "Z scheme" of photosynthesis. Attempts to achieve photochemical splitting of water in vitro based (see, for example, Ref. *33*) on the series scheme have not thus far been brought to fruition.

Current efforts in this laboratory are concerned with determining the quantum requirement for the in vitro water-splitting reaction and the power efficiency of the photochemical conversion. The effects of oxygen on the photovoltaic activity of the platinized Chl a electrode shown in Figure 2 can be related to similar effects observed earlier (*1*). Molecular oxygen appears to exert an inhibitive role in Chl a · H_2O light reactions. This role can be responsible, in part at least, for the observed acceleration of the apparent photolytic rate at temperatures at which oxygen is insoluble in water. It is known that the manganese ion plays a part in the evolution of oxygen in plant photosynthesis (*34*). It may be possible to find the in vitro analog for the in vivo manganese ion effect. It appears reasonable to suppose that success in achieving highly efficient photochemical splitting of water ultimately can depend on the extent to which we are capable of understanding and duplicating the in vivo light reaction in the laboratory.

Acknowledgment

This research was supported by the Alcoa Foundation and by the National Science Foundation. Assistance by R. G. Cooks and A. B. Coddington in the measurement of mass spectra is gratefully acknowledged.

Literature Cited

1. Fong, F. K., Polles, J. S., Galloway, L., Fruge, D. R., *J. Am. Chem. Soc.* (1977) **99**, 5802.
2. Fong, F. K., Hoff, A. J., Brinkman, F. A., *J. Am. Chem. Soc.* (1978) **100**, 619.
3. Fujishima, A., Honda, K., *Nature* (1972) **238**, 37.
4. Wrighton, M. S., Wolczansdi, P. T., Ellis, A. B., *J. Solid State Chem.* (1977) **22**, 17.
5. Mavroides, J. G., Katalas, J. A., Kolesar, D. F., *Appl. Phys. Lett.* (1976) **28**, 241.
6. Tang, C. W., Albrecht, A. C., *Mol. Cryst. Liq. Cryst.* (1974) **25**, 53.
7. Fetterman, L. M., Galloway, L., Winograd, N., Fong, F. K., *J. Am. Chem. Soc.* (1977) **99**, 653.
8. Fong, F. K., Koester, V. J., *J. Am. Chem. Soc.* (1975) **97**, 6888.
9. Fong, F. K., Koester, V. J., *Biochim. Biophys. Acta* (1976) **423**, 52.

10. Galloway, L., Roettger, J. D., Fruge, D. R., Fong, F. K., *J. Am. Chem. Soc.* (1978) **100**, 4635.
11. Beynon, J. H., Cooks, R. G., Caprioli, R. M., *Proc. R. Soc. London, Ser. A* (1972) **327**, 1.
12. Papp, N., Kerwin, L., *Phys. Rev. Lett.* (1969) **22**, 1343.
13. Fong, F. K., Galloway, L., *J. Am. Chem. Soc.* (1978) **100**, 3594.
14. Galloway, L., Fruge, D. R., Fong, F. K., *J. Am. Chem. Soc.* unpublished data.
15. Hersch, P., *Proc. Int. Symp. Microchem. 1958* (1959) 141–150.
16. Fong, F. K., Winograd, N., *J. Am. Chem. Soc.* (1976) **98**, 2287.
17. Tang, C. W., Albrecht, A. C., *J. Chem. Phys.* (1975) **62**, 2139.
18. Ibid. (1975) **63**, 953.
19. Ohnishi, T., Nakato, Y., Tsubomura, H., *Ber. Bunsenges. Phys. Chem.* (1975) **79**, 523.
20. Strouse, C. E., *Proc. Nat. Acad. Sci. USA* (1974) **71**, 325.
21. Chow, H.-C., Serlin, R., Strouse, C. E., *J. Am. Chem. Soc.* (1975) **97**, 7230.
22. Boxer, S. G., Closs, G. L., *J. Am. Chem. Soc.* (1976) **98**, 5406.
23. Boxer, S. G., Ph.D. dissertation, Department of Chemistry, University of Chicago (1976).
24. Winograd, N., Shepard, A., Karweik, D. H., Koester, V. J., Fong, F. K., *J. Am. Chem. Soc.* (1976) **98**, 2369.
25. Fong, F. K., Koester, V. J., Polles, J. S., *J. Am. Chem. Soc.* (1976) **98**, 6406.
26. Fong, F. K., Koester, V. J., Galloway, L., *J. Am. Chem. Soc.* (1977) **99**, 2372.
27. Shipman, L. L., Cotton, T. M., Norris, J. R., Katz, J. J., *Proc. Nat. Acad. Sci. USA* (1976) **73**, 1791.
28. Wasielewski, M. R., Studier, M. H., Katz, J. J., *Proc. Nat. Acad. Sci. USA* (1976) **73**, 4282.
29. Cotton, T. M., Ph.D. dissertation, Department of Chemistry, Northwestern University (1976).
30. Clarke, R. H., Hobart, D. R., *FEBS Lett.* (1977) **82**, 155.
31. Fong, F. K., *Proc. Nat. Acad. Sci. USA* (1974) **71**, 3692.
32. Fong, F. K., *Appl. Phys.* (1975) **6**, 151.
33. Bolton, J. R., *J. Solid State Chem.* (1977) **22**, 4.
34. Cheniae, G. M., *Annu. Rev. Plant Physiol.* (1970) **21**, 467.

RECEIVED February 22, 1978.

19

Further Studies of the Spectroscopic Properties and Photochemistry of Binuclear Rhodium(I) Isocyanide Complexes

KENT R. MANN and HARRY B. GRAY

Arthur Amos Noyes Laboratory of Chemical Physics, California Institute of Technology, Pasadena, CA 91125

The position of the lowest allowed electronic transition ($1a_{2u} \rightarrow 2a_{1g}$ in D_{4h}) in $Rh_2L_4^{2+}$ (L = binucleating isocyanide) depends on the rotameric configuration of the complex [553 nm in $Rh_2(bridge)_4^{2+}$ (eclipsed); 515 nm in $Rh_2(TM4\text{-}bridge)_4^{2+}$ (30° staggered)]; the Rh(I)–Rh(I) distances are 3.26 and 3.25 Å, respectively. Reaction of $Rh_2(bridge)_4^{2+}$ (or Rh_2^{2+}) with H_2 in dilute aqueous acidic solutions yields $Rh_4H_2^{4+}$ ($\lambda_{max} = 780$ nm). The $Rh_4H_2^{4+}$ species also is produced by low-temperature ($-60°C$) irradiation ($\lambda > 520$ nm) of 6M HCl solution of HRh_2Cl^{2+} ($\lambda_{max} = 578$ nm). The possible role of $Rh_4H_2^{4+}$ as an intermediate in the H_2-producing photoreaction of HRh_2Cl^{2+} ($HRh_2Cl^{2+} \xrightarrow[12M\ HCl]{546\ nm} Rh_2Cl_2^{2+} + H_2$) is discussed.

Our interest in the electronic structures and spectroscopic properties of square planar complexes of d^8 ions with π-acceptor ligands dates back 15 years (1). The work reported in this chapter grew out of studies of the spectroscopic properties of Rh(I) isocyanides which we initiated in 1973. One discovery we made early in these studies was that Rh(CNR)$_4^+$ complexes aggregate in solution, yielding discrete binuclear, trinuclear, and even higher oligomers (2, 3). We became intrigued with the spectroscopic properties of these oligomeric Rh(I) species and decided to prepare a particular binuclear complex, Rh$_2$(bridge)$_4^{2+}$ (bridge = 1,3-diisocyanopropane), for detailed study (4). Both the spectroscopic properties and the redox chemistry of $Rh_2(bridge)_4^{2+}$ turned out to be interesting (4, 5, 6, 7), and we have continued our investigations of this system as described herein.

Structural and Spectroscopic Properties

The prototypal complex, $Rh_2(bridge)_4^{2+}$, and similar complexes are synthesized easily by slow addition of the binucleating isocyanide ligand to a dichloromethane solution of $[Rh(COD)Cl]_2$ to give, in the case of

$$[Rh(COD)Cl]_2 + 4\ bridge \rightarrow Rh_2(bridge)_4 Cl_2 + 2COD \qquad (1)$$

bridge, a dark blue precipitate, $Rh_2(bridge)_4 Cl_2$. Other salts can be made by metathetical reactions. The x-ray crystal structure analysis of $Rh_2(bridge)_4(BPh_4)_2$ revealed the presence of discrete $Rh(bridge)_4^{2+}$ cations and BPh_4^- anions (8). A view of the structure of $Rh_2(bridge)_4^{2+}$ is shown in Figure 1. Each RhC_4 unit is very closely square planar, and the ligand bond angles and bond lengths are quite close to values found for other Rh(I) isocyanide complexes (3). The Rh–Rh distance is 3.264 Å.

The close contact of the two planar $Rh(CNR)_4^+$ units results in some very interesting spectroscopic properties. The electronic absorptions of lowest energy can be interpreted satisfactorily using the molecular orbital

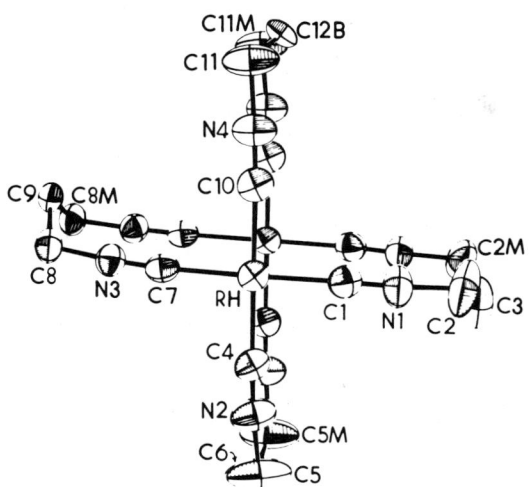

Figure 1. View of the structure of $Rh_2(bridge)_4^{2+}$; the Rh–Rh distance is 3.264 Å (from a crystal structure analysis of $Rh_2(bridge)_4(BPh_4)_2$) (7)

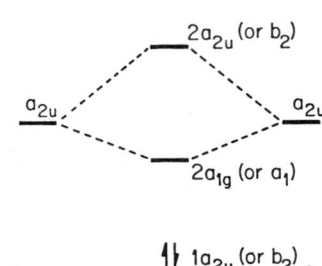
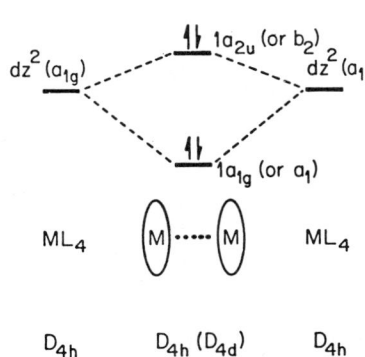

Figure 2. Relative energies of selected molecular orbitals in binuclear Rh(I) complexes of D_{4h} or D_{4d} symmetry (2)

energy level diagram shown in Figure 2. The strongest intermonomer electronic interactions involve the orbitals that extend perpendicular to the molecular plane, namely, $a_{1g}(d_{z^2})$ and $a_{2u}[p_z, \pi^*(CNR)]$.

In the monomeric units a_{1g} is the HOMO and a_{2u} the LUMO. Interaction of the two a_{1g} orbitals causes their splitting into a bonding ($1a_{1g}$) and antibonding ($1a_{2u}$) set. Similar splitting of the upper set occurs to give bonding ($2a_{1g}$) and antibonding ($2a_{2u}$) orbitals. In both cases the upper and lower sets contain orbitals of the same symmetry, allowing considerable mixing that leads to stabilization of the lower set ($1a_{1g}$, $1a_{2u}$) relative to the upper set ($2a_{1g}$, $2a_{2u}$). Since the lower set is filled, a partial bonding interaction exists between the two $Rh(CNR)_4^+$ fragments.

The lowest allowed electronic transition in the binuclear complex is $1a_{2u} \rightarrow 2a_{1g}$, which is predicted to be at substantially lower energy than the corresponding transition in $Rh(CNR)_4^+$ monomers. Electronic absorption and emission spectra for $Rh_2(bridge)_4^{2+}$ in CH_3CN solution are shown in Figure 3. The intense ($\epsilon = 14{,}500$) band at 553 nm in the absorption spectrum is assigned $^1A_{1g} \rightarrow {}^1A_{2u}$ ($1a_{2u} \rightarrow 2a_{1g}$), and the 656 nm emission peak is $^1A_{2u} \rightarrow {}^1A_{1g}$ ($\tau \leqslant 2$ nsec).

Spectral data for several binuclear Rh(I) complexes are given in Table I. The data show that lengthening the carbon chain in the ligand back bone shifts the band to higher energy. Our recent determination (8) of the structure of $Rh_2(TM4\text{-}bridge)_4^{2+}$ (TM4-bridge = 2,5-dimethyl-2,5-diisocyanohexane) shows that the rotameric configuration of the

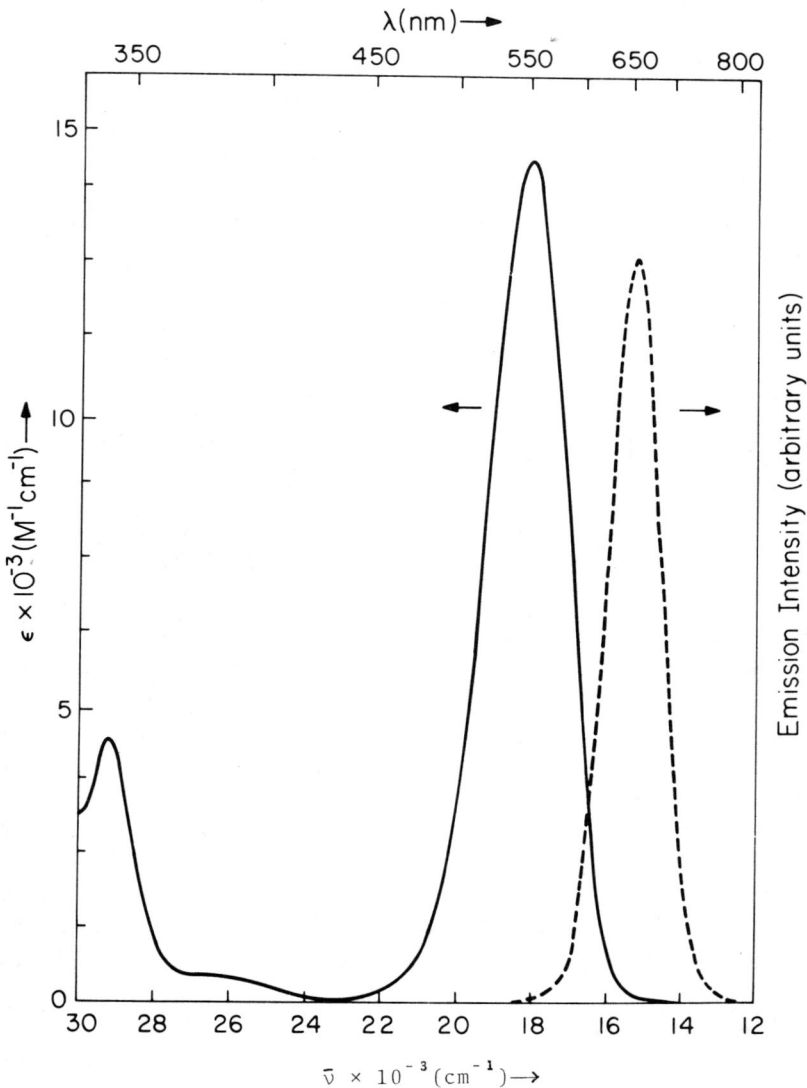

Figure 3. Absorption (———) and emission (- - -) spectra of $Rh_2(bridge)_4$-$(BPh_4)_2$ in acetonitrile solution at $25 \pm 2°C$ (6)

ligands is an important influence on the position of the lowest allowed electronic transition. In this complex the two ligand planes are rotated away from an eclipsed configuration by about 30° (Figure 4), but the Rh–Rh distance is nearly the same as found in $Rh_2(bridge)_4^{2+}$ (3.254 Å). The blue shift of the $1a_{2u} \rightarrow 2a_{1g}$ transition in $Rh_2(TM4\text{-bridge})_4^{2+}$ to 515

Table I. Absorption and Emission Spectral Data[a]

Complex	Lowest Absorption Band (nm)	Emission (nm)
$Rh_2(bridge)_4^{2+}$	553	656
$Rh_2(4\text{-bridge})_4^{2+}$	526	634
$Rh_2(TM4\text{-bridge})_4^{2+}$	515	614
$Rh_2(CYCLO5\text{-bridge})_4^{2+}$	423	583

(4-bridge)   (TM4-bridge)   (CYCLO5-bridge)

[a] 25 ± 2°C in acetonitrile solution.

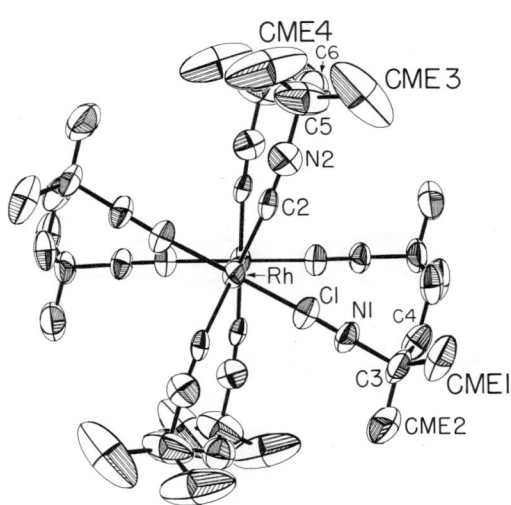

Figure 4. View of the structure of $Rh_2(TM4\text{-bridge})_4^{2+}$; the Rh–Rh distance is 3.254 Å (from a crystal structure analysis of $Rh_2(TM4\text{-bridge})_4\text{-}(PF_6)_2$)

nm is best explained in terms of the smaller splitting of the $2a_{1g}$, $2a_{2u}$ set. Less splitting would be expected because of the much smaller overlap of the CNR π^* components of $2a_{1g}$ and $2a_{2u}$ in the observed configuration.

Thermal Reactions

The $Rh_2(bridge)_4^{2+}$ complex (abbreviated Rh_2^{2+}) and its derivatives undergo a variety of very interesting thermal reactions, e.g.,

$$Rh_2^{2+} \xrightarrow{Cl_2} \underset{\text{yellow}}{ClRh_2Cl^{2+}} \qquad \lambda_{max} = 338 \text{ nm} \qquad (2)$$

$$Rh_2^{2+} \xrightarrow{HCl} \underset{\text{blue}}{HRh_2Cl^{2+}} \qquad \lambda_{max} = 574 \text{ nm} \qquad (3)$$

$$2HRh_2Cl^{2+} \xrightarrow{H_2} \underset{\text{super reduced}}{Rh_4H_2^{4+} + 2HCl} \qquad \lambda_{max} = 780 \text{ nm} \qquad (4)$$

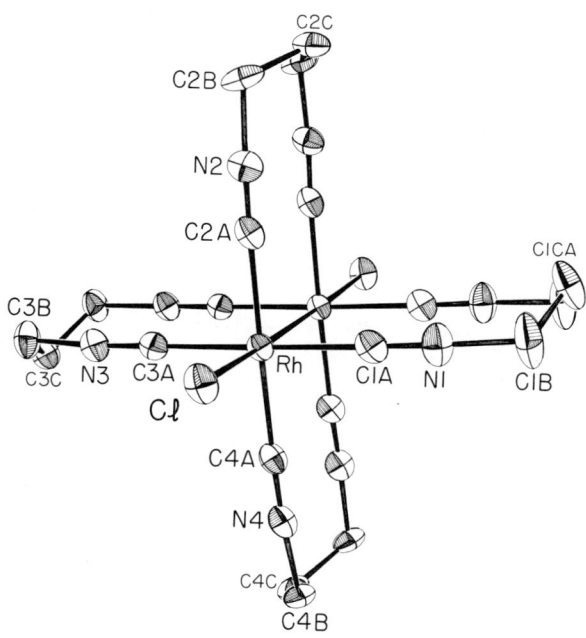

Figure 5. View of the structure of $Rh_2(bridge)_4Cl_2^{2+}$; the Rh–Rh distance is 2.84 Å

In Reaction 2, both Rh(I) atoms are oxidized to a species that contains a Rh–Rh single bond, with the addition of two ligands in a trans arrangement. Several X_2- and XY-type molecules oxidatively add to $Rh_2(bridge)_4^{2+}$ in this way (4). The $Rh_2(bridge)_4Cl_2^{2+}$ complex has been crystallized from aqueous HCl solution as a Cl^- salt and structurally characterized by x-ray techniques (8). A view of the complex cation is shown in Figure 5. The four bridge ligands are again rigorously eclipsed, as in $Rh_2(bridge)_4^{2+}$, but a shorter Rh–Rh distance (2.84 Å) is observed, as expected in a complex possessing a Rh–Rh single bond (9).

Reaction 3 could also be formulated as an oxidative addition although earlier we pointed out (5) that electronic spectroscopic properties of the dark blue HRh_2Cl^{2+} species are noticeably different from those of $Rh_2Cl_2^{2+}$ and related Rh(II)–Rh(II) complexes. The main point was that the intense visible absorption band (λ_{max} = 574 nm, ϵ = 52,700) in the spectrum of HRh_2Cl^{2+} falls at lower energy than most $\sigma \rightarrow \sigma^*$ transitions in d^7–d^7 bonded complexes (e.g., the $\sigma \rightarrow \sigma^*$ transition in $Rh_2Cl_2^{2+}$ is at 338 nm). The temperature dependence of the 574-nm absorption band, however, is very striking (Figure 6). As can be seen, the absorption system blue shifts and sharpens tremendously on cooling to 15°K, which

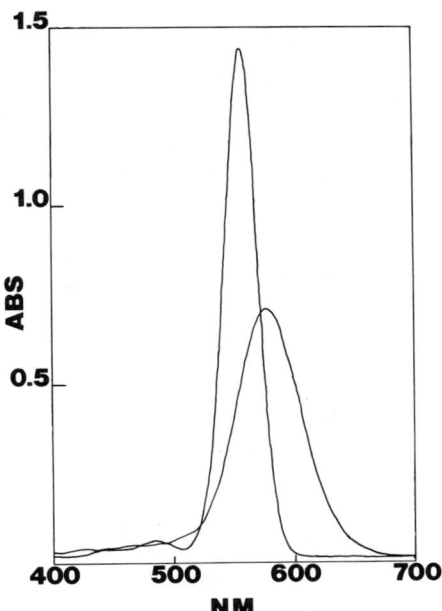

Figure 6. Electronic absorption spectra of HRh_2Cl^{2+} in 6M HCl/6M LiCl at 300 (broad, lower absorption) and 15°K (sharp, higher absorption)

is an established (10) property of a band attributable to a $\sigma \rightarrow \sigma^*$ transition. Thus there is at least some evidence that supports a "chlorohydridodirhodium(II)" formulation of HRh_2Cl^{2+} although it is not clear why the $\sigma \rightarrow \sigma^*$ transition energy is so low in this case.

We have found that molecular hydrogen reacts with HRh_2Cl^{2+} in dilute aqueous acidic solutions to give a tetranuclear species, $Rh_4H_2^{4+}$, that we call "super reduced" (Reaction 4). This species exhibits an intense absorption band at 780 nm. The $Rh_4H_2^{4+}$ complex in aqueous acidic solution has virtually the same spectrum as the "Rh_4^{4+}" oligomer that is obtained (4) by dissolving $Rh_2(bridge)_4Cl_2$ in CH_3OH (λ_{max} for "Rh_4^{4+}" = 778 nm). Thus we presume that Rh_2^{2+} is reduced in CH_3OH to $Rh_4H_2^{4+}$ and to higher oligomers.

Photochemistry

$$HRh_2Cl^{2+} \xrightarrow[12M\ HCl]{546\ nm} Rh_2Cl_2^{2+} + H_2 \qquad (5)$$

$$\text{blue} \qquad\qquad\qquad \text{yellow}$$

Irradiation (546 nm) of HCl solutions of the blue HRh_2Cl^{2+} complex leads to the production of $Rh_2Cl_2^{2+}$ and H_2 gas (Reaction 5). The back thermal reaction is relatively slow, taking several days to return the system to the initial state. When O_2 is present, $Rh_2Cl_2^{2+}$ is produced with higher quantum yields (Table II), indicating preferential reduction of O_2 (rather than reduction of $2H^+$ to H_2). The nature of the dependence of the quantum yield for H_2 production in degassed HCl solutions on a_{H^+}

Table II. Quantum Yields for the Photooxidation of $Rh_2(bridge)_4^{2+}$ in HCl Solutions at 29°C

[HCl] (M)	a_{H^+}	$10^2\ \Phi$ (degassed)[a]	$10^2\ \Phi$ (air saturated)[a]
12.8	6.0×10^4	0.79	[b]
12.1	2.9×10^4	0.56	4.3
11.1	1.1×10^4	0.26	[b]
10.1	4.3×10^3	0.083	[b]
9.1[c]	1.8×10^3	0.028	5.2
8.1	7.8×10^2	0.010	[b]
6.0	1.3×10^2	< 0.002	[b]
1.0	1.6×10^0	<< 0.001	2.2

[a] Based on measurements of the appearance of $Rh_2(bridge)_4Cl_2^{2+}$.
[b] Not measured.
[c] For comparison, data for the photoreaction in 9M HBr (giving H_2 + Rh_2-$(bridge)_4Br_2^{2+}$ in degassed solution) are as follows: $a_{H^+} = 7.8 \times 10^3$, $10^2\ \Phi$ (degassed) = 4.4, and $10^2\ \Phi$ (air saturated) = 8.8.

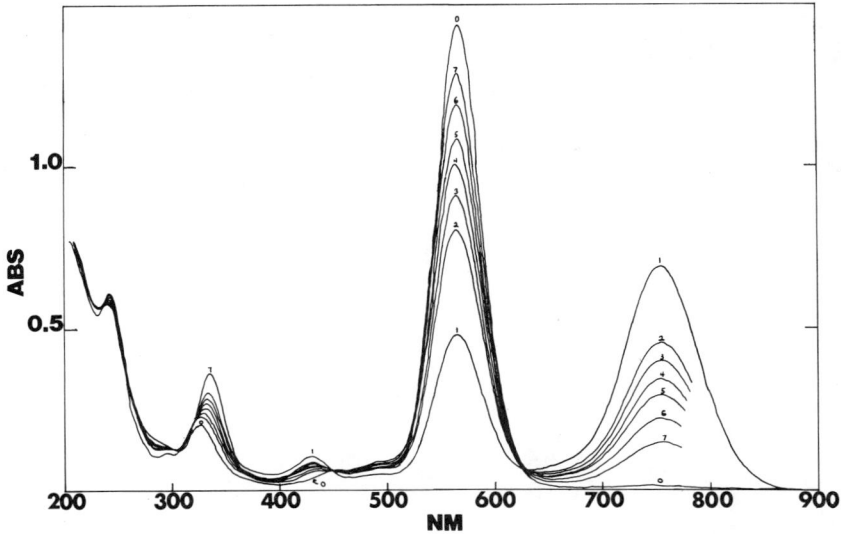

Figure 7. Electronic absorption spectra of HRh_2Cl^{2+} in 6M HCl solution upon irradiation (> 520 nm) at $-60°C$; 0, initial spectrum of HRh_2Cl^{2+}; 1, after 20 min irradiation; 2–7, dark reaction in intervals (total time of ca. 1 hr)

(Table II) suggests that the mechanism of Reaction 5 is quite complicated, perhaps involving several species not observed in room temperature steady state irradiation experiments.

Low temperature ($-60°C$) irradiation (> 520 nm) of degassed 6M HCl solutions (no net production of H_2 is observed in 6M HCl at 25°C) of HRh_2Cl^{2+} yields the spectral changes shown in Figure 7. It is clear that a photochemically induced disproportionation reaction occurs, Reaction 6:

$$3HRh_2Cl^{2+} \xrightarrow[6M\ HCl(-60°C)]{\lambda > 520\ nm} Rh_2Cl_2^{2+} + HCl + Rh_4H_2^{4+} \quad (6)$$

blue → yellow + super reduced

The stoichiometry of Reaction 6 was determined from the known values of the molar extinction coefficients of the rhodium-containing species. In the dark at $-60°C$, the yellow species ($Rh_2Cl_2^{2+}$) reacts slowly (hours) with $Rh_4H_2^{4+}$ to regenerate the initial complex (HRh_2Cl^{2+}). The dark reaction is kinetically first order in each of the rhodium-containing reactants with $k \cong 10^{-1}\ M^{-1}\ sec^{-1}$ at $-60°C$ in 6M HCl.

Flash kinetic spectroscopic experiments on 0.1M HCl solutions of HRh_2Cl^{2+} at 25°C are now being performed. Initial results clearly show that $Rh_4H_2^{4+}$ and $Rh_2Cl_2^{2+}$ are formed several milliseconds after the flash;

these species back react to yield the initial blue complex in a few seconds ($k \cong 10^4$ M^{-1} sec^{-1} at 25°C) (11). Several shorter-lived transients are observed in the flash experiments, and these are probably the precursors of $Rh_4H_2^{4+}$ and $Rh_2Cl_2^{2+}$. The role of these shorter-lived intermediates in the photochemistry of HRh_2Cl^{2+} is now being studied.

The photoreactions of HRh_2Cl^{2+} with H^+ and O_2 likely proceed as

$$3HRh_2Cl \xrightarrow{h\nu} Rh_4H_2^{4+} + Rh_2Cl_2^{2+} + HCl \quad (7)$$

$$Rh_4H_2^{4+} + Rh_2Cl_2^{2+} + HCl \xrightarrow{\Delta} 3HRh_2Cl^{2+} \quad (8)$$

$$Rh_4H_2^{4+} + 2HCl \rightleftharpoons 2HRh_2Cl^{2+} + H_2 \quad (9)$$

$$Rh_4H_2^{4+} + O_2 + 3HCl \rightarrow HRh_2Cl^{2+} + Rh_2Cl_2^{2+} + 2H_2O \quad (10)$$

outlined in Reactions 7, 8, 9, and 10. The photo-induced redox disproportionation (Reaction 7) yields $Rh_2Cl_2^{2+}$, which is the final rhodium-containing product, and the reactive, reducing intermediate $Rh_4H_2^{4+}$. The fate of $Rh_4H_2^{4+}$ is determined by the nature of the reaction medium; in degassed solution at low a_{H^+}, $Rh_4H_2^{4+}$ back reacts with $Rh_2Cl_2^{2+}$ to regenerate the starting complex (Reaction 8); at high a_{H^+}, $Rh_4H_2^{4+}$ reacts with protons to give H_2 and two molecules of HRh_2Cl^{2+} (Reaction 9), but if a_{O_2} is also high, oxidation occurs mainly by pathway (10), owing to the very rapid reaction of $Rh_4H_2^{4+}$ with O_2.

Acknowledgment

This research was supported by the National Science Foundation (CHE75-19086). Matthey–Bishop, Inc. is acknowledged for a generous loan of rhodium chloride.

Literature Cited

1. Gray, H. B., Ballhausen, C. J., *J. Am. Chem. Soc.* (1963) **85**, 260.
2. Mann, K. R., Gordon, J. G., II, Gray, H. B., *J. Am. Chem. Soc.* (1975) **97**, 3553.
3. Mann, K. R., Lewis, N. S., Williams, R. M., Gray, H. B., Gordon, J. G., II, *Inorg. Chem.* (1978) **17**, 828.
4. Lewis, N. S., Mann, K. R., Gordon, J. G., II, Gray, H. B., *J. Am. Chem. Soc.* (1976) **98**, 746.
5. Mann, K. R., Lewis, N. S., Miskowski, V. M., Erwin, D. K., Hammond, G. S., Gray, H. B., *J. Am. Chem. Soc.* (1977) **99**, 5525.
6. Miskowski, V. M., Nobinger, G. L., Kliger, D. S., Hammond, G. S., Lewis, N. S., Mann, K. R., Gray, H. B., *J. Am. Chem. Soc.* (1978) **100**, 485.
7. Gray, H. B., Mann, K. R., Lewis, N. S., Thich, J. A., Richman, R. M., Adv. Chem. Ser. (1978) **168**, 44.

8. Mann, K. R., Thich, J. A., Bell, R. A., Coyle, C. L., Gray, H. B., unpublished data.
9. Norman, J. G., Jr., Kolari, H. J., *J. Am. Chem. Soc.* (1978) **100**, 791.
10. Levenson, R. A., Gray, H. B., *J. Am. Chem. Soc.* (1975) **97**, 6042.
11. Miskowski, V. M., Milder, S., Mann, K. R., unpublished data.

RECEIVED March 20, 1978. Contribution No. 5812 from the Arthur Amos Noyes Laboratory of Chemical Physics.

20

Scope and Applications of Light-Induced Electron-Transfer Reactions of Metal Complexes for Energy Conversion and Storage

PATRICIA J. DE LAIVE and DAVID G. WHITTEN[1]

Department of Chemistry, University of North Carolina, Chapel Hill, NC 27514

CHARLES GIANNOTTI

Institut De Chimie Des Substances Naturelles, CNRS, 91190 Gif Sur Yvette, France

Excited states of transition metal complexes are often quenched by electron donors or electron acceptors in redox processes that can involve efficient conversion of light energy into high energy products. We have examined ways to circumvent the energy-wasting back reactions by modification of the complex, quencher, and medium or by adding reactive substrates to intercept products generated by electron transfer. For oxidative quenching, a variety of substrates are rapidly oxidized by certain RuL_3^{3+} complexes such that RuL_3^{2+} is regenerated but Ox^{red} survives. For reductive quenching processes a combination of relatively slow rates of back reaction and rapid reaction of Red^{ox} enables the isolation of the high energy RuL_3^+ as a stable but highly reactive product.

The oxidation and reduction reactions of excited states of dyes, aromatic hydrocarbons, and metal complexes have been studied widely over the last several years (1–25). Although there are still several aspects of this group of reactions that have not been resolved, it is clear that in many cases quenching of excited states concurrent with one-electron

[1] Author to whom inquiries should be directed.

transfer can be a prominent, rapid, and efficient process for a variety of systems. The one-electron transfer reactions of transition metal complexes have been especially well studied, and both the initial photoreaction and subsequent "dark" processes have been well-characterized for several systems (11–25).

The polypyridyl complexes of a number of metal ions have been used extensively in these investigations because of their prominent luminescence in several cases and of their relative dark and photostability (see, however, Ref. 26, 27, and 28). For example, the 2,2'-bipyridine complexes, $Ru(bipy)_3^{2+}$, $Os(bipy)_3^{2+}$, and $Ir(bipy)_3^{3+}$ (where bipy = 2,2'-bipyridine) all have highly luminescent excited states of moderate lifetime such that dynamic quenching processes can be investigated readily by a variety of techniques. While these complexes are not especially reactive in the ground states towards oxidants and reductants, addition of the excitation energy to the redox couples as illustrated in Scheme 1 can make the excited states very strong potential reductants and oxidants. Much recent work directed at measuring excited-state redox potentials indicates the relationships illustrated in Scheme 1 are approximately correct for a variety of metal complexes such that rapid electron transfer in either direction can occur as an excited-state quenching process. For example, a study of quenching of the MLCT excited state of $Ru(bipy)_3^{2+}$ by a series of nitroaromatics having similar structure but varying reduction potential indicates that the effective reducing power of the excited state is increased on excitation by 2.10 V; this compares closely with the spectroscopically estimated excitation energy of 2.18 (13). Quenching of $Ru(bipy)_3^{2+}$, $Cr(bipy)_3^{3+}$, and $Ir(Me_2phen)_2Cl_2^+$ (where Me_2phen is 5,6-dimethyl-1,10-phenanthroline) excited states by a series of aromatic amine and methoxybenzenes has been found to give values of 0.7, 1.3, and 1.25 V respectively for reduction potentials of these

Scheme 1. *Formal Redox Relationships for Ground and Excited States of Ruthenium Complexes: Left Side, $Ru(bipy)_3$; Right Side, Hydrophobic Complexes*

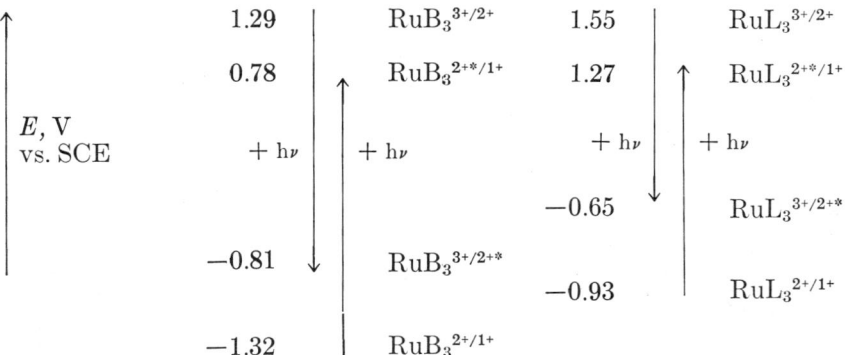

excited states (*29*); the corresponding values estimated from the addition of excited-state energies and ground-state-reduction potentials are 0.79, 1.45, and 1.38, respectively. Thus for these three complexes having different types of excited states, the effective oxidizing power is increased by ca. 90% of the estimated spectroscopic excitation energy. The results cited above are for neutral quenchers in each case and although the separated redox products can be detected in the case of the reductive quenching, quenching of Ru(bipy)$_3^{2+*}$ by the nitroaromatics does not result in the formation of separated redox products detectable by flash photolysis. A reasonable explanation for this is that the ion pair formed in the quenching process has a very low escape probability attributable to attraction of the unlike charges; in contrast, quenching of Ru(bipy)$_3^{2+*}$ by positive electron acceptors results in easily detectable formation of separated redox products. While the polypyridyl complexes have been perhaps most thoroughly studied, a variety of other luminescent metal complexes having excited states of different orbital origin also have been shown to be quenched rapidly and efficiently, yielding separated redox products (*18*). Recently, it has been demonstrated that even relatively short-lived nonluminescent excited states can be quenched in one-electron transfer processes using moderate quencher concentrations (*30*).

In most cases the process of excited state quenching by one-electron transfer (Reactions 1 and 3) is followed by a rapid back-electron-transfer process (Reactions 2 and 4) in which the ground states of the starting material are regenerated. Thus while Reactions 1 and 3 may involve

$$MC^{n*} + Ox \rightarrow MC^{n+1} + Ox^{red} \tag{1}$$

$$MC^{n+1} + Ox^{red} \rightarrow MC^{n} + Ox \tag{2}$$

$$MC^{n*} + Red \rightarrow MC^{n-1} + Red^{ox} \tag{3}$$

$$MC^{n-1} + Red^{ox} \rightarrow MC^{n} + Red \tag{4}$$

efficient energy conversion, the back-electron-transfer reactions are energy-wasting processes which hinder the use of these reactions for production or storage of energy. While the rates of the back-electron-transfer reactions are frequently close to diffusion-controlled, it has been found that introduction of a nonlight-absorbing oxidant or reductant can result in interception of the primary redox products. For example, in oxidative quenching of Ru(bipy)$_3^{2+*}$ by Paraquat (PQ^{2+})(1,1'-dimethyl-4,4'-bipyridinium ion) it has been found that low concentrations of triphenylamine can be oxidized by the Ru(bipy)$_3^{3+}$ generated in the quenching steps so that a sequence given by Reactions 5, 6, and 7 occurs (*17*). Here again there is no net permanent chemical change induced

$$\text{Ru(bipy)}_3^{2+*} + \text{PQ}^{2+} \rightarrow \text{Ru(bipy)}_3^{3+} + \text{PQ}^{\cdot +} \qquad (5)$$

$$\text{Ru(bipy)}_3^{3+} + \phi_3\text{N:} \rightarrow \text{Ru(bipy)}_3^{2+} + \phi_3\text{N}^{\cdot +} \qquad (6)$$

$$\phi_3\text{N}^{\cdot +} + \text{PQ}^{\cdot +} \rightarrow \text{PQ}^{2+} + \phi_3\text{N:} \qquad (7)$$

by irradiation but the back reaction detected by flash photolysis (Reaction 7) is one which involves two species not excited by the irradiating light (17). From the above outlined results, it seems clear that electron-transfer quenching could be an efficient way of generating potentially "stable and storeable" high energy products if it were possible to circumvent the wasting of energy by back electron transfer (31). In evaluating the possibilities for doing this in solution, it appears reasonable that a combination of two approaches might be feasible. First, it would be desirable to selectively retard the initial back-electron-transfer steps (Reactions 2 and 4) to prolong the lifetimes of the high-energy redox products. Secondly, it would be necessary to provide a path whereby one of the initial products is rapidly removed from the system by conversion into a secondary product which would be stable and not redox active. Thus while the sequence shown in Reactions 5, 6, and 7 is not promising since the relay product, $\phi_3\text{N}^{\cdot +}$, is itself reactive, an alternate scheme in which a "redox-inert" product is formed could be effective. At first inspection, neither of these approaches seem particularly simple. While it might be possible to retard the back reaction by a variety of means, in most cases modification of molecular structure to retard the back reaction also would result in a slowing down of the quenching step. However, as will be shown later, a retardation of the quenching can produce no difficulties if an overall high efficiency of the quenching process can be maintained. The rapid and efficient conversion of one of the initial products to a redox inactive species appears to offer more difficulties since in most cases this would involve conversion of a radical or radical ion into an even electron species, a process which would most likely involve radical combination or subsequent electron-transfer steps in competition with the back electron transfer.

While the major portion of this chapter deals with electron-transfer reactions in solution, it is worth discussing briefly the possibilities for using organized or heterogeneous systems to modify the course or rates of light-induced electron-transfer reactions. In a number of studies it has been found that incorporation of one of the two reagents undergoing an excited-state electron-transfer process (Reactions 1 or 3) into a charged micellar environment can result in significant alteration of rates of both quenching and back reactions (32, 33, 34, 35). We have investigated light-induced electron-transfer reactions of surfactant analogs of Ru-

(bipy)$_3^{2+}$ in organized monolayer assemblies (*36, 37*). These investigations indicated that the monolayer-bound surfactant complex is efficiently quenched by a surfactant analog of Paraquat incorporated in the assemblies, provided the Paraquat was positioned in near molecular contact to the ruthenium complex. However, in the assemblies, as well as in solution, no net photoreaction was observed, presumably because of efficient back electron transfer (*37*). Initial experiments on irradiation of monolayer assemblies of the surfactant ruthenium complex alone (deposited on layers of cadmium arachidate) in the presence of water led to the production of hydrogen in an apparent light-induced cleavage of water by the monolayer-bound complex (*36*). Subsequent experiments have shown that this reaction does not occur with the pure surfactant ruthenium complex in monolayer assemblies (*37*). The highly purified complex is not only inactive but it rapidly degraded upon irradiation in the presence of water (*37*). The original preparation was found to contain several impurities including some other surfactant Ru(II) complexes, and it appears possible that the net process of water cleavage in the layers may require at least two different complexes. One possibility for explaining the results obtained in the monolayers can be that a complex–complex electron-transfer process producing two active species can occur (Reaction 8). (This possibility has been suggested by a number of investigators including Sutin and Creutz (*38*).) The solution results that are reported in this chapter point to possible reaction paths which, taken together, could lead to water cleavage among other effective light-mediated redox processes.

$$RuL_3^{2+*} + RuL'_3^{2+} \rightarrow RuL_3^{3+} + RuL'_3^{+} \tag{8}$$

Most of the results reported in the present study deal with the solution-phase reactions of hydrophobic Ru(II) complexes which were synthesized as an outgrowth of our monolayer studies of surfactant complexes (*37, 39, 40*). With these complexes we have found that both quenching and back reactions can be retarded in several cases, allowing new reactions to occur in competition with back electron transfer. Thus both oxidative and reductive quenching processes with these complexes can lead to the generation of isolable but energetic products with some overall promise for use in energy conversion and storage.

Experimental Section

Preparation and Purification of Materials. Spectroquality acetonitrile was dried by distillation from anhydrous P_2O_5 immediately prior to use; handling and transfer prior to degassing was minimized to avoid absorption of water. Isobutyronitrile was distilled from potassium permanganate and used within one month of purification. The various amines used in these studies were purified by vacuum distillation or simple distillation from potassium hydroxide. Triethylamine was purified by distillation from sodium after this treatment. N,N'-dimethyl-4,4'-bipyridine hexafluorophosphate was prepared by the method of R. Young (41). The hydrophobic ruthenium complexes were prepared according to procedures previously described (37, 39, 40).

Electronic Absorption and Emission Spectroscopy. UV and visible spectra were recorded on Cary 14, Cary 171, or Perkin–Elmer 576 ST spectrophotometers. Luminescence excitation and emission spectra were recorded on an Hitachi–Perkin–Elmer MPF-2A spectrofluorimeter equipped with a red-sensitive Hamamatsu R-446 photomultiplier tube. Conventional flash photolysis experiments were performed as described previously (41). The samples were degassed by several cycles of freeze–pump–thaw and sealed under vacuum.

Quantum Yield Determinations. All quantum yields were obtained by irradiating degassed samples in borosilicate glass test tubes in a merry-go-round apparatus. The ruthenium complex samples were irradiated with the 436-nm line of a Hanovia medium pressure (450 W) mercury lamp; Corning filters 3-73 and 5-58 were used to isolate this line. Light intensities were measured using the Reinecke's Salt Actinometer.

Photochemical-ESR Experiments. Solutions for investigations were placed in a 0.1-mm quartz ESR flat cell in the dark or in room light and deoxygenated by bubbling a slow stream of argon through the solution for 20 min. The cell was placed in an ER 400X-RL cavity of a Bruker ER 420 spectrometer equipped with B-ST 100/700, B-MN-12, and B-16 accessories for variable temperature control magnetic field calibration, and frequency measurements, respectively. The samples were irradiated in the cavity with a Hanovia 977 B0090 1000 W mercury–xenon arc lamp in a model LH 15 1H Schoeffel lamp housing. The incident light was focused through a 15-cm flowing water filter and a Corning 3-73 cut-off filter.

Results and Discussion

Reductive Quenching of Ruthenium(II) Complex Excited States. As outlined in the introduction, quenching of excited states of complexes such as $Ru(bipy)_3^{2+}$ by reductants such as triphenylamine in a one-electron transfer step (Reaction 3) is a well-established process. With most complexes examined to date, back-electron transfer is both rapid and efficient so that net chemical change is seldom observed. The hydrophobic Complexes 1, 2, and 3 were found to have absorption spectra, luminescence spectra, and excited-state lifetimes similar to those of $Ru(bipy)_3^{2+}$; however, their solubility and redox behavior is somewhat different (39, 40). While $Ru(bipy)_3^{2+}$ is soluble in water and a few polar organic solvents, Complexes 1, 2, and 3 are water insoluble but rather widely soluble in a variety of nonaqueous solvents. Redox potentials of Complexes 1 and 2 are shifted more anodic as shown in Scheme 1; the changes can be easily rationalized as being attributable to the electron-withdrawing properties of the *p* and *p'*-carboxyester groups on the 2,2'-bipyridine ligands. Complex 3 was found to be electrochemically inactive towards a platinum electrode under conditions where polypyridyl complexes of ruthenium and other metals as well as 1 and 2 are easily oxi-

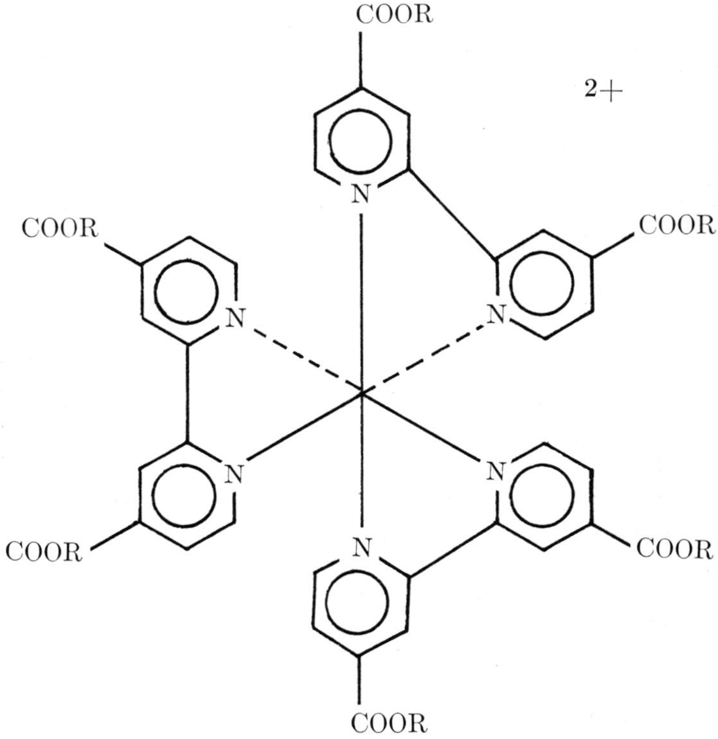

1   R = $(CH_3)_2CH-$

2   R =

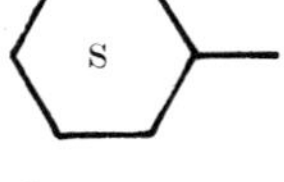

3   R =

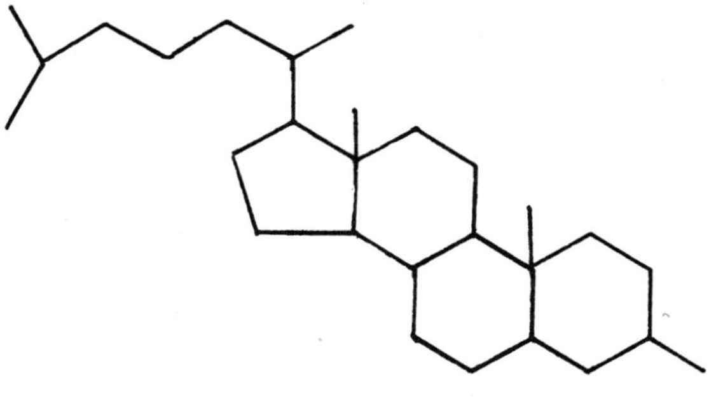

dized and reduced (*39*, *40*). This result, coupled with similar behavior from a series of symmetrically substituted complexes where other bulky hydrophobic groups are attached to the bipyridine ligands, suggests that the bulky groups hinder approach to the electrode of the electrochemically active portion of the complex sufficiently to prevent redox processes. This observation encouraged us to examine the light-induced redox behavior of these complexes since it appeared possible that similar effects might occur in bimolecular electron-transfer reactions.

A study of both oxidative and reductive quenching processes with **1, 2,** and **3** and other hydrophobic complexes indicated that both quenching and back-electron-transfer reactions could be retarded in certain cases (*39*, *40*). Considering that the hydrophobic complexes are dications, it was not surprising to find that electron-transfer processes involving cationic donors and acceptors were retarded more than those involving neutral species. For several quenchers, the behavior observed in dry acetonitrile solution was analogous to that observed with other metal complexes; quenching and back-electron-transfer processes as outlined in Reactions 1, 2, 3, and 4 occurred, although at reduced rates, with the result of no net photochemical change. However, with Complexes **1, 2,** and **3** and amines such as N,N-dimethylaniline and triethylamine, it was found that irradiation resulted in rapid conversion of the starting complex into a "permanent" photoproduct (*39*, *40*). For solutions of **1, 2,** or **3** with triethylamine the changes were consistent with a clean permanent reduction of the Ru(II) complex dication to the one-electron-reduced Ru(II)$^+$ ion; the spectrum obtained for **1** on irradiation with triethylamine in several solvents is identical to that obtained for electrochemical reduction of **1**.

The changes are most simply accounted for by a sequence as outlined in Reactions 9, 10, and 11. Here the combination of a relatively

$$\text{RuL}_3^{2+*} + \text{Et}_3\text{N}: \rightarrow \text{RuL}_3^+ + \text{E}_3\text{tN} \cdot^+ \quad (9)$$

$$\text{RuL}_3^+ + \text{Et}_3\text{N} \cdot^+ \rightarrow \text{RuL}_3^{2+} + \text{Et}_3\text{N}: \quad (10)$$

$$+ \text{Et}_3\text{N} \cdot^+ \rightarrow \text{products} \quad (11)$$

slow back reaction (Reaction 10) and a rapid removal of one of the initial products (Reaction 11) allows the buildup of one of the products. That both processes are important is indicated by the finding that irradiation of solutions of Ru(bipy)$_3^{2+}$ and triethylamine produces no detectable permanent product on moderate irradiation. In this case, a flash photolysis study indicates that redox products are formed but that rapid back electron transfer occurs. A point of major interest in the overall process is the identity of the step or steps which rapidly deplete the triethylamine radical cation. A study of the reaction in acetonitrile at 190 K reveals only a single signal in the ESR spectrum ($g = 1.9962 \pm 0.0003$) characteristic of the reduced complex without any hyperfine splitting. This indicates that any radicals formed from the triethylamine radical cation must be rapidly converted to nonradical products. Possible reaction paths could include those outlined in Reactions 12, 13, 14, 15, and 16. The sequence given by Reactions 12 and 13 has been proposed in a related situation (42, 43), while that given by Reactions 14, 15, and 16 is involved in the photoreduction of benzophenone and other ketones by amines (44, 45). In the latter case the predominance of Reaction 15 results in reduction of a second ketone in a dark process with a limiting

$$\cdot NEt_3^+ + CH_3CN \rightarrow \cdot CH_2CN + H\overset{\oplus}{N}Et_3 \qquad (12)$$

$$2 \cdot CH_2CN \rightarrow \underset{\underset{CH_2-CN}{|}}{CH_2-CN} \qquad (13)$$

$$\cdot NEt_3^+ + :NEt_3 \rightarrow CH_3\overset{\cdot}{C}H-NEt_2 + H\overset{\oplus}{N}Et_3 \qquad (14)$$

$$CH_3\overset{\cdot}{C}H-\overset{..}{N}Et_2 \xrightarrow{-e} CH_3CH\overset{\oplus}{=}NEt_2 \qquad (15)$$

$$2CH_3\overset{\cdot}{C}H-\overset{..}{N}Et_2 \rightarrow \underset{\underset{CH_3-CH-NET_2}{|}}{CH_3CH-NEt_2} \qquad (16)$$

quantum yield of two for disappearance of ketone. If Reaction 14 were to occur in the present case, Reaction 15 could occur with reduction of a second molecule of complex, giving rise to a limiting quantum yield of two for the process.

Irradiation of 1 in moist acetonitrile containing triethylamine leads to no net reduction of the Ru(II) complex; addition of water to solutions of reduced 1 in acetonitrile produce a rapid regeneration of the Ru(II) complex. While the products have not yet been determined, it is evident that a net redox reaction between water and the reduced ruthenium complex is occurring. Under these conditions (water, acetonitrile, and triethylamine), sustained irradiation of 1 can be carried out so that appreciable conversion of the triethylamine occurs; acetaldehyde, which presumably arises through hydrolysis of the Schiff base produced in Reaction 15, is easily detected by VPC as a major product. A trace of product having a retention time identical to succinonitrile is also detectable by VPC, but it appears that Reactions 14, 15, and 16 are the predominant paths for rapid depletion of the triethylamine radical cation.

The photoreduction of **1** by triethylamine occurs in a variety of solvents including isobutyronitrile, THF, acetonitrile–isopropyl alcohol, methylene chloride, and chloroform although the reaction does not appear to be clean or to go to completion in the latter two solvents. Quantum yields measured thus far are 0.35, 0.2, and 0.05 in dry acetonitrile, isobutyronitrile, and THF, respectively. Although the reduced ruthenium complex, RuL_3^+, is "stable" on a time scale of minutes to hours, we find that the pre-irradiation spectrum of RuL_3^{2+} is slowly regenerated on a time scale of hours to days in degassed solutions allowed to stand in the dark at 20°–25°C. Admission of air to the samples results in instantaneous regeneration of the spectrum of RuL_3^{2+}. As mentioned above, the overall reaction sequence appears best described by Reactions 9, 14, and 16 as outlined below. This predicts a limiting quantum yield of two for

$$RuL_3^{2+*} + Et_3N: \rightarrow RuL_3^+ + Et_3N\cdot^+ \tag{9}$$

$$Et_3N\cdot^+ + Et_3N: \rightarrow Et_3\overset{+}{N}H + Et_2\ddot{N}-\dot{C}H-CH_3 \tag{14}$$

$$Et_2\ddot{N}-\dot{C}H-CH_3 + RuL_3^{2+} \rightarrow Et_2N^+=CH-CH_3 + RuL_3^+ \tag{16}$$

production of RuL_3^+, a value substantially higher than the maximum obtained thus far in our investigations. The source of the inefficiency has not been determined although several reasonable possibilities, including some back reaction (Reaction 10) or a less-than-unit efficiency of generation of separated electron-transfer products in the quenching step (Reaction 9), exist. A thermochemical analysis of the reaction by Reactions 17, 18, 19, and 20 indicates that the process involves considerable

$$2(e^- + RuL_3^{2+} \rightarrow RuL_3^+) \tag{17}$$

$$2(H^+ + NEt_3 \rightarrow HN^+Et_3) \tag{18}$$

$$HN^+Et_3 \rightarrow H_2 + Et_2N^+=CH-CH_3 \tag{19}$$

$$H_2 \rightarrow 2H^+ + 2e^- \tag{20}$$

storage of energy; while not enough data are available to permit exact evaluation of the energetics of Reactions 18 and 19, it is clear that Reactions 17 and 19 are energetically uphill as written while Reaction 18 is downhill. It appears that Reactions 18 and 19 should approximately offset one another so that the energy stored can be estimated in terms of the yield of RuL_3^+ produced; the $RuL_3^{2+/1+}$ couple has a potential of -0.9 V or 20.7 kcal/mol (*39, 40*).

If the reductive quenching of complexes such as RuL_3^{2+} by amines is to be used in an energy-conversion process, it will be necessary to develop subsequent chemical or electrochemical steps to convert these primary products back to starting materials concurrent with production of usable products or energy. Although relatively little has been done in this area, it appears that several promising approaches could be developed. As mentioned above, the products appear to slowly revert to the starting materials (as evidenced by the spectroscopically observable $RuL_3^+ \to RuL_3^{2+}$ interconversion) in the dark in degassed solutions. Perhaps the simplest approach for a cyclic use of the system would be to develop catalysts that would accelerate thermal reversion to the starting materials; if this were done it would be possible to transport the photoproducts to a site where an immobilized heterogeneous catalyst could promote the energy-releasing regeneration of the starting materials. An alternative would be a coupling of chemical and electrochemical reactions to bring about a net regeneration of the starting materials with release of energy. Since 1 is electrochemically active, it should be possible to couple Reactions 17 and 20 electrochemically in reverse to regenerate RuL_3^{2+} and hydrogen since the driving force is nearly a volt. The hydrogen produced could then be used to regenerate triethylamine so that no net consumption of any of the reagents would occur. There are, of course, numerous ways in which a high energy species such as RuL_3^+ could be used. It is of interest to note in this regard the recent report of Lehn and Sauvage (46) in which $Ru(bipy)_3^{2+}$ was irradiated in the presence of a complex mixture including the potential reductant, triethanolamine, with a resultant "catalytic" generation of hydrogen. Although the mechanism for this process has not been elucidated, a reasonable possibility appears that a reduced ruthenium complex mediates subsequent electron-transfer steps.

Oxidative Quenching of Ruthenium(II) Complex Excited States. Oxidative quenching of complexes such as $Ru(bipy)^{2+}$, and back electron transfer (Reactions 1 and 2) occurs with a variety of reagents. One of the most widely used oxidants in these studies is Paraquat (PQ^{2+}). For $Ru(bipy)_3^{2+}$ rate constants for the quenching and back-electron-transfer reactions in acetonitrile are, respectively, 2.8×10^9 and 8.1×10^9 M^{-1} sec^{-1} (39, 40). When the hydrophobic complexes are used as substrate, both forward and back reactions are retarded; for example with 1 the corresponding rate constants are 1.2×10^8 and 1.8×10^9 M^{-1} sec^{-1} (39, 40). As pointed out previously with $Ru(bipy)_3^{2+}$ and PQ^{2+}, the electron-transfer products can be intercepted as illustrated in Reactions 5, 6, and 7, but in this case no permanent chemistry occurs. When solutions of 1 and Paraquat are irradiated in dry acetonitrile, we also find that reverse electron transfer is efficient enough so that no permanent chemistry occurs. Recently, we have found that for 1 and several other hydrophobic complexes and in some cases for $Ru(bipy)_3^{2+}$, oxidative quenching by Paraquat can be coupled with the use of the products, in this case the Ru(III) species, to effect a permanent redox change with an added substrate. The process appears to be fairly general and in several cases reasonably efficient such that it offers the promise of considerable utility.

As outlined previously, the photoreaction between excited Ru(II) complexes and Paraquat can be followed by flash spectroscopy (12, 39, 40). The products of the reaction, RuL_3^{3+} and $PQ\cdot^+$, can be detected by transient appearance of new absorbances at 395 and 605 nm, where $PQ\cdot^+$ absorbs, and a transient bleaching of the RuL_3^{2+} absorbance in the 400 to 500-nm range. In dry acetonitrile solutions no permanent buildup of $PQ\cdot^+$ occurs. In contrast, we find that addition of several organic substrates to acetonitrile solutions of 1 and PQ^{2+} results in a rapid buildup of $PQ\cdot^+$ without a concurrent net decomposition of 1 (Figure 1). Since

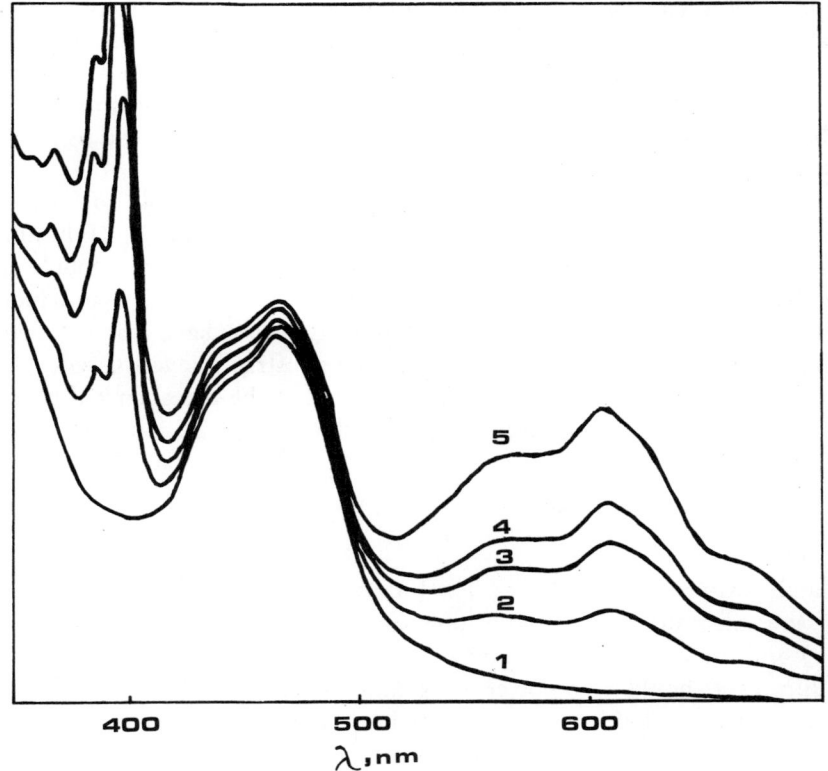

Figure 1. Spectral changes occurring in the visible–near-UV on irradiation of an acetonitrile-2,6-lutidine (2:1) mixture containing PQ^{2+} and 1 with visible light. Curve 1: sample prior to irradiation. Curves 2, 3, ,4 and 5 are from the same sample after progressive irradiations.

these substrates do not themselves quench the excited state of 1 or are used at levels such that quenching of the excited state of 1 by Paraquat is dominant, the observed results are most consistent with a sequence given by Reactions 21, 22, and 23. Thus the draining of oxidized complex

$$RuL_3^{2+*} + PQ^{2+} \rightarrow Ru_3^{3+} + PQ\cdot^+ \tag{21}$$

$$RuL_3^{3+} + \text{Substrate} \rightarrow RuL_3^{2+} + \text{Substrate}^{ox} \tag{22}$$

$$RuL_3^{3+} + PQ\cdot^+ \rightarrow RuL_3^{2+} + PQ^{2+} \tag{23}$$

by Reaction 22 allows the accumulation of PQ^+ as a "permanent" photoproduct while RuL_3^{2+} is regenerated. The light-induced changes also can be conveniently monitored by ESR; in these experiments irradiation of samples carefully dried and degassed with argon leads to the same result: no paramagnetic species are detectable prior to irradiation but photolysis with visible ($\lambda > 420$ nm) light leads to a rapid generation of a single paramagnetic species. Both the envelope and hyperfine splitting pattern are consistent with identification of this species as PQ^+. Thus a pattern similar to that observed in the reductive quenching is occurring here in that Reaction 22 must involve or be followed by steps leading to the depletion of free radicals formed in the initial oxidation.

Substrates which have been found thus far to be able to intercept RuL_3^{3+} in a light-mediated redox process include pyridine, 2,6-lutidine, N,N-dimethylformamide, acetone, triethylamine, isopropyl alcohol, and water. In each case, we have determined that there is "permanent" buildup of PQ^+ upon steady-state illumination; we have not yet determined the products formed for any of the substrates investigated. Condensing Reactions 21, 22, and 23, it is clear that the net chemical change is given by Reaction 24 while the true oxidant is RuL_3^{3+}. That oxidized 1

$$PQ^{2+} + \text{Substrate} \xrightarrow[RuL_3^{2+}]{h\nu} PQ^+ + \text{Substrate}^{ox} \tag{24}$$

should oxidize the substrates listed above is not in itself surprising; all of the substrates have anodic limits in the range $+1.6$ V or lower (47, 48) and the potential of $+1.59$ V for the $RuL_3^{3+/2+}$ couple for 1 (39, 40) indicates it should be a powerful oxidant. What is remarkable is that net conversion of what must be an initial one-electron oxidation product of the substrate to nonradical products can compete with reduction by PQ^+. In fact, the quantum efficiency of net product formation varies quite widely for the different substrates. With 1, "reactive" substrates such as triethylamine and 2,6-lutidine give values $\phi = 0.4$ and 0.07, respectively, while other substrates give considerably lower quantum efficiencies. The high efficiency obtained with triethylamine is perhaps not surprising in view of the results obtained in reductive quenching (vide supra) and with Reactions 14 and 15 occurring, could involve a net of two Paraquat molecules reduced for every molecule of amine oxidized.

The pattern of chemistry observed is not restricted to the hydrophobic Ru(II) complexes. We find that irradiation of Ru(bipy)$_3^{2+}$ with Paraquat and 2,6-lutidine or triethylamine leads to similar qualitative results but with lower quantum efficiencies (0.02 and 0.17, respectively) than for 1. Evidently related phenomena have been observed by Matsuo and co-workers (49) and by Bolton and co-workers (50) for Ru(bipy)$_3^{2+}$ and Paraquat with EDTA in aqueous solutions; the Bolton group has found that organic dyes such as proflavine and acridine orange behave similarly on irradiation under the same conditions (50). An interesting aspect of the work by Bolton and co-workers is the finding that the PQ$^+$ produced in these reactions can be used to produce molecular hydrogen in the presence of the enzyme hydrogenase as a catalyst (50). Thus it is possible to use both initial products of the electron-transfer step for carrying out potentially useful reactions.

Returning to the oxidations mediated by the Ru(III) species, the finding that both the ruthenium complex and Paraquat can be recycled and that a variety of substrates can be used suggests that this process can be quite general and of considerable utility in carrying out oxidation in nonaqueous medium. Thus irradiation of 1 in the presence of Paraquat with visible light generates a usable oxidant, RuL$_3^{3+}$, in neutral, nonaqueous media which is as potent an oxidant as Ce(IV) in nitric acid (51). The mild conditions used suggest that this could be a useful method for the oxidation of even fairly fragile organic substrates; it is interesting to note that the pyridines, which are poorly behaved in electrochemical oxidations, appear to react well in these light-mediated reactions. Greater selectivity or reactivity should be obtainable in these processes by variation of ligand, metal, or solvent; the possibilities for achieving substantial differences are suggested by the rather pronounced changes in redox potential, solubility, and reactivity obtained by the small changes in ligand structure on going from Ru(bipy)$_3^{3+}$ to 1.

The results described above, as well as those obtained in a number of other investigations, indicate rather clearly that light-driven electron-transfer reactions offer a versatile means for efficient energy conversion and storage. Since the process of electron transfer is a general one which is not much limited to specific molecular structures, the range of possible donors and acceptors for use in these processes is virtually unlimited. Although work in this area is clearly only beginning, it should not be too difficult to construct systems which have optimized redox properties for different spectral regions. The main limitation on effective use of these reactions until now has been the inability to avoid energy loss through back-electron transfer reactions. The present investigation and other studies previously mentioned indicate that there are probably a number of ways of getting around the problem which will almost certainly be expanded and optimized through future work. A major problem that remains is the efficient coupling of light-driven one-electron-transfer steps with net two-electron-redox processes or the development of effective reagents which themselves undergo two-electron redox changes upon excitation. Although reactions in homogeneous solution offer considerable promise, it is reasonable to expect that even greater possibilities may exist for useful reaction involving immobilized reagents and interfacial phenomena.

Acknowledgment

This work was supported by a grant from the National Science Foundation (CHE-76-01074). We are also grateful to W. R. Grace and Co. for a grant in support of this work. The authors thank T. J. Meyer for helpful discussion and gratefully acknowledge valuable collaboration with Professor Meyer and his co-workers in portions of these investigations. D. G. Whitten is grateful for an award by the Photochemistry group of the Foundation, "Electricite de France," which made possible the collaboration between the two laboratories.

Literature Cited

1. Leonhardt, H., Weller, A., *Ber. Bunsenges. Phys. Chem.* (1963) **67**, 791.
2. Knibbe, H., Rehm, D., Weller, A., *Ber. Bunsenges. Phys. Chem.* (1968) **72**, 257.
3. Rehm, D., Weller, A., *Ber. Bunsenges. Phys. Chem.* (1969) **73**, 834.
4. Rehm, D., Weller, A., *Isr. J. Chem.* (1970) **8**, 259.
5. Grellmann, K. H., Watkins, A. R., Wellers, A., *J. Phys. Chem.* (1972) **76**, 469.
6. Ibid., 3132.
7. Kawai, K., Yamamoto, N., Tsubomura, T., *Bull. Chem. Soc. Jpn.* (1969) **42**, 369.
8. Yamashita, H., Kokubun, H., Kuizumi, M., *Bull. Chem. Soc. Jpn.* (1968) **41**, 2312.
9. Bonneau, R., Farnier-de-Violet, P., Joussot-Dubien, J., *Photochem. Photobiol.* (1974) **19**, 129.
10. Vogelmann, E., Kramer, H. E. A., *Photochem. Photobiol.* (1976) **23**, 383.
11. Gafney, H. D., Adamson, A. W., *J. Am. Chem. Soc.* (1972) **94**, 8238.
12. Bock, C. R., Meyer, T. J., Whitten, D. G., *J. Am. Chem. Soc.* (1974) **96**, 4710.
13. Ibid. (1975) **97**, 2909.
14. Navon, G., Sutin, N., *Inorg. Chem.* (1974) **13**, 2159.
15. Lawrence, G. S., Balzani, V., *Inorg. Chem.* (1974) **13**, 2976.
16. Harbour, J. R., Tollin, G., *Photochem. Photobiol.* (1974) **19**, 147.
17. Young, R. C., Meyer, T. J., Whitten, D. G., *J. Am. Chem. Soc.* (1975) **97**, 4781.

18. Ibid. (1976) **98**, 286.
19. Creutz, C., Sutin, N., *Inorg. Chem.* (1976) **15**, 496.
20. Lin, C. T., Böttcher, W., Chou, M., Creutz, C., Sutin, N., *J. Am. Chem. Soc.* (1976) **98**, 6536.
21. Lin, C. T., Sutin, N., *J. Am. Chem. Soc.* (1975) **97**, 3543.
22. Creutz, C., Sutin, N., *J. Am. Chem. Soc.* (1977) **99**, 241.
23. Toma, H. E., Creutz, C., *Inorg. Chem.* (1977) **16**, 545.
24. Balzani, V., Moggi, L., Manfrin, M. F., Bolletta, F., Lawrence, G. S., *Coord. Chem. Rev.* (1975) **15**, 321.
25. Juris, A., Gandolfi, M. T., Manfrin, M. F., Balzani, V., *J. Am. Chem. Soc.* (1976) **98**, 1047.
26. Gillard, R. D., *Coord. Chem. Rev.* (1975) **16**, 67.
27. Gillard, R. D., Lyons, J. R., *Chem. Commun.* (1973) 585.
28. Gillard, R. D., Hughes, C. T., private communication.
29. Ballardini, R., Varani, G., Indelli, M. T., Scandola, F., Balzani, V., unpublished manuscript. We thank Prof. Balzani for a preprint of this work.
30. Young, P. C., Nagle, J. K., Meyer, T. J., Whitten, D. G., *J. Am. Chem. Soc.* (1978) **100**, 4773.
31. Balzani, V., Moggi, L., Manfrin, M. F., Bolleta, F., Gleria, M., *Science* (1975) **189**, 852.
32. Scheerer, R., Gratzel, M., *J. Am. Chem. Soc.* (1977) **99**, 865.
33. Maestri, M., Gratzel, M., *Ber. Bunsenges. Phys. Chem.* (1977) **81**, 504.
34. Meisel, D., Matheson, M. S., Rabani, J., *J. Am. Chem. Soc.* (1978) **100**, 117.
35. Kano, K., Takuma, K., Ikeda, T., Nakajima, D., Tsutsui, Y., Matsuo, T., *Photochem. Photobiol.*, in press. We thank Prof. Matsuo for a preprint of this work.
36. Sprintschnik, G., Sprintschnik, H. W., Kirsch, P. P., Whitten, D. G., *J. Am. Chem. Soc.* (1976) **98**, 2337.
37. Ibid. (1977) **99**, 4947.
38. Sutin, N., Creutz, C., ADV. CHEM. SER. (1978) **168**, 1.
39. DeLaive, P. J., Lee, J. T., Sprintschnik, H. W., Abruna, H., Meyer, T. J., Whitten, D. G., *J. Am. Chem. Soc.* (1977) **99**, 7094.
40. DeLaive, P .J., Lee, J. T., Sprintschnik, H. W., Abruna, H., Meyer, T. J., Whitten, D. G., ADV. CHEM. SER. (1978) **168**, 28.
41. Young, R. C., Ph.D. Dissertation, University of North Carolina, 1977.
42. Russell, C. P., *Anal. Chem.* (1963) **35**, 1291.
43. Smith, P. J., Manny, C. K., *J. Org. Chem.* (1969) **35**, 1821.
44. Cohen, S. G., Parola, A., Parsons, G. H., Jr., *Chem. Rev.* (1973) **73**, 141.
45. Cohen, S. G., Baumgarten, R. J., *J. Am. Chem. Soc.* (1965) **87**, 2996.
46. Lehn, J. M., Sauvage, J. P., *Nouv. J. Chim.* (1977) **1**, 449.
47. Weinberg, N. L., Ed., "Techniques of Electroorganic Synthesis," Part II, p. 567, Wiley, New York, 1975.
48. Lund, H., Iverson, P., "Organic Electrochemistry," M. M. Balzer, Ed., p. 210, M. Dekker, New York, 1973.
49. Takuma, K., Kajivava, M., Matsuo, T., *Chem. Lett.* (1977) 1199.
50. Markiewiez, S., Chan, M. S., Sparks, R. H., Evans, C. A., Bolton, J. R., *Int. Conf. Photochem. Conversion and Storage of Solar Energy, London, Canada, 1976*, Abstr. E-7.
51. Richardson, W. H., "Oxidation in Organic Chemistry, Part A," K. B. Wiberg, Ed., p. 244, Academic, New York, 1965.

RECEIVED February 22, 1978.

21

Tungsten(IV) Chelates—Potential Energy Transfer Complexes

RONALD D. ARCHER, CRAIG J. DONAHUE,
WILLIAM H. BATSCHELET, and DAVID R. WHITCOMB

Department of Chemistry, Graduate Research Tower A,
University of Massachusetts, Amherst, MA 01003

Inert eight-coordinate W(IV) chelates have been synthesized in our laboratories and have properties analogous to the ruthenium complexes that are currently an energy-transfer focus. Substituted tetrakis(8-quinolinolato)tungsten(IV) complexes (1), substituted tetrakis(picolinato)tungsten(IV) complexes (2), tetrakis(pyrazinecarboxylato)tungsten(IV) (3), tetrakis(isoquinoline-1-carboxylato)tungsten(IV) (4), and bis(N,N'-disalicylidene-1,2-phenylenediaminato)tungsten(IV) (5) have been synthesized. The ML_4 chelates are substitution-inert d^2 chelates with low energy metal-to-ligand charge-transfer transitions based on relative energies and photooxidation to d^1-W(V) analogs, which possess ligand-to-metal charge-transfer transitions in the visible region. We have initiated the synthesis of analogous polymeric complexes as well.

A sizable number of low-spin inert WL_4 complexes have been synthesized in our laboratories, where L is a bidentate ligand chelating through a heterocyclic aromatic nitrogen donor and a negatively charged oxygen donor (1, 2, 3, 4). Analogous mixed ligand $WL_nL'_{4-n}$ complexes also have been isolated (5, 6). Ligand abbreviations are given in Table I. The characterization of the previously reported (1, 2, 4) substituted tetrakis(8-quinolinolato)tungsten(IV) complexes (**1**, where X = H, Br, Cl, COCH$_3$, or CH$_3$ and Y = H, Br, or Cl) has included a single crystal x-ray study (7) which has verified the eight coordination of these chelates. Furthermore, the structure is one expected from Orgel's rule (8) for d^2

Table I. Ligand Abbreviations

acq⁻	=	5-acetyl-8-quinolinolato
bmq⁻	=	7-bromo-5-methyl-8-quinolinolato
bq⁻	=	5-bromo-8-quinolinolato
cnq⁻	=	7-chloro-5-nitro-8-quinolinolato
cq⁻	=	5-chloro-8-quinolinolato
dbq⁻	=	5,7-dibromo-8-quinolinolato
dcq⁻	=	5,7-dichloro-8-quinolinolato
diq⁻	=	5,7-diiodo-8-quinolinolato
dsp^{2-}	=	N,N'-disalicylidene-1,2-phenylenediaminato
epic⁻	=	5-ethylpicolinato
hpic⁻	=	3-hydroxypicolinato
hqa⁻	=	8-hydroxyquinaldinato
iqc⁻	=	1-isoquinolinecarboxylato
mpic⁻	=	5-methylpicolinato
mq⁻	=	5-methyl-8-quinolinolato
mqq^{2-}	=	methylenebis(quinolin-8-ol-5-ylato)
nd^{2-}	=	1,5-naphthyridine-4,8-diolato
nq⁻	=	5-nitro-8-quinolinolato
pic⁻	=	picolinato
pzc⁻	=	2-pyrazinecarboxylato
pzd^{2-}	=	2,3-pyrazinedicarboxylato
q⁻	=	8-quinolinolato
qd^{2-}	=	5,8-quinoxalinediolato
qq^{2-}	=	bis(quinolin-8-ol-5-ylato)
tbq⁻	=	5-*tert*-butyl-8-quinolinolato
tsb^{4-}	=	N,N',N'',N'''-tetrasalicylidene-1,2,4,5-tetraaminobenzenato

eight-coordinate complexes; i.e., the ground state should have the four π acceptors (the unsaturated nitrogen donors) in the foreshortened tetrahedral array or dodecahedral B positions and the π donors (the oxygen donors) in the elongated tetrahedral or dodecahedral A positions, where A and B are from the nomenclature developed by Hoard and Silverton (9). The B positions have favorable π overlap with the filled b_1 orbital (Scheme 1). This rule reduces the 93 possible dodecahedral isomers for such WL$_4$ systems with inequivalent donors (10) to four isomers (Figure 1), the most symmetrical of which is observed for W(bq)$_4$ · C$_6$H$_6$. Steric considerations can reduce this number even further. (For example, the N and O positions should be reversed in d^0 complexes so that the oxygen π donors can overlap with the empty b_1 orbital (11). But steric considerations caused us to predict (7) that Zr(q)$_4$ would have to be one of the isomers with g edges; this prediction has been verified by Lewis and Fay (12).

For the d^2 complexes under discussion, this rule appears to play a very dominant role, such that slow exchange NMR spectra are observed at elevated temperatures for the W(dcq)$_n$(mpic)$_{4-n}$ complexes (5, 6), even though the iso-electronic and iso-structural W(CN)$_8^{4-}$ ion has been

Scheme 1

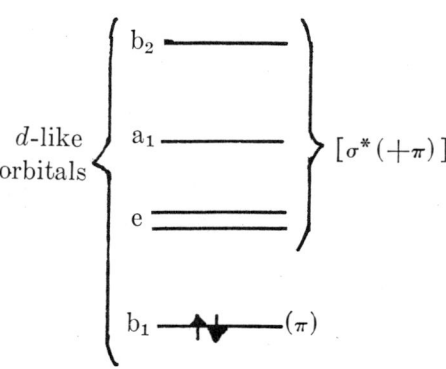

reported to be nonrigid at very low temperatures ($-150°C$) (*13*). Furthermore, two geometrical isomers of $W(dcq)_2(mpic)_2$ have been isolated and characterized (*5, 6*).

Extension of these WL_4 complexes to complexes not involving 8-quinolinol derivatives has included a number of substituted tetrakis-(picolinato)tungsten(IV) complexes (**2**, where Z = H, CH_3, or C_2H_5 and the analogous 3-hydroxypicolinato complex) (*3*). Other species reported previously include the 2-pyrazinecarboxylato and 1-isoquinolinecarboxylato species, $W(pzc)_4$ (**3**) and $W(iqc)_4$ (**4**) (*3*), and the spectrally supported $W(nq)_4$ and $W(hqa)_4$ (*2*). Complexes not previously reported include $W(tbq)_4$ and the quadridentate Schiff base chelate $W(dsp)_2$ (**5**), for which at least two isomers have been observed.

The entire WL_4 series appear to be substitution-inert species, with negligible ligand exchange at room temperature over periods of hours. However, at elevated temperatures ligand exchange can be obtained. Low-spin eight-coordinate dodecahedral d^2 complexes (Scheme 1) are analogous to low-spin six-coordinate octahedral d^6 complexes (Scheme 2) in terms of bonding. That is, the d orbitals that are orthogonal to sigma interactions (the b_1 and t_{2g} orbitals, respectively) are doubly occupied and the sigma antibonding orbitals are empty. Hence, analogous inertness is logical.

Intense charge-transfer transitions observed at the red end of the visible spectrum (14,000–17,000 cm^{-1}) for the tungsten(IV) chelates exhibit energy shifts characteristic of metal-to-ligand charge-transfer species and are analogous to those observed for octahedral d^6 ions with unsaturated nitrogen donor atoms; e.g., Fe(II) and Ru(II) diimines.

The W(IV) chelates are capable of being oxidized to analogous W(V) cations, WL_4^+. A strong oxidant such as chlorine or bromine is required to oxidize the tungsten-picolinato chelates, whereas excess

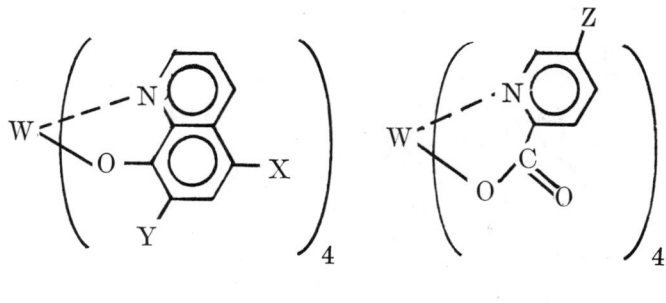

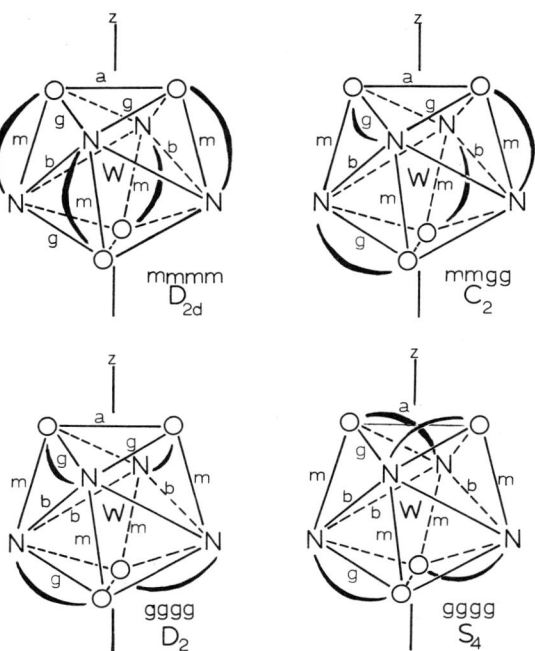

Figure 1. The positions of the chelate rings in the WL_4 isomers allowed by Orgel's rule (8) that acceptors should occupy the B positions in eight-coordinate complexes. The $W(bq)_4 \cdot C_6H_6$ structure is the D_{2d} (mmmm) isomer (7).

Scheme 2

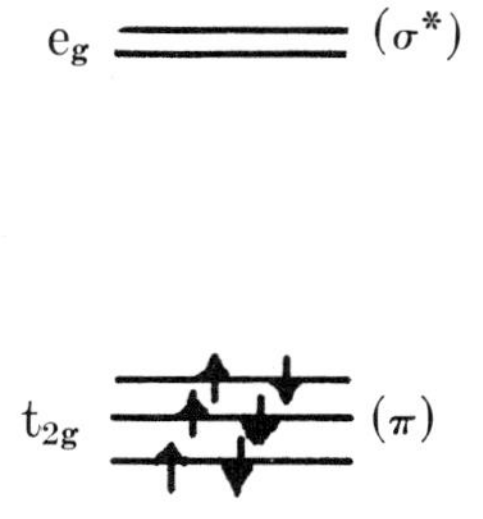

$$e_g \quad \underline{} \quad (\sigma^*)$$

$$t_{2g} \quad \underline{\uparrow \, \uparrow\downarrow \, \uparrow} \quad (\pi)$$

ligand can oxidize the quinolinolato species at elevated temperatures (14). These WL_4^+ chelates, which are d^1 complexes, appear to have moderately intense ligand-to-metal charge-transfer transitions (3, 14) (analogous to octahedral d^5 ions) as a result of the hole in the lowest energy $d\pi$ level; i.e., the b_1 (and t_{2g}) level noted above. The quinolinolato WL_4^+ ions disproportionate in strong alcoholic NaOH solutions to WL_4 and WO_2L_2 type species, with the WL_4 complexes inert to the base. On the other hand, the WL_4^+ picolinato ions are decomposed under these conditions.

These eight-coordinate d^2 and d^1 chelates are analogous to the octahedral d^6 and d^5 ruthenium diimine species that are currently a focus of coordination compound energy transfer. The tungsten species offer both metal and ligand components which are appreciably less expensive than the ruthenium complexes.

Syntheses

Our standard synthetic method for these chelates is the elevated temperature decarbonylation of $W(CO)_6$, either by melt reactions containing excess ligand (2–6, 14) or by the use of high boiling solvents, usually either 2,4,6-trimethylpyridine (2–6, 14) or mesitylene (3). Evidence for 1,2,3,4-tetrahydro-8-quinolinol has been obtained (14); therefore, the reaction has been formulated as:

$$W(CO)_6 + 5HL \rightarrow WL_4 + HLH_4 + 6CO \quad (1)$$

The analogous reaction with the $W_2Cl_9^{3-}$ ion, which we initially used (1), is formulated as:

$$2W_2Cl_9^{3-} + 17HL \rightarrow 4WL_4 + HLH_4 + 12HCl + 6Cl^- \quad (2)$$

In dilute solution hydrogen production can occur, as noted by Dorsett and Walton (15), who have independently produced $W(pic)_4$ by a similar decarbonylation reaction. We also have synthesized these chelates through a stepwise route which does not require elevated temperatures:

$$W(CO)_6 \xrightarrow[-78°C]{Cl_2} W(CO)_4Cl_2 \xrightarrow[0°C]{HL(Cl_2)} WL_4 \quad (3)$$

When all traces of chlorine are removed from the $W(CO)_4Cl_2$, elevated temperatures are required for the second step of Reaction 3. Another indirect route involves an intermediate phosphine:

$$W(CO)_4Cl_2 \xrightarrow{P\phi_3} W(CO)_3(P\phi_3)_2Cl_2$$

$$\downarrow HL$$

$$W(CO)_2(P\phi_3)_2ClL$$

$$W(CO)_3(P\phi_3)ClL \xrightarrow{HL} WL_4$$

$$W(CO)_2(P\phi_3)L_2 \xrightarrow{HL'} WL_nL'_{4-n} \quad (4)$$

With some ligands, further oxidation to the W(V) ions occur under extended reaction conditions.

Synthesis of the previously unreported W(tbq)₄ chelate initially involved a Schraup synthesis of Htbq. Previous attempts in other laboratories had failed because of the acid cleavage of the *tert*-butyl group in the arsenic acid oxidation step (16). We have successfully used picric acid and 4-*tert*-butyl-2-nitrophenol as oxidants with yields of 5–10%. The procedure is similar to that used for Hmq (3, 6). The W(tbq)₄ chelate was prepared by the decarbonylation of W(CO)₆ in mesitylene heated under reflux, followed by sublimation of unreacted material and crystallization from chloroform. Chromatographic separation on silica gel with 1:1 v/v CHCl₃:hexane produced an analytically pure W(tbq)₄ chelate.

Two isomers of W(dsp)₂ have been isolated from the reaction of 1 mmol W(CO)₆, 2 mmol H₂dsp (which had been prepared by a normal Schiff base condensation from salicylaldehyde and 1,2-diaminobenzene), and 2 mmol Hpic in mesitylene heated under reflux conditions for 48 hr. Earlier attempts for shorter time periods had indicated incomplete reaction. Preparative chromatographic separation on silica gel with CHCl₃ as both solvent and eluant gave two products with analyses consistent with W(dsp)₂, plus several mixed complexes. The charge transfer maxima for the W(dsp)₂ species are at 22,500 and 22,700 cm⁻¹. Further work is in progress to elucidate the stereochemistry of these complexes.

Low-Energy Charge-Transfer Transitions

Strong ($\epsilon > 10^4$) electronic transitions at low energies (13,000–17,000 cm⁻¹) are observed for all of the WL₄ tungsten(IV) chelates that we have synthesized (Table II). We (2, 3) have assigned these as metal-to-ligand charge-transfer bands because: (1) the bands are too intense to

Table II. WL₄ Low-Energy Charge-Transfer Bands[a]

C_6H_6 Solutions (cm⁻¹)

W(nq)₄	W(Rq)₄[b]	W(bq)₄	W(bmq)₄	W(dbq)₄
13,400	13,800	14,100	14,100	14,200
W(cq)₄	W(dcq)₄	W(q)₄	W(hqa)₄	W(pic)₄
14,200	14,300	14,300	14,300	16,700

$CHCl_3$ Solutions (cm⁻¹)

W(iqc)₄	W(dcq)₄	W(pzc)₄	W(pic)₄	W(mpic)₄
13,600	14,200	16,200	16,600	16,900
W(epic)₄	W(hpic)₄		{ Mo(pic)₄	}
16,900	16,900		{ > 20,000	}

[a] $\epsilon \geq 10^4$.
[b] R = 5-methyl or 5-tert-butyl.

be simple *d–d* transitions; (2) the bands are at lower energy than the first charge-transfer transitions of the analogous W(V) complexes, whereas ligand-to-metal transitions are lower in energy for higher oxidation states; (3) the filled b_1 level would require a π to a_1 or e level for ligand-to-metal transitions in these species but not for the oxidized form; (4) the intra-ligand transitions are still observable near their free molecule energies; (5) the quinoline derivatives have the transitions at lower energies than the pyridine derivatives; and (6) electron withdrawing groups tend to lower the energies of the transitions. See Scheme 3 for the levels thought to be involved in the low-energy transition and the apparent relative energies of the higher occupied orbitals and lower unoccupied orbitals.

Scheme 3

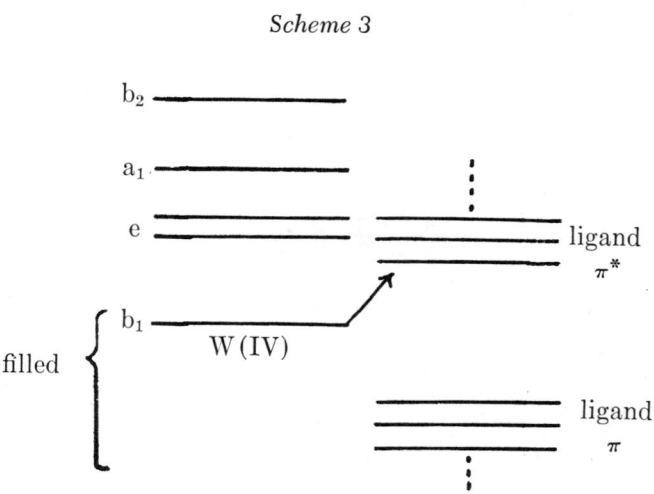

Table III. [WL$_4$]Cl Low-Energy Charge-Transfer Bands[a]

CH$_2$Cl$_2$ Solutions (cm^{-1})

W(dcq)$_4^+$	W(diq)$_4^+$	W(bmq)$_4^+$	W(dbq)$_4^+$
18,300	18,300	18,300	18,500
W(cnq)$_4^+$	W(bq)$_4^+$	W(acq)$_4^+$	W(pic)$_4^+$
18,700	18,900	19,200	23,400[b]

[a] $\epsilon \geq 10^4$.
[b] Acetone.

The analogous W(V) chelates have intense ($\epsilon \geq 10^4$) electronic transitions at somewhat higher energies ($> 18,000$ cm^{-1}) than the charge-transfer transitions in the W(IV) chelates (Table III). The higher energies of analogous W(V) chelates with electron withdrawing groups, W(cnq)$_4^+ >$ W(dcq)$_4^+$ and W(acq)$_4^+ >$ W(bq)$_4^+$, shows that electron withdrawing groups increase the energies of these transitions, in contrast to the charge-transfer bands for the W(IV) chelates. A shift to lower energies for all of the seven-halo species suggests that the nonbonding electrons on the oxygen are involved in the donation from the quinolinol species. The higher energy of the W(pic)$_4^+$ ion is consistent with the electron withdrawing nature of the carboxylate group. The inverse charge-transfer noted for the W(IV) chelates should be at appreciably higher energies in the W(V) chelates because of the lack of electron–electron repulsions in the half-filled b_1 level in the W(V) complexes and of the general lowering of the d orbitals (such as the b_1 level) with increased metal oxidation states. Both effects increase the separation between the b_1 orbital and the ligand π^* orbitals. See Scheme 4.

Scheme 4

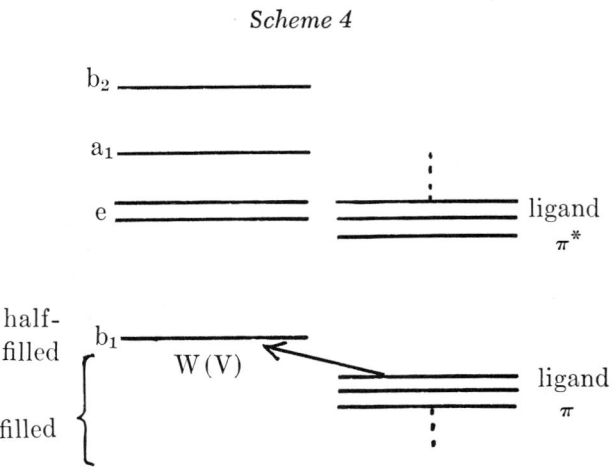

Photosensitivity

Several of the quinolinolato-tungsten(IV) chelates are quite sensitive to oxygen and light during chromatography on silica gel. The product is the W(V) chelate. The W(q)$_4$ chelate is more sensitive than is W(cq)$_4$, which is more sensitive than W(dcq)$_4$. That species is, in turn, more sensitive than any of the picolinato-derivative chelates. Also, the quinolinol derivatives are somewhat photosensitive in air-saturated solutions. Some oxidize in the dark, too, but at a slower rate. The rates also appear to be chelate, solvent, atmosphere, and impurity dependent. Quantification is in progress.

The observed photosensitive oxidation of the tungsten(IV) chelates is further evidence that the low energy bands are metal-to-ligand charge-transfer. Transfer of the π^* electron from the photo-excited state to silica gel leaves the W(V) chelate. Ligand-to-metal charge transfer would tend to give reduced metal or oxidized ligand—contrary to experimental observations.

Polymerization

A current emphasis is the preparation of linear polymers with eight-coordinate tungsten centers. The qd^{2-} biheaded ligand (6) has shown an ability to form a polymeric W(IV) material with a charge transfer band at 12,600 cm^{-1}, but the lower nucleophilicity of this ligand does not allow for polymerization by ligand exchange with the WL$_4$ chelates under conditions that allow ligand exchange between the simpler ligands. The reaction of W(CO)$_4$Cl$_2$ with H$_2$qd followed by Hpic produces a blue product with an electronic transition at 14,300 cm^{-1}. A similar product has been obtained through the reaction performed in the reverse order, but purification has proven difficult. The use of W(CO)$_2$(Pϕ_3)L$_2$ intermediates to form WL$_2$(qd) polymers shows promise. The analogous dione (qd°) provides the two electrons needed to produce the two-electron oxidation. The more nucleophilic ligands, mqq^{2-}, nd^{2-}, and qq^{2-}, also are being investigated. Low-energy transitions have been observed in some of the incompletely characterized species.

Acknowledgment

We wish to acknowledge the support of the Army Research Office for much of this research. Some aspects were supported by the National Science Foundation through the University of Massachusetts Materials Research Laboratory. We appreciate the constructive comments of the reviewers and editors.

Literature Cited

1. Archer, R. D., Bonds, W. D., Jr., *J. Am. Chem. Soc.* (1967) **89**, 2236.
2. Bonds, W. D., Jr., Archer, R. D., *Inorg. Chem.* (1971) **10**, 2057.
3. Donahue, C. J., Archer, R. D., *Inorg. Chem.* (1977) **16**, 2903; cf. Archer, R. D., Donahue, C. J., "Abstracts of Papers," 169th National Meeting, ACS, Philadelphia, PA, 1975, INOR 056.
4. Pribush, R. A., Ph.D. dissertation, University of Massachusetts, 1972.
5. Archer, R. D., Donahue, C. J., *J. Am. Chem. Soc.* (1977) **99**, 269.
6. Donahue, C. J., Archer, R. D., *J. Am. Chem. Soc.* (1977) **99**, 6613.
7. Bonds, W. D., Jr., Archer, R. D., Hamilton, W. C., *Inorg. Chem.* (1971) **10**, 1764.
8. Orgel, L. E., *J. Inorg. Nucl. Chem.* (1960) **14**, 136.
9. Hoard, J. L., Silverton, J. V., *Inorg. Chem.* (1963) **2**, 235.
10. Bennett, W. E., *Inorg. Chem.* (1969) **8**, 1325.
11. Clark, R. J. H., Lewis, J., Nyholm, R. S., Pauling, P., Robertson, G. B., *Nature (London)* (1961) **192**, 222.
12. Lewis, D. F., Fay, R. C., *J. Chem. Soc., Chem. Commun.* (1974) 1046.
13. Muetterties, E. L., *Inorg. Chem.* (1973) **12**, 1963.
14. Archer, R. D., Bonds, W. D., Jr., Pribush, R. A., *Inorg. Chem.* (1972) **11**, 1550.
15. Dorsett, T. E., Walton, R. A., *J. Chem. Soc., Dalton Trans.* (1976) 347.
16. Woodcock, D., *J. Chem. Soc.* (1955) 4391.

RECEIVED February 22, 1978.

22

Stereochemical Aspects of Expanded Coordination Spheres: Seven-Coordinate Tungsten Complexes

JOSEPH L. TEMPLETON

W. R. Kenan, Jr. Laboratory, Department of Chemistry,
University of North Carolina, Chapel Hill, NC 27514

Tricarbonylbis(N,N-dimethyldithiocarbamato)tungsten, $(W(CO)_3(dmtc)_2)$, has been synthesized, and two distinct intramolecular dynamic processes have been identified by variable temperature ^{13}C NMR studies of this seven-coordinate molecule. The sodium salt of dimethyldithiocarbamate reacts with tetracarbonyldiiodotungsten to form the above product, $W(CO)_3(dmtc)_2$. Analytical, IR and NMR data confirm this formulation. The ^{13}C NMR spectrum at $-110°C$ has three distinct resonances, two of which initially coalesce independently of the third ($\Delta G^{\ddagger} = 8.1$ kcal mol^{-1}) as the temperature is increased. All three carbon monoxide signals are averaged at higher temperatures ($\Delta G^{\ddagger} = 9.0$ kcal mol^{-1}). Reversible loss of one CO occurs upon heating to form an insoluble blue $W(CO)_2(dmtc)_2$ compound.

A substantial number of stable seven-coordinate metal complexes have now been isolated and structurally characterized. A comprehensive review by Drew (1) systematizes the structural data that has been accumulated for seven-coordinate compounds. More recently Wreford has contributed to unraveling the details of dynamic processes which can occur for such complexes (2). In one case, dynamic NMR studies of $TaCl(\eta^4-C_{10}H_8)[(CH_3)_2PC_2H_4P(CH_3)_2]_2$ were interpreted in terms of an interconversion of the ground-state pentagonal bipyramid to an intermediate monocapped trigonal prism (3). In another study, site exchange in seven-coordinate species of the type $MX(CO)_2(L-L')_2$ was consistent with a nondissociative mechanism involving a polytopal rearrangement (4).

Although the nature of seven-coordinate compounds is of intrinsic interest, perhaps of greater significance is the role of expanded coordination spheres as intermediates in the chemistry of octahedral complexes. The preparation of the thermodynamically unfavorable trans-$Mo(CO)_2$-(diphos)$_2$ isomer (5) via a seven-coordinate intermediate is an exemplary case of stereochemical control. Plausible seven-coordinate intermediates in the reactions of olefins with pentacarbonyltungstenphenylcarbene (6) can exert stereochemical control of the product distributions.

Although Mo(II) and W(II) exhibit the most extensive seven-coordinate chemistry yet known (consistent with the application of the effective atomic number rule to these d^4 systems), a survey of the substituted metal carbonyl complexes of molybdenum and tungsten reveals no structural data for compounds of the type $M(CO)_3(B-B)_2$ where B-B is a bidentate monoanionic ligand. No tungsten tricarbonyls of this type have been reported to date, but McDonald and co-workers have synthesized and studied the closely related $W(CO)_2(PPh_3)(B-B)_2$ compounds where B-B is a chelating dithiocarbamate (7), xanthate, or dithiophosphate (8). For molybdenum, the compounds $Mo(CO)_3(dtc)_2$ (dtc = $R_2NCS_2^-$) (9) are examples of the $M(CO)_3(B-B)_2$ type as is $Mo(CO)_3$-$[S_2P(i-Pr)_2]_2$ (10), which has been well characterized in solution.

Colton's work with the tricarbonylbisdithiocarbamatomolybdenum series (9) and McDonald's success in studying the triphenylphosphine-substituted tungsten analogs (8) led us to attempt the synthesis and characterization of an analogous tungsten tricarbonyl complex since: (i) the molybdenum derivatives reversibly lose carbon monoxide (11); (ii) the tungsten analogues seemed likely to exhibit different behavior than the molybdenum series (earlier preparative attempts suggested that tungsten dithiocarbamates should be capable of existence but indicated that they would be difficult to characterize (7, 12); and (iii) the seven-coordinate features of $M(CO)_3(B-B)_2$ complexes could be probed via NMR techniques at each of the ligand sites. A synthetic route to $W(CO)_3[(CH_3)_2NCS_2]_2$ and relevant spectroscopic data are reported here.

Experimental

Materials. Tungsten hexacarbonyl, iodine, triphenylphosphine, CO gas, and sodium N,N-dimethyldithiocarbamate were obtained from commercial sources and used without further purification. Solvents were purged with a stream of purified nitrogen prior to use. All solution manipulations were performed under a nitrogen atmosphere using Schlenk techniques.

Physical Measurements. IR spectra were recorded on a Beckman 4250 IR Spectrophotometer. Solid spectra were obtained as Nujol mulls on CsI plates or as KBr pellets. Solution spectra in the region 2400–1600 cm^{-1} were obtained in sealed KBr cells of 0.10 mm pathlength. Polystyrene was used as a calibration marker in all cases. NMR spectra were recorded with a Varian XL-100 Fourier transform spectrometer. Carbon-13 spectra were obtained at a spectrometer frequency of 25.16 MHz with the ^{2}H signal of deuteromethylene chloride serving as an internal lock. Broad band proton decoupling was routinely used for ^{13}C measurements. Spectra were recorded at various temperatures down to −110°C with a solvent mixture consisting of CD_2Cl_2:$CDCl_3$:CCl_4 in a ratio of 60:27:13 with a trace of Me$_4$Si added to serve as an internal reference. Trisacetylacetonatochromium(III) (10 mg) was added to each sample as a paramagnetic relaxation agent in view of the long T_1 values characteristic of metal carbonyl carbon atoms. Analyses were performed by Galbraith Laboratories, Knoxville, Tennessee.

Preparative Procedures. TRICARBONYLBIS(N,N-DIMETHYLDITHIOCARBAMATO)TUNGSTEN(II). The W(CO)$_4$I$_2$ reagent was prepared photochemically according to the method of Colton and Rix (13). The solid reactants W(CO)$_4$I$_2$, 1.50 g (2.72 mmol), and Na(dmtc) (Eastman Chemicals), (0.80 g, 5.60 mmol) were placed in a flask under nitrogen prior to adding 30 mL of THF which had been freshly distilled from calcium hydride under a nitrogen atmosphere. The solids dissolved to form a red solution, and gas evolution commenced. Within 30 min the reaction was judged complete based on the cessation of gas evolution, with the amount of gas collected approximately equal to that expected for the loss of one CO per mole of W(CO)$_4$I$_2$ initially present. The solution was filtered prior to removal of the THF solvent by vacuum evaporation. The solid that remained was chromatographed on an alumina column using benzene as the eluent. Vacuum evaporation of the benzene solution produced a solid which was purified by recrystallization from toluene by cooling a solution that had been saturated at 50°C. The initial yield was 50% after chromatographing the crude material. The final yield was 30% after recrystallization. Analysis: Calcd for C$_9$H$_{12}$N$_2$O$_3$S$_4$W: C, 21.27; H, 2.38; N, 5.51; S, 25.23; W, 36.17. Found: C, 21.97; H, 2.53; N, 5.57; S, 23.92; W, 35.63.

DICARBONYL BIS(N,N-DIMETHYLDITHIOCARBAMATO)TUNGSTEN(II). A solid sample of W(CO)$_3$(dmtc)$_2$, 50 mg (0.1 mmol), was placed in a tube and heated under vacuum. A temperature of 100°C was sufficient to cause a color change within minutes. The initial orange crystals became dark blue-green and the IR spectrum of the new material showed new bands growing in the carbonyl stretching region at 1930 and 1803 cm^{-1}, but the bands attributed to the starting material were of almost equal intensity. Heating at 170°C for 2 hr produced a solid that displayed only the two carbonyl absorptions at 1930 and 1803 cm^{-1}. The blue-green solid resulting from thermal loss of CO was insoluble in common organic solvents, and quantitative conversion of W(CO)$_3$(dmtc)$_2$ to W(CO)$_2$-(dmtc)$_2$ was not accomplished without some accompanying decomposition to material with no carbonyls.

TRIPHENYLPHOSPHINEDICARBONYLBIS(N,N-DIMETHYLDITHIOCARBAMATO)TUNGSTEN(II). (Method A). Equimolar amounts of W(CO)$_3$(dmtc)$_2$,

250 mg (0.5 mmol) and $P(C_6H_5)_3$, 130 mg (0.5 mmol) were placed in a flask followed by addition of 10 mL of CH_2Cl_2 as a solvent. After stirring at room temperature for 12 hr the solution was filtered. The solid product was characterized by IR after removal of the CH_2Cl_2 solvent under vacuum. The solid state IR spectrum of $W(CO)_2(PPh_3)(dmtc)_2$ showed two strong absorptions attributed to CO stretching frequencies at 1910 and 1820 cm^{-1} in agreement with the values reported previously by Chen, Yelton, and McDonald (8), who synthesized this and related compounds from substitution reactions of $W(CO)_3(PPh_3)_2Cl_2$ with uni-negative chelating ligands. (Method B) Thermal loss of CO was induced to form the blue-green $W(CO)_2(dmtc)_2$ from $W(CO)_3(dmtc)_2$, 50 mg (0.1 mmol). Addition of $P(C_6H_5)_3$, 30 mg (0.1 mmol) and toluene (5 mL) caused the blue-green solid to dissolve, and the solution IR spectrum showed only two bands in the carbonyl region: 1930 and 1840 cm^{-1}, as found previously for $W(CO)_2(PPh_3)(dmtc)_2$ prepared according to Method A.

TRICARBONYLBIS(N,N-DIMETHYLDITHIOCARBAMATO)TUNGSTEN(II), ^{13}CO ENRICHED. A toluene solution of the isotopically normal $W(CO)_3(dmtc)_2$ was stirred at 50°C for 45 min under a ^{13}CO atmosphere. The solution was allowed to cool and cyclohexane was added to precipitate the enriched ^{13}CO sample.

Results and Discussion

The preparation of seven-coordinate W(II) complexes of the type $W(CO)_3(B-B)_2$ was pursued via a route similar to Colton's scheme for isolating $Mo(CO)_3(dtc)_2$ (9). We selected N,N-dimethylthiocarbamate (dmtc) since it provides a bidentate ligand with a charge of -1, and it was known to form seven-coordinate compounds of the type we sought. The ease of preparation of $W(CO)_4I_2$ by photolysis of $W(CO)_6$ + I_2 (13) dictated that $W(CO)_4I_2$ be used initially as the W(II) halocarbonyl reactant of choice. IR analysis of the reaction mixture consisting of $W(CO)_4I_2$ and Na(dmtc) was encouraging in view of the tricarbonyl pattern observed, and when coupled with the CO evolution and the formation of insoluble NaI that occurred upon stirring the THF solution, the data suggested that indeed $W(CO)_3(dmtc)_2$ had formed. Minor carbonyl containing impurities were evident in the solution IR spectrum between 2200 and 1700 cm^{-1}, but the three strong bands at 2020, 1944, and 1926 cm^{-1} dominated. Chromatography on alumina affected purification such that only these three bands were observed in solution. A further recrystallization of the material eliminated a small amount of insoluble blue-green material that formed after elution. The orange $W(CO)_3(dmtc)_2$ could be dried under vacuum and dissolved in organic solvents with no loss of carbonyl ligands. The stability of the tricarbonyl in solution provided an opportunity to observe both the solution IR and NMR behavior of the complex. Similar solution stability has been reported for $Mo(CO)_3(dtc)_2$ in organic solvents in a recent publication (14).

The IR spectrum of W(CO)$_3$(dmtc)$_2$ was typical of substituted cis tricarbonyls in the absence of threefold symmetry. The A_1 and E normal modes generated by a C_3 axis produce two allowed absorptions. Departure from the threefold symmetry splits the degenerate E band into two components. Figure 1 reproduces this simple pattern that was observed in the carbonyl region for W(CO)$_3$(dmtc)$_2$. The solid state IR spectrum obtained as a KBr pellet for purposes of comparison with similar molybdenum data displayed four distinct absorptions (2010, 1932, 1909, and 1880 cm^{-1}) while only three were reported for Mo(CO)$_3$(dmtc)$_2$ (2020, 1920, and 1882 cm^{-1}) (12). The additional solid-state absorption band is probably the result of solid-state splitting of the three fundamental modes, but the possibility of more than one isomer also exists.

Heating solid W(CO)$_3$(dmtc)$_2$ in vacuo resulted in an abrupt color change from orange to blue-green. An IR spectrum, after heating at 100°C for 2 hr, indicated that new absorptions had grown in at 1930 and 1803 cm^{-1} while much of the original tricarbonyl was still unreacted. Prolonged heating at 170°C produced an insoluble solid with only the two new absorptions in the carbonyl region. This was interpreted as conversion to W(CO)$_2$(dmtc)$_2$, analogous to the molybdenum dicarbonyl derivatives characterized previously where ν(CO) for Mo(CO)$_2$(dmtc)$_2$ was observed at 1930 and 1840 cm^{-1} (12).

The loss of CO from W(CO)$_3$(dmtc)$_2$ was reversible, as evidenced by the dissolution of the insoluble blue-green dicarbonyl in toluene when CO was added to the reaction vessel. The solution spectrum which resulted was that of the tricarbonyl while no dissolution of the dicarbonyl

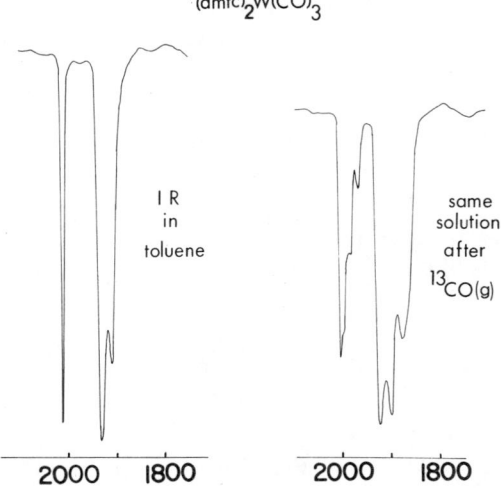

Figure 1. *IR spectrum of W(CO)$_3$(dmtc)$_2$ from 1800 to 2100 cm^{-1}*

occurred in the absence of CO. The insolubility of $W(CO)_2(dmtc)_2$ in organic solvents is somewhat surprising if indeed the complex is monomeric since $W(CO)_3(dmtc)_2$ has excellent solubility and stability in a variety of solvents. The possibility of dimer or polymer formation through bridging sulfur atoms by dative bond formation using one of the sulfur lone pairs (15) could maintain an effective atomic number of 18, as occurs in $[W(CO)_4I_2]_2$ via iodine bridges. If indeed such an interaction occurs, it must be weak and unfavorable relative to CO coordination, as evidenced by the rapid uptake of CO that occurs almost instantaneously when the dicarbonyl solid interacts with a solution containing dissolved CO. A monomer–dimer equilibrium has been proposed by McDonald and co-workers for $Mo(CO)_2(dtc)_2$ in solution based on electronic spectra as a function of concentration (14). The remaining IR absorptions of solid $W(CO)_3(dmtc)_2$ were consistent with the presence of bidentate dithiocarbamate ligands (16), and the $\nu(C-N)$ frequency of 1523 cm^{-1} reflects the partial double bond character of the CN bond typical in this delocalized ligand system.

The ^1H NMR of $W(CO)_3(dmtc)_2$ showed only a singlet at room temperature ($\delta = 3.20$ ppm) as would be appropriate for the methyl protons of dmtc (17), and further ^1H NMR investigations were not pursued because of the greater sensitivity of ^{13}C NMR to small differences in environment. The initial ^{13}C NMR spectrum at room temperature displayed three resonances in addition to those attributable to solvent and Me_4Si. The methyl groups were easily assigned to the singlet at 39.2 ppm while the central carbons of the dithiocarbamate ligands (referred to as C carbons throughout the discussion) were assigned at 208.7 ppm and the carbonyl carbons at 233.1 ppm in view of typical values for these ligands (18, 19). The facile interconversion routes available to seven-coordinate compounds suggested that a dynamic process was averaging the environments of the three carbonyl ligands on the NMR time scale since a structure with three equivalent CO ligands seemed highly improbable.

A temperature-dependent ^{13}C NMR study was undertaken. The original data were obtained on natural abundance ^{13}C and are reproduced in Figure 2. Indeed the carbonyl signal at 233.1 ppm broadened substantially at $-27°C$ and no carbonyl resonance was observed at either $-37°$ or $-74°C$. Cooling to $-104°C$ produced a total of three resonances in the low field region: the 248.7 ppm resonance clearly assignable to CO, which had now reappeared in the slow exchange limit; the 206.1 ppm resonance interpretable as a slight negative temperature dependence for the central dithiocarbamate carbon (C); and the 203.7 ppm resonance which would be assigned to a unique carbonyl position in order that the weighted average of the carbonyl chemical shifts reproduce the room

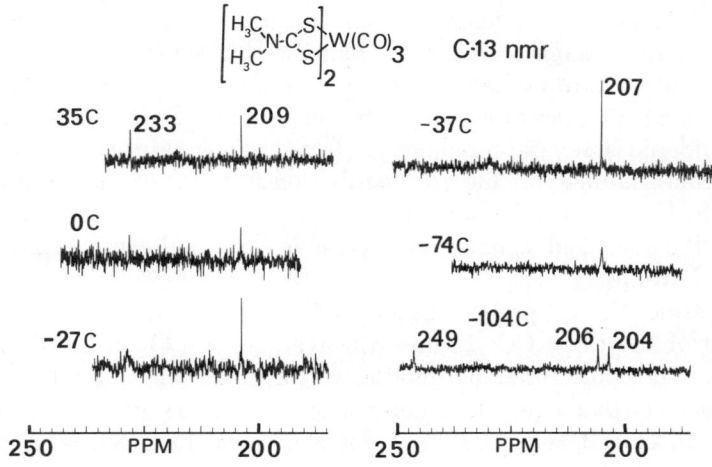

Figure 2. Variable temperature ^{13}C-$\{^1H\}$ NMR spectra of natural abundance $W(CO)_3(dmtc)_2$ in the low-field region displaying resonances attributed to ^{13}CO and $R_2N^{13}CS_2^-$

temperature value ($\frac{1}{3}(2 \times 248.7 + 203.7) = 233.7$, cf. 233.1 observed). The methyl singlet remained sharp at $-74°C$ but at $-104°C$ two signals had replaced the original singlet.

One can interpret the above data, admittedly limited by the small number of different temperatures and the signal-to-noise ratio present for exchanging ^{13}C sites with no attached protons to enhance these resonance signals, as indicative of two unique sites for the carbonyls in a ratio of 2:1, equivalent sites for both central dmtc carbons, and two sites for the four dmtc methyls in a 1:1 ratio. Although the ^{13}CO-enriched NMR study contradicts each of these conclusions for the low temperature structure, vide infra, the above case offers a convenient point of departure for analyzing the stereochemical nonrigidity of these complexes.

The lack of structural data for the ground state of molecules of the type $M(CO)_3(B-B)_2$ necessitates that reasonable postulates regarding ligand distribution serve as a basis for the interpretation of the NMR data. The preference exhibited by carbonyl ligands which favor cis geometries to enhance the π-acceptor properties of these ligands is well documented (20). The observed pattern and intensities of the IR absorptions for the three carbonyls are consistent with such a cis arrangement, and we suggest that a useful starting point for structural considerations is the 4:3 geometry. The cis arrangement of the three CO ligands will provide the planar group of three donor atoms by definition, regardless of the details of their location. The two bidentate ligands are then required to furnish the four remaining donor atoms in a plane parallel to that of the three carbon donor atoms if the strict 4:3 definition is to apply. In fact, there is no reason to assume planarity for the four S atoms, but as pointed out by Drew, the 4:3 geometry is sufficiently close to both the capped octahedron and the capped trigonal prism (CTP) that it need not be considered a separate geometry. Thus, in our case, we can use the conceptual simplicity of the 4:3 distribution in terms of the natural division of $S_4:C_3$ and alter the bidentate ligand positions accordingly to realize any of the idealized geometries accessible within the constraint of cis carbonyl groups.

Consider the 4:3 projections shown below as A and B. In both cases, rotation of the three CO ligands will average CO(1) with CO(2) and CO(3). The same rotational motion will average all CH_3 groups when one considers that equivalent minima are accessible after each rotation of 60°. In case B, where C(1) is not equivalent to C(2), averaging of

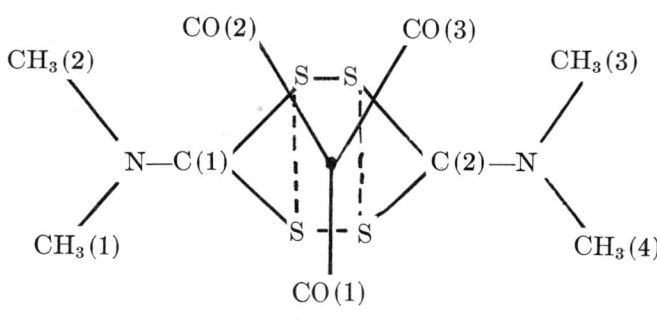

A

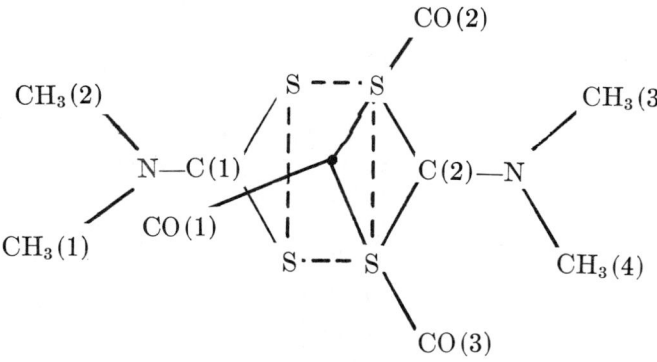

B

these signals will also occur. Hence, only one signal will be seen for each of the three distinct types of carbons in the fast exchange limit for both A and B. Upon cooling the sample to obtain the limiting spectrum, both case A and B will produce two separate CO signals in a ratio of 2:1 and two CH_3 signals in a ratio of 1:1, but case B would predict two different C environments and hence two signals while A has both C atoms in magnetically equivalent sites in the frozen geometry. Although other possible geometries could be analyzed, suffice it to note that case A provides a geometry that is consistent with the possible assignments discussed with respect to the natural abundance ^{13}C data above.

The incorporation of ^{13}CO into the sample proceeded as expected in view of the lability of the CO in the tricarbonyl when heated under vacuum or when reacted with triphenylphosphine. The IR spectrum comparison in Figure 1 clearly reflects substantial ^{13}CO enrichment, probably in the neighborhood of 15 to 20%, an estimated based on NMR data.

The room-temperature spectrum of the isotopically enriched $W-(CO)_3(dmtc)_2$ provided definitive evidence that the exchange process did not involve a dissociative loss of CO to form a six-coordinate intermediate. The presence of ^{183}W (I = 1/2) in 14.4% natural abundance produced a doublet with $J(^{183}W-^{13}C)$ = 119 Hz, indicating that any exchange was necessarily intramolecular in order that the coupling be retained. The intensity of the doublet relative to the total ^{13}CO intensity was approximately 15%, and the coupling constant of 119 Hz is in the range typical of W–C coupling constants (21), thus confirming the origin of these sidebands. The enriched ^{13}CO NMR spectra as a function of temperature for $W(CO)_3(dmtc)_2$ are illustrated in Figures 3, 4, and 5.

The improved signal-to-noise ratio in the labelled species allowed us to obtain useful data over a broad temperature range. The ^{13}CO signal could be seen at each of the 11 temperatures chosen for study, and the additional data dictated that the dynamic processes were more complex than originally formulated. After initial broadening to an unsymmetrical doublet at $-65°C$, two signals with an intensity ratio of 1:2 were observed at $-78°C$, with the less intense signal at low field corresponding to the 248.7 ppm signal frozen out in the earlier study. The $-78°C$ spectrum also showed broadening of the dithiocarbamate carbon C resonance. Further cooling produced a doublet for the C atoms (the high field signal of this pair had earlier been incorrectly attributed to the unique carbonyl ligand). Again two signals of approximately equal intensity appeared for the methyl groups in the low temperature region.

The unique CO resonance at low field continued to sharpen at temperatures below $-78°C$. Coupling to both tungsten ($J(^{183}W-^{13}C) \approx 100$ Hz) and carbon ($J(^{13}C-^{13}C) \approx 10$ Hz) was observed. The upfield

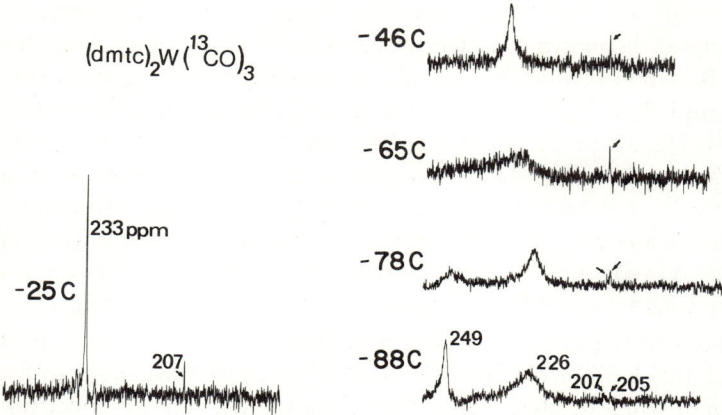

Figure 3. Variable temperature $^{13}C\text{-}\{^1H\}$ NMR spectra of ^{13}CO-enriched $W(CO)_3(dmtc)_2$ from $-25°C$ to $-88°C$

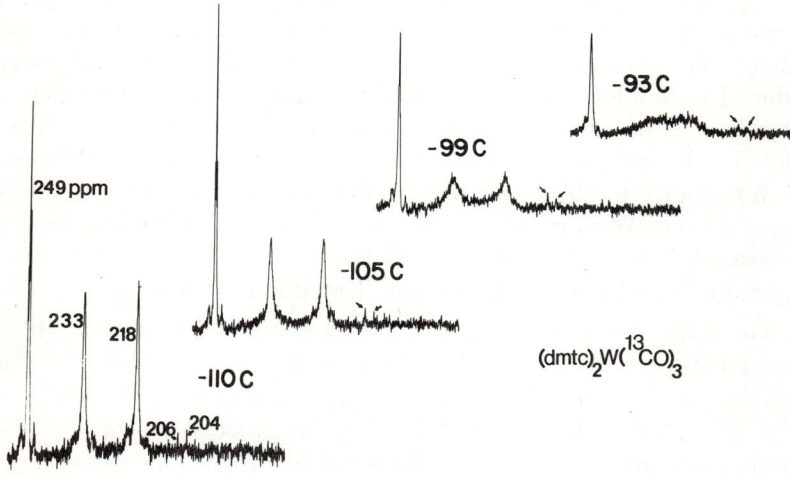

Figure 4. Variable temperature $^{13}C\text{-}\{^1H\}$ NMR spectra of ^{13}CO-enriched $W(CO)_3(dmtc)_2$ from $-93°$ to $-110°C$

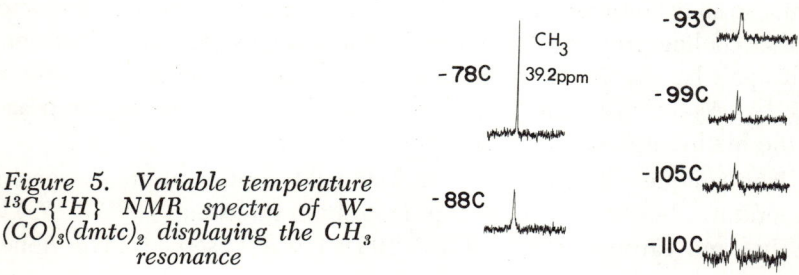

Figure 5. Variable temperature $^{13}C\text{-}\{^1H\}$ NMR spectra of $W(CO)_3(dmtc)_2$ displaying the CH_3 resonance

CO resonance with twice the intensity of the lowfield signal underwent a second broadening process at lower temperatures, however. While the lowest temperature attained ($-110°C$) failed to produce a linewidth sufficiently narrow to measure coupling constants for these two carbonyl ligands, the chemical shifts were accurately determined, and rate constants were extracted from the coalescence temperature and linewidth measurements. Close inspection of the methyl region suggests that the two signals observed at $-99°C$ may be broadening again at lower temperatures, indicating that more than two sites can exist for the four methyl groups.

In summary then, three carbonyl ligands undergo two distinct exchange processes while the dithiocarbamate ligands provide two additional sites at which to accumulate exchange data, carbons C and CH_3. Table I lists the activation barrier, $\Delta G^{\ddagger}$, calculated from the Eyring equation (1), where k_{ex} was calculated according to the Gutowsky–Holm equation at the coalescence temperature (22), and the fast exchange and

Table I. Activation Barriers

C	$T(°C)^a$	$k_{ex}(sec^{-1})^b$	$\Delta G^{\ddagger}(kcal\ mol^{-1})$
CH_3	-88	70	9.1
	$-91*$	50	9.1
	-99	28	8.9
			9.0 ± 0.1^c
R_2NCS_2	-65	1300	9.1
	-78	290	9.1
	$-83*$	141	9.1
	-93	60	8.9
	-99	25	8.9
			9.0 ± 0.1
CO 1,2,3	$-65*$	1290	9.1
	-78	310	9.0
	-88	110	8.9
			9.0 ± 0.1
CO 2,3	$-91*$	840	8.0
	-93	760	8.2
	-99	220	8.1
	-105	110	8.0
	-110	70	8.0
			8.1 ± 0.1

[a] The * identifies the coalescence temperature, T_c.
[b] Rate constants were calculated as indicated in the text.
[c] The $\pm$ values of $\Delta G^{\ddagger}$ listed serve only as an indication of the agreement of these determinations and do not represent error limits.

$$k_{ex} = \frac{kT}{h} \exp(-\Delta G^{\ddagger}/RT) \quad (1)$$

$$T > T_c \qquad k_{ex} = \frac{\pi(\Delta\nu_{AB})^2}{2\delta\nu} \quad (2)$$

$$T = T_c \qquad k_{ex} = \frac{\pi(\Delta\nu_{AB})}{\sqrt{2}} \quad (3)$$

$$T < T_c \qquad k_{ex} = \pi(\delta\nu) \quad (4)$$

slow exchange approximations were used above and below T_c, respectively (23). Here $\Delta\nu_{AB}$(Hz) is the difference in the chemical shift of the exchanging sites, and $\delta\nu$(Hz) is the linewidth caused by exchange phenomena. For the CH_3 and C analyses, all three approximations are valid, but for the carbonyl processes only the coalescence point and the slow exchange treatments are applicable since the three-site exchange which occurs initially complicates the analysis until the first carbonyl T_c is reached. For large $\Delta\nu_{AB}$, as in the three-site CO case, the Gutowsky–Holm equation is still applicable even though the system is not a simple two-site problem (24). Below the first T_c the unique lowfield CO resonance will display a linewidth reflecting the first-order decay which determines the lifetime at this site. This will be equal to twice the decay rate at the remaining two equivalent CO sites, but since the second exchange process begins to influence this portion of the spectrum below T_c, the slower rates required for interchange of the two equivalent sites with the third CO could not be independently determined. In a similar manner, the two-site CO exchange could be independently analyzed only at temperatures below the region where the higher energy three-site process was not influencing line shapes.

The results tabulated in Table I can be interpreted as a function of two exchange processes. The similar values of $\Delta G^{\ddagger}$ for the high temperature CO process and the C and CH_3 interconversion barriers suggest that one intramolecular rearrangement is responsible for all three of these averaging processes. The previous analysis of the natural abundance ^{13}C NMR data can be used again, but now the limiting spectrum requires a distinct site for each dithiocarbamate C carbon, consistent with projection B rather than projection A. All other details of the rotation of the pseudo C_3 axis vs. the two bidentate ligand sites are in agreement with the experimental data. The barrier of 9.0 kcal mol^{-1} is similar to values found for other $M(CO)_3$ moieties which are fluxional (25).

The low temperature CO exchange is definitely not responsible for averaging either the CH_3 or C since the activation energy differs by nearly 1 kcal mol^{-1}. A more vivid statement of this fact is that T_c for the CH_3 groups is approximately equivalent to T_c for the low temperature CO exchange, so the rates at $-91°C$ have a ratio equal to the ratio of $\Delta\nu_{AB}(CO)$ to $\Delta\nu_{AB}(CH_3)$ and thus differ by a factor of 13. It seems likely that this second fluxional process results from the facile interconversion of seven-coordinate geometries. A direct exchange of the two carbonyls is unlikely since the path of least motion is not available in the simple two-site case where the ligands would be required to pass through one another.

We now refine our previous model in light of the requirement that each carbonyl environment be unique. A further consideration involves the possibility that more than two sites exist for the CH_3 groups but that the low temperature process responsible for averaging $CO(2)$ and $CO(3)$ also averages the CH_3 sites to produce only two signals. Assuming $\Delta\nu_{AB}$ for the CH_3 sites to be 10 Hz or less, and certainly one expects the methyl environments will differ at most slightly, it is possible to calculate a second T_c for the methyls based on $\Delta G^{\ddagger} = 8.1$ kcal mol^{-1}. The resultant prediction for T_c has an upper limit of $160°K$ which is below the lowest temperature accessible in our experiments. In fact, the methyl signals seem to broaden slightly compared with the C carbons which become increasingly narrow at $-110°C$. We interpret this as indicating nonequivalent chemical shifts for the four methyl carbons in the final low temperature limit.

The above criteria lead us to postulate a ground state structure possessing C_1 symmetry. Consider the idealized pentagonal bipyramid (PB) substituted as shown, so no symmetry other than the identity is present. The numbering scheme serves to label the three COs (1, 2, and 3), the two Cs (45 and 67), and the four CH_3s (4, 5, 6, and 7) for our discussion. A small rotation of the equatorial bidentate dmtc ligand converts the PB to a capped trigonal prism (CTP) with $C(1)$ the capping ligand, $C(2)$, $C(3)$, $S(6)$, and $S(7)$ the quadrilateral face ligands, and

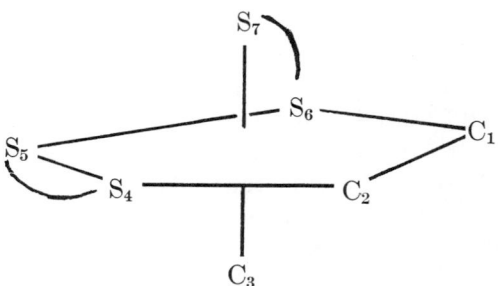

Table II. ^{13}C NMR Data
Chemical Shifts in ppm Relative to Me_4Si

^{13}C	$35°C$	$-78°C$	$-110°C$
CH_3	39.2	39.2	38.4
			39.4
R_2NCS_2	208.7	206	203.7
			206.3
CO	233.1	226	218.2
		248	233.2
			248.8

S(4) and S(5) in the unique edge positions. From this C_s form, the seven-coordinate species can equilibrate C(2) and C(3) while C(1) remains unique. According to both paths 1 and 2, the two C positions (45 and 67) remain unique in agreement with the observed data. The above rearrangements both exchange $CH_3(6)$ and $CH_3(7)$, but only Scheme 1 interconverts $CH_3(4)$ and $CH_3(5)$. The CH_3 data suggests that indeed some CH_3 exchange is occuring at $-110°C$, but in the absence of a true limiting spectrum to identify the four distinct CH_3 shifts, one cannot eliminate the possibility that $CH_3(4)$ and $CH_3(5)$ remain distinct while only $CH_3(6)$ and $CH_3(7)$ exchange. The slight asymmetry of the two CH_3 signals observed between $-99°C$ and $-110°C$ and the broadening evident at $-110°C$ could be reconciled with either of the above interconversions.

While detailed geometrical discussions must await crystallographic data regarding the solid state structure, the only assumption crucial to the above analysis is that the carbonyls occupy adjacent vertices of the coordination sphere. The general 4:3 geometry or some distortion thereof can then be applied to analysis of the high energy exchange phenomenon, and the lack of symmetry in the true ground state can be represented by deviation from an idealized pentagonal bipyramid with the axial sites occupied by one sulfur and one carbon.

Recognition that the postulated C_s CTP intermediate has the essential features of the 4:3 geometry discussed earlier allows us to consider the higher energy process with $\Delta G^{\ddagger} = 9.0$ kcal mol^{-1} in terms of this effective geometry in which C(2) and C(3), $CH_3(4)$ and $CH_3(5)$, and $CH_3(6)$ and $CH_3(7)$ are equivalent. The rotational process described earlier then interchanges all three carbonyl ligands as well as the two dithiocarbamate carbons and the four methyl groups. The combined effects of these two distinct dynamic processes are consistent with the observed temperature-dependent NMR properties.

22. TEMPLETON   Seven-Coordinate Tungsten Complexes   277

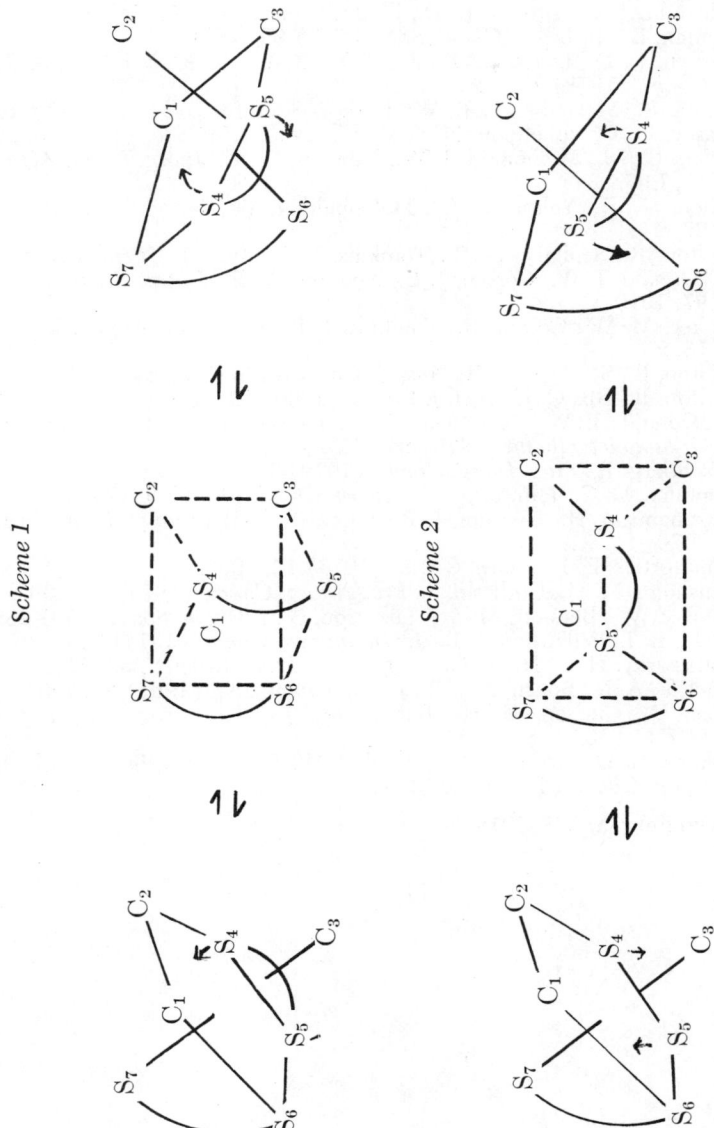

Literature Cited

1. Drew, M. G. B., *Prog. Inorg. Chem.* (1977) **23**, 67.
2. Datta, S., Wreford, S. S., *Inorg. Chem.* (1977) **16**, 1134.
3. Albright, J. O., Brown, L. D., Datta, S., Kouba, J. K., Wreford, S. S., Foxman, B. M., *J. Am. Chem. Soc.* (1977) **99**, 5518.
4. Brown, L. D., Datta, S., Kouba, J. K., Smith, L. K., Wreford, S. S., *Inorg. Chem.* (1978) **17**, 729.
5. Datta, S., McNeese, T. J., Wreford, S. S., *Inorg. Chem.* (1977) **16**, 2661.
6. Casey, C. P., Polichnowski, S. W., *J. Am. Chem. Soc.* (1977) **99**, 6097.
7. Chen, G.J.-J., McDonald, J. W., Newton, W. E., *Inorg. Chim. Acta* (1976) **19**, L67.
8. Chen, G.J.-J., Yelton, R. O., McDonald, J. W., *Inorg. Chim. Acta* (1977) **22**, 249.
9. Colton, R., Scollary, G. R., Tomkins, I. B., *Aust. J. Chem.* (1968) **21**, 15.
10. McDonald, J. W., Corbin, J. L., Newton, W. E., *J. Am. Chem. Soc.* (1975) **97**, 1970.
11. Anker, M. W., Colton, R., Tomkins, I. B., *Rev. Pure Appl. Chem.* (1968) **18**, 23.
12. Colton, R., Scollary, G. R., *Aust. J. Chem.* (1968) **21**, 1427.
13. Colton, R., Rix, C. J., *Aust. J. Chem.* (1969) **22**, 305.
14. McDonald, J. W., Newton, W. E., Creedy, C. T. C., Corbin, J. L., *J. Organomet. Chem.* (1975) **92**, C25.
15. Eisenberg, R., *Prog. Inorg. Chem.* (1970) **12**, 295.
16. Pantaleo, D. C., Johnson, R. C., *Inorg. Chem.* (1970) **9**, 1248.
17. Abrahamson, H., Heiman, J. R., Pignolet, L. H., *Inorg. Chem.* (1975) **14**, 2070.
18. Muetterties, E. L., *Inorg. Chem.* (1973) **12**, 1963.
19. Chisholm, M. H., Godleski, S., *Prog. Inorg. Chem.* (1976) **20**, 299.
20. Abel, E. W., Bennett, M. A., Wilkinson, G., *J. Chem. Soc.* (1959) 2323.
21. Todd, L. J., Wilkinson, J. R., *J. Organomet. Chem.* (1974) **77**, 1.
22. Gutoswsky, H. S., Holm, C. H., *J. Chem. Phys.* (1956) **25**, 1228.
23. Anet, F. A. L., Bourn, A. J. R., *J. Am. Chem. Soc.* (1967) **89**, 760.
24. Kost, D., Carlson, E. H., Raban, M., *J. Chem. Soc., Chem. Commun.* (1971) 656.
25. Adams, R. D., Cotton, F. A., Cullen, W. R., Hunter, D. L., Mihichuk, L., *Inorg. Chem.* (1975) **14**, 1395.

RECEIVED February 22, 1978.

Effect of Aluminum Additions on the Thermodynamic and Structural Properties of LaNi$_{5-x}$Al$_x$ Hydrides

MARSHALL H. MENDELSOHN and DIETER M. GRUEN

Chemistry Division, Argonne National Laboratory, Argonne, IL 60439

AUSTIN E. DWIGHT[1]

Materials Science Division, Argonne National Laboratory, Argonne, IL 60439

Desorption isotherms for the hydrides of LaNi$_{4.6}$Al$_{0.4}$ and LaNi$_{4.5}$Al$_{0.5}$ are presented and values for the enthalpy and entropy changes of the hydriding reactions are calculated from the van't Hoff plots of log P vs. 1/T. A crystallographic model of LaNi$_4$Al is shown and consideration of the nearest neighbor atom distribution leads to a rationalization of the observed linear relationship between the enthalpy change, ΔH, and the aluminum composition. Brief discussions of methods to predict dissociation pressures or interstitial site occupation are included. The cubic and hexagonal AB$_5$ phases are compared and, finally, the application of these alloys in chemical heat pump systems is noted.

A great deal of interest exists in alloys of general composition AB$_5$, mainly because of their remarkable ability to absorb hydrogen rapidly and reversibly at moderate pressures near room temperature. LaNi$_5$, which crystallizes with the CaCu$_5$ structure, has been thoroughly investigated. Although substitutions of 20% of other lanthanides for lanthanum or other transition metals for nickel have been known for some time to change the equilibrium hydrogen pressure by a factor of

---

[1] Present address: Department of Physics, Northern Illinois University, Dekalb, IL 60115

about four (1), it has been shown recently that aluminum substitutions are particularly effective in lowering the hydrogen pressure by a factor of about 300 in going from $LaNi_5$ to $LaNi_4Al$ (2, 3).

Aluminum substitutions have been studied in four other AB_5 alloys in addition to $LaNi_5$. The hexagonal $CaCu_5$ structure of $ThNi_5$ is maintained up to the composition $ThNi_2Al_3$ without undergoing a phase transition (4). Cubic UCu_5 transforms to the hexagonal $MgZn_2$ structure at the composition $UCu_{4.5}Al_{0.5}$ and to the hexagonal $CaCu_5$ structure at the composition $UCu_{3.5}Al_{1.5}$ (5). Cubic $ZrNi_5$ and cubic UNi_5 both form single-phase solid solutions up to the compositions $ZrNi_4Al$ and UNi_4Al (6, 7). Of these systems, only the hydrogen absorption properties of $Th(Ni,Al)_5$ ternaries have been studied recently (8). As with $La(Ni,Al)_5$ ternaries, there appears to be a large decrease in the equilibrium plateau pressure for $Th(Ni,Al)_5$, except that $ThNi_2Al_3$ did not appear to absorb any hydrogen (8).

The present work reports hydrogen dissociation pressures obtained as a function of temperature on homogeneous samples of $LaNi_{5-x}Al_x$ ternary alloys. The derived thermodynamic values are considered to be more accurate than those reported by us on less well-characterized samples (2). Some apparent relationships of the structure of the alloys to the chemical properties of the hydrides are also discussed.

Experimental

$LaNi_{4.6}Al_{0.4}$ and $LaNi_{4.5}Al_{0.5}$ were both prepared by the Denver Research Institute under a contract to Argonne National Laboratory. The alloys were prepared by standard inert atmosphere, arc melting procedures. After melting, the alloys were heat treated at 1050°C for 2 hr in a sealed quartz tube in an inert atmosphere. Weight losses during melting were on the order of 0.1%. The samples were characterized metallographically as a single-phase material and were also checked at Argonne by x-ray diffraction.

Samples, weighed to ± 0.0001 g, were placed in an all-316 stainless steel reactor equipped with a welded 1μ porous stainless steel filter disk. The reactor was connected to an all-316 stainless steel high pressure manifold with connections to 0–1000 psia and 0–100 psia Sensotec pressure transducers (accuracy ± 0.1%), a vacuum pump, and a high pressure hydrogen gas cylinder. Hydrogen was Matheson's ultra-high purity grade (99.999% min). The reactor was immersed in a bath of dixylylethane at temperatures up to 130°C and Dow Corning 710 silicone oil above 130°C. Temperatures were maintained to ± 0.2°C at the lower temperatures to ± 0.8°C at the higher temperatures with YSI models 71 and 72 temperature controllers. Equilibration was considered achieved when the pressure remained constant for a period of about 15 min. Equilibration times were from 30 min to 16 hr. To obtain an alloy that rapidly absorbs and desorbs hydrogen, it is necessary in most cases to use an activation procedure.

Activation is usually accomplished by exposing the alloy to higher pressures and/or temperatures than subsequently required for hydrogen absorption. In many cases these conditions can only be determined by trial and error. $LaNi_{4.6}Al_{0.4}$ was activated by simply evacuating the reactor containing the alloy for several minutes and exposing the sample to 250 psi H_2 for 1 hr. The $LaNi_{4.5}Al_{0.5}$ sample was slightly more difficult to activate. This sample was heated to about 90°C in 200 psi H_2, then evacuated and pumped on for about 20 min while cooling, and finally exposed to 300 psi H_2 for about 2 hr. Desorption isotherms were determined by removing measured amounts of hydrogen and equilibrating the system.

Results

One method of determining the thermodynamic properties of alloy–hydrogen systems is to measure their hydrogen equilibrium pressures as a function of the hydrogen contained in the solid phases. These pressure–composition isotherms generally can be divided into three regions. Starting with the hydrided alloy, there is first a region of rapid pressure decrease without much change in hydrogen composition. This is the β phase or hydride region. Next there is a plateau region of constant or relatively constant pressure where the two solid phases ($\alpha + \beta$) co-exist. Finally, there is again a region of rapid decrease in pressure, this being the α phase or solid solution region of hydrogen in the alloy.

Hydrogen dissociation pressures for the plateau region of $LaNi_{4.6}Al_{0.4}H_y$ are shown in Figure 1a at two temperatures comparing data from the homogeneous samples to earlier results. The reproducibility of a desorption isotherm is indicated by the data in Figure 1b taken from two separate runs. The error in β-phase composition appears to be $\pm 2\%$ while the error in the plateau pressure region appears to be $\pm 1\%$. Tables I and II give the experimental pressure–composition data for $LaNi_{4.6}Al_{0.4}H_y$ at six temperatures and $LaNi_{4.5}Al_{0.5}H_y$ at five temperatures, respectively. Figures 2 and 3 show a plot of these data for three temperatures. Figure 4 shows a plot of log P vs. $1/T$ for both alloy hydrides. For $LaNi_{4.6}Al_{0.4}H_y$, the pressures shown in Figure 4 were obtained at a composition $y = 2.75$ while for $LaNi_{4.5}Al_{0.5}H_y$, the pressures were obtained at $y = 2.50$. The errors introduced into the determination of ΔH resulting from the somewhat arbitrary decision to use pressures near the middle of the plateau are greatly reduced in the present work over the earlier data (2) because of the relative "flatness" of the plateaus. From a least squares fit of the log P vs. $1/T$ plot, the enthalpy (ΔH) and the entropy (ΔS) of transition were calculated from the slope and intercept, respectively. These values are given in Table III and for $LaNi_{4.6}Al_{0.4}H_y$ are compared with an earlier reported value. The errors are

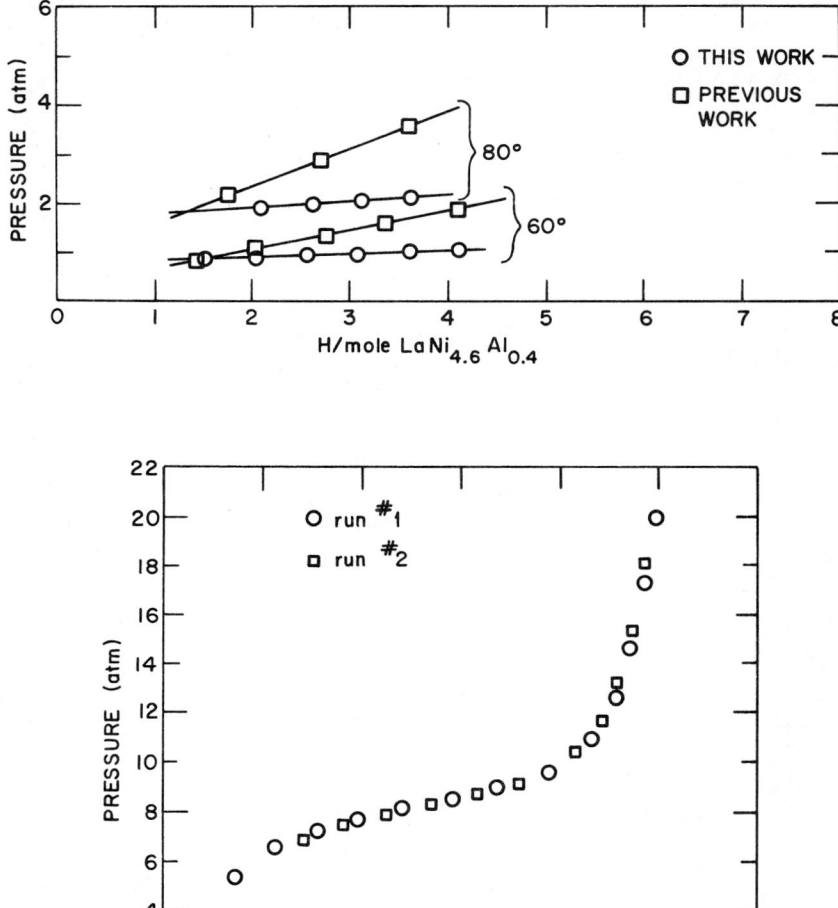

Figure 1. (a) Comparison of plateau pressure data; (b) reproducibility of two separate runs

Table I. Pressure–Composition Data for LaNi$_{4.6}$Al$_{0.4}$

Temperature (°C)	Composition (H:M)	Pressure (Atm)	Temperature (°C)	Composition (H:M)	Pressure (Atm)
30° ± 0.2°	5.87	7.72	80° ± 0.2°	5.50	12.1
″	5.76	4.18	″	5.42	9.69
″	5.59	1.91	″	5.34	7.75
″	5.18	0.89	″	5.23	6.09
″	4.93	0.54	″	5.07	4.56
″	4.48	0.34	″	4.85	3.37
″	3.94	0.30	″	4.57	2.63
″	3.38	0.29	″	4.14	2.27
″	2.81	0.28	″	3.63	2.16
″	1.73	0.26	″	3.13	2.08
″	1.22	0.24	″	2.62	2.01
″	0.83	0.22	″	2.11	1.93
48° ± 0.2°	5.78	7.79	″	1.58	1.88
″	5.69	5.34	″	1.12	1.67
″	5.60	3.89	″	0.64	1.47
″	5.46	2.52	″	0.40	0.67
″	5.22	1.48	100° ± 0.5°	4.99	12.3
″	4.79	0.86	″	4.88	10.4
″	4.20	0.71	″	4.73	8.23
″	3.66	0.66	″	4.54	6.54
″	3.13	0.63	″	4.47	5.41
″	2.59	0.61	″	4.18	5.03
″	2.11	0.58	″	3.70	4.54
″	1.57	0.55	″	3.41	4.46
″	1.04	0.51	″	3.04	4.23
″	0.55	0.43	″	2.74	4.22
″	0.29	0.22	″	2.41	4.07
60° ± 0.2°	5.60	9.78	″	1.82	3.82
″	5.55	7.89	″	1.25	3.64
″	5.47	6.12	″	0.73	3.27
″	5.35	4.13	″	0.43	2.46
″	5.16	2.89	″	0.26	0.95
″	4.90	1.87	″	0.15	0.26
″	4.56	1.27	126° ± 0.8°	4.90	16.2
″	4.10	1.07	″	4.75	13.8
″	3.59	1.02	″	4.51	11.5
″	3.07	0.98	″	4.24	10.3
″	2.56	0.95	″	3.87	9.59
″	2.03	0.92	″	3.35	9.13
″	1.51	0.87	″	2.77	8.74
″	1.01	0.80	″	2.19	8.42
″	0.52	0.68	″	1.67	8.06
″	0.28	0.28	″	1.26	7.50
			″	0.75	6.53
			″	0.56	4.52
			″	0.38	1.39
			″	0.27	0.24

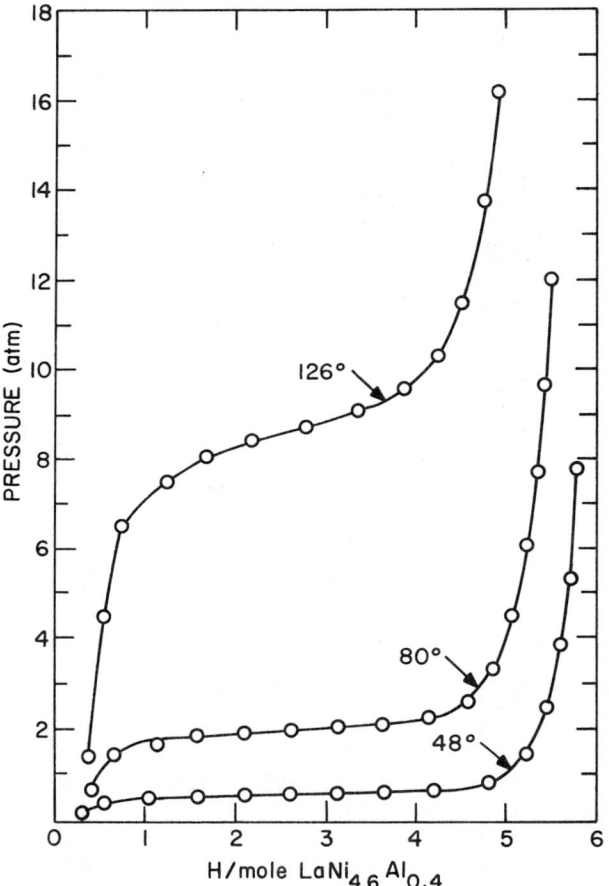

Figure 2. Desorption isotherms for $LaNi_{4.6}Al_{0.4}$

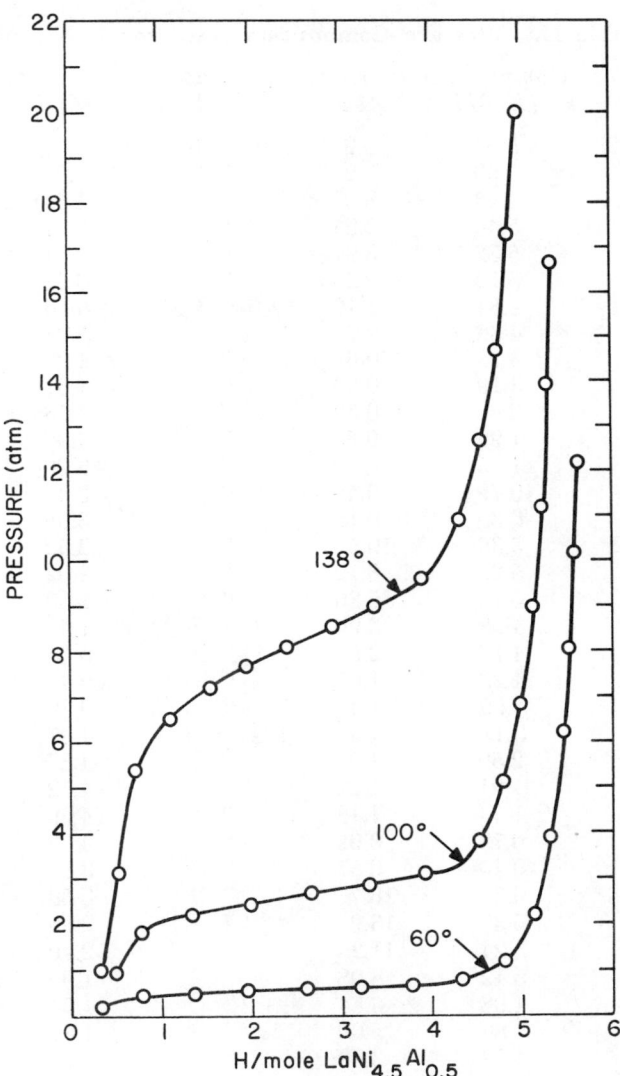

Figure 3. Desorption isotherms for $LaNi_{4.5}Al_{0.5}$

Table II. Pressure–Composition Data for LaNi$_{4.5}$Al$_{0.5}$

Temperature (°C)	Composition (H:M)	Pressure (Atm)	Temperature (°C)	Composition (H:M)	Pressure (Atm)
60° ± 0.2°	5.63	12.2	100° ± 0.6°	2.64	2.69
"	5.59	10.2	"	1.97	2.43
"	5.53	8.07	"	1.34	2.23
"	5.46	6.24	"	0.77	1.84
"	5.32	3.91	"	0.51	0.94
"	5.12	2.23	"	0.38	0.17
"	4.81	1.19	120° ± 0.8°	5.16	19.3
"	4.38	0.78	"	5.07	16.3
"	3.78	0.67	"	4.96	13.4
"	3.19	0.63	"	4.81	10.6
"	2.60	0.59	"	4.58	8.13
"	1.96	0.55	"	4.30	6.60
"	1.35	0.50	"	3.96	5.84
"	0.79	0.44	"	3.48	5.43
"	0.35	0.18	"	3.08	5.17
80° ± 0.2°	5.36	10.5	"	2.52	4.90
"	5.28	8.12	"	2.02	4.63
"	5.15	5.86	"	1.50	4.29
"	4.98	3.99	"	1.04	3.88
"	4.73	2.55	"	0.65	3.22
"	4.22	1.65	"	0.44	1.76
"	3.69	1.47	"	0.32	0.63
"	3.11	1.38	138° ± 1°[a]	4.97	20.0
"	2.50	1.31	"	4.87	17.3
"	1.91	1.22	"	4.72	14.7
"	1.33	1.12	"	4.56	12.6
"	0.79	0.99	"	4.32	10.9
"	0.40	0.57	"	3.89	9.59
100° ± 0.6°	5.35	16.7	"	3.38	8.97
"	5.30	13.9	"	2.91	8.54
"	5.24	11.2	"	2.40	8.12
"	5.12	8.98	"	1.95	7.69
"	4.98	6.84	"	1.54	7.20
"	4.80	5.09	"	1.11	6.56
"	4.52	3.82	"	0.72	5.40
"	3.96	3.12	"	0.53	3.14
"	3.30	2.86	"	0.38	0.97

[a] Run #1.

Table III. Derived Thermodynamic Data

Alloy	$-\Delta H \left(\dfrac{kcal}{mol\ H_2}\right)$	$-\Delta S \left(\dfrac{cal}{deg - mol\ H_2}\right)$	Ref.
$LaNi_{4.6}Al_{0.4}$	8.7 ± 0.1	26.1 ± 0.3	This work
$LaNi_{4.6}Al_{0.4}$	9.1 ± 0.2	28.1 ± 0.7	2
$LaNi_{4.5}Al_{0.5}$	9.21 ± 0.04	26.6 ± 0.1	This work
$LaNi_5$	7.2 ± 0.1	26.1 ± 0.4	2

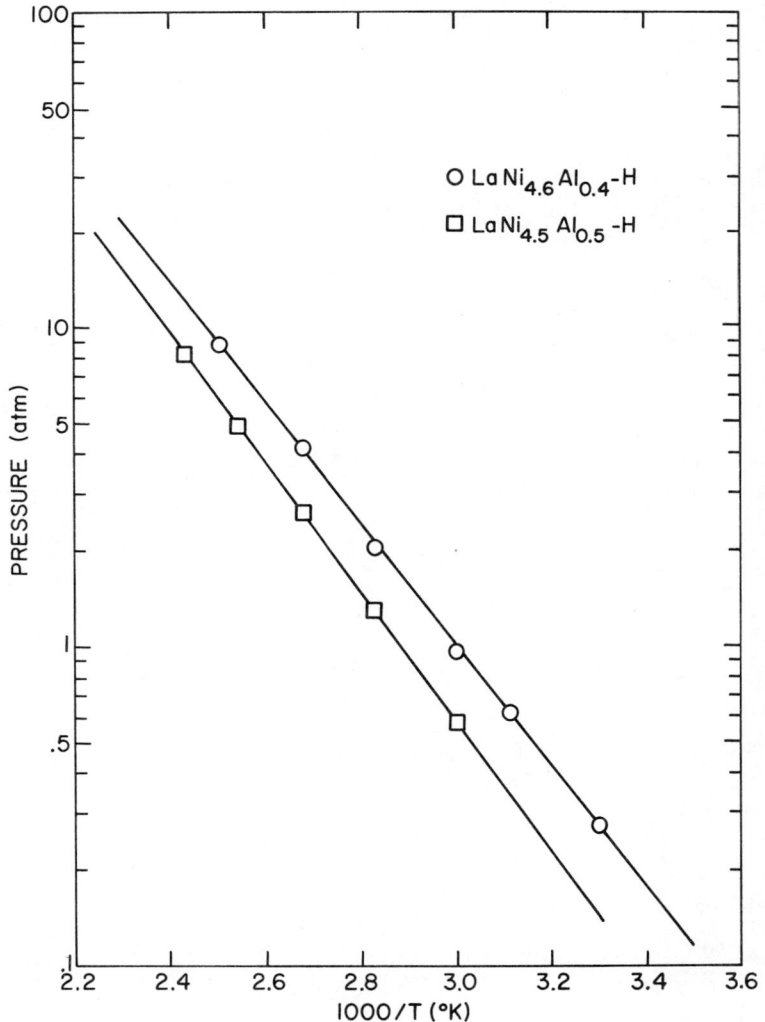

Figure 4.   Log $P_{plateaus}$ vs. 1000/temperature

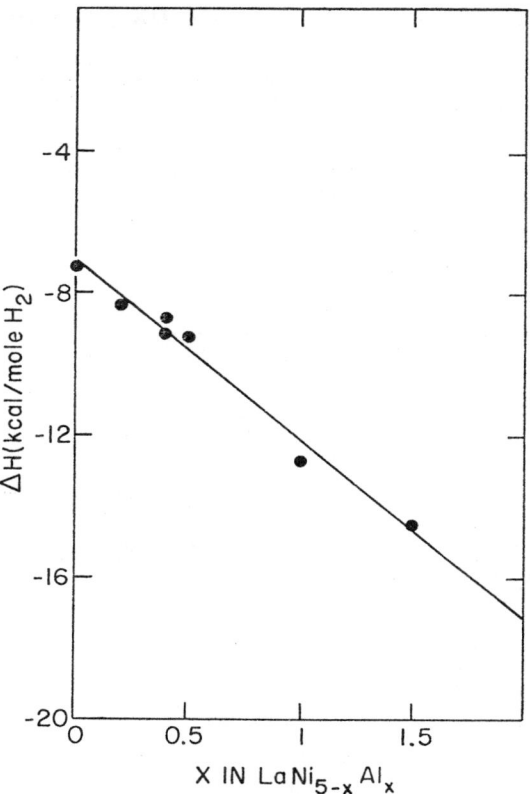

Figure 5. ΔH vs. x in $LaNi_{5-x}Al_x$

Table IV. Hysteresis in $LaNi_{5-x}Al_x$ Alloys

	Temp. (°C)	H:M	P_{abs} (atm)	P_{des} (atm)	P_A:P_D
$LaNi_{4.6}Al_{0.4}$	30	3.09	0.31	0.28	1.11
″	48	3.10	0.71	0.64	1.11
″	60	3.10	1.16	1.00	1.16
″	80	2.84	2.41	2.05	1.18
″	100	2.75	4.77	4.20	1.14
$LaNi_{4.5}Al_{0.5}$	60	2.64	0.65	0.59	1.10
″	80	2.48	1.36	1.29	1.05
″	100	2.83	2.99	2.73	1.10
″	120	2.40	5.27	4.80	1.10
$LaNi_5$[a]	20	3.0	2.0	1.6	1.25

[a] From Ref. *9*.

Table V. Crystallographic Data

Compound	a_o (Å)	c_o (Å)	V (Å^3)	ΔV/V
LaNi$_{4.6}$Al$_{0.4}$	5.018	4.030	87.89	—
	(5.037)[a]	(4.017)	(88.24)	—
LaNi$_{4.6}$Al$_{0.4}$H$_5$	5.358	4.207	104.6	0.19
	(5.343)	(4.233)	(104.7)	(0.19)
LaNi$_{4.5}$Al$_{0.5}$	5.049	4.021	88.77	—
	(5.037)	(4.032)	(88.59)	—
LaNi$_{4.5}$Al$_{0.5}$H$_{4.5}$	5.340	4.201	103.7	0.17

[a] Values in parenthesis from Ref. 21.

deviations from a least squares fit of the data. A plot of the available data for ΔH vs. the aluminum composition is shown in Figure 5. The apparent linear relationship will be discussed later.

In general, absorption pressures do not appear to be as reproducible as desorption pressures, and no extensive data of the absorption isotherms were obtained. However, for practical applications absorption pressures need to be known at least approximately. Therefore, single data points on the absorption pressure–composition diagram were taken at several temperatures for both alloys and are listed in Table IV. As can be seen from this table, the magnitude of the hysteresis as measured by the ratio $P_{absorption}:P_{desorption}$ is less than that observed for LaNi$_5$. Crystallographic data obtained on both alloys and both corresponding hydrides are collected in Table V and are compared with previous results where available.

Discussion

The conclusion that sloping plateaus are attributable to compositional inhomogeneities in the initial alloy sample (10) seems reasonable since carefully prepared alloys with different La:Ni ratios, but within the CaCu$_5$ phase field, display different plateau pressures (10). Therefore, the "flatter" plateaus of the present series of samples (shown in Figure 1a) are indicative of a better degree of homogeneity than the earlier samples.

There have been several discussions in the literature of the relationships of structure to hydride stability (11, 12). It is of interest to discuss the influence of aluminum on the structure and on the plateau pressure of AB$_5$ alloys. Up to the composition LaNi$_{3.5}$Al$_{1.5}$, the LaNi$_{5-x}$Al$_x$ alloys crystallize with the CaCu$_5$ structure. Two views of a model of this structure for LaNi$_4$Al are shown in Figure 6. Figure 6a is a view looking down the c axis and shows the configuration of nickel and aluminum atoms around the central lanthanum atoms. Figure 6b shows the layer-

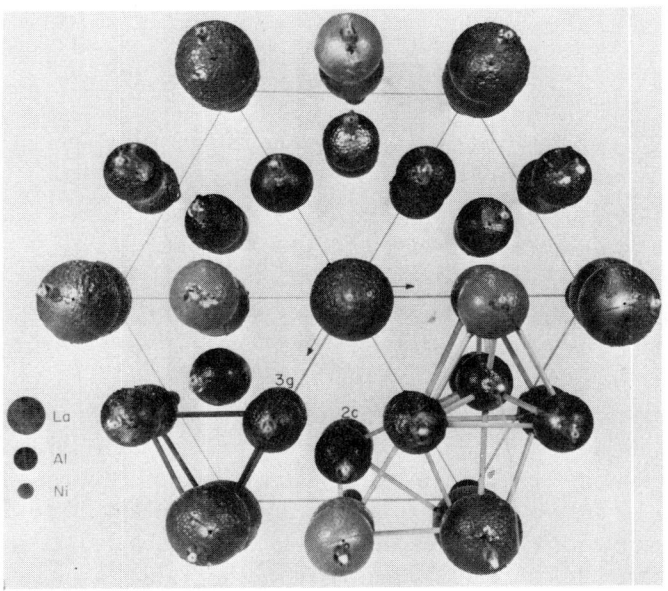

Figure 6. Two views of a model of $LaNi_4Al$

type structure parallel to the basal plane. This view allows one to discern the distinct $2c$ and $3g$ sites of nickel. On the basis of x-ray diffraction intensities (2), it has been shown that the aluminum atoms preferentially occupy the $3g$ sites. In the LaNi$_4$Al model shown in Figure 6, the aluminum atoms also have been ordered in the $3g$ sites without experimental evidence. On the basis of this model, a comparison of nearest neighbor distributions is given in Table VI for six different types of interstitial sites.

Several workers recently have reported results on neutron diffraction investigations of LaNi$_5$D$_6$ (13, 14, 15). In each of these papers, it is concluded that the symmetry of the alloy is lowered to space group $p31m$ on deuteration and that the deuterium atoms occupy $3c$ and $6d$ tetrahedral sites. These sites in the deuteride are closely related spatially to the $12n$ and $12o$ tetrahedral sites of the alloy having P6/mmm symmetry. Concentrating on the n and o sites only, one notes that the fraction of those sites containing one aluminum nearest neighbor is one-half for LaNi$_4$Al and one-fourth for LaNi$_{4.5}$Al$_{0.5}$. Thus, if one assumes the aluminum atoms to replace nickel in the ordered manner depicted in Figure 6, then the fraction of n and o sites with one aluminum nearest neighbor is a linear function of the aluminum composition. If one further assumes that hydrogens in n and o sites are bound with energies averaged over all available sites, then the "average" interstitial site bonding energy also would be expected to be proportional to the aluminum composition, thus providing a rationale for the observed linear relationship between ΔH and aluminum composition shown in Figure 5.

Shinar et al. (16) have proposed that specific interstitial site occupations can be determined by associating different binding energies with

Table VI. Nearest Neighbor Distribution around Alloy Interstitial Sites

Site Designation	LaNi$_5$			LaNi$_4$Al			
	Neighbors		Number of Sites	Neighbors			Number of Sites
	La	Ni		La	Ni	Al	
b	2	6	1	2	4	2	1
f	2	4	3	2	2	2	1
				2	4	0	2
h	0	4	4	0	3	1	4
m	2	2	6	2	2	0	2
				2	1	1	4
n	1	3	12	1	2	1	4
				1	3	0	8
o	1	3	12	1	2	1	8
				1	3	0	4

hydrogen atoms in different interstitial sites. The binding energies would be proportional to ΔH, the sum of the heats of formation of (imaginary) binary hydrides formed with the A and B atoms surrounding a particular site. On the basis of this scheme, one would predict that the hydrogen atoms prefer sites having an aluminum nearest neighbor since the heat of formation of aluminum hydride is considered to be more negative than that of nickel hydride (17). However, because the most recent measured values of ΔS for the aluminum-containing alloys are almost identical to those of $LaNi_5$ (see Table III), configurational entropy calculations (12) indicate that the distribution of occupied sites is the same for $LaNi_5$ as for the $LaNi_{5-x}Al_x$ alloys. It should be noted that although the present ΔS values are considered to be more accurate than those previously reported (2), more precise ΔS values should be available soon from precision calorimetric experiments being performed in collaboration with the Argonne National Laboratory Chemical Engineering Division calorimetry group.

Calculations from Shinar et al. for $LaNi_5H_6$ show that the octahedral $3f$ and tetrahedral $6m$ sites would be preferred for hydrogen occupation, and contention is made that this result is supported by neutron diffraction data. They state that the $3c$ and $6d$ tetrahedral interstitial sites for the deuteride "are analogous to the $3f$ and $6m$ sites in the ($LaNi_5$) mother compound (P6/mmm space group)" (16). However, the occupied $3c$ sites in the deuteride are displaced ~ 0.08 Å from $3f$ interstitial sites in $LaNi_5$ and are, in fact, closer to $12n$ interstitial sites (15), as already noted. Furthermore, $6d$ sites are never equivalent to $6m$ interstitial sites except in one special case which does not pertain to the deuteride, but they are closely related to the $12o$ interstitial sites. It seems therefore more reasonable to conclude that the hydrogens occupy tetrahedral sites in $LaNi_5H_6$ and that the neutron diffraction results support an "average" interstitial site bonding energy model over models which assume discrete bonding energies associated with every crystallographically distinct site occupied by hydrogen. In any event, the question of site occupation of hydrogen in the AB_5 hydrides remains an intriguing one and will require more precise structural and thermodynamic measurements on which to base more accurate theoretical calculations for its solution.

A scheme for correlating hydride stabilities, the so called "rule of reversed stability" (see e.g., 18, 19), states that for a series of analogous alloys, the more stable the alloy, the less stable (i.e., higher dissociation pressure) the corresponding hydride. Using Miedema's formula (20), the calculated heat of formation for $LaNi_5$ is -11.2 kJ/mol and for $LaAl_5$ is -42.1 kJ/mol. Since $LaAl_5$ is more stable (more negative ΔH) than $LaNi_5$, the rule of reversed stability predicts the $LaNi_{5-x}Al_x$ hydrides to be less stable than $LaNi_5H_6$ contrary to observation. Similarly, Shinar et al. (*16*) have found disagreement with the rule of reversed stability for the $LaNi_{5-x}Cu_x$ hydrides. It appears that in the case of the aluminum and copper ternaries, as in many other AB_5 hydride systems, the factors determining the hydrogen dissociation pressures are closely correlated with cell volumes (*2*) or interstitial hole sizes (*11*).

The two most commonly observed structure types for compounds having the composition AB_5 are the cubic UNi_5 structure and the hexagonal $CaCu_5$ structure. The relationships these structures bear to one another are important because, to our knowledge, alloys of composition AB_5 which have been reported to absorb large quantities of hydrogen are exclusively of the $CaCu_5$ structure type, although up to now it appears that the only cubic compound tested for hydrogen absorption is YNi_4Mn (*21*). The $CaCu_5$ hexagonal phase, in accordance with Dwight's empirical rule (*22*), is only stable if the radius ratio $r_A:r_B > 1.30$. Compounds with $r_A:r_B < 1.30$ crystallize in the cubic UNi_5 structure. However, it has been found that the limit 1.30 is not critical for the cubic structure, and indeed cubic compounds are known with radius ratios up to 1.42 (*23*). If one examines the ionic crystal radii for lanthanum, neodymium, thorium, erbium, and uranium, and the structures and volumes of their corresponding ANi_5 compounds, as shown in Table VII, the volume of UNi_5 seems to be anomalously small. However, the metallic radius of uranium is smaller than all of the rare earth metallic radii. This implies that uranium is not tri-valent in this metallic compound but rather of higher valency. Lacking an estimate of the metallic radius of uranium in UNi_5, advantage has been taken of the apparent linear relationship between ionic crystal radii and alloy cell volume for several ANi_5 compounds. In fact, by extrapolating such a plot, one would expect the ionic radius of uranium to be about 0.98 Å for the cell volume to be 77.9 Å^3. Thus, uranium could be either tetra-valent (CR = 1.03, Ref. *24*) or penta-valent (CR = 0.90, Ref. *24*) or possibly of intermediate valency.

Table VII.

Ion	Crystal Radius[a] (Å)	ANi_5 Structure	ANi_5 Volume (Å^3)
La^{3+}	1.172	hexagonal	86.7[b]
Nd^{3+}	1.123	hexagonal	84.3[b]
Th^{4+}	1.08	hexagonal	84.1[c]
Er^{3+}	1.030	hexagonal	81.0[b]
U^{+3}	1.165	cubic	77.9[d]

[a] All values from Ref. *24*.
[b] From Ref. *25*.
[c] From Ref. *4*.
[d] From Ref. *6*.

Finally, a two-metal hydride concept operating as a chemical heat pump for storage and recovery of thermal energy for heating, cooling, and energy conversion has been proposed (26) and is currently being tested (27). Hydrogen gas is transferred from one hydride bed by thermal energy input at a characteristic temperature to a second bed where hydrogen is absorbed, and thermal energy is released at another characteristic temperature. This concept is illustrated diagrammatically in Figure 7. The heat pump application becomes most efficient for the case when the values of ΔS for the hydriding reactions of the metal hydride pair are equal (28), a property closely approximated by the $LaNi_{5-x}Al_x$ alloys. This property together with the ability to vary the hydrogen dissociation pressure over wide ranges make the $LaNi_{5-x}Al_x$ alloy system of particular interest for chemical heat pump applications.

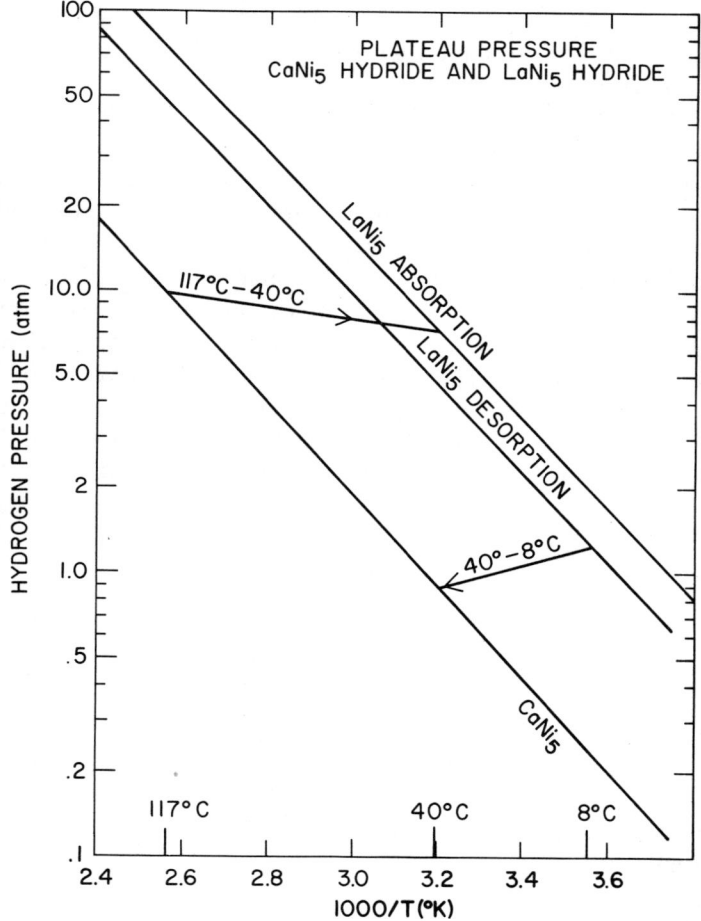

Figure 7. *Representation of a two metal hydride chemical heat pump*

Acknowledgment

We wish to thank S. Peterson for helpful discussions.

Literature Cited

1. van Mal, H. H., Buschow, K. H. J., Miedema, A. R., *J. Less-Common Met.* (1974) **35**, 65.
2. Mendelsohn, M. H., Gruen, D. M., Dwight, A. E., *Nature* (1977) **269**, 45.
3. Achard, J. C., Percheron-Guegan, A., Diaz, H., Briancourt, F., Denany, F., *Int. Congr. on Hydrogen in Metals, 2nd, Paris, France, 1977.*
4. Ban, Z., Sikirica, M., Raseta, R., *J. Less-Common Met.* (1967) **12**, 478.
5. Blazina, Z., Ban, Z., *Z. Naturforsch.* (1973) **28b**, 561.
6. Blazina, Z., Ban, Z., *J. Less-Common Met.* (1973) **33**, 321.
7. Blazina, Z., Ban, Z., *Croat. Chem. Acta* (1971) **43**, 59.
8. Takeshita, T., Wallace, W. E., *J. Less-Common Met.* (1977) **55**, 61.
9. Kuijpers, F. A., van Mal, H. H., *J. Less-Common Met.* (1971) **23**, 395.
10. Buschow, K. H. J., van Mal, H. H., *J. Less-Common Met.* (1972) **29**, 203.
11. Lundin, C. E., Lynch, F. E., Magee, C. B., *J. Less-Common Met.* (1977) **56**, 19.
12. Gruen, D. M., Mendelsohn, M. H., *J. Less-Common Met.* (1977) **55**, 149.
13. Andresen, A. F., *Proc. Int. Symp. on Hydrides for Energy Storage, Geilo, Norway, August 1977.*
14. Fischer, P., Furrer, A., Busch, G., Schlapbach, L., *Proc. Int. Symp. on Hydrides for Energy Storage, Geilo, Norway, August, 1977.*
15. Bowmann, A. L., Anderson, J. L., Nereson, N. G., *Proc. Rare Earth Res. Conf., 10th, Carefree, Arizona, 1973.*
16. Shinar, J., Jacob, I., Davidov, D., Shaltiel, D., *Proc. Int. Symp. on Hydrides for Energy Storage, Geilo, Norway, August, 1977.*
17. "Metal Hydrides," W. M. Mueller, J. P. Blackledge, G. G. Libowitz, Eds., Academic, New York, 1968.
18. van Mal, H. H., Buschow, K. H. J., Miedema, A. R., *J. Less-Common Met.* (1974) **35**, 65.
19. Buschow, K. H. J., van Mal, H. H., Miedema, A. R., *J. Less-Common Met.* (1975) **42**, 163.
20. Miedema, A. R., *J. Less-Common Met.* (1976) **46**, 67.
21. Mendelsohn, M. H., Gruen, D. M., Dwight, A. E., *Proc. Rare Earth Res. Conf., 13th, Wheeling, West Virginia, October, 1977.*
22. Dwight, A. E., *Trans. ASM* (1961) **53**, 479.
23. Buschow, K. H. J., van der Goot, A. S., Birkham, J., *J. Less-Common Met.* (1969) **19**, 433.
24. Shannon, R. D., *Acta Crystallogr.* (1976) **A32**, 751.
25. Wernick, J. H., Geller, S., *Acta Crystallogr.* (1959) **12**, 662.
26. Gruen, D. M., et al., *Proc. Intersoc. Energy Conver. Engin. Conf., 11th, Stateline, Nevada, 1976.*
27. Gruen, D. M., et al., Argonne National Laboratory Report, **ANL-77-39**, June, 1977.
28. Gruen, D. M., Mendelsohn, M. H., Sheft, I., *Sol. Energy* (1978) **21**(2).

RECEIVED March 7, 1977. Work performed under the auspices of the Division of Basic Energy Sciences of the Department of Energy.

24

Sunlight Engineering Efficiency of Thin-Layer Iron–Thiazine Photogalvanic Cells

Evidence That Surface-Induced Back Reaction Is a Key Limiting Factor

NORMAN N. LICHTIN[1], PETER D. WILDES, and TERRY L. OSIF
Department of Chemistry, Boston University, Boston, MA 02215

DALE E. HALL[2]
Exxon Research & Engineering Co., Linden, NJ 07036

Absorption spectra and redox potentials in acid solution limit sunlight engineering efficiency (S.E.E.) of unsensitized iron–thiazine photogalvanic cells to ~ 2%. The highest S.E.E. value obtained with totally illuminated single thin-layer (TI-TL) iron–thionine cells with SnO_2 anodes and Pt cathodes, .036%, corresponds to $V_{power\ point}$ ~ 35% of theoretical limit. Potentials at the selective anode are dominated by the dye–leucodye couple. Potentials at the poorly selective cathode are dominated by the iron couple. I_{sc} varies linearly with photostationary concentration of leucothionine and, with electrode spacing $\leqslant 50\mu m$, is not limited by solution lifetime of charge carriers. Inefficient electron transfer at the electrodes is believed to reduce S.E.E. by a factor of ~ 5, possibly because of surface-promoted back reaction on SnO_2.

A photogalvanic cell is a system of many components that must be carefully matched to each other and, for use as a solar transducer, must be matched to the insolation spectrum. No practical photogalvanic cell has yet been achieved. Rational optimization of photogalvanic con-

---

[1] Senior author.
[2] Present address: Paul D. Merica Research Laboratory, The International Nickel Co., Suffern, NY 10901

version requires understanding of all aspects of the system. These aspects include photochemistry, electrochemistry, ground-state solution chemistry, device design, and device materials.

Iron–thiazine photogalvanic cells use the photoredox reactions of Fe(II) with thiazine dyes, represented for thionine by Reactions 1, 2, 3, 4, and 5, to convert photon energy into chemical potential. The spontaneous ground state reactions represented by Reactions 6, 7, 8, and 9 also occur in homogeneous solution during illumination. Photogalvanic action results when homogeneous Reactions 7, 8, and 9 are replaced by anodic oxidation of TH_4^{2+} and $TH_2\cdot^+$ coupled with cathodic reduction of Fe(III). The free energy change for the oxidation of leucothionine, TH_4^{2+}, to thionine, TH^+, by Fe(III) in aqueous sulfuric acid at pH = 2 corresponds to $E^{\circ\prime} = 0.28$ V (1).

Established Elementary Steps in the Iron–Thionine Photoredox Reaction in Acid Solution

$$TH^+(S_0) \xrightarrow[\sim 600 \text{ nm}]{h\nu} TH^+(S_1) \tag{1}$$

$$TH^+(S_1) \to TH^+(S_0) \tag{2}$$

$$TH^+(S_1) \to TH^+(T_1) \xrightarrow{H^+} TH_2^{2+}(T_1) \tag{3}$$

$$TH_2^{2+}(T_1) \to TH^+(S_0) + H^+ \tag{4}$$

$$TH_2^{2+}(T_1) + Fe(II) \to TH_2\cdot^+ + Fe(II) \tag{5}$$

$$2TH_2\cdot^+ + H^+ \rightleftarrows TH^+ + TH_4^{2+} \tag{6}$$

$$TH_4^{2+} + Fe(III) \rightleftarrows \text{"Complex"} \tag{7}$$

$$\text{"Complex"} \to TH_2\cdot^+ + 2H^+ + Fe(II) \tag{8}$$

$$TH_2\cdot^+ + Fe(II) \to TH^+ + H^+ + Fe(II) \tag{9}$$

$TH^+ =$

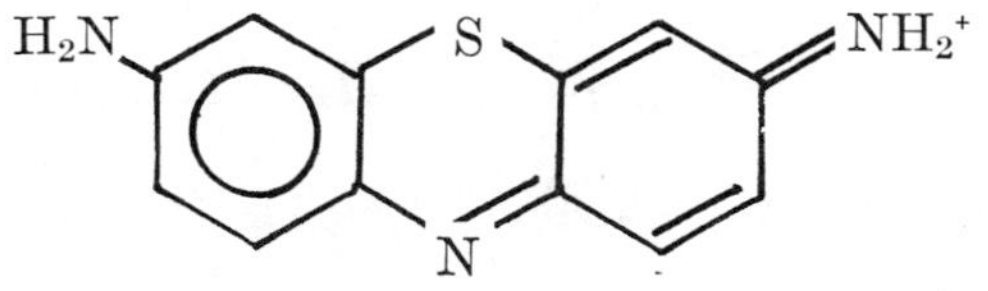

Thionine

λ_{max} 601 nm

The pioneering work of Rabinowitch (2) achieved photogalvanic transduction by means of a cell in which a platinum electrode in contact with illuminated solution served as the anode while a similar electrode in contact with solution maintained in the dark served as the cathode. We have been concerned with understanding and optimizing an area device, the totally illuminated thin-layer (TI-TL) photogalvanic cell, first described by Clark and Eckert (3). In this device photogalvanic transduction has been achieved by using one transparent electrode, usually n-type SnO_2, which is more reversible to the dye:reduced dye couple than to the Fe(III):Fe(II) couple, together with a second relatively nonselective electrode, usually either platinum or indium tin oxide (ITO). In such a cell, the SnO_2 electrode is the anode. In principle, it would be desirable to replace the nonselective electrode by one which is much more reversible to the Fe(III):Fe(II) couple than to the dye couple.

Practical conversion efficiency of a solar transducer is measured by the sunlight engineering efficiency (S.E.E.):

$$\text{S.E.E.} = \frac{100 \times \text{electrical energy or power delivered to load at the power point}}{\text{incident sunlight energy or power}}$$

The best S.E.E. values that we have obtained with the single layer TI-TL iron–thionine cell with SnO_2 anode and Pt cathode have been in the range .02–.036%. These values have been obtained using 50 v/v % aqueous CH_3CN as solvent and solutions containing $\sim .001M$ dye, .01M sulfuric acid, and Fe(II), with SO_4^{2-} and HSO_4^- as the only anions, Fe(III) present initially at its impurity level ($\sim 5 \times 10^{-5}M$), and electrodes spaced 80 μm apart.

The use of transparent nonselective electrodes, e.g., indium tin oxide, makes possible multi thin-layer (MTL) cells in which the layers are in series optically and in parallel electrically. A sunlight efficiency of .063% was obtained with a MTL cell constructed of four 80 μm layers (4). Such cells are capable of absorbing a larger proportion of incident light while maintaining the electrochemical properties of thin-layer cells.

Processes in the TI-TL Iron–Thionine Photogalvanic Cell

Photogalvanic transduction in the TI-TL iron–thionine cell or similar cells using other thiazines (4) can be analyzed in terms of five basic processes (5): (1) absorption of incident light; (2) conversion of absorbed radiant energy into chemical potential of charge carriers; (3)

diffusion of charge carriers to electrode surfaces; (4) transfer of charge to electrode surfaces; and (5) delivery of electrical current to the external circuit.

The theoretical maximum S.E.E. of an iron–thionine cell at pH = 2 is ~ 2%, based on ability to absorb up to 15% of the insolation flux, 100% quantum efficiency in photoredox reactions, and conversion of 2 V photons to 0.28 V electrical potential. The best S.E.E. achieved with a single element TI-TL cell at pH = 2 has been less than 1/50 of the theoretical maximum. Furthermore, the theoretical limit is low compared with S.E.E. values already achieved with many other solar transduction devices. It is thus necessary to find means of both increasing the theoretical limit and coming close to actually reaching this limit if photogalvanic transduction is to remain a significant option for use of solar energy. We have performed a variety of studies of each of the five basic steps of photogalvanic transduction so that reasons for losses of energy could be understood and steps taken to eliminate or reduce these losses.

Absorption of Light

Efficient absorption of light in a thin-layer cell requires 10^{-2} to $10^{-3} M$ thionine (ϵ_{max} ~ 6×10^4 M^{-1} cm^{-1}) or other thiazine dye. At such concentrations, association of these dyes to dimers and higher oligomers is very extensive in aqueous solution. Since only monomeric dye contributes to transduction of light to electricity and the absorption spectrum of dimeric dye extensively overlaps that of monomer, it is necessary to suppress association (6). This can be done in various ways, e.g., by incorporation of surfactants. Greatest improvement in cell performance has been achieved by using organic co-solvents to suppress association. The most satisfactory solvent yet identified is 50 v/v % aqueous CH_3CN in which $1 \times 10^{-3} M$ solutions of thionine are virtually completely monomeric.

One approach to increasing limiting theoretical efficiency is to use mixtures of photoredox dyes which can absorb a larger portion of the insolation flux than can only a single dye. Thionine and methylene blue can together absorb about 25% of the insolation spectrum at Air Mass – 1. The theoretical maximum S.E.E. for an iron–thionine–methylene blue cell at pH = 2 is ~ 4%. Additivity in output of TI-TL SnO_2/Pt cells containing both dyes at $10^{-4} M$ concentration has been approached but, with both dyes $10^{-3} M$, cell output was no more than that obtained with methylene blue alone and only about 80% of the output with $10^{-3} M$ thionine. The reason for this unsatisfactory result still is not established. Roughly additive output has been achieved with 1 mM thionine and 4 mM methylene blue incorporated in different elements of a two-layer cell (4).

Sensitization of photogalvanic action by dyes which are themselves not capable of photoredox action has been demonstrated (7). Action spectra of solutions containing rhodamine 6G and two coumarin dyes in addition to thionine and methylene blue closely parallel absorption spectra, corresponding to the possible use of about 50% of the insolation spectrum and a theoretical maximum sunlight engineering efficiency of ~ 7%. It has been demonstrated that rhodamine 6G does, in fact, increase the power output of iron–thionine (or other thiazine) cells under white-light illumination; an approximately 40% increase has been observed under illumination with 35 mWcm^{-2} (8).

Formation and Decay of Charge Carriers in Solution

Previous workers have shown that, for thionine, competition of Reaction 2 with Reaction 3 is essentially negligible, i.e., the quantum yield for intersystem crossing from S_1 to T_1 is close to unity (9). We have studied the dependence of the rates of intrinsic decay of triplet thionine to the ground state, Reaction 4, and of Reaction 5, the reduction of the triplet by Fe(II), upon pH and the nature of solvent and anions using flash photolytic techniques (10). These measurements have shown that reduction by Fe(II) is greatly favored by use of 50 v/v % aqueous CH_3CN rather than pure water and by use of sulfate rather than $F_3CSO_3^-$. Under the favored conditions, .01M Fe(II) reduced 97% of triplet thionine at pH = 2. In a related study, it was found that the logarithm of the specific rate of disproportionation of semithionine, the forward direction of Reaction 6, varies linearly with the value of Kosower's Z for the solvent in which the reaction occurs (11). This relationship was observed over three orders of magnitude in rate constant and variation of Z by ~ 17.

The kinetics and mechanism of the bulk reaction of leucothionine, TH_4^{2+}, with Fe(III), also has been studied by flash photolytic technique (12). These experiments have shown that the reaction proceeds via reversible formation of a 1:1 association complex, Reaction 7, and have explored dependence of equilibrium and rate constants on pH, ionic strength, and nature of solvent and anions. The product of the association constant and the electron transfer rate constant, K_7k_8, corresponds to a second-order rate constant which is rather small, ~ 350 to 1800M^{-1} sec^{-1} under the range of conditions used. The magnitudes of intracomplex electron transfer rate constants, k_8, are of the order of 1 sec^{-1}. Formation of relatively long-lived complexes of leucothionine with Fe(III) in bulk solution is consistent with a mechanism for loss of charge carriers at the SnO_2 electrode that is suggested below.

A great deal of useful information has been obtained by direct measurement of the composition and kinetics of relaxation of the iron–thionine photostationary state (*13*). The measured dependence of photostationary state composition on the initial solute concentrations and on the nature of solvent has been found to agree to within experimental error with values calculated from measured rate constants for the relevant elementary reactions in the system. Interestingly, under illumination with 100 mWcm^{-2} of light from a Xenon lamp (essentially "1 sun"), a solution in 50 v/v aqueous CH_3CN initially .001M in thionine, .01M in acid and Fe(II), and .00006M in Fe(III), with sulfate as anion, is 48% bleached. This degree of bleaching is nearly optimal since it provides a high concentration of both absorbing dye and charge-carrying reduced species.

The absence of detectable absorption at 390 and 770 nm in the photostationary state shows that less than 10% of the reduced dye is present as semithionine under conditions of observation involving 35–95% photobleaching of 10^{-5}–$10^{-3}M$ thionine (*14*). Calculations based on kinetic data show that no more than $\sim 0.3\%$ of the photobleached dye is present as semithionine in a solution initially $10^{-3}M$ in thionine and with other aspects of initial composition and illumination as given above. From a comparison of such results with measured monochromatic quantum yields for current generation, e.g., $\sim 7\%$ at 578, 589, and 620 nm in 80 μm cells under similar conditions (*4*), it can be concluded that leucothionine is the principal charge carrier derived from the dye under these conditions.

Kinetic analysis of the observed linear dependence of the pseudo first-order rate of relaxation of the photostationary state upon initial concentrations of thionine and Fe(III) (*13, 14*) gives a completely independent determination of the apparent second-order rate constant for the reaction of Fe(III) with leucothionine, K_7k_8. The resulting values are in excellent agreement with those determined by flash photolysis (*12*). The same analysis leads to evaluation of the rate constant for synproportionation of leucothionine to give semithionine, k_{-6}. Knowledge of the rate constants for both the forward and reverse directions of Reaction 6 makes possible evaluation of the equilibrium constant for this reaction, K_6, under a variety of conditions (*14*). The resulting values, e.g., $K_6 =$ [Semi]^2/[Leuco][Thionine] $= (.6 \pm .2) \times 10^{-6}$ in $10^{-2}M$ aqueous sulfuric acid, are less by between four and five orders of magnitude than a value estimated by Michaelis (*15*), which has led a number of workers to conclude that semithionine is the principal reduced form of the dye in iron–thionine photogalvanic cells. A knowledge of K_6 also has made possible evaluation of the potentials for the two one-electron redox equilibria of thionine, e.g., $E^{\circ\prime}_{T/S} = .196 \pm .004$ V and $E^{\circ\prime}_{S/L} = .570 \pm .005$ V vs. NHE in $10^{-2}M$ aqueous sulfuric acid. In $10^{-2}M$ solutions of sulfuric acid in 50 v/v % aqueous CH_3CN, $E^{\circ\prime}_{T/S} = .176 \pm .005$ V and

$E°_{S/L} = .530 \pm .005$ V vs. NHE. These and related values have been used in constructing an energy diagram for the SnO_2^--electrolyte interface that is described below in the section on electron transfer at the electrode–solution interface.

Diffusion of Charge Carriers to the Electrodes

Substitution of experimentally determined lifetimes for relaxation of the photostationary state (13), 1.6 sec under the conditions indicated above, into the diffusion equation gives 50 μm as the average diffusion length for charge carriers in solution under these conditions. Thus, loss of charge carriers by recombination in bulk solution cannot be a major process with the 25 and 80 μm electrode separations used in most of our TI-TL cells.

Transfer of Charge to the Electrodes

Short circuit current of TI-TL cells varies linearly with concentration of photobleached dye, mostly leucothionine (16). Current per unit area of electrode per unit concentration of leucothionine was found to be the same for TI-TL cells with 25 and 80 μm spacings of a given set of electrodes and increased slightly with the proportion of CH_3CN in aqueous CH_3CN solutions. The best efficiencies obtained with SnO_2 anodes in 50 v/v % aqueous CH_3CN with sulfate as anion corresponded to about 130 μA/mM-cm². Substitution of this value in the Nernst diffusion-layer equation indicates a diffusion-layer thickness of 100 μm. The fact that essentially identical current densities were obtained with 25 and 80 μm electrode separations indicates, however, that the diffusion layer was no greater than 25 μm in thickness and could have been significantly less. This suggests that at least 75% of charge carriers that reached those particular electrodes did not produce current in the external circuit and were wasted. Little of this loss is incurred within the electrodes after electron transfer, as reverse biasing of the cell increases the current by only 25–50%. Apparently, most of this loss results from inefficiency of electron transfer between solution charge carriers and electrode(s).

Single electrode potentials have been measured at SnO_2 anodes and ITO cathodes for solutions in 50 v/v % aqueous CH_3CN, .01M in H_2SO_4, in which the photostationary ratio of TH^+ to TH_4^{2+} varied by a factor of more than 10 and the Fe(III):Fe(II) ratio varied by a factor of approximately six (1). The potential at the SnO^2 anode paralleled the potential calculated for the $TH^+:TH_4^{2+}$ couple from known compositions with the aid of the Nernst potential equation. Measured values were, however, more positive (for the particular samples of SnO_2 used) than calculated values by ~ 100 mV. Measured potentials at the ITO cathode paralleled those calculated for the Fe(III):Fe(II) couple but were about 70 mV more negative (for the particular samples of ITO used). Cell voltages were thus about 170 mV smaller than would be expected from a cell with an anode fully selective for the $TH^+:TH_4^{2+}$ couple and a cathode fully selective for the Fe(III):Fe(II) couple.

Electron transfer processes at the SnO_2 electrode have been studied extensively by cyclic voltammetry and other electrochemical techniques (*4, 6, 17*). The tin oxide electrodes were characterized by determining charge carrier densities and flat-band potentials by means of Schottky–Mott plots of $1/C^2$ vs. E. The highly conductive (30–100 ohms/sq) defect-structure SnO_2 had a charge carrier density in the range $(4-7) \times 10^{20}$ cm^{-1}. The flat band potential was $+.05$ to $+.25$ relative to NHE and essentially independent of the solvent in contact with the SnO_2. Thus, $E°'_{pH=2}$ of the thionine–semithionine couple falls within the flat band potential range with either .01M aqueous sulfuric acid or .01M sulfuric acid in 50 v/v % aqueous CH_3CN as solvent. $E°'_{pH=2}$ of the semithionine–leucothionine couple is at least .28 V positive with respect to the flat-band potential in aqueous CH_3CN and more positive by about .04 V in the neat aqueous solvent while $E°'$ of the Fe(III):Fe(II) couple is at least .4 V positive to the flat band potential in both media. Figure 1, taken from Ref. *17*, summarizes interfacial energies. These data suggest (*18*) that the SnO_2 electrode should be relatively poorly selective since leucothionine is, as shown above, the principal reduced form of the dye at the photostationary state. The results also suggest that selectivity of the SnO_2 anode with respect to the leucothionine–semithionine couple should be better with 50 v/v % aqueous CH_3CN as solvent than with neat water.

Results of cyclic voltammetric measurements at the SnO_2 electrode (*4, 6, 17*) are summarized below.

(1) The mechanism of reduction of thionine is EE, i.e., no discernible chemical process takes place between the two electron-transfer steps.

(2) Reactions of the thionine–leucothionine couple are kinetically controlled. Rectification significantly reduces efficiency of oxidation of leucothionine at the SnO_2 electrode.

(3) Contrary to what might be expected by considering only the interfacial energy diagram, reversibility is greater for the dye couple with neat H_2O as solvent than with 50 v/v % aqueous CH_3CN.

(4) Leucothionine is strongly adsorbed on SnO_2 while thionine is weakly adsorbed. Adsorption is greater from solutions in pure water than from 50 v/v % aqueous CH_3CN.

(5) The Fe(III):Fe(II) couple is somewhat more reversible at SnO_2 with 50 v/v % aqueous CH_3CN as solvent than with pure water, but with either solvent it is much less reversible at SnO_2 than the dye couple is.

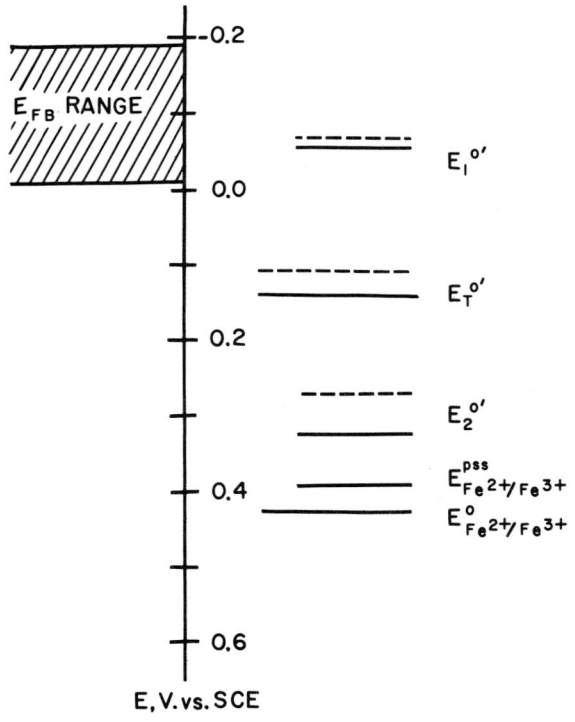

Journal of the Electrochemical Society

Figure 1. Interfacial energy diagram for the SnO_2 anode in contact with the components of the iron–thionine photogalvanic cell at pH = 2. Solid lines: aqueous sulfuric acid. Dashed lines: sulfuric acid in 50 v/v % aqueous CH_3CN. Taken from Ref. 17.

Summary and Speculation

The fifty-fold to one hundred-fold factor by which the best observed values of S.E.E. for the iron–thionine TI-TL cell with a SnO_2 anode and a Pt or ITO cathode fall short of the theoretical maximum can be ascribed, in part, to two relatively simple factors. One of these is inefficient absorption of light. At the photostationary state, the concentration of unbleached dye in optimized cells was $\sim 5 \times 10^{-4}M$, so that only $\sim 25\%$ of the incident light in the wavelength region of the thionine band was absorbed in the 80 μm-thick cell. If absorption by SnO_2 and ITO in the four-element MTL cell is ignored, approximately two-thirds of incident light in the thionine band is absorbed by the dye. The difference is, in fact, similar to the difference in S.E.E. of single-element TI-TL and four-element

MTL cells. The other simple factor is the relatively low cell potential. Cell potentials at the power point were $\sim$ 100 mV, i.e., about 35% of $E^{\circ'}{}_{pH=2}$ of the cell. Most of this deficiency can be ascribed to selectivity of electrodes being only partial. The two loss-factors together account for reduction of S.E.E. by a factor of $\sim$ 12.

Taken together, five lines of evidence suggest that the remaining four-fold to eight-fold loss factor can be ascribed to inefficient transfer of charge between solution charge carriers and SnO_2 anodes. One pertinent piece of evidence is the relatively long diffusion length calculated for charge carriers in solution. This length is sufficient to allow most charge carriers to reach the electrodes. Second is the contrast between data which suggest that the thickness of the Nernst diffusion layer is 100 μm and evidence that it cannot be more than 25 μm in thickness. Third is the cyclic-voltammetric evidence that electron transfer at the SnO_2–solution interface is controlled kinetically. Fourth is the cyclic-voltammetric evidence that leucothionine is strongly adsorbed on SnO_2. Fifth is the evidence that the reaction of leucothionine with Fe(III) in bulk solution proceeds via a relatively long-lived ($\tau \sim$ 1 sec) complex.

At least three paths by which leucothionine can be wasted at the anode suggest themselves. (1) Adsorbed leucothionine is oxidized at the interface by Fe(III). The resulting semithionine would be expected to rapidly transfer an electron to SnO_2. Thus, this path could introduce a loss-factor of only two. (2) Adsorbed leucothionine complexes with Fe(III) at the interface. The resulting complex then reacts in the adsorbed state. This path also would be expected to introduce a loss-factor of only two. (3) Adsorbed leucothionine complexes with Fe(III) at the interface. The resulting complex desorbs and diffuses back into bulk solution before undergoing intracomplex electron transfer. This path can lead to complete wastage of charge carriers. If the analysis presented in this chapter is correct, path (3) plays an important role in reducing the efficiency of charge transfer at the SnO_2 electrode.

Acknowledgment

This work, a joint project of the Department of Chemistry of Boston University and the Solar Energy Conversion Unit of Exxon Research and Engineering Co., was supported by the National Science Foundation Research Applied to National Needs Program under Grant No. SE/AER/72-03579. Some of the research on solution kinetics was supported by Energy Research and Development Administration Contract No. EY-76-S-02-2889. Many valuable discussions of various aspects of this work have been held with Morton Z. Hoffman.

Literature Cited

1. Wildes, P. D., Brown, K. T., Lichtin, N. N., *J. Am. Chem. Soc.* (1978) **100**, "Abstracts of Papers," *National Meeting, ACS, Aug. 28–Sept. 2, 1977*, PHYS 117.
2. Rabinowitch, E., *J. Chem. Phys.* (1940) **8**, 560.
3. Clark, W. D. K., Eckert, J. A., *Sol. Energy* (1975) **17**, 147.
4. Hall, D. E., Eckert, J. A., Lichtin, N. N., Wildes, P. D., *J. Electrochem. Soc.* (1976) **123**, 1705.
5. Lichtin, N. N., "Photogalvanic Processes," *in* "Solar Power and Fuels," J. Bolton, Ed., pp. 119–142, Academic, New York, 1977.
6. Hall, D. E., Clark, W. D. K., Eckert, J. A., Lichtin, N. N., Wildes, P. D., *Am. Ceram. Soc., Bull.* (1977) **56**, 408.
7. Wildes, P. D., Hobart, D. R., Lichtin, N. N., Hall, D. E., Eckert, J. A., *Sol. Energy* (1977) **19**, 567.
8. Lichtin, N. N., Wildes, P. D., U.S. Patent **4,052,536**, "Electrolytes Which Are Useful in Solar Energy Conversion," 1977.
9. Havemann, R., Reimer, K. G., *Z. Phys. Chem. (Leipzig)* (1961) **216**, 334, and earlier papers.
10. Wildes, P. D., Lichtin, N. N., Hoffman, M. Z., Andrews, L., Linschitz, H., *Photochem. Photobiol.* (1977) **25**, 21.
11. Wildes, P. D., Lichtin, N. N., Hoffman, M. Z., *J. Am. Chem. Soc.* (1975) **97**, 2288.
12. Osif, T. L., Lichtin, N. N., Hoffman, M. Z., "Abstracts of Papers," *National Meeting, 174th, ACS, Aug. 28–Sept. 2, 1977*, PHYS 17.
13. Wildes, P. D., Lichtin, N. N., Hoffman, M. Z., "Application of Solution and Photo Dynamics to the Optimization of Output of Iron–Thiazine Photogalvanic Cells," *in* "Solar Energy," J. B. Berkowitz, I. A. Lesk, Eds., p. 128–138, The Electrochemical Society, 1976.
14. Wildes, P. D., Lichtin, N. N., *J. Phys. Chem.* (1978) **82**, 981.
15. Michaelis, L., Schubert, M. P., Granick, S., *J. Am. Chem. Soc.* (1940) **62**, 204.
16. Wildes, P. D., Brown, K. T., Hoffman, M. Z., Lichtin, N. N., Hall, D. E., *Sol. Energy* (1977) **19**, 579.
17. Hall, D. E., Wildes, P. D., Lichtin, N. N., *J. Electrochem. Soc.* (1978) **125**, in press.
18. Gerischer, H., *in* "Semiconductor Electrochemistry in Physical Chemistry," L. Eyring, D. Henderson, W. Jost, Eds., pp. 463–542, Academic, New York, 1970.

RECEIVED February 22, 1978.

25

Nickel(0) Catalyzed Reactions of Strained Ring Systems

R. NOYORI

Department of Chemistry, Nagoya University, Nagoya, Japan

In connection with the activation of saturated hydrocarbons via homogeneous catalysis, we have examined transition metal catalyzed reactions of various strained hydrocarbon systems that have unique steric and electronic properties. Strained carbon-to-carbon single bonds have considerable π-bonding character. The chemistry of these substrates should be intermediate between well-documented transition metal chemistry of alkenes and rather unclarified alkane chemistry (1, 2, 3). Our attention has been focused particularly on the stereoselectivity, regioselectivity, and periselectivity of the Ni(0)-catalyzed reactions (4–14).

Reactive organometallic intermediates that play a crucial role in the catalytic reactions are difficult to isolate. We have therefore intended to analyze the reactions on the basis of well-defined basic principles of organic chemistry.

Nickel(0)-Catalyzed Valence Isomerization of Quadricyclane to Norbornadiene Catalytic Trapping of the Organometallic Intermediate

Quadricyclane (1) undergoes both intramolecular and intermolecular 2 + 2 reaction with the aid of a Ni(0) catalyst (4). When the hydrocarbon 1 was treated with a catalytic amount of bis(acrylonitrile)nickel(0) [Ni(an)$_2$] or bis(1,5-cyclooctadiene)nickel(0) [Ni(cod)$_2$], norbornadiene (2) was obtained in a high yield. The reaction in the presence of an excess of electron-deficient olefins such as acrylonitrile or acrylates gave rise to the cycloaddition products of type 3. The reaction using methyl α,*cis*-β-di-deuterio-acrylate has revealed that the catalytic cycloaddition goes in a stereospecific manner with retention of configuration. Furthermore, this catalytic reaction is in marked contrast with the corresponding uncatalyzed reaction which leads to the 2 + 2 + 2 cycloadduct 4.

Various evidence goes to support the operation of the mechanism outlined below. The strained σ bond of 1 undergoes oxidative addition to form the organonickel intermediate 6. Ligand isomerization followed by extrusion of the Ni(0) catalyst produces the isomeric diene 2 whereas insertion of the coordinated olefin into the Ni–C bond in 6 gives 8, which

Scheme 1

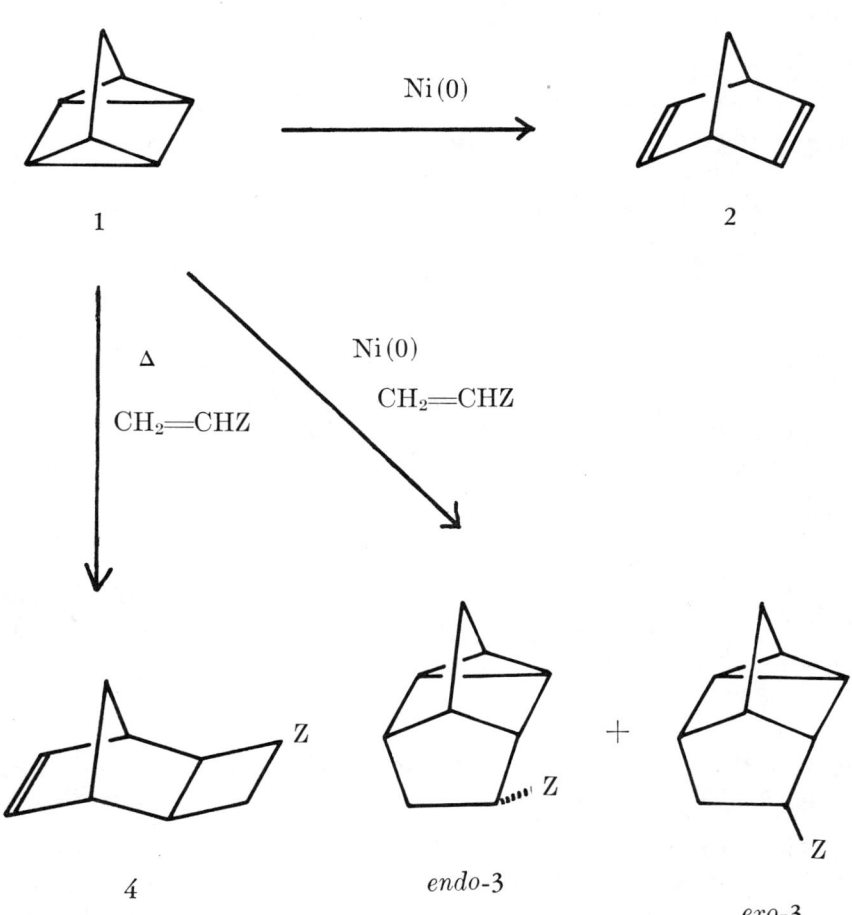

Z = CN, COOR.

Scheme 2

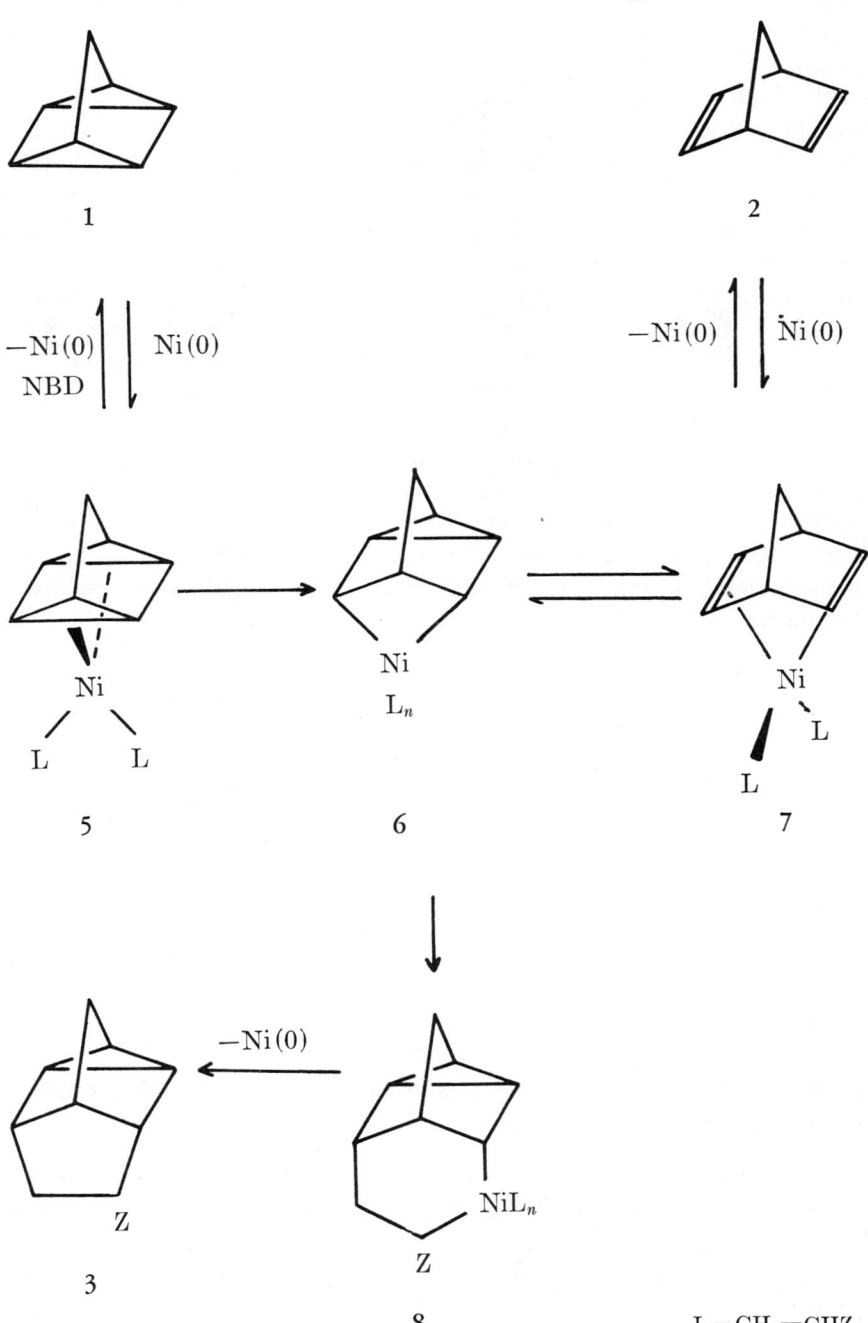

$L = CH_2=CHZ$.

through elimination of the Ni(0) complex results in the cycloaddition products 3. In the presence of Ni(0) catalysts, the diene 2 also forms the olefin addition product 3 (15). Careful comparison of the catalytic behavior of 1 and 2 showed that: (1) the saturated hydrocarbon 1 reacted with acrylates much faster than the diene 2; (2) the co-existence of 2 lowers the rate of the catalytic isomerization of 1 to 2 and the cycloaddition of 1 across acrylates giving 3; and (3) the product distribution of the cycloaddition, *endo*-3:*exo*-3 ratio, does not depend on whether the starting hydrocarbon is 1 or 2 but only on the reaction conditions. These observations imply that both catalyzed reactions involve the common intermediate 6 in the product-determining step. The transformations 5 (or the unidentately coordinated complex) → 6 → 7 and 5 → 6 → 8 are facile because of the apparent relief of strain while the conversion of 7 to 6 is unfavorable because of the significant increase in strain energy which accounts for the difference in reactivities of 1 and 2.

The intermediacy of rhodium complex 9 structurally similar to 6 has been proposed in the stoichiometric formation of the acylrhodium complex 10 from the hydrocarbon 1 and di-μ-chloro-tetracarbonyldirhodium(I) (16). Thus both stoichiometric and catalytic trapping experiments have demonstrated the valence isomerization 1 → 2 is going in a stepwise fashion via an organometallic intermediate.

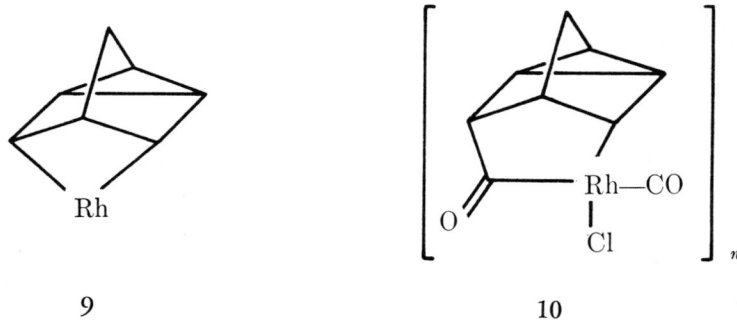

9   10

Nickel(0)-Catalyzed Cycloaddition between Bicyclo[2.1.0]pentane with Electron-Deficient Olefins

The title reaction exemplifies intriguing features of transition metal catalysis. In the absence of transition metal catalysts, reaction between bicyclo[2.1.0]pentane (11) and electron-poor olefins requires extremely forcing conditions and produces a very complex mixture. For example, because of the molecular orbital (MO) restrictions, *trans*-1,2-dicyanoethylene enters into the 2 + 2 cycloaddition across the central σ bond, giving bicyclo[2.2.1]heptane derivatives in a poor yield and with lack of stereospecificity (*17*). The most characteristic feature is the fact that maleic anhydride approaches from the endo side of the bicyclo envelope of 11 in the cycloaddition.

These features of the reaction are dramatically changed by the added Ni(0) catalysts (*10, 11*). When a mixture of 11 and acrylonitrile or methyl acrylate was treated with Ni(an)$_2$ at 40°–70°C, the 2 + 2 cycloadducts 12 and 13 and some substitutive addition product 14 were obtained in good yields. Under these catalytic conditions, dimethyl fumarate or maleate undergoes the 2 + 2 cycloaddition in a highly stereospecific manner with retention of configuration.

The catalytic reaction is most reasonably explained by a mechanism involving initial oxidative addition of the central bond of 11 onto Ni(0) catalyst. Results of the deuterium labeling experiments were in full accord with this catalytic cycle. Olefin insertion in the complex 15 forms the ring-expanded organonickel 16, from which the organic moiety eliminates to give the formal 2 + 2 product 17. The monocyclic 1:1 adduct 19 can be derived from the common intermediate 16 through the intramolecular β metal hydride elimination, forming 18, followed by the reductive elimination.

Scheme 3

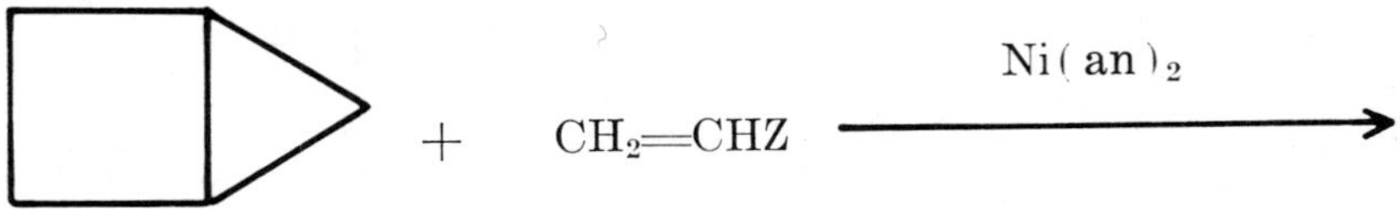

11 + CH$_2$=CHZ $\xrightarrow{\text{Ni(an)}_2}$

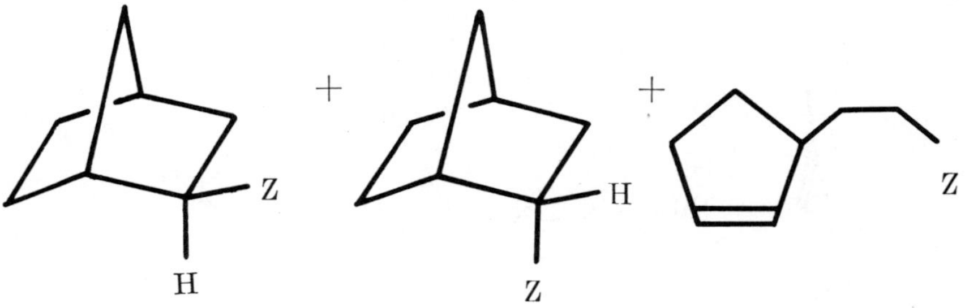

12   13   14

Z = CN, COOCH$_3$.

Scheme 4

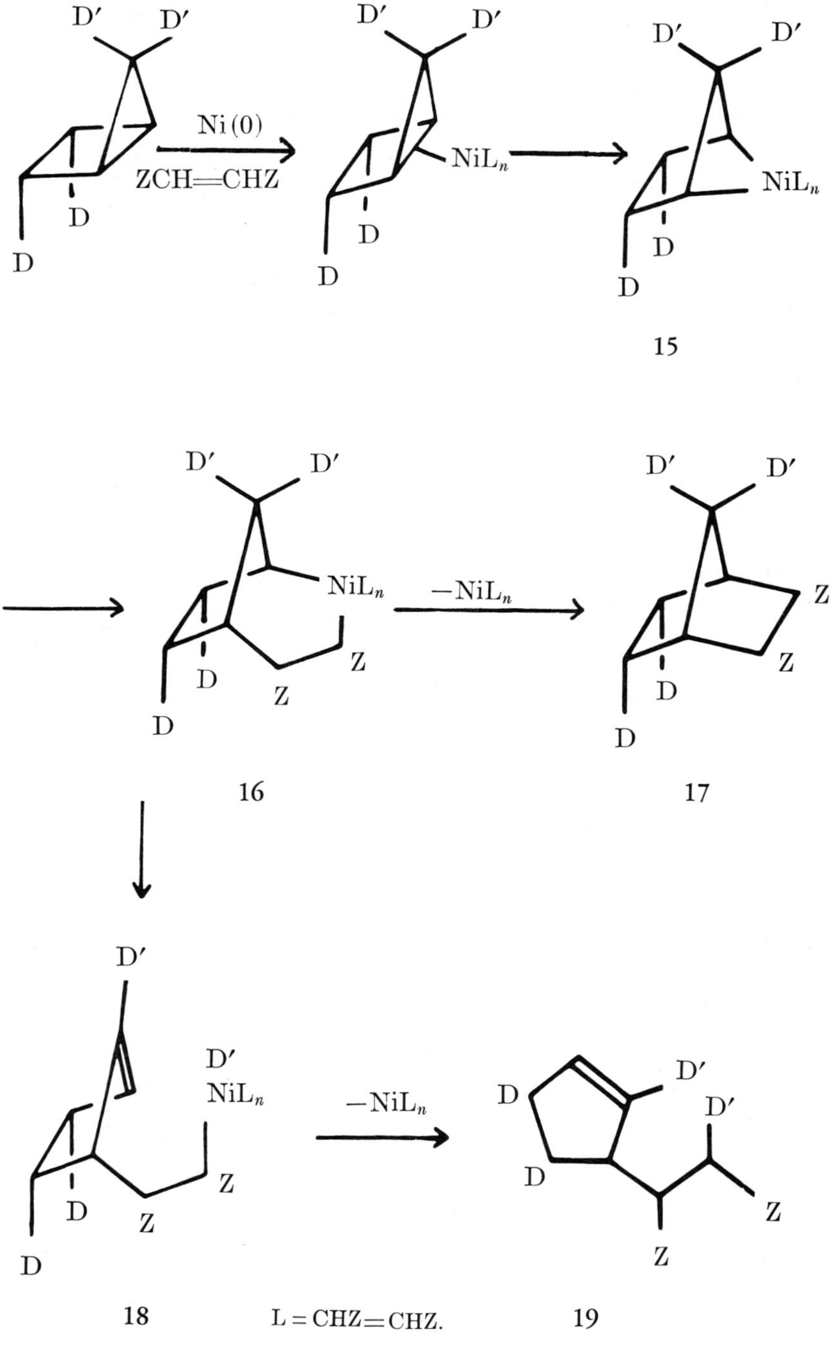

15

16   17

18   L = CHZ=CHZ.   19

Scheme 5

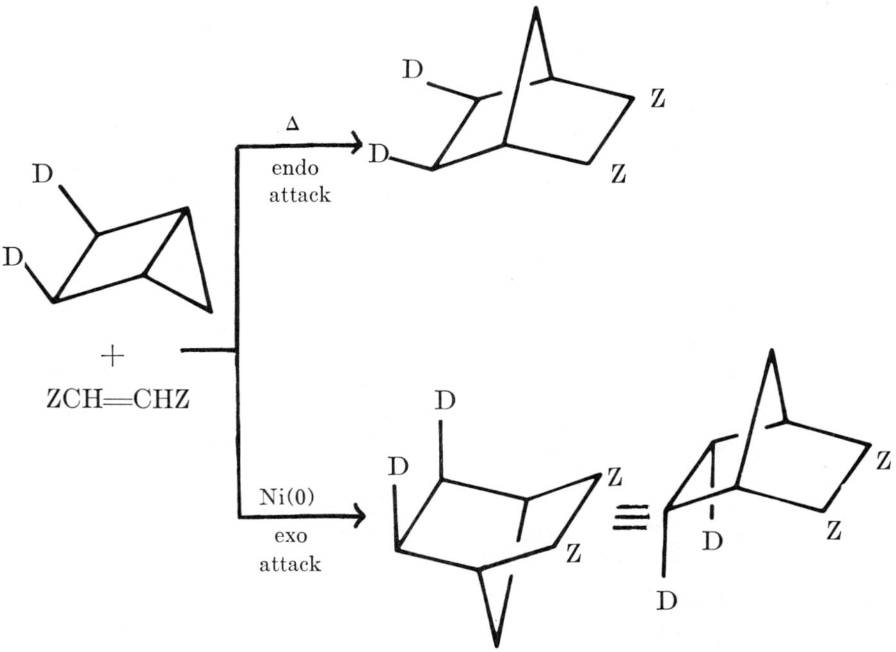

Z = elctron-withdrawing group.

It should be noted that the stereochemistry of the catalyzed cycloaddition is virtually the reverse of that encountered in the uncatalyzed reaction. In the absence of the transition metal complex, the hydrocarbon 11 reacts with an electron-deficient olefin with double inversion of stereochemistry at the angular positions while in the presence of a Ni(0) complex the cycloaddition proceeds with retention of original configuration.

MO theory provides a powerful tool in understanding the origin of selectivities of the catalytic reaction. The principal bonding interaction between a transition metal and a saturated hydrocarbon molecule is based on overlapping of a high-lying occupied MO of the saturated molecule (σ donation) and LUMO of the metal and a low-lying (π back donation) (Figure 1). As to the coordination mode, transition metals can interact with saturated hydrocarbons in an edge-on, face-on, or corner-on manner if the phase of the MOs coincides each other. Here preference in the coordination mode as well as relative importance of the electron donation and back donation in stabilization of the complex highly depends on the nature of the metals and organic molecules. The exo approach of the

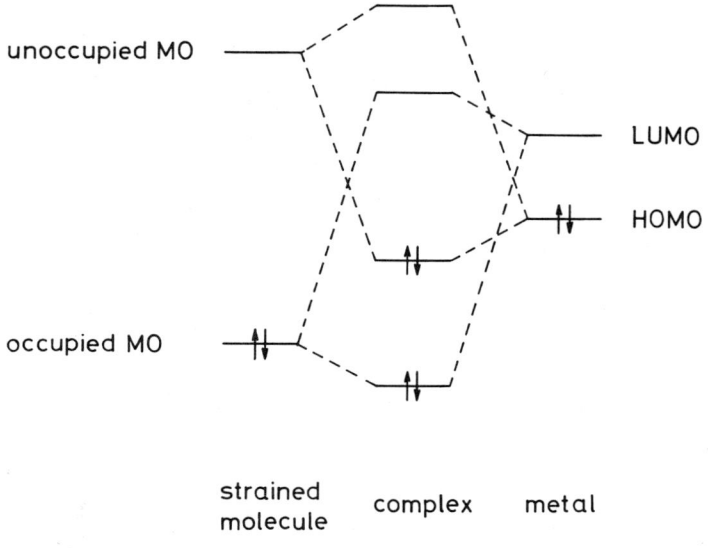

Figure 1. Correlation digram of the energy levels of a strained molecule and a transition metal

electropositive Ni(0) complex to bicyclopentane central bond is easily rationalized by such treatment. As can be seen in Figure 2, the MO contribution of the C–C bond that participates in the edge-on coordination must be strongly bonding in nature in any appropriate high-lying filled orbital and, at the same time, must have an antibonding character in a low-lying vacant orbital. An MO calculation of 11 indicated that the exo region of the envelope has a strongly symmetric, bonding character while the endo side has an antisymmetric, antibonding character (*18*). Therefore, the Ni(0) atom approaches the σ bond preferentially

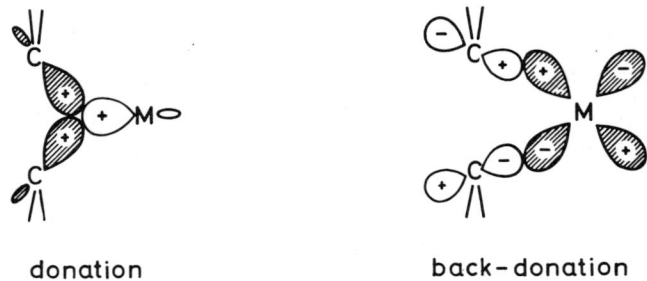

Figure 2. The edge-on interaction of a strained C–C σ bond and a transition metal

from the exo side to form the edge complex. In the absence of transition metals, however, exo 2 + 2 pericyclic interaction of an electron-poor olefin (2π-electron system) is highly unfavorable, leading to the opposite stereochemical result (17).

Molecular Orbital Control of the Selectivities in the Transition Metal Catalyzed Reactions of 1,8-Bis-Homocubane System

The catalytic behavior of 1,8-bis-homocubane (20) and its derivatives is quite metal dependent. Ni(0) and Rh(I) complexes promoted the cleavage of a cyclobutane ring to give the dienes of type 21 (5, 6, 19, 20). By contrast, Ag⁺ caused the bicyclo[2.2.0]hexane to dicyclopropyl conversion, 20 → 22 (21, 22). The reaction courses with Pd(II) complex were highly influenced by the ligand attached to the metal (19). Even in the Ni(0) catalysis system, addition of tetracyanoethylene caused a considerable change in the product, and the dicyclopropyl derivative of type 22 became the major product (6). Apparently the 20 → 21 transformation was achieved by metals that have strong oxidative-addition ability and the 20 → 22 rearrangement with metals having an eminent σ-accepting ability.

Scheme 6

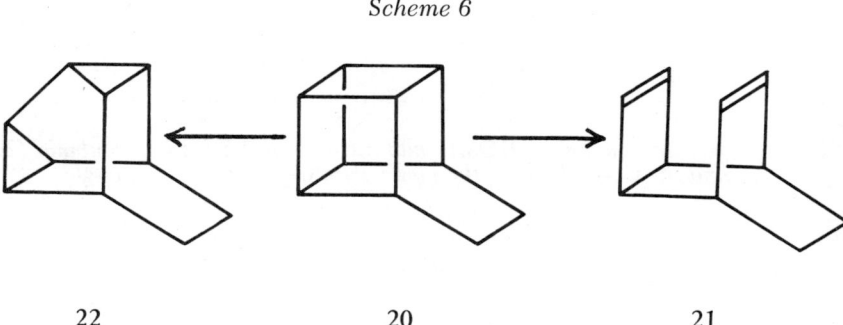

22   20   21

The extended Hückel calculation of 20 suggests that only C_2–C_5 and the structurally equivalent bonds have MO shapes suitable for the edge-on coordination. Only these bonds are bonding in nature in the HOMO (and next HOMO) and antibonding in the LUMO. This situation is clearly indicated in the detailed contour maps of these MOs around the C_2–C_5 bond (Figure 3). The orbital symmetries rule out positive interactions of Ni(0) or Rh(I) catalyst with other C–C bonds. Thus the bis-homocubane ligand in the initially formed complex 23 undergoes oxidative addition onto the π-donating metal to produce the metallocyclopentane 24 which in turn experiences the second C–C bond breakage, yielding the final diene product 25 in a regiospecific fashion. Judging

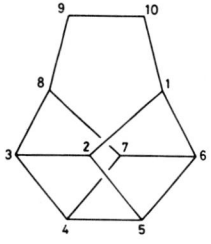
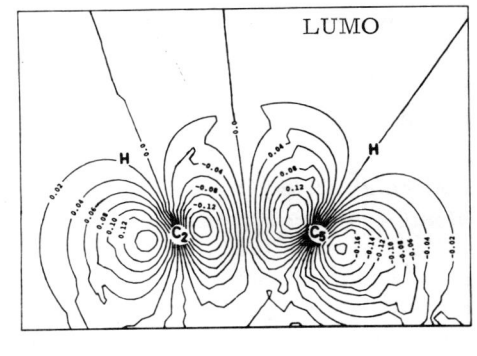
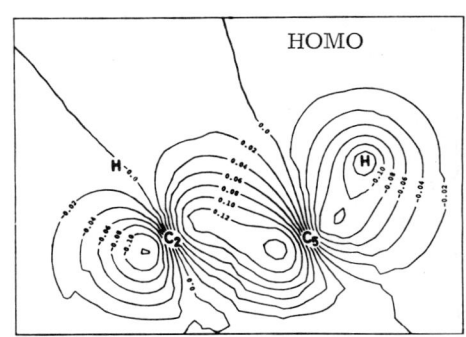

Figure 3. Contour maps of HOMO and LUMO of 1,8-bis-homocubane (20) around the C_2–C_5 bond in the plane bisecting the C_4–C_5–C_6 angle

Scheme 7

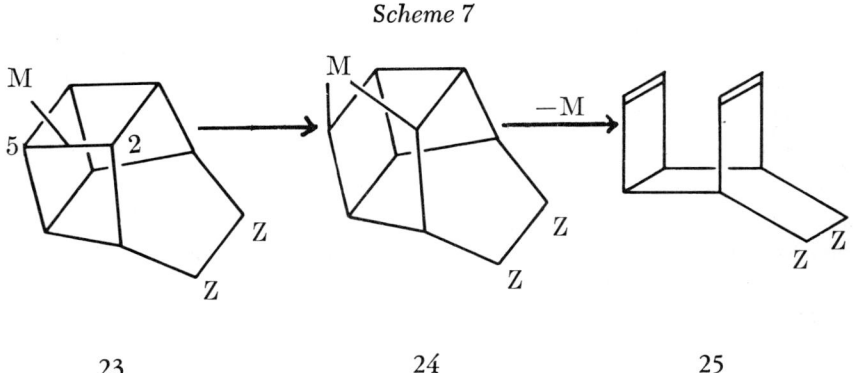

M = Ni, Rh.   Z = COOCH$_3$.

from the MO calculation, the orbitals appear to spread in directions appropriate to only unidentate, edge-on coordination, and any concerted mechanisms involving a face-on coordinated complex (23) are not likely to be operating.

Corner-on metalation to strained molecules is feasible with strongly σ-accepting metals (24, 25, 26). To obtain such interaction, only hydrocarbon-to-metal electron donation is significant. Thus the Ag⁺-promoted rearrangement can also be understood by looking at the shapes of the frontier MOs of 20. The Ag⁺ catalyst can attack the C_2 corner from the back-side along the C_2–C_5 axis where the HOMO is developing greatly. Such electrophilic corner-on interaction (26) leads to the site-selective generation of a cyclobutyl or cyclopropylcarbinyl cation 27. Subsequent ejection of Ag⁺ species produces the formal bicyclo[2.2.0]hexane–dicyclopropyl rearrangement product 28.

Scheme 8

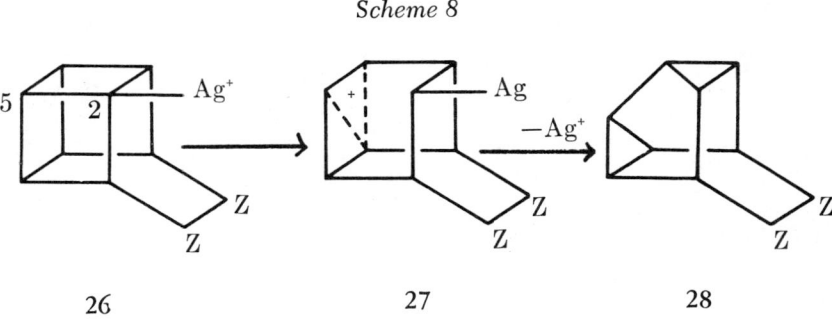

Nickel(0)-Catalyzed Reaction of Bicyclo[1.1.0]butanes Regioselective Carbene Generation and the Trapping with Electron-Deficient Olefins

This type of reactions provide a good model of olefin metathesis reactions. When bicyclo[1.1.0]butane (29) and electron-deficient olefins such as methyl acrylate or acrylonitrile was exposed to Ni(0) catalysts, Ni(cod)₂ or Ni(an)₂, the allylcyclopropane derivatives (30) were obtained in good yields (7, 8, 9). The labeling experiments have revealed that the reaction involves cleavage of the central bond and one of the peripheral bonds (geminal two-bond cleavage) and proceeds in a stereospecific manner with respect to the olefinic substrates. Thus this coupling reaction is viewed in a formal sense as an intramolecular retrocarbene addition of a bicyclobutane followed by an intermolecular olefin trapping of the resulting allylcarbene intermediate.

The Ni(0)-catalyzed reaction is considered to proceed according to the following scheme (M = Ni). The central bond of bicyclobutanes shows a symmetric shape in the HOMO and antisymmetric nature in the

Scheme 9

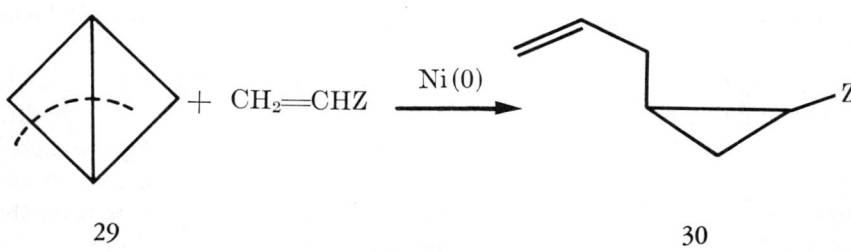

29   30

Z = COOR, CN.

Scheme 10

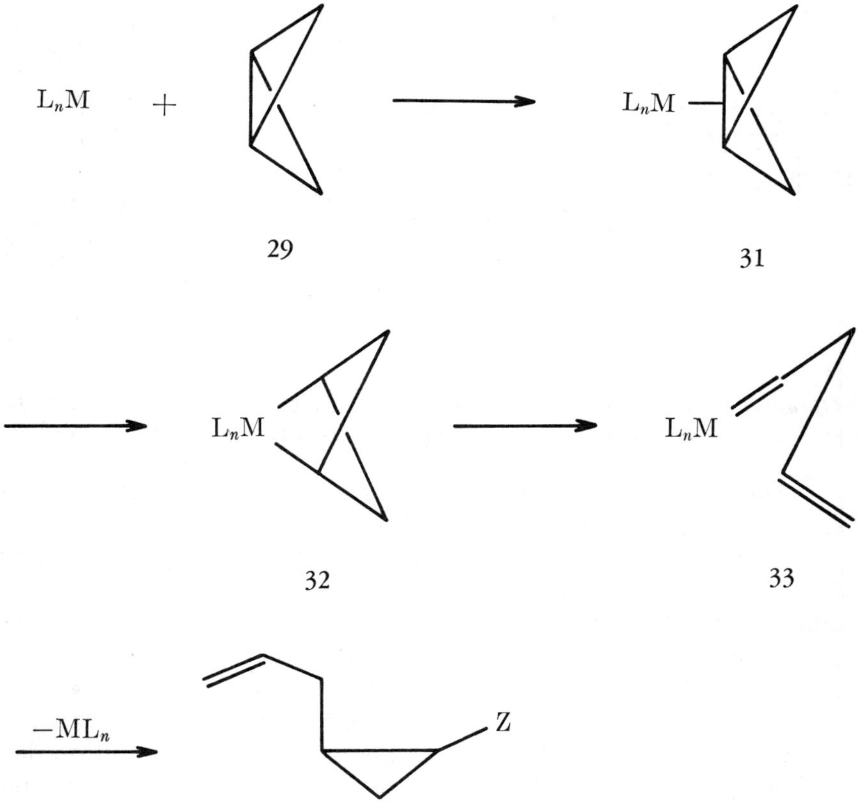

L_nM + [29] ⟶ L_nM—[31]

⟶ L_nM—[32] ⟶ L_nM [33]

$\xrightarrow{-ML_n}$ [30]

L = olefinic ligand.
Z = COOR, CN.

LUMO and consequently can form an edge-on transition metal π complex (31). Subsequent oxidative addition of the strained bond onto the metal produces the metallocyclic intermediate 32 which in turn undergoes metathetic bond cleavage to give the carbene complex 33. Finally, interligand reaction between the carbenic and olefinic ligands produces the cyclopropane adduct 30.

If one assumes the metathetic breakage 32 → 33 as proceeding along a reaction coordinate through a rather late transition state, its electronic structure could be approximated by that of 33. Namely, regioselectivity of the reaction of unsymmetrically substituted bicyclobutanes can be determined by the relative stabilities of the Ni(0) carbenoid intermediates of type 33. The MO calculation can shed light on the substituent effect on the stability of the carbene complexes. Figure 4 illustrates the correlation diagram of methylene–Ni(0) complex calculated by the extended Hückel method. The methylene σ orbital is interacted with the $(3d_{z^2})$ $(4s)$-hybridized orbital of Ni(0), and the methylene p_π (p_x) orbital with the Ni(0) d_{zx} orbital. The stabilization energy of the metal complex was found to arise through such HOMO–LUMO interactions. Here substituents on the carbenic carbon affect to a little or moderate extent the stabilization energy, ΔE_σ, gained through the σ donation. On the other hand, the effect by π back donation, ΔE_π, greatly depends on the nature of the substituents. In general, electron-withdrawing groups increase much ΔE_π value compared with that of the parent methylene complex, whereas methyl or phenyl substituent decreases this value. Thus the calculations suggest that the relative stabilities of the Ni(0) complexes are controlled principally by ΔE_π term and decrease in the order of $CHCOOCH_3 \sim CHCN > CH_2 > CHCH_3 \sim CHC_6H_5$. This sequence does not seem unreasonable in view of the electropositive property of Ni(0) atom.

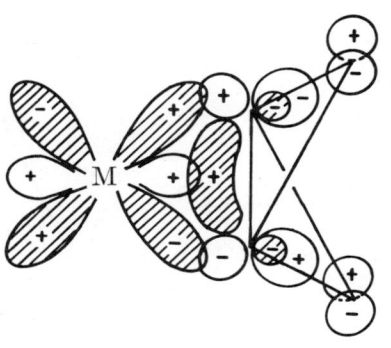

32

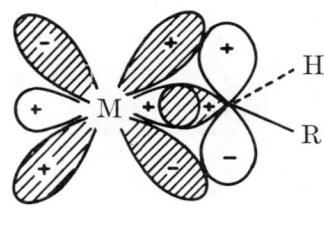

34

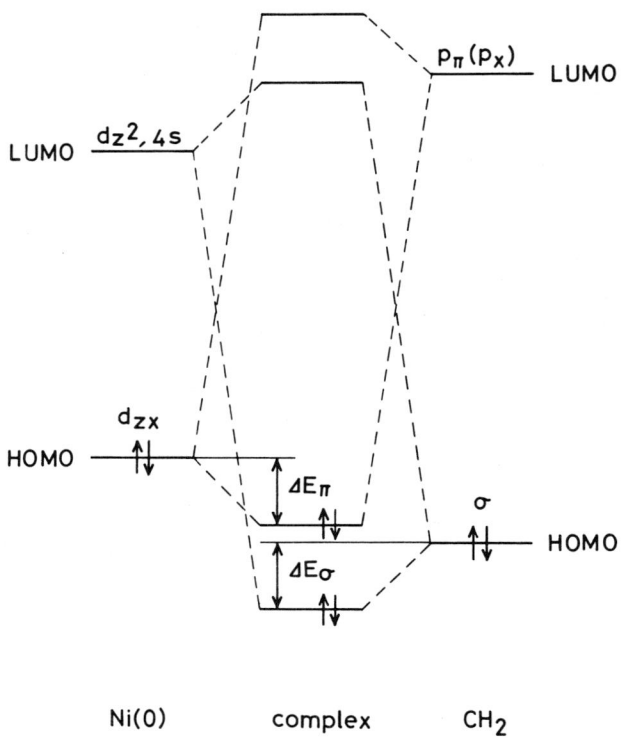

Figure 4. Correlation diagram of Ni(0)–CH$_2$ complex

The experimental results were fully consistent with this consideration (27). The angular carbon substituted by an electron-donating methyl group or conjugative phenyl group avoids becoming the carbenic carbon. In the reaction of carbomethoxy derivatives, two possible pathways compete with the route via carbomethoxy-substituted carbenes predominating.

Scheme 11

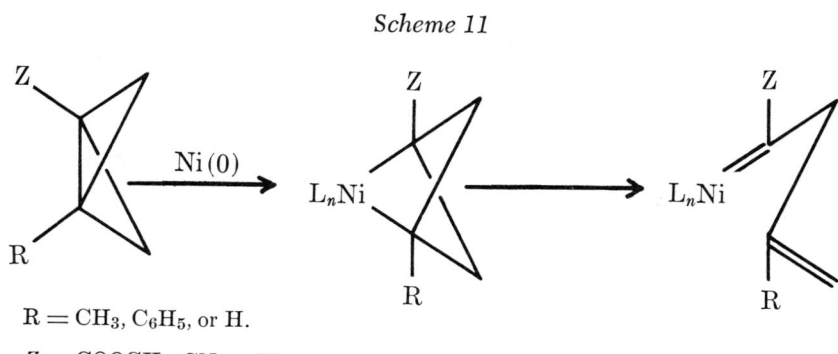

R = CH$_3$, C$_6$H$_5$, or H.

Z = COOCH$_3$, CN, or H.

Scheme 12

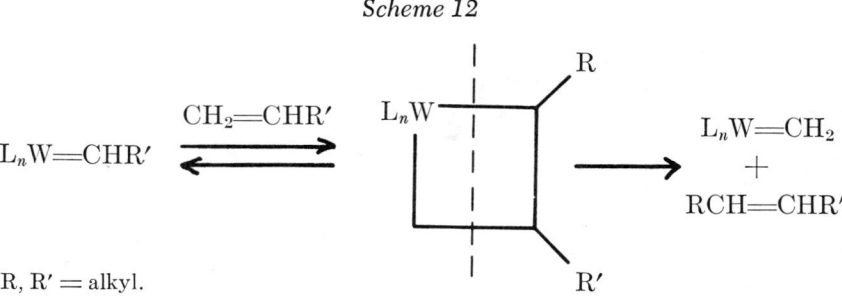

R, R' = alkyl.

The directing effect displayed by the cyano group is the most enlightening. The carbon bearing the electron-accepting cyano group is incorporated exclusively as a carbenic center in the Ni(0) carbenoid intermediate. Thus the unique regioselectivity in the geminal bond cleavage of bicyclobutanes is primarily controlled by electronic properties of the metal and bicyclobutane substituents and is reasonably rationalized by the MO treatment. Notably, closely related selectivity has been observed in the tungsten-catalyzed metathesis of olefins (28).

Regioselectivity of the Catalytic Hydrogenolysis of Strained Molecules

Striking similarities in behavior of strained hydrocarbons displayed in heterogeneous and homogeneous conditions have promoted us to analyze the selectivity in the heterogeneously catalyzed hydrogenolysis (29) in terms of homogeneous chemistry. Hydrogenolysis of bicyclo[2.1.0]pentane to cyclopentane (30), reductive cleavage of 1,8-bis-homocubane at the C_2–C_5 bond (31), and two-bond cleavage of bicyclo[1.1.0]butanes to butanes (32) all are interpreted by assuming that metallocyclic intermediates form by oxidative addition of the strained bond edgewise coordinated on the metal or metal (poly)hydride. Hydrogenolysis of cyclopropane derivatives also can be considered to proceed by way of metallocyclobutane intermediates of type 35.

Scheme 13

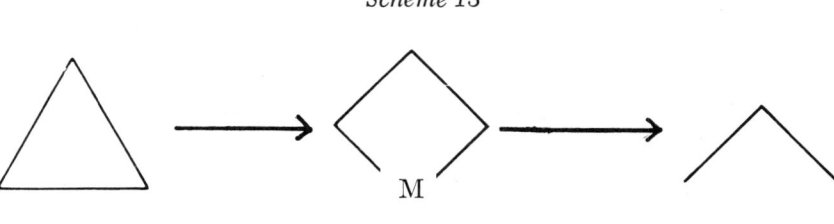

35

In the actual heterogeneous system, M would constitute a part of metal cluster structures. Electronic and steric effects of the substituents are known to perturb significantly the rate and reaction course of the cyclopropane hydrogenolysis. Here any postulated reaction mechanisms must account for the generally observed regioselectivity; mono- or 1,1-dialkylated cyclopropanes prefer the cleavage of the C_2–C_3 bond whereas introduction of an electronegative group causes breakage at the C_1–C_2 bond. Monoalkylated cyclopropanes react faster than parent cyclopropane although further alkylation retards the hydrogenolysis. The reaction of the substrates bearing an electron-withdrawing group requires rather forcing reaction conditions.

MOs of 1,1-dimethylcyclopropane and 1,1-dicyanocyclopropane as models have been calculated by the extended Hückel method. Interestingly, the HOMOs of these substrates (Figure 5 and 6) are not nicely set up for edge-on, π coordination at the bonds to be broken catalytically. If the frontier MO consideration can apply safely here, the calculation suggests the initiation of such hydrogenolysis through corner-on metal coordination (24). Electrophilic attack of metallic species can occur at the C_2 corner of alkylated cyclopropanes from outside where HOMO is developing greatly while the catalyst approaches from the inside to the

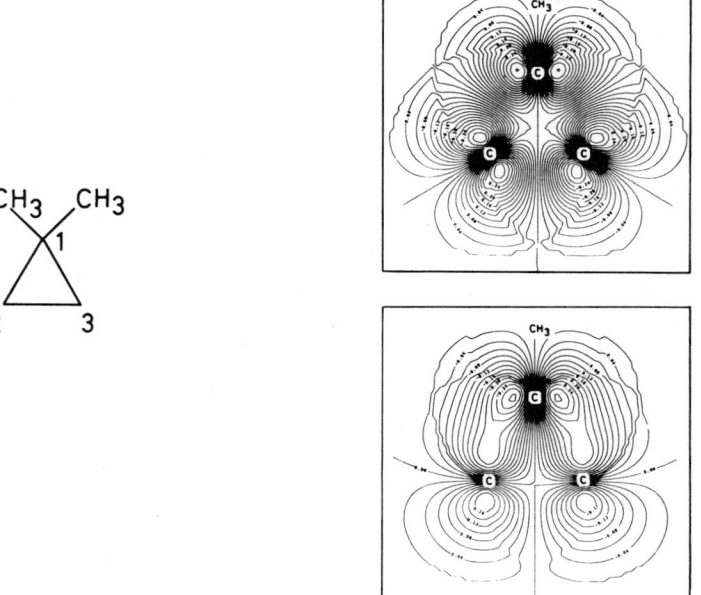

Figure 5. HOMO (lower) and LUMO (upper) of 1,1-dimethylcyclopropane

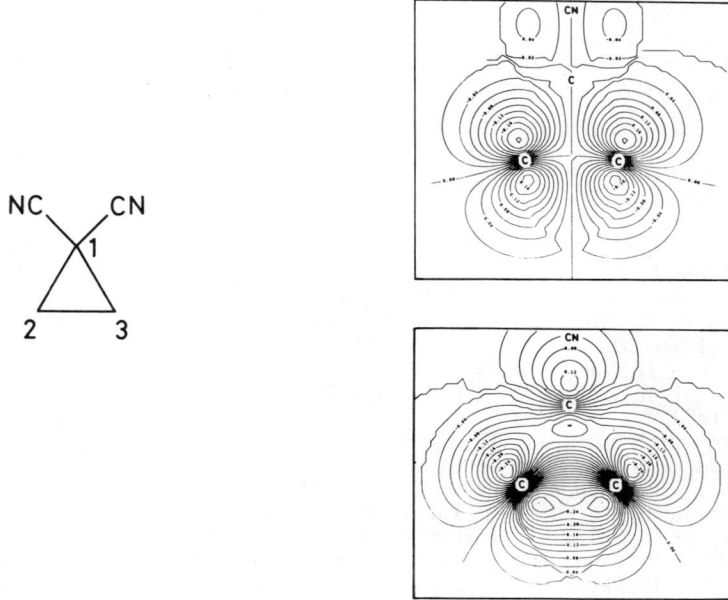

Figure 6. HOMO (lower) and LUMO (upper) of 1,1-dicyanocyclopropane

C_2 corner of electronegatively substituted cyclopropanes. The coordination complexes thus formed, 36 and 37, then will rearrange to the metallocyclobutane intermediates of type 35. Energy levels of HOMOs would be associated with the ease with which the hydrogenolysis takes place. Further detailed experimental and theoretical scrutinies are required on this problem.

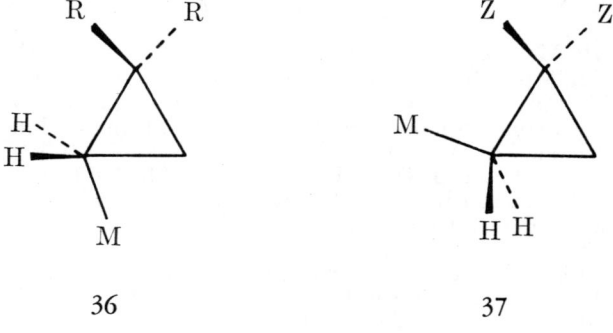

R = alkyl groups. Z = electron-withdrawing groups.

Acknowledgment

We are grateful for the important contribution of our co-workers at Nagoya, particularly of H. Takaya. We thank M. Yamakawa for performing the MO calculations described herein.

Literature Cited

1. Bishop, K. C., III, *Chem. Rev.* (1976) **76**, 461.
2. Mango, F. D., *Coord. Chem. Rev.* (1975) **15**, 109.
3. Halpern, J., *Org. Synth. Met. Carbonyls* (1977) **2**, 705.
4. Noyori, R., Umeda, I., Kawauchi, H., Takaya, H., *J. Am. Chem. Soc.* (1975) **97**, 812.
5. Takaya, H., Yamakawa, M., Noyori, R., *Chem. Lett.* (1973) 781.
6. Noyori, R., Yamakawa, M., Takaya, H., *J. Am. Chem. Soc.* (1976) **98**, 1471.
7. Noyori, R., Suzuki, T., Kumagai, Y., Takaya, H., *J. Am. Chem. Soc.* (1971) **93**, 5894.
8. Noyori, R., *Tetrahedron Lett.* (1973) 1691.
9. Noyori, R., Kawauchi, H., Takaya, H., *Tetrahedron Lett.* (1974) 1749.
10. Noyori, R., Suzuki, T., Takaya, H., *J. Ann. Chem. Soc.* (1971) **93**, 5896.
11. Noyori, R., Kumagai, Y., Takaya, H., *J. Am. Chem. Soc.* (1974) **96**, 634.
12. Noyori, R., Odagi, T., Takaya, H., *J. Am. Chem. Soc.* (1970) **92**, 5780.
13. Noyori, R., Kumagai, Y., Umeda, I., Takaya, H., *J. Am. Chem. Soc.* (1972) **94**, 4018.
14. Noyori, R., Umeda, I., Takaya, H., *Chem. Lett.* (1972) 1189.
15. Schrauzer, G. N., *Adv. Catal.* (1968) **18**, 373.
16. Cassar, L., Halpern, J., *Chem. Commun.* (1970) 1082.
17. Gassman, P. G., Mansfield, K. T., Murphy, T. J., *J. Am. Chem. Soc.* (1969) **91**, 1684.
18. Collins, F. S., George, J. K., Trindle, C., *J. Am. Chem. Soc.* (1972) **94**, 3732.
19. Dauben, W. G., Kielbania, A. J., Jr., *J. Am. Chem. Soc.* (1971) **93**, 7345.
20. Paquette, L. A., Boggs, R. A., Farnham, W. B., Beckley, R. S., *J. Am. Chem. Soc.* (1975) **97**, 1112.
21. Paquette, L. A., Stowell, J. C., *J. Am. Chem. Soc.* (1970) **92**, 2584.
22. Dauben, W. G., Schallhorn, C. H., Whalen, D. L., *J. Am. Chem. Soc.* (1971) **93**, 1446.
23. Mango, F. D., Schachtschneider, J. H., "Transition Metals in Homogeneous Catalysis," Schrauzer, G. N., Ed., p. 223, Marcel Dekker, New York, 1971.
24. DePuy, C. H., McGirk, R. H., *J. Am. Chem. Soc.* (1974) **96**, 1121.
25. Hogeveen, H., Nusse, B. J., *J. Am. Chem. Soc.* (1978) **100**, 3110.
26. Lehn, J.-M., Wipff, G., *J. Chem. Soc., Chem. Commun.* (1973) 747.
27. Noyori, R., "Organotransition-Metal Chemistry," Y. Ishii, M. Tsutsui, Eds., p. 231, Plenum, New York, 1975.
28. Gassman, P. G., Johnson, T. H., *J. Am. Chem. Soc.* (1977) **99**, 622.
29. Kieboom, A. P. G., van Rantwijk, F., "Hydrogenation and Hydrogenolysis in Synthetic Organic Chemistry," p. 102, Delft University Press, Delft, 1977.
30. Criegee, R., Rimmelin, A., *Chem. Ber.* (1957) **90**, 414.
31. Sasaki, N. A., Zunker, R., Musso, H., *Chem. Ber.* (1973) **106**, 2922.
32. Wiberg, K. B., *Adv. Alicyclic Chem.* (1968) **2**, 185.

RECEIVED February 22, 1977.

26

Sensitization of Olefin Photoreactions by Copper(I) Compounds

CHARLES KUTAL and PAUL A. GRUTSCH

Department of Chemistry, University of Georgia, Athens, GA 30602

> *Copper(I) compounds accelerate the rates of a diverse assortment of olefin photoreactions, including rearrangement, oligomerization, and molecular fragmentation. This chapter seeks to provide a basis for understanding this sensitization effect in terms of the ground- and excited-state properties of Cu(I). Thus the ability to generate vacant coordination sites and form stable olefin complexes, as well as the absence of low-lying vacant d orbitals, are two characteristics of the metal which have potentially significant consequences for sensitization. A review of Cu(I)-sensitized olefin photoreactions is presented and a general classification of sensitization processes is proposed. The potential sensitization behavior of some previously unstudied Cu(I) systems also is explored.*

Olefins undergo a fascinating assortment of photochemical transformations (1, 2, 3). During the past 15 years there has been an increasing awareness that the presence of certain transition metal compounds can influence the course of these processes, in some cases resulting in the formation of novel or otherwise difficult-to-synthesize products. In the most general sense, the role of the metal is to render the olefin sensitive to the action of the irradiating light. Consequently, in this chapter we use the term sensitization as a generic label to denote a process in which a transition metal compound (the sensitizer) accelerates the rate of an olefin photoreaction with at least one photon being required per product molecule formed.

Particularly prominent in this role are Cu(I) compounds whose sensitization of a variety of olefin rearrangement (4–13), oligomerization (14, 15, 16, 17, 18), and molecular fragmentation (19) processes is well documented. While the majority of systems examined to date have featured CuX (where X is Cl, Br, and F_3CSO_3) salts as sensitizers, recent results from our laboratory have established that a considerably broader range of Cu(I) compounds can function in this capacity. Such compounds possess a potentially significant advantage over simple CuX salts in that they provide a vehicle for "tailoring" the sensitization properties of Cu(I) by selection of the appropriate combination of coordinated ligands. This approach recently has been used in attempts to develop Cu(I) sensitizers that absorb strongly in the visible wavelength region since this would facilitate the use of sunlight as a photoexcitation source (20, 21).

In searching for effective Cu(I) sensitizers, however, some rational basis for selecting the most promising candidates is needed. Thus any consideration of sensitization behavior should be predicated upon a fundamental understanding of the ground- and excited-state properties of Cu(I) compounds. This viewpoint has been adopted in the present chapter. In subsequent sections we survey the structural and chemical properties of Cu(I) compounds and then consider some pertinent features of their excited-state behavior. In this manner we hope to delineate the characteristics of Cu(I) which contribute to its effectiveness as a sensitizer. Within this framework the results of several studies dealing with Cu(I) sensitization of olefin photoreactions are reviewed, and some mechanistic generalizations are proposed. Finally, the potential sensitization behaviors of some novel Cu(I) systems are considered.

Important Features of Cu(I) Chemistry

There exists an extensive literature on the preparation and characterization of Cu(I) compounds (22, 23); a representative sampling of reactions is presented in Scheme 1. Although Cu^+ disproportionates to Cu^{+2} and Cu^0 in aqueous solution, the $+1$ oxidation state can be stabilized via coordination of strongly binding ligands. Thus the soluble complexes CuI_2^- and $Cu(NH_3)_2^+$ are stable in aqueous solution, and insoluble salts such as the Cu(I) halides exist in the solid state or in contact with water. Ligands possessing low lying, vacant π^* orbitals which are capable of accepting electron density from the metal also tend to stabilize the $+1$ state; common examples are 1,10-phenanthroline(phen), 2,2'-bipyridine (bipy), phosphines, arsines, and olefins.

While discrete two- and three-coordinate Cu(I) compounds have been reported, the favored coordination number of the metal is four. Quite frequently, two or more copper atoms will share ligands, resulting in the formation of polynuclear clusters. Complexes of stoichiometry $Cu_4X_4L_4$ (X = Cl, Br, I; L = PR_3, AsR_3) for example, exist as tetrameric units whose structure depends upon the steric demands of X and R. In

Scheme 1

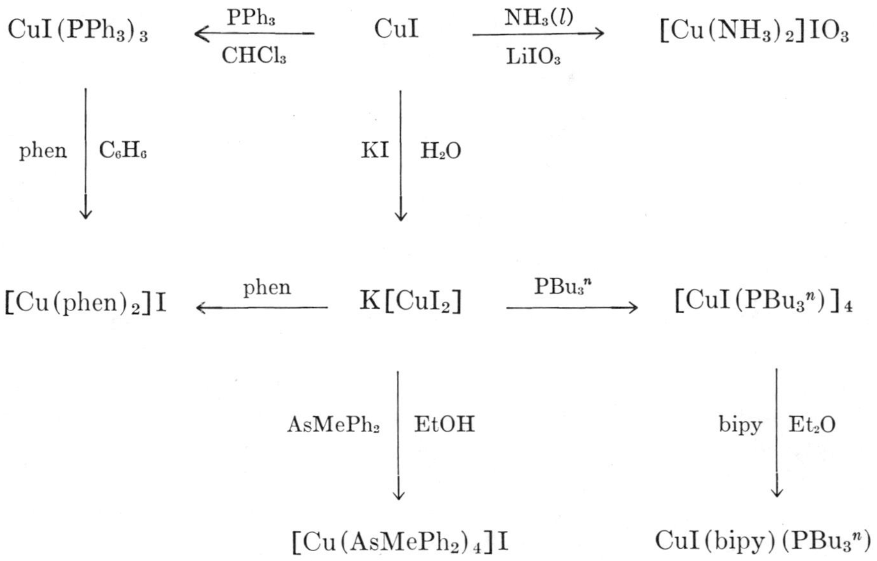

$$CuI(PPh_3)_3 \xleftarrow[CHCl_3]{PPh_3} CuI \xrightarrow[LiIO_3]{NH_3(l)} [Cu(NH_3)_2]IO_3$$

phen | C_6H_6 ↓   KI | H_2O ↓

$$[Cu(phen)_2]I \xleftarrow{phen} K[CuI_2] \xrightarrow{PBu_3^n} [CuI(PBu_3^n)]_4$$

AsMePh$_2$ | EtOH ↓   bipy | Et$_2$O ↓

$$[Cu(AsMePh_2)_4]I \qquad\qquad CuI(bipy)(PBu_3^n)$$

[CuClPPh$_3$]$_4$ the geometry of the Cu$_4$Cl$_4$ core has been loosely described (24) as "cubane-like" (Figure 1A) while increased anion crowding forces [CuBrPPh$_3$]$_4$ to adopt a "step" structure (25) containing both three- and four-coordinate Cu(I) (Figure 1B). The long Cu–Cu distances and ready deformability of the Cu$_4$X$_4$ framework in these clusters suggest the absence of strong metal–metal bonding. Indeed the proposal has been advanced that the stabilities of copper clusters arise from the presence of bridging ligands rather than from direct metal–metal bonds (26, 27).

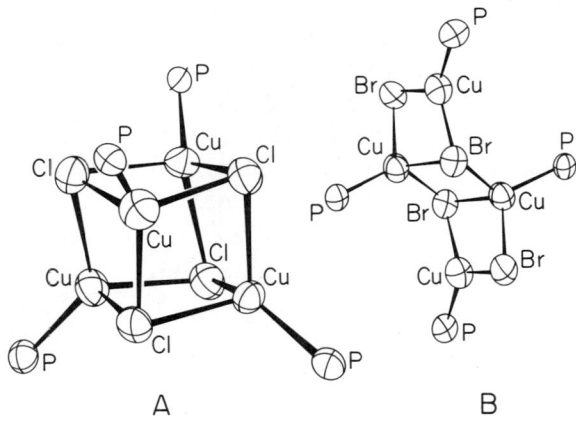

Figure 1. Structures of $Cu_4X_4L_4$ cluster compounds

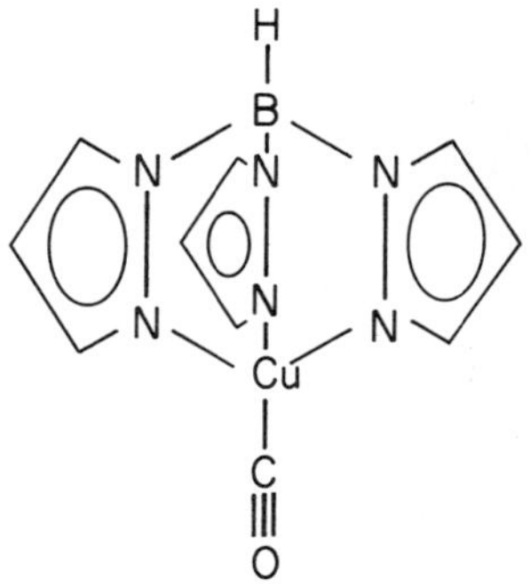

Figure 2. Structure of Cu-[HB(pz)$_3$]CO

Cu(I) forms stable complexes with a variety of cyclic and acyclic olefins (22). The bonding in such complexes is adequately described in terms of the synergistic model proposed by Dewar and by Chatt and Duncanson (28, 29). Thus σ donation from a filled olefin π orbital to an empty metal orbital and π back donation from a filled metal d orbital to an empty π^* olefin orbital both contribute to the stability of the Cu(I)–olefin bond (30, 31). Contrary to its behavior towards olefins, Cu(I) forms few stable, isolable complexes with CO. This difference reflects the inability of Cu → CO backbonding to offset the poor σ-donor characteristics of the ligand. The degree of backbonding overlap can be improved, however, if the electron density on copper is enhanced by the incorporation of a strongly electron-donating group into the complex. Thus tert-BuOCu(CO) can be sublimed (60°C, 1 mm) with no apparent decomposition (32), and Cu[HB(pz)$_3$]CO (Figure 2; HB(pz)$_3$ is hydrotris(1-pyrazolyl)borate) is both air and heat stable in the solid state (33, 34, 35). In solution the latter complex reacts with a variety of donor ligands L, including olefins, to produce Cu[HB(pz)$_3$]L with the concomitant evolution of CO.

Based upon simple ligand field considerations (i.e., d^{10} complexes have zero ligand field contribution to the activation energy), we would expect uncharged, nonchelating ligands (such as CO) to undergo facile exchange between the first coordination sphere of Cu(I) and bulk solution. Direct support for this premise emerges from the observation that

Table I. Spectral Characteristics

Compound[a]	Solvent	λ_{max} (nm)
ClCu–NBD	Ethanol	248
Cu(2,9-Mephen)$_2^+$	Isoamyl alcohol	454
CuCl$_3^{-2}$	HCl/H$_2$O	273
Cu(PPh$_3$)$_2$BH$_4$	Benzene	264

[a] NBD is norbornadiene; 2,9-Mephen is 2,9-dimethyl-1,10-phenanthroline; PPh$_3$ is triphenylphosphine.

the exchange of uncoordinated and Cu(I) bound phosphines (36) or olefins (30, 31) is rapid on the NMR time scale. This ability of certain Cu(I) compounds to readily generate a vacant site can be profitably incorporated into useful catalytic schemes (vide infra).

Excited State Characteristics of Copper(I) Compounds

Because of the presence of a completely filled metal $3d$ subshell, Cu(I) compounds possess no low-lying ligand field excited states; furthermore, the metal localized $3d \rightarrow 4p$ transition is predicted (37) to lie at inconveniently high energies ($> 60,000$ cm^{-1}). Thus the absorption characteristics of Cu(I) compounds can be attributed to one or more of the following transitions: (1) ligand-to-metal charge transfer; (2) metal-to-ligand charge transfer; (3) charge transfer to solvent (CTTS); and (4) intraligand. Some repersentative spectral assignments are listed in Table I.

Different types of behavior are expected depending upon the identity of the excited state populated. The photo-induced redistribution of electron density in a ligand-to-metal CT state formally generates a reduced metal and oxidized ligand radical; the latter can be susceptible to nucleophilic attack by solvent or other species in solution. For a metal-to-ligand CT state, promotion of electron density to the periphery of the complex can favor bimolecular electron transfer. Thus McMillin and co-workers recently have reported that visible irradiation of Cu(2,9-Mephen)$_2^+$ in the presence of cis-bis(iminodiacetato)cobaltate(III) ion in a water–alcohol mixture results in the production of Cu(II) and Co(II) (40). This photo-induced redox behavior mimics that of Ru(bipy)$_3^{+2}$ (41) and suggests that the Cu(I) compound functions as an excited-state reductant. Qualitatively similar behavior has been found by Hurst and co-workers (42, 43) in the binuclear bridged systems (NH$_3$)$_5$Co–L–Cu (L is $^-$O$_2$C(CH$_2$)$_n$CH=CHR or NH$_2$(CH$_2$)$_n$CH=CH$_2$) where irradiation induces Cu(I)-to-Co(III) electron transfer mediated by the π-delocalized orbitals of the bridging ligands. Stevenson and Davis have reported that population of a CTTS state in CuCl$_3^{-2}$ can result in production of a

of Some Cu(I) Compounds

ϵ (M^{-1} cm^{-1})	Assignment	Ref.
6.3×10^3	NBD $\rightarrow$ Cu or Cu $\rightarrow$ NBD	11
7.8×10^3	Cu $\rightarrow$ 2,9-Mephen	38
3.3×10^3	CTTS	39
1.9×10^4	$n(P) \rightarrow \pi^*$	13

solvated electron which, in turn, can be scavenged by H⁺ (producing hydrogen) or other facile electron acceptors (*44*). Intraligand π—π^* or n—π^* excited states are expected to exhibit the reactivity patterns characteristic of the uncoordinated molecule. The possibility exists, however, that coordination to the metal can orient the ligand in a particularly reactive (or nonreactive) conformation or perhaps induce a high degree of stereospecificity in product formation. Furthermore, the metal can perturb the radiative and nonradiative processes of the ligand excited states.

Additional possibilities arise when considering the behavior of copper cluster systems. Thus transitions which involve the stabilizing bridging ligands can result in one or more of the following pathways: (1) multiple ligand loss; (2) structural changes in the cluster framework; and (3) declusterification.

Relatively few cases of luminescence from Cu(I) compounds have been reported. Ziolo, et al. observed emission from $(PPh_3)_m Cu_n X_n$ (X is a halogen) in solid-state and low-temperature glasses (*45*). Hardt reported that cuprous halides react with various nitrogen bases (e.g., pyridine, picolines) to form compounds whose emission maxima are strongly temperature dependent (so-called fluorescence thermochromism) (*46, 47*). Some 1:1 cuprous halide complexes with methyl or ethyl isonicotinate have been noted to emit while the corresponding 1:2 species do not (*48*). McMillin and Buckner have observed emission from Cu(bipy)-$(PPh_3)_2$ both in the solid state and in a glass at 77°K and suggest that it originates from a metal-to-bipy charge transfer state (*49*). Recently we reported some rare examples of luminescence from Cu(I) complexes in room-temperature fluid solution (*13*). Thus $Cu(PPh_3)_2BH_4$ and Cu-(diphos)BH₄ (diphos is 1,2-bis(diphenylphosphino)ethane) both exhibit a broad emission in the visible region which is tentatively assigned as ligand localized.

Survey of Cu(I)-Sensitized Olefin Photoreactions

In a series of papers (*14, 15, 16*) Trecker et al. reported their detailed studies of the cuprous halide-sensitized dimerization of norbornene. Compelling evidence was presented for the involvement of a preformed metal–olefin complex in the mechanism of sensitization. Thus ether solutions of norbornene and CuBr separately displayed no maximum in their electronic spectra down to 220 nm whereas a mixture of the two exhibited an intense absorption at 239 nm. The appearance of this band was attributed to the formation of a 1:1 BrCu–norbornene complex. Upon 254-nm irradiation, predominant production of the exo-trans-exo dimer **1** (Reaction 1) ensued. The observed second-order dependence of the quantum yield for this process upon the norbornene concentration was interpreted

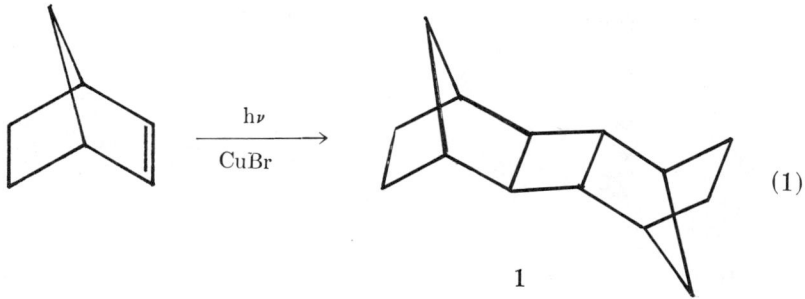

(1)

1

in terms of a termolecular interaction between two ground-state olefin molecules and the photo-excited BrCu–norbornene complex.

Salomon and Kochi reexamined this dimerization process (among others) by using the very attractive sensitizer, $Cu(F_3CSO_3)$ (17). Earlier work (30, 31) by these authors had established that Cu(I) complexes containing one, two, three, and even four coordinated olefinic bonds could be formed by displacement of the weakly coordinating F_3CSO_3 anion. In contrast, cuprous halides exhibit little tendency to coordinate more than one olefinic bond (30, 50). Irradiation of solutions containing $Cu(F_3CSO_3)$ and norbornene resulted in preferential formation (~90%) of dimer 1 but with a quantum yield that displayed a first-order dependence upon olefin concentration. While no completely satisfactory explanation for this discrepancy with the results of Trecker, et al. (vide supra) could be offered, it was noted that solubility considerations allowed the quantum yield dependence to be tested over a much wider olefin concentration in the $Cu(F_3CSO_3)$ system. The photochemical data were explicable in terms of the mechanistic sequence outlined in Scheme 2. Thus while both 1:1 and 1:2 Cu(I)–norbornene complexes are present in solution, dimer 1 is produced exclusively by photoexcitation of the latter species.

Minimally, ground-state complex formation between Cu(I) and norbornene facilitates the absorption of light by the otherwise weakly absorbing olefin. There is, in addition, the possibility that the close proximity of two (or more) norbornene molecules in the complex predisposes them toward dimerization. Ample precedent exists for this template effect of transition metals in olefin cycloaddition reactions (51, 52, 53). Another, quite intriguing role which has been proposed for Cu(I) involves the photo-induced formation of a carbenium ion intermediate 2 from the initially formed complex (Reaction 2). Such "photocupration" recently has been implicated in a series of novel Cu(I)-sensitized rearrangement and fragmentation processes of 7-methylenenorcarane (19). Additional studies are clearly warranted to test the generality of this process.

332

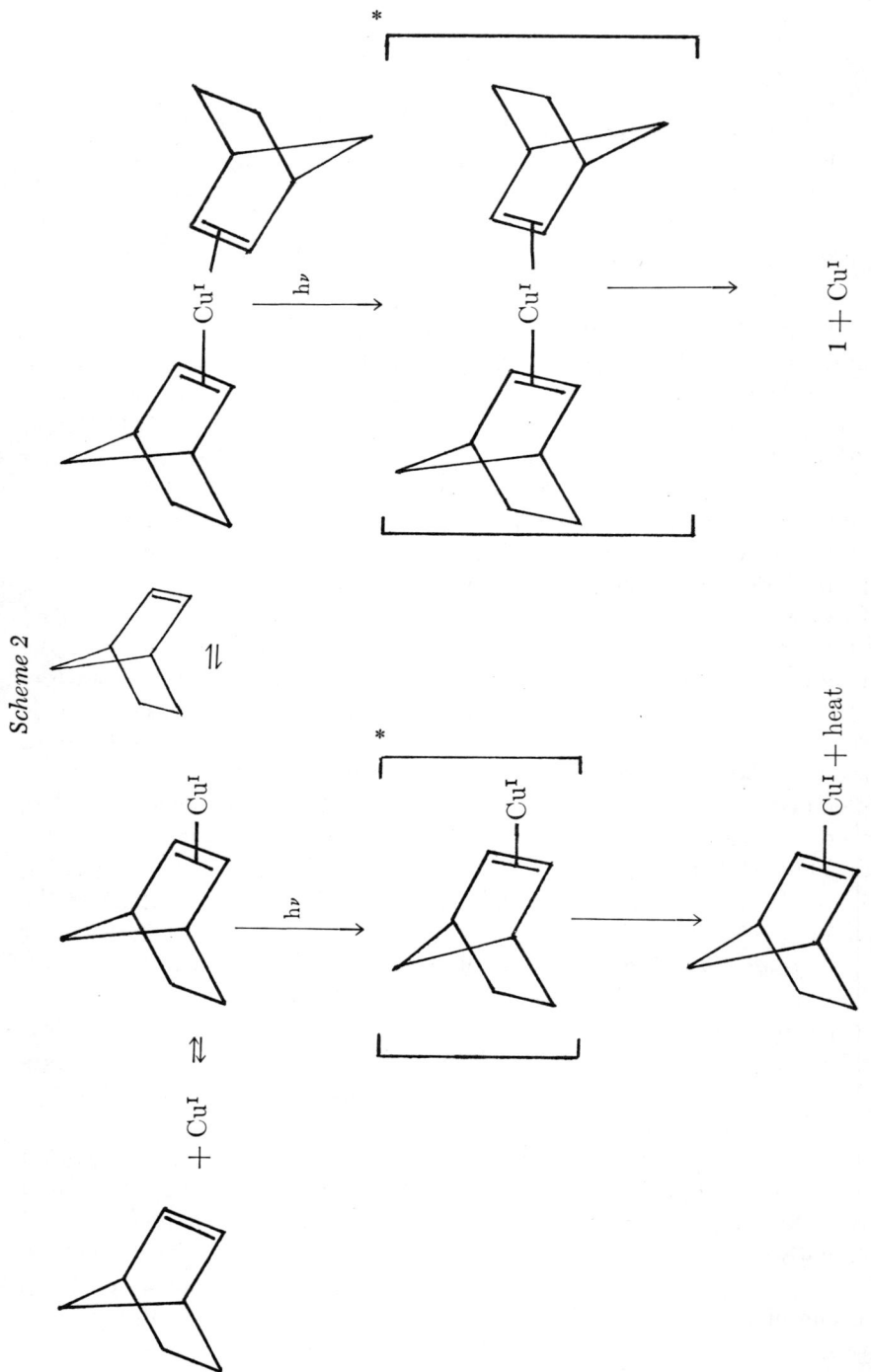

Scheme 2

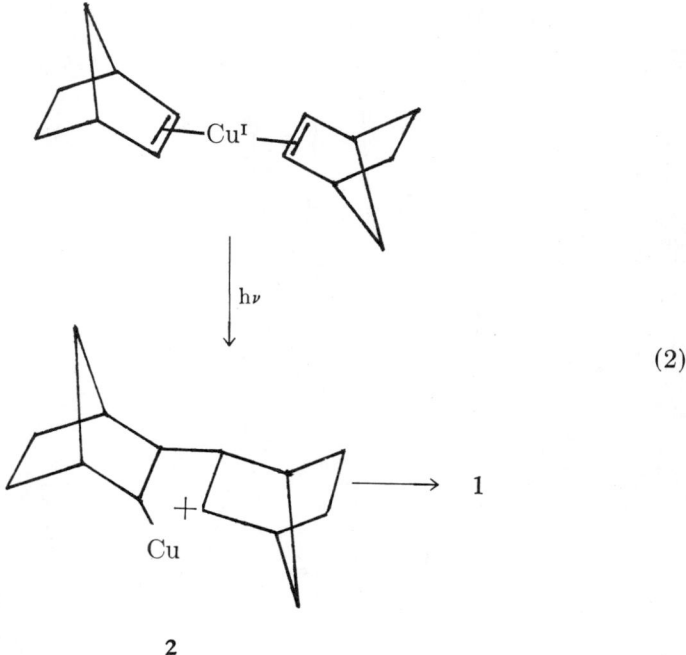

(2)

Recent work (*20, 21*) in our laboratory has focused upon the use of transition metal compounds to sensitize the energy-storing valence isomerization of norbornadiene, NBD, to quadricyclene, Q (Reaction 3). In particular we have found that a catalytic amount of CuCl functions as an effective and quite specific sensitizer for this transformation. Conversions of greater than 90% have been achieved since Cu(I) is ineffective as a catalyst for the energy-releasing reverse reaction. Spectral and photochemical evidence support a mechanism which features a 1:1 ClCu–NBD complex as the photoactive species. As illustrated in Figure 3, an obvious consequence of complexation is a shift of the absorption spectrum of the system into a region accessible to the 313-nm irradiation used. Possible pathways by which the photo-excited complex relaxes to Q have been discussed (*12*).

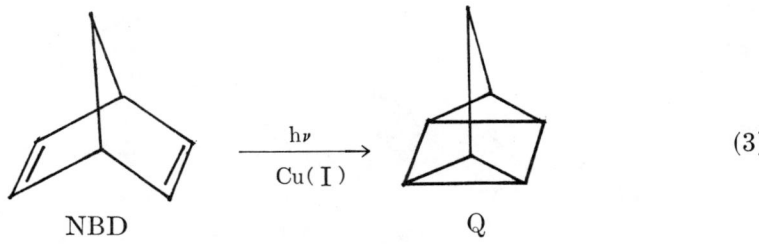

(3)

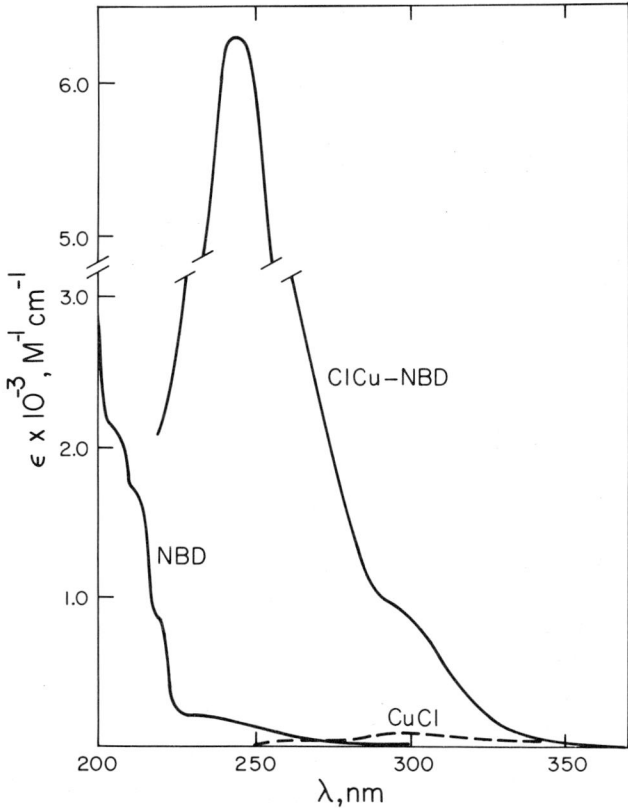

Figure 3. Spectral evidence for complex formation between CuCl and NBD in ethanol (11)

The Cu(I) sensitized rearrangement of *cis,cis*-1,5-cyclooctadiene, COD, to tricyclo[$3.3.0.0^{2,6}$]octane, **3** (Reaction 4), represents an interesting contrast in behavior to the systems considered thus far. In his study of this process (*4, 5*), Srinivasan noted that the spectrum of an ether solution containing both CuCl and the olefin is essentially the sum of the individual component spectra, suggesting (although not conclusively proving) the absence of appreciable complex formation. Since *cis,cis*-1,5-COD was present in considerable excess in the photochemical studies, it constituted the major absorbing species. In fact, a reasonable value (< 1) for the quantum yield of rearrangement could only be calculated on the basis of the light absorbed by all (coordinated and uncoordinated) of the olefin in the system. The proposed mechanism (Reaction 5) involves the interaction of CuCl with a photo-excited olefin molecule. The exact nature of this interaction, however, was not specified.

$$\text{COD} \xrightarrow{h\nu} \text{COD}^* \xrightarrow{\text{CuCl}} \mathbf{3} \cdot \text{CuCl} \quad (5)$$

In a subsequent study (9), Whitesides, et al. examined the photobehavior of Cu_2Cl_2(*cis,cis*-1,5-COD), **4**, in pentane. While precise quantum yield measurements were precluded by the unknown, though apparently appreciable extent of ligand dissociation and the presence of a heterogeneous suspension, several noteworthy observations were reported. Thus 254-nm irradiation of the system produced, in addition to **3**, significant quantities of insoluble CuCl complexes of *cis,trans*-1,5-COD and *trans,trans*-1,5-COD. The initial rate of formation of **3** was less than that of *cis,trans*-1,5-COD but became greater at higher conversions. While irradiation of *cis,cis*-1,5-COD or *cis,trans*-1,5-COD in the absence of CuCl afforded no **3**, irradiation of *trans,trans*-1,5-COD yilded **3** with high efficiency.

These observations suggest that **3** is not formed directly from *cis,cis*-1,5-COD but instead involves the intermediacy of *cis,trans*-1,5-COD and perhaps *trans,trans*-1,5-COD. A general mechanism which incorporates these features is outlined in Scheme 3. While the detailed role of CuCl was not definitively established, it was suggested that the ability of Cu(I) to bind strongly to ground-state trans olefinic bonds shifts the position of the photoequilibria between the three 1,5-COD isomers toward the sterically strained but strongly complexing cis-trans- and trans,trans-dienes at the expense of the relatively strain-free but weakly coordinating *cis,cis*-1,5-COD.

A related study (18) by Salomon, et al. concerning the Cu(F_3CSO_3)-sensitized dimerization of cyclohexene and cycloheptene also emphasized the importance of photo-generated trans olefinic bonds. Thus the proposed mechanism involves the photo-induced isomerization of a preformed Cu(I)–olefin complex to a highly strained and reactive *trans*-cyclohexene or *trans*-cycloheptene followed by a thermal cycloaddition of a second olefin molecule (Reaction 6).

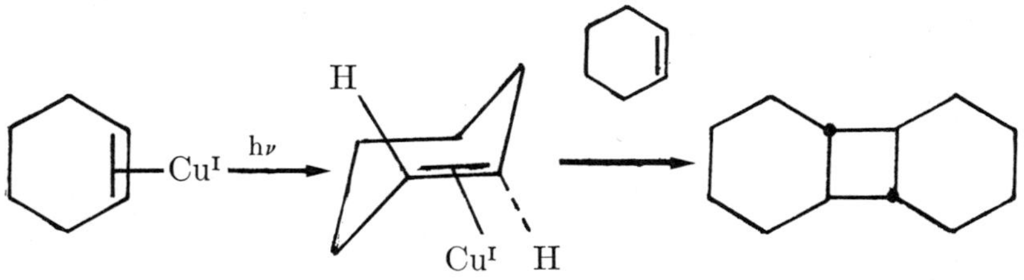

$$ (6) $$

In a recent photochemical study (13) of the NBD to Q conversion (Reaction 3) in the presence of $Cu(P)_2BH_4$ (where P is a tertiary phosphine), we presented evidence for a previously unreported mechanistic role of Cu(I) compounds in sensitized olefin photoreactions. The salient features of our results can be summarized by reference to the behavior of $Cu(PPh_3)_2BH_4$, the structure of which is shown in Figure 4 (54, 55). While molecular weight measurements indicate that this coordinatively saturated compound is undissociated in solution, ^{31}P NMR studies reveal that the bound phosphines are labile and undergo rapid exchange with added PPh_3. Nevertheless, experiments designed to detect the displacement of bound PPh_3 by NBD to form a logical Cu(I)–NBD precursor to sensitization invariably have yielded negative results. Consequently, the direct coordination of NBD to the metal does not represent a viable pathway for sensitization in this system. The alternative mechanism involving interaction of photo-excited $Cu(PPh_3)_2BH_4$ with ground-state olefin thus appears to be operative. Current work in our laboratory is directed toward defining the nature of this excited-state interaction as well as discovering additional examples of this type of sensitization.

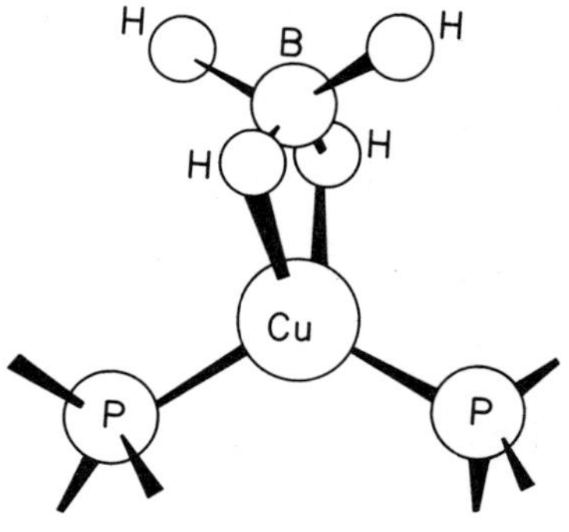

Figure 4. Structure of Cu-$(PPh_3)_2BH_4$

Scheme 3

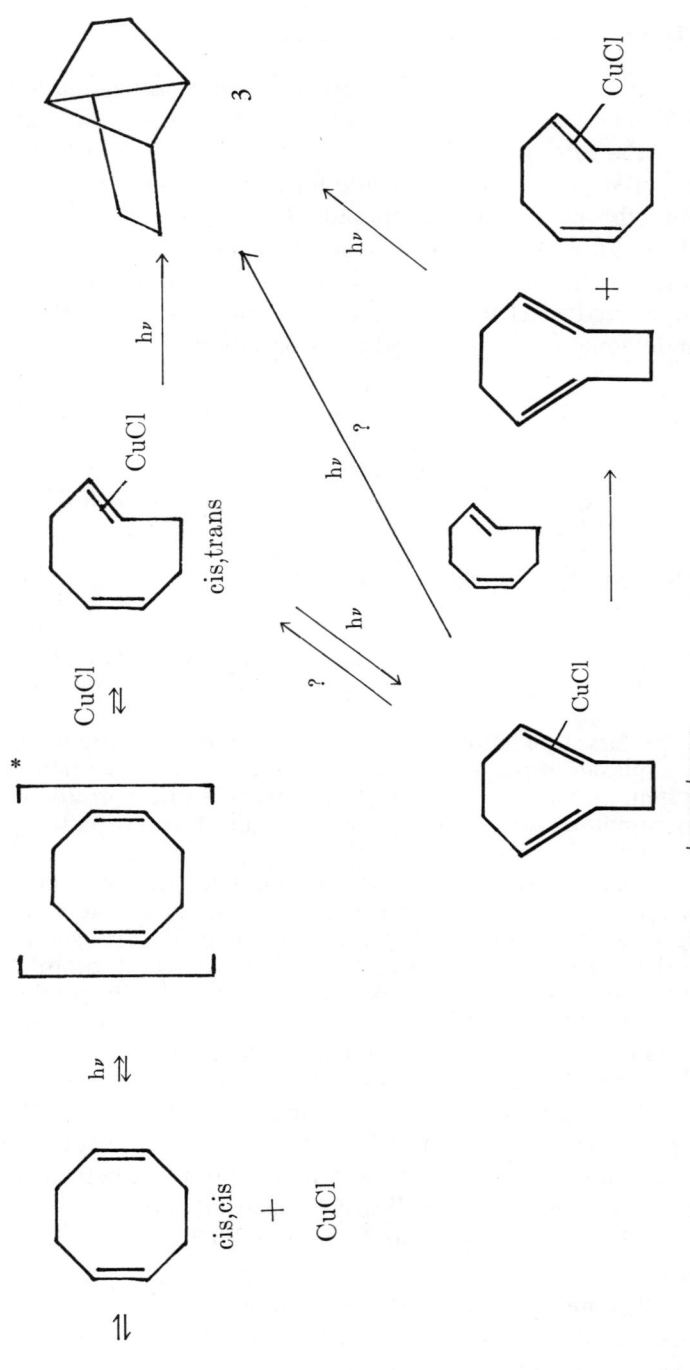

Classification of Cu(I) Sensitization Processes

In this chapter we have used the term sensitization in a generic sense to denote a process in which a Cu(I) compound (the sensitizer) accelerates the rate of an olefin photoreaction with at least one photon being required per product molecule formed. Within this rather broad definition, the preceding section provides examples of three general classes of sensitization process; these are depicted in Scheme 4, where the symbol Cu(I) is taken to represent some Cu(I) compound, an asterisk signifies an excited electronic state, and a prime indicates that the olefin has undergone a photo-induced transformation.

Scheme 4

Class 1 $Cu^I + olefin \rightleftharpoons Cu^I - olefin \xrightarrow{h\nu} [Cu^I - olefin]^* \rightarrow Cu^I + olefin'$

Class 2 $olefin \xrightarrow{h\nu} olefin^* \xrightarrow{Cu^I} [Cu^I + olefin^*] \rightarrow Cu^I + olefin'$

Class 3 $Cu^I \xrightarrow{h\nu} Cu^{I*} \xrightarrow{olefin} [Cu^{I*} + olefin] \rightarrow Cu^I + olefin'$

Class 1—Photoexcitation of a preformed copper–olefin complex. Complexation plays a key role by shifting the absorption spectrum of the system into a wavelength region accessible to the irradiating light and/or by providing a sterically or electronically favored pathway to the photoproduct.

Class 2—Bimolecular interaction between an excited state of the olefin and the ground-state Cu(I) compound. The role of Cu(I) can be to stabilize the excited state of the olefin via complexation (9) or perhaps to form an intermediate (such as that postulated for photocupration (19)) which is particularly amenable to product formation.

Class 3—Bimolecular interaction between an excited state of the Cu(I) compound and ground-state olefin. Possible interactions include electronic energy transfer and exciplex formation (56, 57).

Although most reported examples of Cu(I) sensitization can be accommodated within a single class (Table II), a more complicated situation may result in systems where multiple photochemical steps are involved. Thus the overall CuCl-sensitized conversion of *cis,cis*-1,5-COD to 3 (Scheme 3) appears to feature both Class 1 and Class 2 sensitization processes.

The most obvious distinction among the three classes rests upon the identity of the light-absorbing species. Our choice in organizing the experimental results along these lines rather than, for example, on the basis of the identities of possible ground- or excited-state intermediates

Table II. Classification of Some Cu(I)-Sensitized Olefin Photoreactions

	Example	Ref.
Class 1	Dimerization of norbornene in presence of CuBr (Reaction 1) or Cu(F$_3$CSO$_3$) (Scheme 2)	16, 17
	Rearrangement and fragmentation of 7-methylenenorcarane in the presence of Cu(F$_3$CSO$_3$)	19
	Valence isomerization of norbornadiene to quadricyclene in presence of CuCl (Reaction 3)	11, 12
	Rearrangement of *cis,trans*-1,5-COD to 3 in presence of CuCl (Scheme 3)	9
Class 2	Isomerization of *cis,cis*-1,5-COD to *cis,trans*-1,5-COD in presence of CuCl (Scheme 3)	9
Class 3	Valence isomerization of norbornadiene to quadricyclene in presence of Cu(PPh$_3$)$_2$BH$_4$ and related compounds	13, 59

formed subsequent to light absorption reflects the presently incomplete understanding of many of the mechanistic details of Cu(I)-sensitized processes. Future work along these lines is clearly needed.

Based upon the general classification in Scheme 4, however, it is possible to delineate some ground- and excited-state characteristics of Cu(I) which can contribute to its effectiveness as a sensitizer. Thus the ability of a Cu(I) compound to generate vacant coordination sites and to form stable olefin complexes is clearly advantageous for Class 1 sensitization; although less certain, it is probably important in Class 2 processes as well. Conversely, Class 3 sensitization will be more prominent among coordinatively saturated Cu(I) compounds whose ligands are not readily displaced by olefins. The absence of low-lying vacant d orbitals on Cu(I) has potentially significant consequences in that the energy of an absorbed photon, which might otherwise be dissipated via the ligand substitution processes characteristic of ligand-field excited states (*58*), can be channeled into chemically more productive pathways. This feature is particularly crucial for Class 3 sensitization which requires that an excited state characteristic of the Cu(I) compound undergo a bimolecular interaction with a ground-state olefin molecule.

Future Directions

The availability of a diverse assortment of reasonably stable and, in many cases, highly colored Cu(I) compounds (*23*) suggests that studies of Cu(I) sensitization need not be limited to simple CuX salts. The novel results obtained for Cu(PPh$_3$)$_2$BH$_4$ (*13*) and related compounds (*59*) reinforce this view. Consequently, in this final section we briefly consider the potential sensitization properties of some previously unstudied, but rather intriguing Cu(I) systems. While the discussion is admittedly speculative, we hope that it is equally suggestive.

We have recently noted (21) the attractive characteristics of the series, $CuXPR_3(N-N)$ (typically X = Cl, Br; R = alkyl or phenyl, N–N = phen or bipy). Depending upon the lability of PR_3, these compounds can conceivably function as either Class 1 or Class 3 sensitizers. Furthermore, their strong absorption in the visible wavelength region raises the possibility of harnessing sunlight to drive useful photochemical reactions.

The ready accessibility of only one coordination (via loss of CO) site about copper makes $Cu[HB(pz)_3]CO$ (Figure 2) potentially attractive as a specific sensitizer for photorearrangements of olefins. For example, endo-dicyclopentadiene in the presence of $Cu(O_3SCF_3)$ undergoes virtually exclusive photodimerization (Reaction 7, path i), presumably via prior formation of a 2:1 olefin:Cu(I) complex (17). Since the analogous bis-olefin complex involving the $Cu[HB(pz)_3]$ moiety is sterically improbable, the most likely photoprocess is internal cyclization (Reaction 7, path ii). Preliminary studies in our laboratory indicate that $Cu[HB(pz)_3]CO$ sensitizes the NBD-to-Q conversion (Reaction 3) with respectable quantum efficiency (21).

The sensitization properties of cluster systems such as $[CuXL]_4$ (Figure 1) are particularly intriguing since the possibility of multicenter sensitization exists. Thus the photochemical or thermal (likely when L is bulky) substitution of two or more L groups by olefins generates a species in which the olefin molecules are situated in close proximity to each other. One likely consequence of this template effect of the cluster is efficient photo-induced olefin oligomerization.

Another interesting cluster system is $[CuHPPh_3]_6$, whose structure contains the copper atoms at the apices of a distorted octahedron (Figure 5) (60). Although the exact locations of the hydride ligands are uncer-

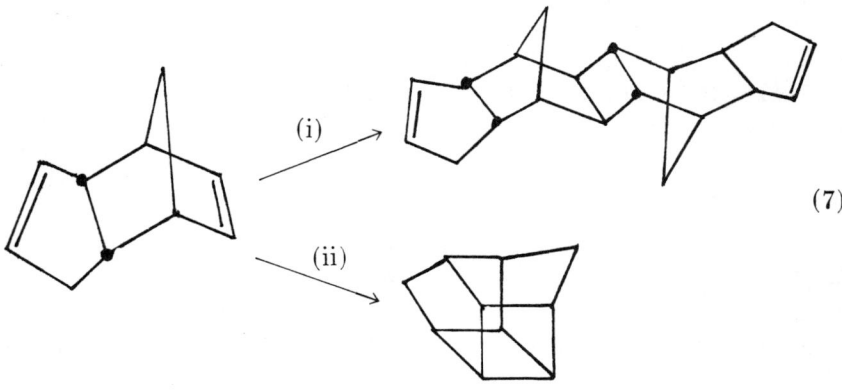

(7)

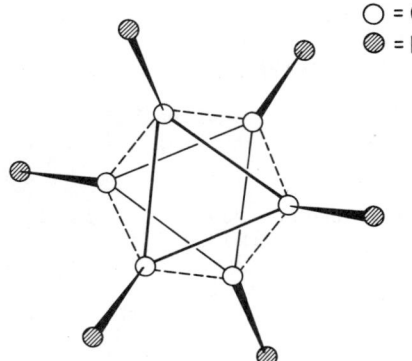

Figure 5. Structure of $[CuH(PPh_3)]_6$ (26)

tain, they most likely bridge the six long Cu–Cu distances indicated by solid lines. Two potential photoreactive pathways of the cluster are elimination of hydrogen (61, 62) and hydrogenation of added unsaturated molecules. The latter process can have some catalytic value if an appropriate source to replenish the hydrogen consumed can be found.

Acknowledgment

We wish to acknowledge the contributions of D. P. Schwendiman and E. M. Sweet. Our studies described herein have been supported by the Department of Energy (EY-76-S-09-0893) and the National Science Foundation (MPS 75-13752).

Literature Cited

1. Fonken, G. J., "Organic Photochemistry," O. L. Chapman, Ed., p. 197, Marcel Dekker, New York, 1967.
2. Kropp, P. J., *Pure Appl. Chem.* (1970) **24**, 585.
3. Coyle, J. D., *Chem. Soc. Rev.* (1974) **3**, 329.
4. Srinivasan, R., *J. Am. Chem. Soc.* (1963) **85**, 3048.
5. Srinivasan, R., *J. Am. Chem. Soc.* (1964) **86**, 3318.
6. Baldwin, J. E., Greeley, R. H., *J. Am. Chem. Soc.* (1965) **87**, 4514.
7. Haller, I., Srinivasan, R., *J. Am. Chem. Soc.* (1966) **88**, 5084.
8. Nozaki, H., Nisikawa, Y., Kawanisi, M., Noyori, R., *Tetrahedron* (1967) **23**, 2173.
9. Whitesides, G. M., Goe, G. L., Cope, A. C., *J. Am. Chem. Soc.* (1969) **91**, 2608.
10. Deyrup, J. A., Betkouski, M., *J. Org. Chem.* (1972) **37**, 3561.
11. Schwendiman, D. P., Kutal, C., *Inorg. Chem.* (1977) **16**, 719.
12. Schwendiman, D. P., Kutal, C., *J. Am. Chem. Soc.* (1977) **99**, 5677.
13. Grutsch, P. A., Kutal, C., *J. Am. Chem. Soc.* (1977) **99**, 6460.
14. Arnold, D. R., Trecker, D. J., Whipple, E. B., *J. Am. Chem. Soc.* (1967) **87**, 2596.
15. Trecker, D. J., Henry, J. P., McKeon, J. E., *J. Am. Chem. Soc.* (1965) **87**, 3261.
16. Trecker, D. J., Foote, R. S., Henry, J. P., McKeon, J. E., *J. Am. Chem. Soc.* (1966) **88**, 3021.

17. Salomon, R. G., Kochi, J. K., *J. Am. Chem. Soc.* (1974) **96**, 1137.
18. Salomon, R. G., Folting, K., Streib, W. E., Kochi, J. K., *J. Am. Chem. Soc.* (1974) **96**, 1145.
19. Salomon, R. G., Salomon, M. F., *J. Am. Chem. Soc.* (1976) **98**, 7454.
20. Kutal, C., Schwendiman, D. P., Grutsch, P. A., *Solar Energy* (1977) **19**, 651.
21. Kutal, C., ADV. CHEM. SER. (1978) **168**, 158.
22. Quin, H. W., Tsai, J. H., *Adv. Inorg. Chem. Radiochem.* (1968) **12**, 217.
23. Jardine, F. H., *Adv. Inorg. Chem. Radiochem.* (1975) **17**, 115.
24. Churchill, M. R., Kalra, K. L., *Inorg. Chem.* (1974) **13**, 1065.
25. Churchill, M. R., Kalra, K. L., *Inorg. Chem.* (1974) **13**, 1427.
26. Cotton, F. A., Wilkinson, G., "Advanced Inorganic Chemistry," 3rd ed., Chap. 25, John Wiley, New York, 1972.
27. Fackler, J. P., *Prog. Inorg. Chem.* (1976) **21**, 55.
28. Dewar, M. J. S., *Bull. Soc. Chim. Fr.* (1951) **18**, C79.
29. Chatt, J., Duncanson, L. A., *J. Chem. Soc.* (1953) 2939.
30. Salomon, R. G., Kochi, J. K., *J. Am. Chem. Soc.* (1973) **95**, 1889.
31. Salomon, R. G., Kochi, J., *J. Organomet. Chem.* (1974) **64**, 135.
32. Tsuda, T., Habu, H., Horiguchi, S., Saegusa, T., *J. Am. Chem. Soc.* (1974) **96**, 5930.
33. Bruce, M. I., Ostazewski, A. P. P., *J. Chem. Soc., Chem. Commun.* (1972) 1124.
34. Bruce, M. I., Ostazewski, A. P. P., *J. Chem. Soc., Dalton Trans.* (1973) 2433.
35. Churchill, M. R., DeBoer, B. G., Rotella, F. J., Salah, O. M. A., Bruce, M. I., *Inorg. Chem.* (1975) **14**, 2051.
36. Muetterties, E. L., Alegranti, C. W., *J. Am. Chem. Soc.* (1970) **92**, 4114.
37. Moore, C. E., "Atomic Energy Levels," NBS Circular 467, p. 117, 1952.
38. Day, P., Sanders, N., *J. Chem. Soc. A* (1967) 1536.
39. Sugasaka, K., Fujii, A., *Bull. Chem. Soc. Jpn.* (1976) **49**, 82.
40. McMillin, D. R., Buckner, M. T., Ahn, B. T., *Inorg. Chem.* (1977) **16**, 943.
41. Creutz, C., Sutin, N., *Inorg. Chem.* (1976) **15**, 496.
42. Hurst, J. K., Lane, R. H., *J. Am. Chem. Soc.* (1973) **95**, 1703.
43. Farr, J. K., Hulett, L. G., Lane, R. H., Hurst, J. K., *J. Am. Chem. Soc.* (1975) **97**, 2654.
44. Stevenson, K. L., Davis, D. D., "Abstracts of Papers," *174th National Meeting, ACS, Chicago, August, 1977*, INOR 121.
45. Ziolo, R. F., Lipton, S., Dori, Z., *J. Chem. Soc., Chem. Commun.* (1970) 1124.
46. Hardt, H. D., *Naturwiss.* (1974) **61**, 107.
47. Hardt, H. D., Pierre, A., *Inorg. Chim. Acta* (1977) **25**, L59.
48. Goher, M. A. S., Drátovský, M., *Naturwiss.* (1975) **62**, 96.
49. Buckner, M. T., McMillin, D. R., "Abstracts of Papers," *174th National Meeting, ACS, Chicago, August, 1977*, INOR 183.
50. Manahan, S. E., *Inorg. Chem.* (1966) **5**, 2063.
51. Schrauzer, G. N., *Adv. Catal.* (1968) **18**, 373.
52. Fraser, A. R., Bird, P. H., Bezman, S. A., Shapley, J. R., White, R., Osborn, J. A., *J. Am. Chem. Soc.* (1973) **95**, 597.
53. Hill, B., Math, K., Pillsbury, D., Voecks, G., Jennings, W., *Mol. Photochem.* (1973) **5**, 195.
54. Lippard, S. J., Melmed, K. M., *Inorg. Chem.* (1967) **6**, 2223.
55. Gill, J. T., Lippard, S. J., *Inorg. Chem.* (1975) **14**, 751.
56. Lamola, A. A., "Techniques of Organic Chemistry," Vol. 14, P. A. Leermakers, A. Weissberger, Eds., Chap. 2, John Wiley, New York, 1969.
57. Balzani, V., Moggi, L., Manfrin, M. F., Bolletta, F., Laurence, G. S., *Coord. Chem. Rev.* (1975) **15**, 321.

58. Zinato, E., "Concepts of Inorganic Photochemistry," A. W. Adamson, P. Fleischauer, Eds., Chap. 4, Interscience, New York, 1975.
59. Bommer, J. C., Morse, K. W., "Abstracts of Papers," *174th National Meeting, ACS, Chicago, August, 1977*, INOR 182.
60. Churchill, M. R., Bezman, S. A., Osborn, J. A., Wormald, J., *Inorg. Chem.* (1972) **11**, 1818.
61. Geoffroy, G. L., Pierantozzi, R., *J. Am. Chem. Soc.* (1976) **98**, 8054.
62. Geoffroy, G. L., Bradley, M. G., *Inorg. Chem.* (1977) **16**, 744.

RECEIVED February 22, 1978.

27

Catalysts for the Isomerization of Quadricyclane to Norbornadiene in a Photochemical Energy Storage System

E. M. SWEET, R. B. KING, R. M. HANES, and S. IKAI

Department of Chemistry, University of Georgia, Athens, GA 30602

The applications of transition metal complexes as catalysts for the conversion of quadricyclane to norbornadiene in the energy release step of a solar energy storage system are discussed. Co(II) tetraarylporphyrins anchored to macroreticular polystyrene through sulfonamide and carboxamide linkages can be prepared which are active catalysts for the conversion of quadricyclane to norbornadiene. Certain polystyrene-supported phosphine Pd(II) chloride derivatives are also active catalysts for this reaction. New homogeneous catalysts for the conversion of quadricyclane to norbornadiene include the triphenylcyclopropenylnickel complexes $[(C_6H_5)_3C_3Ni(CO)X]_2$ (X = Cl and Br) and the metal dithiolenes $[(CF_3)_2C_2S_2]_3MO$ and $[(CF_3)_2C_2S_2]_2Ni$.

The photosensitized conversion of norbornadiene (Structure 1) to quadricyclane (Structure 2) coupled with the transition metal complex-catalyzed reversion of quadricyclane to norbornadiene is a promising system for photochemical solar energy storage (1, 2, 3, 4). A device based on this reaction requires two steps: (a) energy storage through the sensitized photolysis of norbornadiene (1) to quadricyclane (2) in an endothermic reaction; (b) energy release through the catalyzed reconversion of quadricyclane (2) to norbornadiene (1) in an exothermic reaction. The preceding chapter (1) describes some research at the University of Georgia directed towards the development of inorganic sensitizers based on Cu(I) complexes for the energy storage step ((a) above).

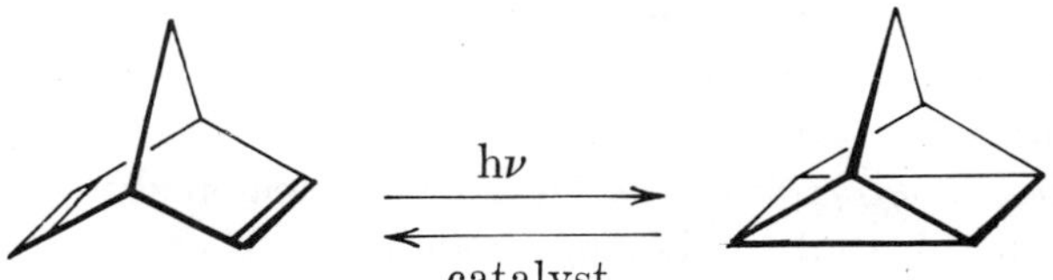

$$\text{1} \quad \xrightleftharpoons[\text{catalyst}]{h\nu} \quad \text{2}$$

This chapter summarizes some key aspects of our work at the University of Georgia directed towards the development of appropriate transition-metal complex catalysts for the energy release step ((b) above).

The catalyst for the energy release step (b) in a photochemical energy storage system based on this principle must be kept away from the site in the system where the energy storage step (a) is taking place. This can be most effectively accomplished by immobilizing the catalytically active structure by chemical bonding to an insoluble polymeric matrix. This chapter discusses some approaches directed towards anchoring catalytically active structures for the conversion of quadricyclane to norbornadiene onto macroreticular polystyrene beads through chemical bonding to the benzene rings of the polystyrene. In addition, some new homogeneous catalysts for the conversion of quadricyclane (2) to norbornadiene (1) are described.

Supported Co(II) Tetraphenylporphyrin Catalysts

The demonstrated high catalytic activity of Co(II) tetraphenylporphine for conversion of quadricyclane to norbornadiene (5, 6) suggested that this would be an attractive transition metal system to incorporate into our supported catalysts. Such a supported catalyst also appeared promising for the following reasons: (1) leaching of the cobalt from the tetradentate porphyrin ligand appeared improbable; (2) even if the cobalt were leached from the porphyrin, as nonporphyrin Co(II), it would be catalytically inactive; and (3) a catalyst based on cobalt would be significantly less expensive than one based on a platinum metal such as rhodium.

The tetraarylporphyrin ligand was bonded to macroreticular polystyrene through carboxamide and sulfonamide linkages. The aminopolystyrene used for the preparation of these materials was obtained from unfunctionalized polystyrene by nitration, followed by reduction of the nitro groups with stannous chloride. This aminopolystyrene was then coupled with the tetra(p-chlorosulfophenyl)porphine and the tetra(p-chlorocarbonyl)-phenylporphine obtained from the respective acids or acid salts to give the sulfonamide-linked and the carboxamide-linked polystyrene-anchored tetraarylporphyrins, respectively. These coupling reactions were effected in chlorinated hydrocarbon solvents using tertiary amines as proton scavengers. Cobalt was incorporated into these polystyrene-anchored tetraarylporphyrins by treatment under nitrogen with Co(II) acetate 4-hydrate in boiling acetic acid. The resulting materials can be schematically represented by Structures 3a and 3b.

Five batches of sulfonamide-linked (3a) and two batches of carboxamide-linked (3b) polystyrene-anchored Co(II) tetra-arylporphyrin catalysts were prepared with cobalt contents ranging from 0.28 to 0.48%. The standard activities with 0.1M quadricyclane in benzene were measured for each batch (Table I). The pseudo first-order rate constants for disappearance of quadricyclane were obtained with 0.1 g of catalyst in 10 mL of solvent. These rates were normalized to standard activity conditions of 1.0 g/L (k_w in Table I) and to an activity based on metal content (k_M in Table I).

No immediate correlation of activity with metal content is apparent for these polystyrene-anchored Co(II) tetraarylporphyrin catalysts. There is some evidence that nonporphyrin cobalt is present, perhaps bound to unreacted pendant amino groups, since the sulfur:cobalt ratios of two batches were 3.9:1 (batch 2A-1) and 2.7:1 (batch 2B-4) as compared with a theoretical minimum of 4:1 when cobalt is bound only to tetra(p-sulfophenyl)porphine units. However, this observation should be interpreted with caution since elemental analyses are notoriously unreliable for polymer systems (7).

The effect of repeated use on the activity of the catalyst was examined. Thus a single 0.1 g charge of catalyst batch 1A-1 was treated repeatedly with 1.0M quadricyclane in xylene. The activity (k_w) showed a steady decrease correlating well with the total volume (V_T) of the

Table I. Standard Activities of the Polystyrene-Anchored Co(II) Tetraarylporphyrin Catalysts for the Conversion of Quadricyclane to Norbornadiene

Batch	Linkage[a]	% Co	mol Co $\times 10^5$/g	$k_w \times 10^4$ in sec^{-1} (g/L)$^{-1}$	k_M in sec^{-1} (g-atom Co/L)$^{-1}$
1A-1	C	0.44	7.5	2.42 (0.23)[b]	3.2 (0.31)[b]
2A-1	S	0.45	7.6	1.95 (0.19)	2.6 (0.25)
2A-2	S	0.34	5.8	2.15 (0.11)	3.7 (0.20)
2B-1	S	0.31	5.3	2.07 (0.13)	3.9 (0.25)
2B-2	S	0.28	4.8	0.88 (0.052)	1.8 (0.11)
2B-3	C	0.45	7.6	1.37 (0.11)	1.8 (0.15)
2B-4	S	0.48	8.2	1.73 (0.13)	2.1 (0.16)

[a] C = carboxamide linked, S = sulfonamide linked.
[b] The figures in parentheses in these columns refer to standard deviations obtained from four or more runs.

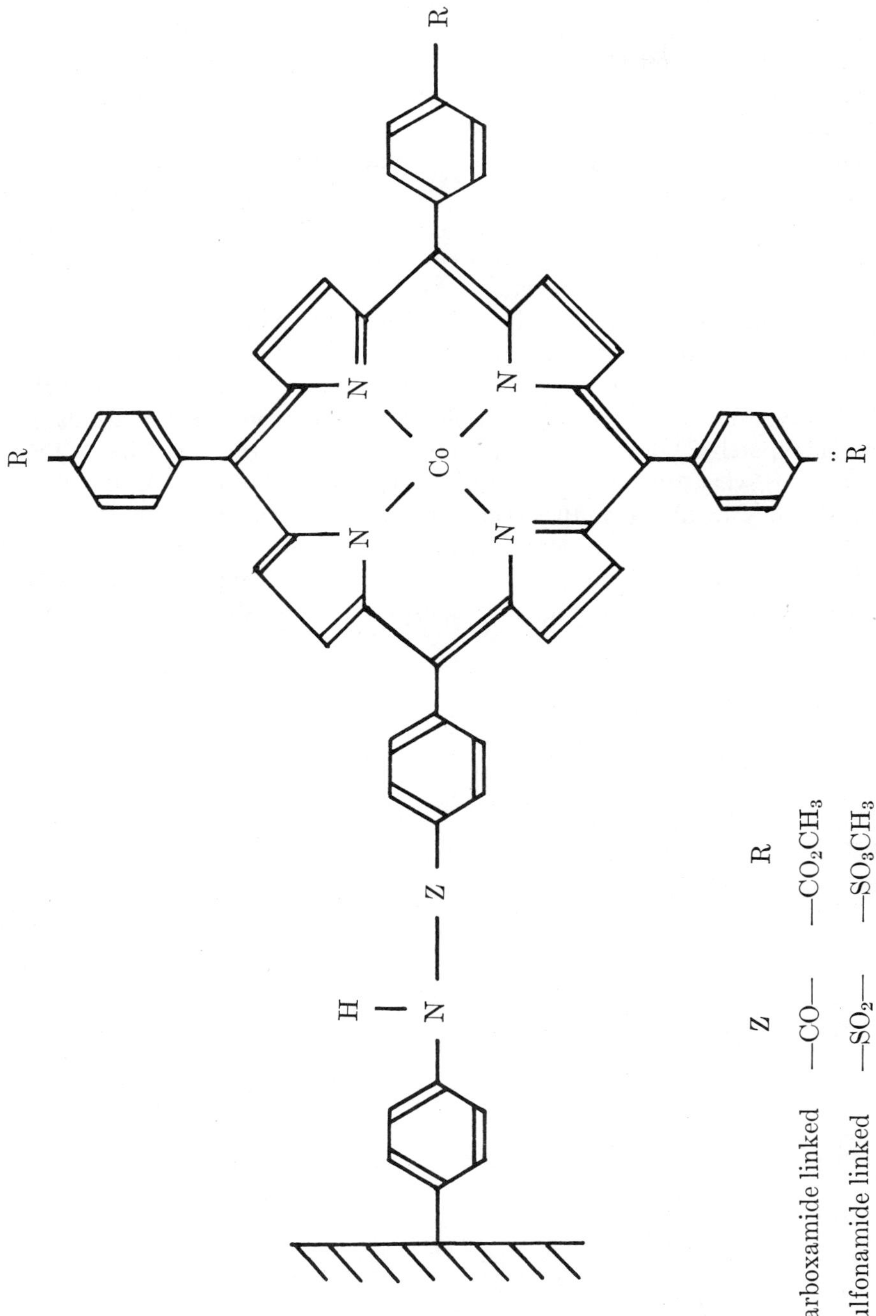

	Z	R
	—CO—	—CO$_2$CH$_3$
	—SO$_2$—	—SO$_3$CH$_3$

3a Carboxamide linked
3b Sulfonamide linked

solution to which the catalyst had been exposed (Figure 1). The extrapolated intercept for zero catalytic activity was 127 mL, corresponding to an average turnover per metal in excess of 15,000 moles of quadricyclane per g-atom of cobalt. In a similar study on batch 2A-1 using $0.1M$ quadricyclane in benzene, the decrease in activity with volume was not as regular (Table II). Some of the lost catalytic activity could be restored by treatment of the catalyst with strong reducing agents, either Ti(III) in organic solvents or potassium benzophenone ketyl in THF. This restoration of activity by treatment with strong reducing agents suggests that the loss of catalytic activity upon repeated use arises from oxidation of Co(II) to Co(III).

Diffusion was shown to reduce the activity of the polystyrene-anchored Co(II) tetraarylporphyrin catalysts. Thus when beads of catalyst batch 2A-1 were ground to a fine powder, the activities in four runs (k_w) were greater by factors of 2–5 than in any runs using the whole bead catalyst. In the studies with the pulverized catalysts, the standard activities were found to increase with an increase in the ratio of weight of catalyst to the volume of the solution, again suggesting that some component of the solution is responsible for the deactivation of the catalyst.

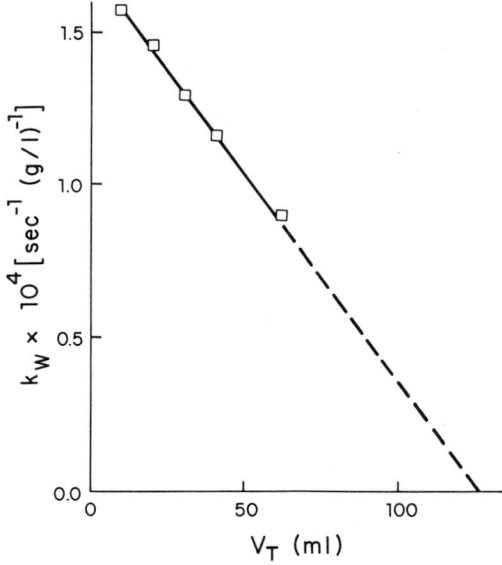

Figure 1. Recycling study of batch 1A-1 of the carboxamide-linked polystyrene-anchored Co(II) tetraarylporphyrin catalyst for the conversion of quadricyclane to norbornadiene; plot of activity (k_w) vs. total volume of solution to which the catalyst was exposed (V_T)

Table II. Studies on the Regeneration of Recycled Catalyst Using Reducing Agents[a]

Run/Operation	$k_w \times 10^4$ in sec^{-1} (g/L)$^{-1}$
1	2.59 (0.042)[b]
2	1.84 (0.047)
3	1.44 (0.019)
4	1.29 (0.013)
5	1.39 (0.012)
6	1.31 (0.011)
Reduction with Ti(III) in methanol (5 mL for 30 min)	
7	1.67 (0.022)
8	1.56 (0.013)
Reduction with Ti(III) in methanol (5 mL for 30 min)	
9	1.71 (0.012)
10	1.55 (0.012)
11	1.46 (0.0090)
12	1.39 (0.0079)
13	1.37 (0.0082)
14	1.34 (0.0099)
15[c]	0.69 (0.0037)
16[c]	0.50 (0.0081)
17	0.53 (0.0075)
Reduction with Ti(III) in methanol (10 mL for 45 min)	
18	0.73 (0.0090)
Reduction with Ti(III) in THF (5 mL for 2 hr)	
19	0.83 (0.0086)
Reduction with Ti(III) in THF	
20	0.98 (0.019)
21	0.87 (0.0097)
Reduction with potassium benzophenone ketyl in THF	
22	1.07 (0.020)
23	1.04 (0.012)

[a] These runs were performed in the indicated sequence using 0.107 g of sulfonamide-linked polystyrene-supported Co(II) porphyrin from batch 2A-1 (Table I). Except where indicated, 10 mL of 0.1M quadricyclane in benzene was used.

[b] The figures in parentheses in this column represent the error in the first-order kinetics plots, volumes of reaction solutions, and initial weight of the catalyst. Variations in temperature ($\pm 0.2°C$) and physical loss of catalyst on workup are not included.

[c] These runs were made using neat quadricyclane rather than 10 mL of 0.1M quadricyclane in benzene.

Supported Palladium Chloride Catalysts

Hogeveen and Volger (8) observed that 1,5-cyclooctadienedichloropalladium and η^3-allylchloropalladium dimer were active catalysts for the conversion of quadricyclane (2) to norbornadiene (1). Furthermore, Dauben and Kielbania (9) have found that [(C$_6$H$_5$)$_3$P]$_2$PdCl$_2$ effects the conversion of tricyclo-[4,1,0,0^{2,7}]-heptane into 3-methylenecyclohexene.

These observations suggested that Pd(II) chloride bound to phosphine-substituted polystyrene might be an effective catalyst for the conversion of quadricyclane to norbornadiene. We therefore investigated the preparation of such a polymer-anchored phosphine Pd(II) chloride catalyst.

Macroreticular polystyrene beads were diphenylphosphinated by lithiation with *n*-butyllithium (*10*) in the presence of tetramethylethylenediamine, followed by addition of diphenylchlorophosphine. Pd(II) chloride was then complexed with this polymer by stirring in acetonitrile solution. The resulting lemon yellow beads were found to be active catalysts for the conversion of quadricyclane (2) norbornadiene (1). Two batches of beads prepared by this general method gave the following elemental analyses: (A) 0.70% Pd, 0.28% P, 0.39% Cl; and (B) 1.36% Pd, 0.97% P, 1.08% Cl, and 0.026% N, corresponding to the elemental ratios Pd:P:Cl:N of 1:1.4:1.7:? and 1:2.4:2.4:0.14, respectively. While $(R_3P)_2PdCl_2$ coordination is not required for bonding of the palladium to the polymer, these elemental ratios, particularly the low N:Pd ratio, indicate that a substantial portion of the palladium may be in such a coordination environment.

Activities of the polymer-bound palladium catalysts were measured for both batches A and B (Table III). Except for the first run with a given charge of catalyst, the disappearance of quadricyclane was found to obey first-order kinetics. However, the first run of a given charge was found to give a first-order rate decreasing in time. Thus with batch A, the activity of the catalyst appears to decrease between the first and second use; however, independent analysis of the final portion of the first run gives an activity quite similar to that of the second run. Unused catalyst that had been pretreated with norbornadiene was found to give an activity for its initial run similar to that of the second run with untreated catalyst. These observations suggest the formation of a polymer-bound palladium–norbornadiene complex.

The catalytic activities of both batches A and B of the polystyrene-supported phosphine Pd(II) chloride catalyst were found to be substantially less than that of the homogeneous analogue $[(C_6H_5)_3P]_2PdCl_2$ and to decrease with extended use. The precise cause of this loss of catalytic activity is unclear. Possible causes include oxidation of the pendant phosphine groups, leading to metal leaching, or polymerization of norbornadiene within the polystyrene matrix initiated by intermediates in such phosphine oxidation.

The catalytic activities of the polystyrene-supported phosphine Pd(II) catalysts (Table III) for the conversion of quadricyclane to norbornadiene were considerably less than those of the polystyrene-supported Co(II) tetraphenylporphyrin catalysts (Table I) under comparable conditions.

Table III. Standard Activities of the Polystyrene-Anchored Pd(II) Chloride Catalysts for the Conversion of Quadricyclane to Norbornadiene

System	Solvent	Quadricyclane Conc. (M)	$k_w \times 10^4$ in $sec^{-1} (g/L)^{-1}$	k_M in sec^{-1} (g-atom Pd/L)$^{-1}$
Polymer Batch A, Charge 1				
Run 1 Average	xylene	1.0	0.20	0.31
Initial			0.31	0.47
Final			0.13	0.20
Run 2	xylene	1.0	0.15	0.23
Run 3	xylene	1.0	0.16	0.24
Run 4	xylene	1.0	0.041	0.062
Run 5	xylene	1.0	0.018	0.028
Polymer Batch A, Charge 2 (pretreated with excess norbornadiene)				
Run 1	xylene	1.0	0.14	0.21
Polymer Batch A, Charge 3				
Run 1	benzene	1.0	0.29	0.44
Polymer Batch B, Charge 1				
Run 1	benzene	1.0	0.94	0.63
Run 2	benzene	1.0	0.36	0.24
Run 3	benzene	1.0	0.19	0.13
Run 4	benzene	1.0	0.10	0.07
[(C$_6$H$_5$)$_3$P]$_2$PdCl$_2$				
Run 1	benzene	0.5		80

Homogeneous Catalysts

Another aspect of our work has involved identification and development of new homogeneous catalysts for the conversion of quadricyclane to norbornadiene. New classes of compounds found to have promising catalytic activity include certain triphenylcyclopropenylnickel halide and metal dithiolene complexes.

Triphenylcyclopropenylnickel Halide Catalysts. Addition of [(C$_6$H$_5$)$_3$C$_3$Ni(CO)X]$_2$ (4: X = Cl or Br) (11) to pure quadricyclane rapidly produces norbornadiene with noticeable evolution of heat. However, addition to the chloride [(C$_6$H$_5$)$_3$C$_3$Ni(CO)Cl]$_2$ (4: X = Cl) of ligands containing functionalities useful for binding the complexes to a polymer support (e.g., tri-valent phosphorus derivatives) was found to produce new complexes, which in similar qualitative studies (Table IV) show generally decreased activity. In practically all cases, the phosphine and phosphite complexes proved difficult to purify, and in some instances,

Table IV. Some Triphenylcyclopropenylnickel Halide Derivatives and Their Catalytic Activities for the Conversion of Quadricyclane to Norbornadiene

Compound[a]	Color	Catalytic Activity[b]
$[Ph_3C_3Ni(CO)Cl]_2$	orange	+++
$[Ph_3C_3Ni(CO)Br]_2$	orange	+++
$Ph_3C_3Ni(bipy)Cl$[c]	dark red	—
$Ph_3C_3Ni(PPh_2Me)Cl$	brown	+
$Ph_3C_3Ni(PPh_2Me)_2Cl$	dark orange	—
$Ph_3C_3Ni(PPh_3)Cl$[c]	dark brown	++
$Ph_3C_3Ni(PPh_3)_2Cl$	dark red	+
$Ph_3C_3Ni[P(OPh)_3]Cl$[c]	brown	+++
$Ph_3C_3Ni[P(OPh)_3]_2Cl$	orange	+
$Ph_3C_3Ni[P(OMe)_3]_2Cl$	orange	—
$Ph_3C_3Ni[P(OPr^i)_3]_2Cl$	orange	+
$Ph_3C_3Ni(PPh_2Cl)_2Cl$	yellow	+
$Ph_3C_3Ni(PPhCl_2)_2Cl$[c]	yellow	+++

[a] bipy = 2,2'-bipyridyl; Me = methyl; Pr^i = isopropyl; Ph = phenyl.
[b] The catalytic activities at 23°C are indicated as follows: (—) no detectable activity; (+) minimum detectable activity (5–10% conversion even upon prolonged standing at room temperature); (++) moderate activity (some conversion in short periods at room temperature; major amounts of conversion at higher temperatures or upon prolonged standing at room temperature); and (+++) high activity resulting in an immediate exothermic nearly complete conversion of quadricyclane to norbornadiene.
[c] The approximate compositions of these previously unreported compounds were verified by carbon and hydrogen analyses.

the "compounds" in Table IV are inferred from the stoichiometry of the reagents (i.e., $[(C_6H_5)_3C_3Ni(CO)Cl]_2$ and the tri-valent phosphorus ligand) used in their preparation. The previously reported triphenylcyclopropenylnickel complexes $(C_6H_5)_3C_3NiC_5H_5$ (12) and $(C_6H_5)_3C_3Ni(NC_5H_5)_2Cl \cdot NC_5H_5$ (13) were found to be catalytically inactive for the conversion of quadricyclane to norbornadiene.

4

The notable exceptions to the decrease in catalytic activity of [(C$_6$H$_5$)$_3$C$_3$Ni(CO)Cl]$_2$ upon addition of the tri-valent phosphorus ligand arise from the complexes of stoichiometries (C$_6$H$_5$)$_3$C$_3$NiP(OC$_6$H$_5$)$_3$Cl and (C$_6$H$_5$)$_3$C$_3$Ni(PCl$_2$C$_6$H$_5$)$_2$Cl (Table IV). Both of the tri-valent phosphorus ligands in these complexes are relatively strong π acceptors and weak σ donors. The relatively low catalytic activity in the complex (C$_6$H$_5$)$_3$C$_3$Ni[P(OC$_6$H$_5$)$_3$]$_2$Cl suggests that ligand dissociation occurs from (C$_6$H$_5$)$_3$C$_3$Ni(PCl$_2$C$_6$H$_5$)$_2$Cl under the reaction conditions to give an active 16-electron species such as (C$_6$H$_5$)$_3$C$_3$Ni(PCl$_2$C$_6$H$_5$)Cl.

A more detailed investigation of the kinetics was made with [(C$_6$H$_5$)$_3$C$_3$Ni(CO)Cl]$_2$ in methylene chloride. Except for very low ratios of quadricyclane to catalyst, where first-order behavior was observed, the time-dependent reaction is of mixed order (Figure 2). At long reaction times the reaction approaches first-order behavior with a rate apparently independent of catalyst concentration. This latter observation could be a fortuitous combination of the effects of metal concentration and of inhibition by the norbornadiene product, with greater concentrations of norbornadiene being present for higher metal concentrations during the period when this behavior is observed. That the [(C$_6$H$_5$)$_3$C$_3$Ni(CO)Cl]$_2$ catalyst is inhibited by norbornadiene is indicated by separate experiments in the presence of added norbornadiene (Figure 3).

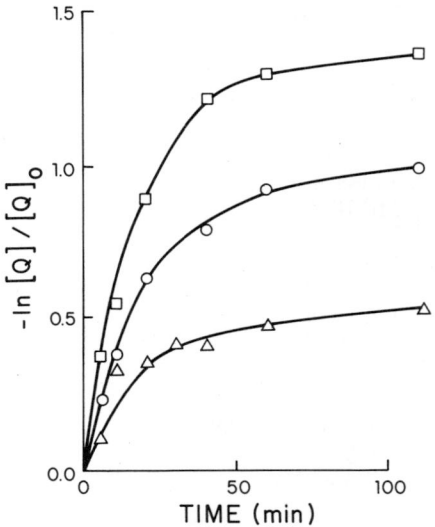

Figure 2. Plot of $-\ln([\text{quadricylane}]/[\text{quadricyclane}]_0)$ vs. time for various [(C$_6$H$_5$)$_3$C$_3$Ni(CO)Cl]$_2$ concentrations. ($\square$) [Ni] = 4.38 mM; ($\bigcirc$) [Ni] = 2.19 mM; ($\triangle$) [Ni] = 1.42 mM.

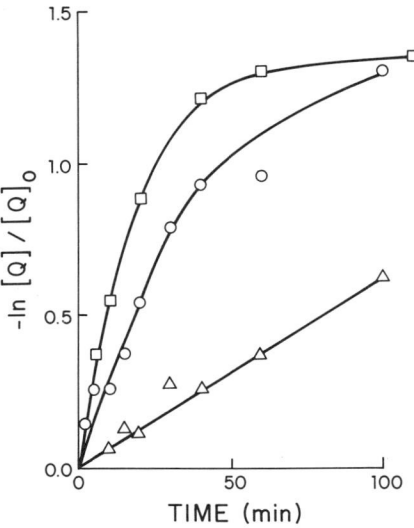

Figure 3. Plot of $-\ln([quadricyclane]/[quadricyclane]_0)$ vs. time for $[(C_6H_5)_3C_3Ni(CO)Cl]_2$ in the presence of added norbornadiene. (□) $[NBD] = 0.0$; (○) $[NBD] = 0.27$M; (△) $[NBD] = 1.1$M.

Table V. Evaluation of Metal Dithiolenes as Catalysts

Catalyst $(mol/L \times 10^3)$	Initial Quadricyclane Conc. (mol/L)
Molybdenum Complexes	
$[(CF_3)_2C_2S_2]_3Mo$ (0.95)	0.42
$[(CF_3)_2C_2S_2]_3Mo$ (0.95)	0.47
$[(CF_3)_2C_2S_2]_3Mo$ (0.95)	0.79
$[(CF_3)_2C_2S_2]_3Mo$ (0.95)	0.77
$[(CF_3)_2C_2S_2]_3Mo$ (0.95)	0.12
$[(CF_3)_2C_2S_2]_3Mo$ (0.95)	0.12
$[(CF_3)_2C_2S_2]_2Mo(Pf-Pf)$ [a] (0.95)	0.64
$[Ph_4As][(CF_3)_2C_2S_2]_3Mo$ (0.95)	0.46
$[Ph_4As]_2[(CF_3)_2C_2S_2]_3Mo$ (0.95)	0.49
$(Ph_2C_2S_2)_3Mo$ (0.95)	0.58
$(Ph_2C_2S_2)_2Mo(CO)_2$ (0.95)	0.78
Nickel Complexes	
$[(CF_3)_2C_2S_2]_2Ni$ (3.78)	2.35
$[Et_4N][(CF_3)_2C_2S_2]_2Ni$ (3.78)	1.10
$[Et_4N]_2[(CF_3)_2C_2S_2]_2Ni$ (3.78)	1.19
$(Ph_2C_2S_2)_2Ni$ (3.78)	1.08

[a] $Pf-Pf = Ph_2PCH_2CH_2PPh_2$.

No detailed mechanistic interpretation of the catalysis of the conversion of quadricyclane to norbornadiene by these triphenylcyclopropenylnickel derivatives is warranted at the present time. However, these reactions appear to be quite similar to isomerizations of quadricyclane to norbornadiene catalyzed by other nickel complexes (14).

Metal Dithiolene Catalysts. The metal dithiolenes (15) $[(CF_3)_2C_2S_2]_3Mo$ and $[(CF_3)_2C_2S_2]_2Ni$, obtained by treatment of bis(trifluoromethyl)dithietene with the metal carbonyls, were found to catalyze the isomerization of quadricyclane to norbornadiene. The molybdenum complex is significantly more catalytically active than the nickel complex (Table V). The catalytic activities of these neutral metal dithiolenes $[(CF_3)_2C_2S_2]_nM$ (M = Mo, n = 3; M = Ni, n = 2) are destroyed upon reduction to the corresponding mono- or dianions by substitution of the CF_3 groups with less electronegative groups such as phenyl and by substitution of even one of the dithiolate ligands by a tri-valent phosphorus ligand. Thus the presence of the highest possible metal oxidation state, the presence of strongly electron-withdrawing substituents such as CF_3, and the exclusive presence of dithiolate ligands all appear to be necessary for metal dithiolenes to have appreciable catalytic activity for the conversion of quadricyclane (2) to norbornadiene (1).

for the Conversion of Quadricyclane to Norbornadiene

Solvent	Other Additive (mol/L)	Norbornadiene Production after 20 hr (as % of total C_7H_8 hydrocarbons)	
benzene	none	98	+ polymer
benzene	H_2O (0.37)	13	+ nortricyclanol
MeOH	none	4.3	
MeOH	H_2O (0.37)	0	
CH_2Cl_2	none	100	+ polymer
CH_2Cl_2	pyridine (0.083)	0.2	
benzene	none	1.5	
benzene	none	4.3	
benzene	none	0.8	
benzene	none	0.1	
benzene	none	0	
CH_2Cl_2	none	21	
CH_2Cl_2	none	2.2	
CH_2Cl_2	none	0	
CH_2Cl_2	none	1.7	

The ability of $[(CF_3)_2C_2S_2]_3Mo$ to isomerize quadricyclane to norbornadiene and the ultimate course of the reaction appear to be rather solvent sensitive (Table V). Thus the catalytic activity of $[(CF_3)_2C_2S_2]_3Mo$ was found to decrease when coordinating solvents were used or when water was added. Unfortunately, in no case was a clean conversion of quadricyclane to norbornadiene observed.

The treatment of quadricyclane with catalytic amounts of $[(CF_3)_2C_2S_2]_3Mo$ in benzene or dichloromethane solution at room temperature was found to produce not only the expected norbornadiene but also a polymer. Control experiments indicate that $[(CF_3)_2C_2S_2]_3Mo$ polymerizes norbornadiene in benzene at the boiling point but not at room temperature. Addition of water to a benzene solution of quadricyclane containing $[(CF_3)_2C_2S_2]_3CMo$ gives small quantities of nortricyclanol in addition to slowing the rate of conversion of quadricyclane to norbornadiene.

Conclusions

The results summarized in this article indicate that a variety of transition metal derivatives can catalyze the conversion of quadricyclane (2) to norbornadiene (1). These include not only the derivatives of cobalt, palladium, nickel, and molybdenum discussed in this chapter but also derivatives of other metals (particularly rhodium (8)) that have been studied in detail by other workers. Of the systems that have been studied up to the present time, the Co(II) tetraarylporphyrins appear to be the most promising ones for use in a practical solar energy storage system for the following reasons: (1) the low cost of cobalt relative to the platinum metals, particularly rhodium; (2) the relative ease of chemically bonding the porphyrin nucleus to an insoluble polymer by conventional chemical reactions; (3) the general stability of porphyrin systems toward prolonged exposure to heat and light (e.g., the stability and ubiquity of chlorophyll in nature); and (4) the requirement of complexation with a porphyrin or similar planar polydentate ligand (5) for cobalt to be an active catalyst for the conversion of quadricyclane to norbornadiene, thereby preventing malfunction of the solar energy storage system by uncomplexed cobalt leached upon prolonged use of the system. For these reasons, further work in this laboratory directed towards the development of catalysts for a practical solar energy storage system based on the interconversion of norbornadiene to quadricyclane will involve mainly the use of Co(II) complexes of porphyrins and related planar macrocyclic ligands such as phthalocyanines.

Acknowledgment

The authors are indebted to the Division of Energy Storage Systems of the U.S. Department of Energy for partial support of this work under contract EY-76-S-09-0893. We also acknowledge frequent helpful discussions with C. R. Kutal and R. R. Hautala of the University of Georgia.

Literature Cited

1. Kutal, C., Grutsch, P., ADV. CHEM. SER. (1978) **173**, 325.
2. Hautala, R. R., Little, J., Sweet, E. M., *Sol. Energy* (1977) **19**, 503.
3. Kutal, C., Schwendiman, D. P., Grutsch, P., *Sol. Energy* (1977) **19**, 651.
4. Schwendiman, D. P., Kutal, C., *J. Am. Chem. Soc.* (1977) **99**, 5677.
5. Manassen, J., *J. Catal.* (1970) **18**, 38.
6. Wilson, H. D., Rinker, R. G., *J. Catal.* (1976) **42**, 268.
7. Card, R. J., Neckers, D. C., *J. Am. Chem. Soc.* (1977) **99**, 7733.
8. Hogeveen, H., Volger, H. C., *J. Am. Chem. Soc.* (1967) **89**, 2486.
9. Dauben, W. G., Kielbania, A., Jr., *J. Am. Chem. Soc.* (1972) **94**, 3669.
10. Evans, D. C., George, M. H., Barrie, J. A., *J. Polym. Chem.* (1974) **12**, 247.
11. Gowling, E. W., Kettle, S. F. A., *Inorg. Chem.* (1965) **4**, 604.
12. Rausch, M. D., Tuggle, R. M., Weaver, D. L., *J. Am. Chem. Soc.* (1970) **92**, 4981.
13. Weaver, D. L., Tuggle, R. M., *J. Am. Chem. Soc.* (1969) **91**, 6505.
14. Noyori, R., Umeda, I., Kawauchi, H., Takaya, H., *J. Am. Chem. Soc.* (1975) **97**, 812.
15. McCleverty, J. A., *Progr. Inorg. Chem.* (1968) **10**, 49.

RECEIVED February 22, 1978.

28

An Entatic State for Copper in Redox Enzymes?

ROBERT H. LANE, NANTELLE S. PANTALEO, JAMES K. FARR, WILLIAM M. CONEY, and M. GARY NEWTON

Department of Chemistry, University of Georgia, Athens, GA 30602

The molecular structure of $[Co(en)_2(SCH_2CH_2NH_2) \cdot Cu\text{-}(CH_3CN)_2]_2(ClO_4)_6 \cdot 2H_2O$ (en = ethylenediamine) contains a unique mercaptide-bridged $Cu(I)_2(SR)_2$ planar unit which is structurally related to proposed models for (reduced) Type 3 (EPR-nondetectable) copper in multicopper oxidases. The Cu–Cu distance of 3.146(2) Å is well within the 5–6 Å estimated as a maximum separation for Type 3 coppers. The S–S distance of 3.557(6) Å rules out a disulfide bond. The S–Cu–S bond angle (97°) indicates that the planar ring imparts, or at least allows, a geometry about the copper atoms intermediate between that preferred by Cu(I) (tetrahedral) and Cu(II) (tetragonal), a fact which may be of importance in the catalysis of biological redox processes by Type 3 copper.

A number of copper metalloproteins which function in either the transport or reduction of molecular oxygen possess the Type 3 copper chromophore which uses two copper atoms as part of a unit that accepts/donates two electrons and which apparently interacts directly with the oxygen molecule (1, 2). Significant spectral features of this chromophore are an unusually intense electronic absorption maximum at 330 nm ($\epsilon \geq 2.7 \times 10^3$) in the oxidized form and the absence of an EPR signal in either the oxidized or reduced form of the chromophore. Redox potentials for Type 3 copper are considerably more positive (0.4–0.8 V vs. SHE) than the standard potential for the Cu(II)/Cu(I) aqua couple (0.167 V) or for most synthetic copper complexes (with the exception of those with highly distorted geometries), indicating that the structural environment of copper in proteins effects a greater stabilization of Cu(I) relative to Cu(II) than is usual for copper complexes (3). Spectral and redox behavior thus strongly suggest that Type 3 copper is structurally atypical when compared with small-molecule copper complexes.

There are three reasonable combinations of metal oxidation states for oxidized Type 3 copper that are consistent with spectral and redox data: (1) Cu(I)–Cu(I) with some other group, e.g., disulfide, functioning as a two-electron acceptor; (2) Cu(I)–Cu(III) where Cu(III) is low spin; and (3) an antiferromagnetically coupled Cu(II)–Cu(II) dimer. Magnetic susceptibility studies on *Rhus vernicifera* laccase have established that the two Type 3 copper atoms in this enzyme are present as an antiferromagnetically coupled Cu(II) dimer (4). The Type 3 copper atoms of hemocyanin and tyrosinase appear to be similarly coupled and separated by 3–5 Å (5, 6, 7). Further structural information on the Type 3 copper chromophore is scanty; neither the identity of the ligands nor the geometry of the site has been ascertained. There is likewise a paucity of literature on binuclear copper complexes that exhibit structural features expected for Type 3 copper.

Although ligands coordinated to the Type 3 copper atoms have not been identified, they are expected to have an affinity for both Cu(II) and Cu(I). On this basis the most attractive potential ligands of biological importance are sulfur from cystine (disulfide), cysteine (sulfhydryl or mercaptide) or methionine (thioether), and the imidazole nitrogen of histidine (8). Several combinations of ligating atoms are, of course, possible. However, models incorporating two mercaptides or a disulfide group as ligands bridging the two copper atoms have received the most attention (1, 2, 9). Sulfur as a ligand bridge is quite appealing since: (1) the bridge would consist of a single atom, allowing the metals to reside in close enough proximity for strong magnetic interaction; (2) sulfur stabilizes Cu(I) relative to Cu(II) which would shift the redox potential of the copper couple in the desired direction; and (3) sulfur is known to mediate facile electron transfer in transition metal systems (10, 11), a fact which may be of importance in multicopper redox enzymes where electron transfer among copper sites has been indicated (12, 13).

A potentially severe structural constraint for the Type 3 copper site, in fact for all redox-active biological copper sites where the Cu(II)/Cu(I) couple is involved, arises from the different structural preferences of Cu(II) (tetragonal) and Cu(I) (tetrahedral or larger bond angles). The requirement of Cu(I) for a nontetragonal environment is apparently quite strict, even with such Cu(I)-stabilizing ligands as sulfur (14). The inability of ligands to distort from tetragonal symmetry is reflected in very negative potentials for reduction of Cu(II) complexes (15). If in the ground states of the oxidized and reduced forms of Type 3 copper the copper atoms reside in their preferred structural environments, then the enthalpic barrier to activation for electron transfer is expected to be large. This barrier would arise from the necessity for attaining in the transition state some configuration intermediate between those of the Cu(II) and Cu(I) ground states. Such a barrier is difficult to reconcile with the high redox rates for copper enzymes (12, 13). If on the other hand the ground state structures of the oxidized and reduced forms of Type 3 copper are constrained to resemble the electron transfer transition state, i.e., they are in a poised or entatic state (16), then little structural reorganization at the copper site is required. The enthalpy of activation would thus be lowered, not by lowering the energy of the transition state, but rather by raising the energies of the respective ground states. Such a poised configuration could serve the additional function of "fine tuning" the redox potential of the metal site to meet the requirements of the particular redox protein. Thus a structure that more nearly resembles that preferred by Cu(I) would have the effect of raising the ground state energy of Cu(II) to a greater extent than that of Cu(I), yielding a more positive potential and vice versa.

Because of the paucity of available data, presently it is not possible to describe the arrangement of metal and ligand atoms at the Type 3 copper site. However, considerable insight into plausible arrangements can be developed from an examination of structural parameters of small-molecule copper complexes. Structural information is available for a number of bridged binuclear complexes containing a variety of ligand types that use oxygen or chlorine as bridging atoms (cf. Table I).

Although considerable effort has gone into the preparation and structural determination of sulfur-containing copper complexes, definitive characterization of binuclear complexes containing a mercaptide bridge has proved exceedingly difficult. Two major problems which manifest this difficulty are the tendency for Cu(II) to readily oxidize mercaptans

Table I. Comparison of Distances and Angles

Complex	Bridging Atoms (X)
$[Cu_2Cl_6]^{2-}$	Cl
$[Cu(OH)EAEP]_2$—[b] $(ClO_4)_2$	O
$[Cu(C_8H_{10}N_2O)]_2$[a]	O
Cu_8S_{12} (Cp(I)-cubane)	S (dithiosquarate, singly bridged)
$Cu(I)_2S_6$	S (thiourea)
$Cu(I)_2(SR)_2$	S (mercaptide)

[a] Copper atoms are doubly bridged except where noted.
[b] EAEP = 2-(2-ethylaminoethyl)pyridine.

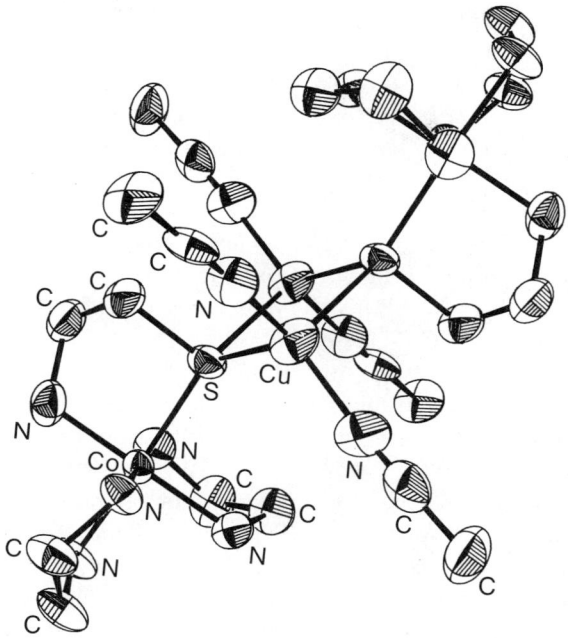

Journal of the American Chemical Society

Figure 1. ORTEP drawing of the $[Co(en)_2(SCH_2\text{-}CH_2NH_2) \cdot Cu(CH_3CN_2]_2^{6+}$ cation. The 50% thermal probability ellipsoids are shown for all nonhydrogen atoms. Unlabeled atoms are related to the labeled ones by the crystallographic center of inversion (25).

for Some Bridged Copper Complexes[a]

X–Cu–X (deg)	Cu–X–Cu (deg)	X–X (Å)	Cu–Cu (Å)	Ref.
86.7	93.3	—	3.355	20
80.2	98.8	—	2.917	21
81.5	99.5			
76.1	103.9	—	3.001	22
116.2	78.6	3.814	2.844	23
107.4	72.6	3.865	2.840	24
97.0	83.0	3.557	3.146	25

[c] $C_8H_{10}N_2O^{2-}$ = Schiff base derived from pyrrole-2-carboxaldehyde and 3-aminopropanol.

to their corresponding disulfides (17, 18) and for Cu(I) to form polymeric species with mercaptans (19). We have been able to circumvent the second problem by complexing Cu(I) to a mercaptide that has undergone prior coordination to substitution-inert Co(III). The resulting [Co(en)$_2$-(SCH$_2$CH$_2$NH$_2$) · Cu(CH$_3$CN)$_2$]$_2^{6+}$ complex cation contains a Cu$_2$(SR)$_2$ planar unit (Figure 1) which is structurally unique among bridged binuclear copper complexes and which has features expected for Type 3 copper in the reduced state. A preliminary report of the structure has appeared (25); the final refinement will be reported elsewhere.

Features of the Cu$_2$(SR)$_2$ unit that suggest its similarity to Type 3 copper include: (1) a Cu–Cu separation (Table II) that is well within the range expected for strong magnetic coupling; (2) biologically relevant sulfur ligands that stabilize Cu(I) relative to Cu(II); and (3) S–Cu–S bond angles that are intermediate between those preferred by Cu(I) and Cu(II).

The third point is especially significant with regard to the possibility of an entatic state for copper at the Type 3 site. Thus in this example the S–Cu–S bond angle is distorted from those preferred by either Cu(I) or Cu(II) toward an intermediate value that might be expected for an electron transfer transition state. The significance of this configuration for a doubly mercaptide-bridged complex is underscored by the fact that in doubly bridged complexes where atoms other than sulfur serve as bridging ligands, the X–Cu–X bond angle is almost invariably less than 90° whereas in singly bridged sulfur complexes and in doubly bridged thiourea complexes, S–Cu–S angles more nearly approach those preferred by Cu(I) (Table I). Furthermore, in the present example an apparent entatic state for copper is attained without the necessity for potentially modifying constraints imparted by a protein molecule. To the extent that a similar Cu–S arrangement results for Type 3 copper, a major role of the protein may well be to constrain non-ring ligands in a configuration

Table II. Bond Distances and Angles for the Cu$_2$(SR)$_2$ Core of [Co(en)$_2$(SCH$_2$CH$_2$NH$_2$) · Cu(CH$_3$CN)$_2$]$_2^{6+}$

Type	Distance (Å)	Type	Angle (deg)
Cu–S	2.406(4)[a] 2.342(5)	S–Cu–S	97.0(1)
Cu–N	1.951(1) 1.951(2)	N–Cu–N	118.3(5)
Cu–Cu	3.146(2)	N–Cu–S	117.1(5) 105.4(5)
S–S	3.557(6)	Cu–S–Cu	83.0(1)

[a] Estimated standard deviation.

that "fine tunes" the redox potential of the site to the appropriate value. Verification or rejection of these hypotheses, however, awaits more definitive structural data on the Type 3 copper chromophore.

Acknowledgment

Acknowledgment is made to the donors of the Petroleum Research Fund, administered by the American Chemical Society, and to the National Institute of General Medical Sciences (Grant No. 22592) for partial support of this work.

Literature Cited

1. Malkin, R., Malmström, B. G., *Adv. Enzymol.* (1970) **33**, 177.
2. Fee, J. A., *Struct. Bonding (Berlin)* (1975) **23**, 1.
3. Moore, G. R., Williams, R. J. P., *Coord. Chem. Rev.* (1976) **18**, 125.
4. Solomon, E. I., Dooley, D. M., Wang, R. H., Gray, H. B., Cerdonio, M., Mongo, F., Romani, G. L., *J. Am. Chem. Soc.* (1976) **98**, 1029.
5. Van Holde, K. E., Van Bruggen, E. F. J., *Biol. Macromol.* (1971) **5**, 1, S. M. Timasheff, G. D. Fasman, Ed.
6. Caughy, W. S., Wallace, W. J., Volpe, J. A., Yoshikawa, S., *Enzymes* (1975) **12**, 299.
7. Lontie, R., Witters, R., "Inorganic Biochemistry," G. L. Eichhorn, Ed., p. 344, Elsevier, New York, 1973.
8. Österberg, R., *Coord. Chem. Rev.* (1974) **12**, 309.
9. Briving, C., Deinum, J., *FEBS Lett.* (1975) **51**, 43.
10. Lane, R. H., Sedor, F. A., Gilroy, M. J., Eisenhardt, P. F., Bennett, Jr., J. P., Ewall, R. X., Bennett, L. E., *Inorg. Chem.* (1977) **16**, 93.
11. Lane, R. H., Bennett, L. E., *J. Am. Chem. Soc.* (1970) **92**, 1089.
12. Holwerda, R. A., Gray, H. B., *J. Am. Chem. Soc.* (1974) **96**, 6008.
13. Ibid (1975) **97**, 6036.
14. Dockal, E. R., Diaddario, L. L., Glick, M. D., Rorabacher, D. B., *J. Am. Chem. Soc.* (1977) **99**, 4530.
15. Patterson, G. S., Holm, R. H., *Bioinorg. Chem.* (1975) **4**, 257.
16. Vallee, B. L., Williams, R. J. P., *Proc. Natl. Acad. Sci. USA* (1968) **59**, 498.
17. Jocelyn, P. C., "Biochemistry of the SH Group," p. 95, Academic, New York, 1972.
18. Friedman, M., "The Chemistry and Biochemistry of the Sulfhydryl Group in Amino Acids, Peptides and Proteins," p. 25, Pergamon, New York, 1973.
19. Vortisch, V., Kroneck, P., Hemmerich, P., *J. Am. Chem. Soc.* (1976) **98**, 2821.
20. Textor, M., Dubler, E., Oswald, H. R., *Inorg. Chem.* (1974) **13**, 1361.
21. Lewis, D. L., Hatfield, W. E., Hodgson, D. J., *Inorg. Chem.* (1972) **11**, 2216.
22. Bertrand, J. A., Kirkwood, C. E., *Inorg. Chim. Acta* (1972) **6**, 248.
23. Hollander, F. J., Coucouvanis, D., *J. Am. Chem. Soc.* (1977) **99**, 6268.
24. Taylor, Jr., I. F., Weininger, M. S., Amma, E. L., *Inorg. Chem.* (1974) **13**, 2835.
25. Lane, R. H., Pantaleo, N. S., Farr, J. K., Coney, W. M., Newton, M. G., *J. Am. Chem. Soc.* (1978) **100**, 1610.

RECEIVED February 22, 1978.

29

Metal Tetrathiolenes: Chemistry, Stereochemistry, Electrochemistry, and Semiconductivity

BOON-KENG TEO

Bell Telephone Laboratories, Murray Hill, NJ 07974

> *A class of organochalcogen compounds containing one or two chalcogen–chalcogen bonds was chosen as ligands for organometallic synthesis. New discrete molecular complexes containing two, four, and six metal atoms were prepared and characterized. These cluster systems exhibit rich electrochemistry as established by cyclic voltammetry and unusual stereochemistry as revealed by x-ray crystallography. A new class of organometallic polymers based on these ligands was synthesized and characterized. Temperature-dependent electrical conductivity measurements revealed semiconductivity consistent with pseudo-one-dimensionality. Electrical conductivity can be correlated with the oxidation potential of the free ligands. These new semiconducting organometallic polymers can be used as reversible anode materials for a rechargeable battery system.*

Recently there has been considerable interest in the molecular design of bimetallic complexes bridged by a quadridentate or a bis-bidentate ligand with highly delocalized π system (1–23). One incentive for such an attempt is to synthesize new bimetallic clusters that are potentially capable of forming multiparallel stacks of pseudo-one-dimensional systems via intermolecular metal and/or ligand orbital overlaps (1–5, 7–9, 24–28).

We have chosen a class of organochalcogen compounds containing one or two chalcogen–chalcogen bonds as potential ligands for organometallic synthesis (1, 2, 3, 4, 5). Typical members are $C_{10}H_6XY$ (where

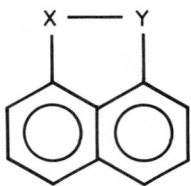

XY = SS, SeSe, TeTe
SSe, STe, SeTe

Figure 1. Dichalcolene ligands involving 1,8-substituted naphthalene systems

XY = S_2, Se_2, Te_2, SSe, STe, and SeTe) (29, 30, 31, 32) which contain one chalcogen–chalcogen bond (cf. Figure 1) and $C_{10}H_4X_4$ (33), $C_{18}H_8X_4$ (34, 35), and $C_{10}Cl_4X_4$ (36) where X= S, Se, and Te) which contain two chalcogen–chalcogen bonds (cf. Figure 2). These organic molecules possess three favorable steric and electronic features which make them excellent ligands. First, they are structurally planar with extensive electron delocalization (29–36). Secondly, the chalcogen–chalcogen bond(s) in each of these molecules are well suited for oxidative addition reactions with low-valent transition metal complexes (1, 2, 3, 4, 5). For the latter class of tetrachalcogen ligands, it may constitute a bridge between two metal atoms (1, 2, 3, 4, 5). Finally, upon coordination to one or two metal atoms, the di- and tetrachalcogen ligands can accommodate a formal charge of −2 and −4, respectively (1, 4). The implication of such a qualitative consideration is that the resulting complexes, which can appropriately be termed as metal di- and tetrachalcolenes, will exhibit unusually rich electrochemistry (4). In this chapter, we will first give a general view on the ligands and then restrict ourselves to the metal tetrathiolenes in the subsequent discussions.

Conceivably, metal tetrachalcolenes of the types exemplified in Figure 3 for metal tetrathiolenes can be synthesized. For discrete polynuclear metal complexes, one can either build up the oligomeric clusters by forming in-plane, metal–ligand σ bonds with tetrathiolene and other

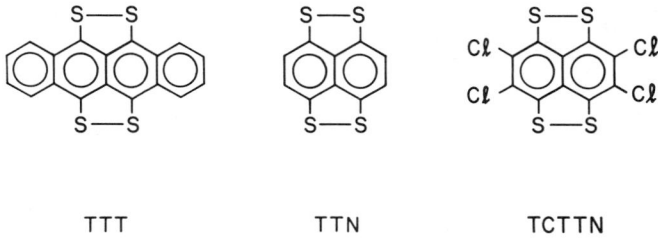

Figure 2. Tetrathiolene ligands (a subclass of tetrachalcolenes) $C_{18}H_8S_4$(TTT), $C_{10}H_4S_4$(TTN), and $C_{10}Cl_4S_4$(TCTTN)

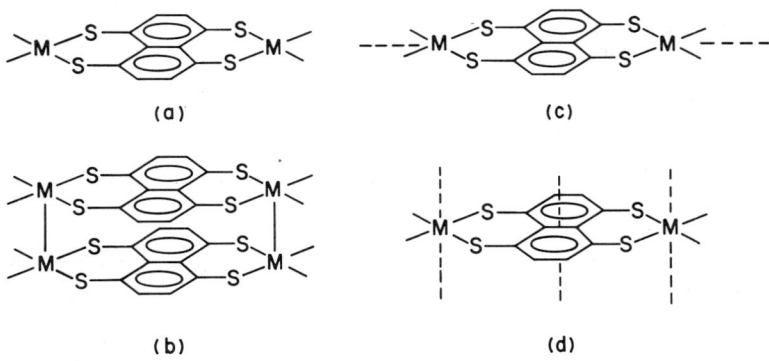

Figure 3. *Examples of molecular and polymeric metal tetrathiolenes, a subclass of metal tetrachalcolenes*

bidentate ligands (Figure 3a) or form small clusters by out-of-plane π-orbital overlaps (Figure 3b). For the polymeric species, one can have pseudo-one-dimensional chains with alternating units of metal atoms (or complexes) and the tetrathiolene ligands (Figure 3c) or multiple columnar stacks of metal complexes and/or the ligands (Figure 3d). It is obvious that Figures 3c and 3d can formally be considered as extensions of Figures 3a and 3b. A further variation of Figures 3b and 3d may involve the corresponding "staggered" structures, with the metal atoms interacting with the sulfurs rather than the metals of the adjacent unit(s) or chain(s) (*37*). It is also readily apparent that for structural types shown in Figures 3b and 3d, it is necessary to minimize the steric requirements of the terminal ligands.

A summary of our recent attempts in the synthesis, characterization, structure, and bonding of new metal tetrathiolene clusters and polymers will be presented. Emphasis will be placed on their novel stereochemistry, their unusually rich electrochemistry, and their interesting physical properties (*1, 2, 3, 4, 5*), with the hope of generating and understanding new materials of technological significance.

Organochalcogen Ligands

A major incentive for the organic synthesis of the ligands to be discussed in this section comes from the recent intense interest in pseudo-one-dimensional "organic metal" (*31, 33, 38–47*). In fact, the electrical conductivity of the tetrathiotetracene (TTT) and its monocation, first synthesized by Marschalk in 1948 (*34, 35*), has been noted for a long time. It is not until recently that its close analogs, tetrachlorotetrathionaphthalene (TCTTN) (36) and tetrathionaphthalene (TTN) (33), were made in 1972 and 1976, respectively. Similarly, while dithionaphthalene (DTN) (29) was made in 1911, the seleno and telluro (as well as the mixed combinations) analogs were reported only recently (1977) (31, 32).

To get an idea why these di- or tetramercapto compounds are prone to oxidation and form stable radical monocations or dications, we need only consider the conversion of naphthalene ($C_{10}H_8$) to dithionaphthalene (DTN, $C_{10}H_6S_2$) to tetrathionaphthalene (TTN, $C_{10}H_4S_4$). All three molecules are strictly planar with highly delocalized π-systems. However, the orbital characters of the highest-occupied molecular orbitals (HOMO) in these molecules change dramatically upon replacement of the hydrogens at 1,8 and 4,5 positions, successively, by the sulfur atoms. It goes from a naphthyl-ring-centered bonding π orbital in $C_{10}H_8$ to a dithio-centered π orbital which is antibonding between the two sulfur atoms in $C_{10}H_6S_2$ to a tetrathio-centered π orbital which is antibonding between the two pairs of sulfur atoms in $C_{10}H_4S_4$, as portrayed in Figures 4a, 4b, and 4c, respectively. As expected, with the increasing degree of antibonding, the orbital energies rise drastically in the same direction: -12.79, -8.53, and -6.17 eV in $C_{10}H_8$, $C_{10}H_6S_2$, and $C_{10}H_4S_4$, respectively. These results were based on nonparameterized molecular orbital calculations performed using an approximate Hartree–Fock–Roothaan SCF–LCAO method developed by Fenske et al. (48). This energetic trend then predicts that it should become increasingly easy to oxidize the series $C_{10}H_{8-n}S_n$ with increasing $n = 0, 2, 4$ as depicted in Figure 5 (the potentials are converted to vs. Ag/0.01M AgNO$_3$ in acetonitrile at a platinum electrode) (30, 33). In fact, it is possible to plot the orbital energy as a function of the one-electron oxidation potential for the con-

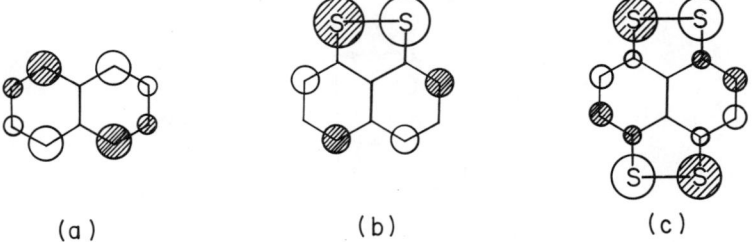

Figure 4. Representations of highest occupied molecular orbital (HOMO) of naphthalene (a), dithionaphthalene (b), and tetrathionaphthalene (c). The shaded and empty circles represent the positive and negative lobes of the atomic π orbitals (perpendicular to the molecular plane).

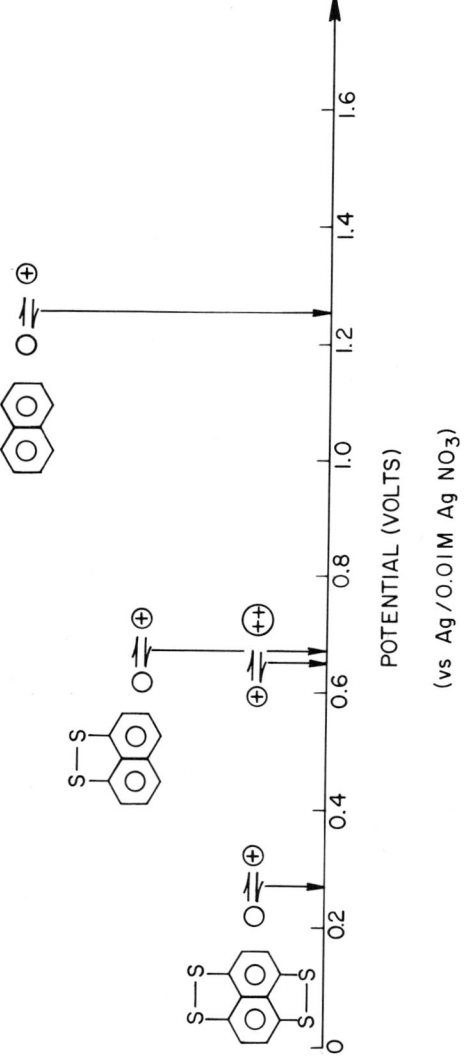

Figure 5. Oxidation potentials of naphthalene $(C_{10}H_8)$, dithionaphthalene $(C_{10}H_6S_2)$, and tetrathionaphthalene $(C_{10}H_4S_4)$ vs. $Ag/0.01M\ AgNO_3$

version of the neutral $C_{10}H_{8-n}S_n$ to its radical cations $[C_{10}H_{8-n}S_n]^+$ as shown in Figure 6. A gratifying linear relationship is clearly observed. Furthermore, the fact that the electron(s) come out of orbitals which are highly antibonding between the sulfur atoms for $n = 2$ and $n = 4$ in $C_{10}H_{8-n}S_n$ suggests that the resulting cations should be quite stable.

An obvious variation of these prototypes of organic molecules is the replacement of one or both of the sulfur atoms by selenium or tellurium. This has been done with $C_{10}H_6S_2$. In fact, the complete series of compounds with the general formula $C_{10}H_6XY$ where XY = SS, SeSe, TeTe, SSe, STe, and SeTe recently has been successfully synthesized and characterized. The oxidation potential is believed to decrease along the series $C_{10}H_6S_2$, $C_{10}H_6Se_2$, and $C_{10}H_6Te_2$. The room-temperature-compressed pellet electrical resistivity of the corresponding monocations as TCNQ⁻ salts also decreases dramatically from 7.2×10^{11} to 1×10^7 to 50 ohm-cm along the same series (30, 31). The temperature (T) dependence of the electrical resistivity (R) follow the linear relationship $\ln R$ vs. $T^{-1/2}$ over a reasonably large temperature range, suggesting pseudo-one-dimensionality. Similar replacement of the sulfur atoms in $C_{10}H_4S_4$ by its heavier congeners to form $C_{10}H_4X_nY_{4-n}$ (where X, Y = S, Se, or Te and $0 \le n \le 4$) or $C_{10}H_4X_lY_mZ_n$ (where X,Y,Z = S, Se, or Te and $l + m +$

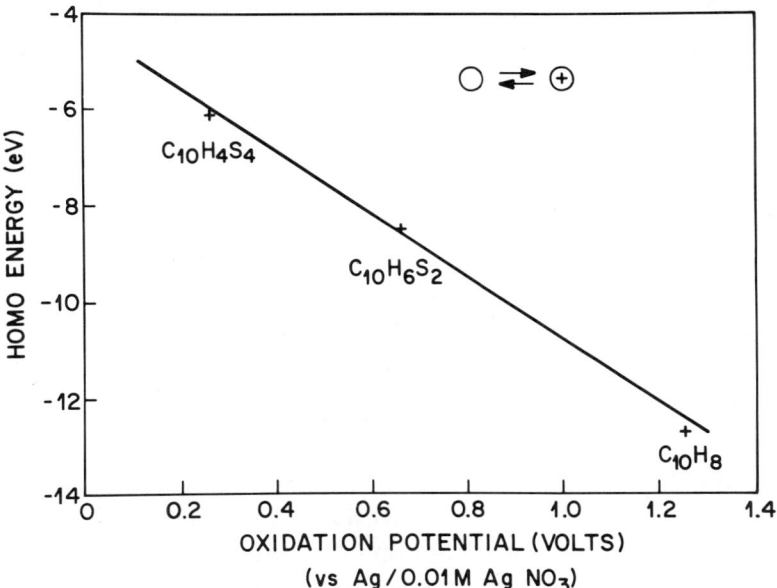

Figure 6. A linear correlation of the first one-electron oxidation potential of naphthalene ($C_{10}H_8$), dithionaphthalene ($C_{10}H_6S_2$), and tetrathionaphthalene ($C_{10}H_4S_4$) with the calculated HOMO energies

$n = 4$) has not been realized. Such a series of closely related compounds undoubtedly will give rise to highly interesting trends of redox potentials, electrical conductivities, as well as other physical or transport properties which will shed light on the chemical bonding and solid-state properties (e.g. columnar stacking interactions via orbital overlaps) of these species.

A different variation of these planar organic molecules can be achieved by ring substitution. For example, formal replacements of the four hydrogens in $C_{10}H_4S_4$ (TTN) by two benzo rings and four chlorines yield $C_{18}H_8S_4$ (TTT) and $C_{10}Cl_4S_4$ (TCTTN), respectively (cf. Figure 2). These compounds, however, were made by different procedures under different conditions. The redox potentials are also drastically different. Each molecule exhibits two reversible oxidations and one (or two) irreversible reduction reaction(s). The first oxidation step occurs at -0.05, $+0.27$, and $+0.64$ V whereas the second oxidation reaction occurs at $+0.44$, $+0.65$, and $+0.97$ V for TTT (49), TTN (33), and TCTTN (4), respectively, as shown schematically in Figure 7. The corresponding irreversible reduction wave occurs at -1.30 (also -1.67), -1.42, and -1.26 V. These values are standardized with respect to Ag/$0.01M$ AgNO$_3$. The electrical conductivity of the monocations are 0.07 to ~ 1 and 40 (ohm-cm)$^{-1}$ for TTT$^+$ (38, 39, 40) and TTN$^+$ (33), respectively. It is clear that there seems to be a correlation between the electrical conductivity and the oxidation potential: viz, as the ease of oxidation declines along the series TTT, TTN, TCTTN, so does the conductivity. This may be related to the bandgap of the semiconducting monocationic species which, qualitatively speaking, increases with the lowering of the energy of the highest occupied molecular orbital which, in turn, correlates with the decline of the ease of oxidation.

It occurs to us that these planar, highly π-delocalized, easily redoxed organochalcogen compounds can function as excellent ligands in organometallic synthesis. The chalcogen–chalcogen bond(s) in these molecules can undergo facile oxidative–addition reactions with a variety of inorganic or organometallic compounds, especially those with low oxidation states, thereby producing oligomeric or polymeric organometallic complexes comprised of chains of transition metals bridged by the above-mentioned tetrachalcogen ligands as well as other bidentate ligands. The free ends of these clusters or chains can be terminated by the above-mentioned dichalcogen ligands, other ligands, or other metal complexes. We expect these new materials to exhibit novel stereochemistry, rich electrochemistry, and unusual transport or catalytic properties. In the following sections, we summarize part of our recent attempts in synthesizing these new materials and in studying their intriguing chemical and physical properties. We will focus on the three tetrathiolene ligands TTT, TTN, and TCTTN. It should be emphasized, however, that a large

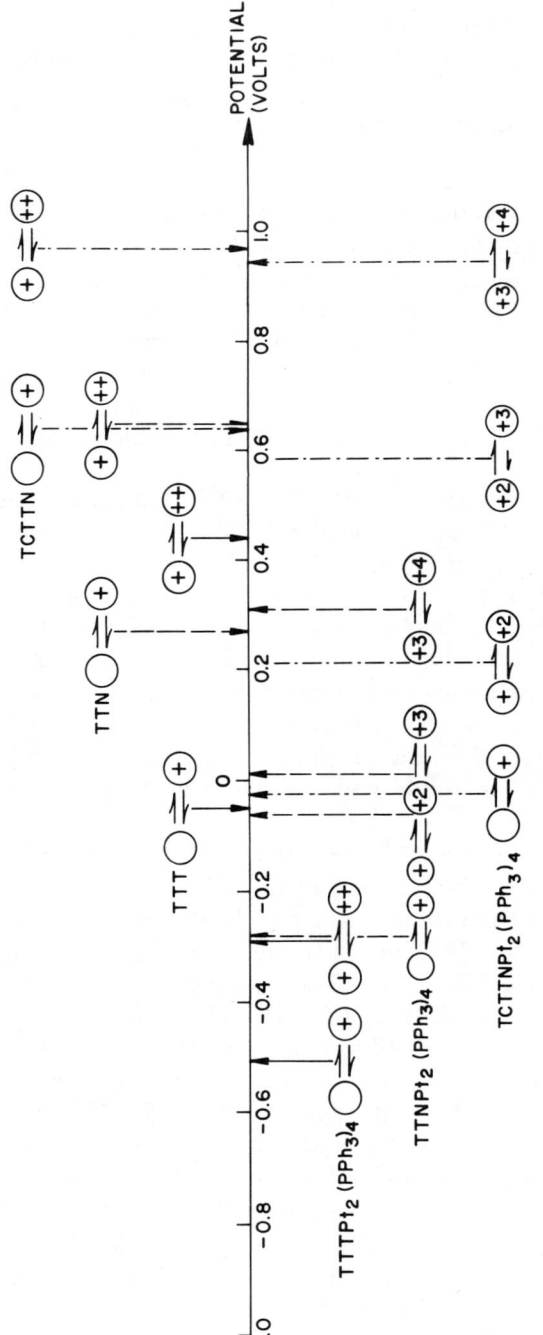

Figure 7. Oxidation potentials (vs. Ag/0.01M AgNO₃) of tetrathiolene ligands TTT, TTN, and TCTTN (top) and diplatinum–tetrathiolene complexes $Pt_2(PPh_3)_4$ TTT, $Pt_2(PPh_3)_4$TTN, and $Pt_2(PPh_3)_4$ TCTTN (bottom)

number of different types of reactions and properties are conceivable for these di- and tetrachalcogen–organometallic compounds. These are currently under investigation and will be subjects of future publications.

Molecular Metal Tetrathiolene Clusters

Bimetallic tetrathiolene complexes of the general type $L_{2n}M_2(TTL)$ can be obtained by reacting two moles of low-valent transition metal complex $L_{n+m}M$ (here L and M refers to the ligands and the metal, respectively; the $m+n$ ligands L need not be identical) with one mole of the TTL (where TTL = TTT, TTN, or TCTTN) ligands via oxidative addition of the two sulfur–sulfur bonds to the two metal atoms (Equation 1). The resulting bimetallic compleses can adopt a molecular

$$2L_{m+n}M + TTL \rightarrow L_{2n}M_2(TTL) + 2mL \qquad (1)$$

structure (3) comprised of a tetrathiolene (TTL) ligand bridging two metal complexes (L_nM) via four metal–sulfur bonds, two on each side of the molecule as exemplified in Figure 3a.

The solubilities of these complexes in common organic solvents are rather limited, being somewhat soluble in N,N'-dimethylformamide, methylene chloride, and chloroform, sparingly soluble in benzene, and insoluble in acetonitrile, acetone, hexane, etc. The limited solubility precludes measurements such as molecular weight determinations. Nevertheless, the products generally precipitate out of solvents such as benzene as microcrystalline solids and can be recrystallized with mixed solvents (1, 3, 4).

The IR spectra of the resulting complexes reveal characteristic bands which are diagnostic of the presence of the TTL ligands. These bands, however, are shifted from those of the free ligands which occur as four strong features at: (a) 1616(m), 1317(m), 1304(s), 968(w), 742(s) (or 714(w)) cm^{-1} for TTT; (b) 1540(s), 1362(s), 1185(vs), 797(vs) cm^{-1} for TTN; and (c) 1528(s), 1428(s), 1299(vs), 854(m) cm^{-1} for TCTTN. Similarly, the UV–visible spectra of these complexes exhibit features characteristic of the TTL ligands. In fact, the colors of these complexes parallel those of the free ligands: the TTT, TTN, and TCTTN complexes are generally green, red, and orange, respectively. This correlates with the visible absorptions of the free ligands which occur at (λ_{max} in nm and ϵ in M^{-1} cm^{-1} in parentheses): 694 (6.39 × 10^3), 637 (4.85 × 10^3), 583 (sh, 3.08 × 10^3), 471 (4.85 × 10^3), 428 (sh, 2.86 × 10^3), and 403 (sh, 1.54 × 10^3) for TTT (green); 420 (1.84 × 10^4), 397 (sh, 1.49 × 10^4), and 377 (sh, 8.60 × 10^3) for TTN (red); and 423 (2.07 × 10^4), 397 (1.58 × 10^4), and 372 (sh, 7.23 × 10^3) for TCTTN (golden yellow).

The reactions of Vaska's compounds, trans-Ir(PPh$_3$)$_2$(CO)X (X = Cl, Br, I), with TTN in benzene under reflux for 3–5 hr in a 2:1 molar ratio gave rise to orange complexes which analyze as (Ph$_3$P)$_2$(CO)$_2$X$_2$Ir$_2$(TTN) (Equation 2) (4). The IR spectra revealed the presence of coordinated

$$2\text{Ir}(\text{PPh}_3)_2(\text{CO})\text{X} + \text{TTN} \rightarrow$$
$$(\text{Ph}_3\text{P})_2(\text{CO})_2\text{X}_2\text{Ir}_2(\text{TTN}) + 2\text{PPh}_3 \quad (2)$$

CO, TTN, and Ph$_3$P ligands. The carbonyl stretching frequency occurs at 2019, 2017, and 2005 cm^{-1} for X = Cl, Br, and I, respectively, with fine structures developing upon refluxing. The four TTN bands, which are analogous for the series, occur at 1532(s), 1340(s), 1190(m), and 813(br,w) cm^{-1}. The intensities of the TTN bands are comparable with those of the Ph$_3$P bands for these complexes with the TTN:Ph$_3$P ratio of 1:2. The electronic spectra of (Ph$_3$P)$_2$(CO)$_2$X$_2$Ir$_2$(TTN) complexes in CH$_2$Cl$_2$ are dominated by two major bands in the visible region. The lowest-energy band, which occurs at 446 (2.42 × 10^4), 452 (2.52 × 10^4), and 438 nm (2.62 × 10^4 M^{-1} cm^{-1}) for X = Cl, Br, and I, respectively, is more or less insensitive to the nature of the halogen. On the other hand, the shoulder-like band at higher energy which occurs at (λ_{max} in nm and ϵ in M^{-1} cm^{-1} in parentheses) 318 (sh, 1.42 × 10^4), 349 (sh, 1.27 × 10^4), and 380 (sh, 1.61 × 10^4) for X = Cl, Br, and I, respectively, is highly halogen sensitive, being shifted to lower energy along the sequence X = Cl, Br, I. The UV bands occur at 270 nm (sh, 2.56 × 10^4, 2.70 × 10^4, and 3.17 × 10^4 for X = Cl, Br, and I, respectively). Preliminary cyclic voltammetry studies revealed that these diiridium clusters undergo one irreversible oxidation and one irreversible reduction reaction.

A series of diplatinum tetrathiolene complexes can be prepared by reacting Pt(PPh$_3$)$_4$ in benzene with the corresponding TTL ligand in a molar ratio of 2:1 (Equation 3) (4). The microcrystalline products,

$$2\text{Pt}(\text{PPh}_3)_4 + \text{TTL} \rightarrow (\text{Ph}_3\text{P})_4\text{Pt}_2(\text{TTL}) + 4\text{PPh}_3 \quad (3)$$

which form as green, red, and orange precipitate for TTL = TTT, TTN, and TCTTN, respectively, have been formulated as (Ph$_3$P)$_4$Pt$_2$(TTL) by elemental analysis. IR spectroscopy revealed the presence of both TTL and Ph$_3$P ligands, with the former being much weaker in band intensity than the latter, attributable to the 1:4 ratio of TTL:Ph$_3$P. The four strong TTL bands occur in the complexes at 1609(w), 1276(s), 954(vw), and 740(m) cm^{-1} for TTL = TTT; at 1530(w), 1346(m), 1178(s), 818(w), and 808(sh) cm^{-1} for TTL = TTN; and at 1465(w), 1392(w), 1239(s), and 836(w) cm^{-1} for TTL = TCTTN. The visible spectra of the (Ph$_3$P)$_4$Pt$_2$TTL complexes have two major bands with varying degrees of unresolved fine structure at (λ_{max} in nm and ϵ in M^{-1} cm^{-1} in parentheses): 720 (1.80 × 10^4), 652 (sh, 1.03 × 10^4), 440 w·sh, 5.54 × 10^3), and 387 (sh, 1.05 × 10^4) for TTL = TTT; 519 (1.44 × 10^4), 490 (1.35 × 10^4), 439 (1.43 × 10^4), and 423 (1.43 × 10^4) for TTL = TTN; and 502 (1.24 × 10^4), 447 (8.50 × 10^3), and 409 (9.44 × 10^3) for TTL = TCTTN. These visible bands are, in general, shifted to lower energies in comparison with the free ligands.

In order to establish unambiguously the stereochemistry of these metal tetrathiolenes as well as the mode of binding of the tetrathiolene ligands to transition metal complexes, single-crystal x-ray structural determinations of (Ph$_3$P)$_4$Pt$_2$(TTN) and (Ph$_3$P)$_2$(CO)$_2$Br$_2$Ir$_2$(TTN) were undertaken (3).

Figures 8 and 9 depict the structure of (Ph$_3$P)$_4$Pt$_2$(TTN) in different views. It involves a tetrathionaphthalene (TTN) ligand bridging two bis-triphenylphosphine platinum moieties with each platinum atom being coordinated to two phosphorus (from two PPh$_3$ ligands) and two sulfur (from the TTN ligand) atoms. The platinum coordination and the bridging TTN ligand (except, perhaps, the sulfur atoms) are close to planarity. The molecule as a whole, however, is by no means planar. The overall distortions from planarity can be visualized as a small rotation of the sulfur atoms about the C(3)–C(3)' bond, followed by a large rotation of each of the two platinum coordination planes about the S···S edge, resulting in the dihedral angles of 12.6° between the average planes formed by naphthalene group and S(1)–C(1)–C(3)–C(2)–S(2) and of 38.4° between the average planes formed by S(1)–C(1)–C(3)–C(2)–S(2) and the PtS$_2$P$_2$ coordination. The resulting molecular geometry is centrosymmetric with the center of symmetry located at the midpoint of C(3) and C(3)'. An interesting observation of the crystal structure of (Ph$_3$P)$_4$Pt$_2$(TTN) is that the intra- and intermolecular Pt···Pt vectors form parallel arrays of zig-zag chains, with the former (9.043(4) Å) being substantially longer than the latter (7.662(4) Å).

On the other hand, the structure of (Ph$_3$P)$_2$(CO)$_2$Br$_2$Ir$_2$(TTN) is a total surprise to us (3). On chemical as well as stereochemical grounds, one might expect two square–pyramidal or trigonal–bipyramidal (Ph$_3$P)-Br(CO)IrS$_2$ complexes bridged by the TTN ligand (cf. (Ph$_3$P)$_4$Pt$_2$-(TTN)). The determined structure of the former compound, however, bears no resemblance to the latter. The two iridiums, instead of oxidatively cleaving the two sulfur–sulfur bonds in TTN, react with only one of them, resulting in a "butterfly" arrangement of the Ir$_2$S$_2$ (from TTN) fragment. The molecule is best described as two (Ph$_3$P)(CO)BrIr moieties bridged by two sulfur atoms (S···S of 3.07 Å) from TTN. This unusual iridium dimer has an Ir–Ir distance of 2.68 Å with the metal–

metal bond conceptually occupying the sixth coordination site of the highly distorted octahedral coordinations.

The significance of the $(Ph_3P)_4Pt_2(TTL)$ complexes lies in their rich electrochemistry (4). Cyclic voltammetry (a $10^{-3}M$ solution in $0.1M$ $(n\text{-}C_4H_9)_4N^+ClO_4^-$ in CH_2Cl_2 using a scan rate of 200 mV/sec, a platinum bead as working electrode, a platinum wire as counter electrode, and

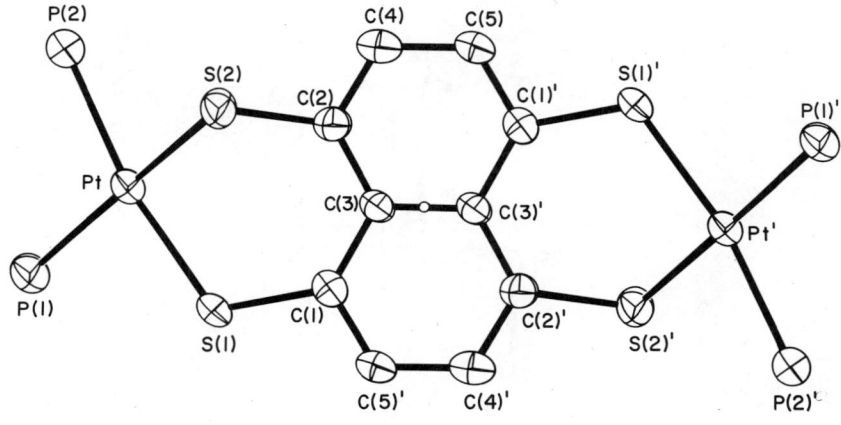

Figure 8. The $P_4Pt_2C_{10}S_4$ core of the $[(C_6H_5)_3P]_4Pt_2(C_{10}H_4S_4)$ molecule (ORTEP diagram, 50% probability thermal ellipsoids, infinity projection) with crystallographic C_i-$\overline{1}$ symmetry located at the midpoint between C(3) and C(3)'. (a) View along the normal of the naphthalene plane and (b) view similar to (a) but rotated 90° about the C(3)–C(3)' bond.

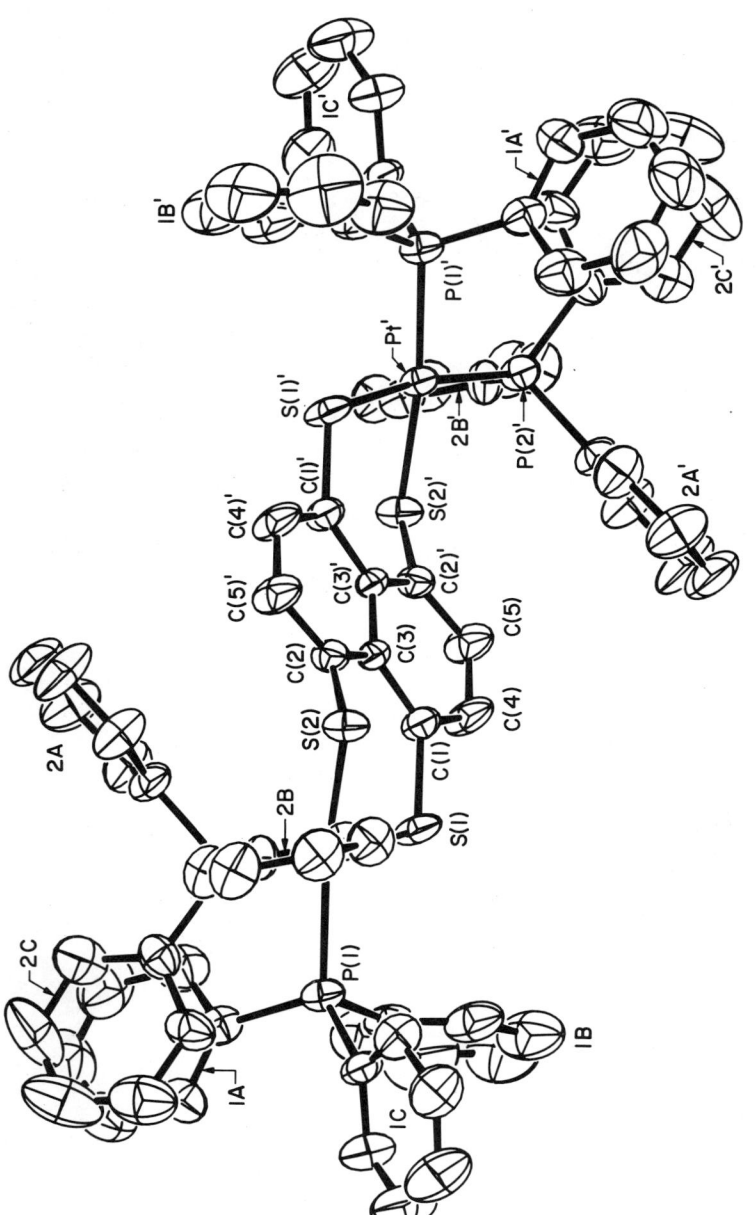

Figure 9. Stereochemistry of the $[(C_6H_5)_3P]_4Pt_2(C_{10}H_4S_4)$ molecule (only nonhydrogen atoms are shown in 50% probability ellipsoids)

Ag/0.01M AgNO$_3$ in CH$_3$CN as reference electrode) revealed two reversible one-electron oxidations at −0.51 and −0.28 V for (Ph$_3$P)$_4$Pt$_2$-(TTT) (at ambient temperature) but four reversible one-electron oxidation waves at −0.28, ∼ −0.05, ∼ 0.01, and 0.31 V for (Ph$_3$P)$_4$Pt$_2$(TTN) (at ambient temperature). The cyclic voltammogram of (Ph$_3$P)$_4$Pt$_2$-(TCTTN) at ambient temperature turns out to be more complicated. However, at dry ice/acetone temperature, it exhibits two reversible one-electron oxidations at −0.02 and +0.21 V and two quasi-reversible oxidations at +0.58 and +0.94 V. These oxidation potentials are shifted dramatically in the negative direction (more easily oxidized) with respect to the free ligands (vide supra) as depicted in Figure 7. This is taken as an indication of a buildup of negative charge in the TTL ligand upon coordination. The EPR spectra of the paramagnetic mono- and trications of (Ph$_3$P)$_4$Pt$_2$(TTN) indicated that in the monocation the spin densities are substantially localized on the TTN ligand with no observable hyperfine interaction(s) with the platinum atoms, whereas in the trication there is a significant amount of spin densities "localized" on one platinum atom (nonequivalent hyperfine interactions) even though the unpaired electron also resides mainly on the bridging ligand (*1*).

The stereochemical novelty and the electrochemical richness of the bimetallic tetrathiolene complexes prompted us to synthesize and study longer chains of oligometallic tetrathiolene clusters. The tetraplatinum clusters (DPPA)$_4$Pt$_4$(TTL)$_3$ (where DPPA = Ph$_2$C ≡ CPh$_2$; TTL = TTT, TTN) can be prepared by reacting stoichiometric equivalents of the DPPA-bridged platinum dimer Pt$_2$(DPPA)$_2$(PPh$_3$)$_4$ with the corresponding tetrathiolene as shown in Equation 4 (*50*). IR and UV–visible

$$2\text{Pt}_2(\text{DPPA})_2(\text{PPh}_3)_4 + 3\text{TTL} \rightarrow (\text{DPPA})_4\text{Pt}_4(\text{TTL})_3 + 8\text{PPh}_3 \quad (4)$$

spectroscopies suggest the presence of the tetrathiolene and the bisphosphine ligands. These compounds are in general more deeply colored and less soluble than the diplatinum complexes. A zig-zag chain-like structure with bridging tetradentate TTL and bidentate DPPA ligands depicted in 1 is expected for these tetraplatinum clusters. Cyclic voltammograms of these tetraplatinum–tetrathiolene clusters indicated a complex manifold of overlapping reversible and quasi-reversible oxidation waves

(TTL)Pt(DPPA)$_2$Pt(TTL)Pt(DPPA)$_2$Pt(TTL)

1

(which amount to a total of approximately 12 electron transfers in the case of $(DPPA)_4Pt_4(TTN)_3$) (50).

Longer homo- or heteronuclear chain-like clusters such as the hexametal complex 2 can be prepared by reacting the corresponding metal clusters with a combination of the tetrathiolenes and other appropriate multidentate ligands as bridges. The metal atoms M, M', and M", the tetrathiolene ligands, TTL, TTL', and the l bidentate LL and l' terminal L ligands need not be identical within each group. We are actively pursuing the synthesis, structure, and electrochemistry of these unusual cluster compounds.

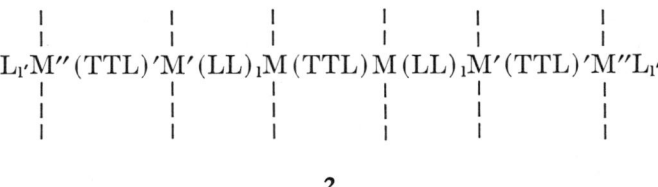

2

Semiconducting Organometallic Polymers

In an attempt to synthesize new planar organometallic complexes which will form multiparallel columnar stacks of square-planar transition metal complexes bridged by a highly delocalized organic π system, tetrathiotetracene (TTT), tetrathionaphthalene (TTN), and tetrachlorotetrathionaphthalene (TCTTN) were reacted with a variety of transition metal complexes with varying degree of steric requirements. Since carbonyls are capable of stabilizing metals in low formal oxidation states and are also sterically quite innocent, metal carbonyls were chosen as starting reactants for such investigations (2, 5). Our initial goal was to prepare bimetallic compounds such as $Ni_2(CO)_4(TTL)$ (cf. Figure 3a) which will conceivably form columnar stacking via overlaps of the π orbitals of the bridging ligand and/or the metal orbitals (either with or without the involvement of the terminal carbonyl ligands). Instead, much to our initial surprise, we obtain a new class of polymeric metal–tetrathiolene compounds formulated as $[Ni(TTL)]_x$ and $[Co_2(CO)_2$-$(TTL)]_x$ from the reaction of the tetrathiolenes with the corresponding metal carbonyl (phosphine) complexes. These new organometallic polymers exhibit interesting semiconducting properties (2, 5). These materials are distinctly different from either the organic conductors such as TTF–TCNQ (51, 52) or the inorganic conductors such as $K_2Pt(CN)_4 \cdot X_{0.3}$ (53, 54, 55) (where X = Cl, Br) in that the chain direction lies, presumably, in the molecular plane (along the long molecular axis) rather than perpendicular to it.

$$Ni(CO)_4 + TTN \rightarrow \frac{1}{x}[Ni(TTN)]_x + 4CO \qquad (5)$$

The compound $[Ni(TTN)]_x$ can be prepared by reacting TTN with excess nickel tetracarbonyl in benzene (Equation 5). The dark brown-red amorphous material, insoluble in common organic solvents, exhibits no carbonyl stretching frequencies but four strong TTN bands at 1528(s), 1338(s), (1320(sh)), 1185(m), and 800(m) cm^{-1} in the IR spectrum. A new band also is observed at 965 cm^{-1} which can be attributed to C $\cdots$ S bonds. The polymeric chain-like structure 3 was proposed for $[Ni(TTN)]_x$.

$$\ldots Ni(TTN)Ni(TTN) \ldots$$

3

Similarly, reaction of TTN with a stoichiometric amount of $Co_2(CO)_8$ in benzene gave the polymer $[Co_2(CO)_2(TTN)]_x$ (Equation 6). IR spec-

$$Co_2(CO)_8 + TTN \rightarrow \frac{1}{x}[Co_2(CO)_2(TTN)]_x + 6CO \qquad (6)$$

troscopy indicated the presence of: (1) terminal carbonyls (broad band at 2010(s)); (2) the coordinated TTN ligand at 1525(m), 1339(m), 1320(sh), 1190(m), and 810(br,w) cm^{-1}; and (3) a new band at 970 cm^{-1} which is probably attributable to C $\cdots$ S stretching frequencies. Again, a polymeric chain-like structure 4 was proposed for $[Co_2(CO)_2(TTN)]_x$.

$$\ldots (TTN)Co_2(CO)_2(TTN) \ldots$$

4

The most intriguing physical property of these polymeric materials is their electrical conductivity. The temperature (T) dependence of the powder resistance (R) can be characterized by the relation

$$\ln \frac{R}{R_o} = \left(\frac{T}{T_o}\right)^{-1/2} \qquad (7)$$

where T_o is the square of the slope of the $\ln R$ vs. $T^{-1/2}$ plot and is inversely proportional to the density of localized states (Equation 8) (56, 57, 58).

$$T_o = \frac{4\alpha}{kAN(\epsilon_F)} \qquad (8)$$

Here α^{-1} is the radius of the localized state wavefunction, $N(\epsilon_F)$ is the density of localized states, A is the area of the compressed pellet, and k is Boltzmann constant. Plots of this type have been observed for a number of known one-dimensional systems and taken as evidence for one-dimensional hopping conductivity between localized states (*38–47, 56–64*). For comparison, the hopping conductivity for a two-dimensional system can be characterized by the relation $\ln R/R_o = (T/T_o)^{-1/3}$ where $T_o = 8\alpha^2/(kDN(\epsilon_F))$ (here D is the thickness), whereas the hopping conductivity for a three-dimensional system follows the relation $\ln R/R_o = (T/T_o)^{-1/4}$ where $T_o = 16\alpha^3/(kN(\epsilon_F))$. This theory has been questioned recently by Mott (*64*). Nevertheless, this type of plot can be used to characterize the conductivity of these materials. The T_o values of 1.7×10^5 K observed for both compounds are, however, significantly higher than that generally observed for one-dimensional conductors or semiconductors (range: $0.5 \sim 5 \times 10^4$ K) (*24–26, 38–47, 56–64*).

A more convenient preparation of the series of polymers $[\text{Ni}(\text{TTL})]_x$ (where TTL = TTT, TTN, or TCTTN) starts with a 2:1 ratio of Ni$(\text{CO})_2(\text{PPh}_3)_2$ and TTL in refluxing benzene (Equation 9) (*5*). Simi-

$$\text{Ni}(\text{CO})_2(\text{PPh}_3)_2 + \text{TTL} \xrightarrow[\text{benzene}]{\text{reflux}} \frac{1}{x}[\text{Ni}(\text{TTL})]_x + 2\text{CO} + 2\text{PPh}_3 \quad (9)$$

larly, the series of cobalt tetrathiolene polymers $[\text{Co}_2(\text{CO})_2(\text{TTL})]_x$ has been prepared from the reaction of $\text{Co}_2(\text{CO})_8$ with TTL in refluxing benzene (Equation 10). All of these materials exhibit IR bands indicative of the coordinated ligands.

$$\text{Co}_2(\text{CO})_8 + \text{TTL} \xrightarrow[\text{benzene}]{\text{reflux}} \frac{1}{x}[\text{Co}_2(\text{CO})_2(\text{TTL})]_x + 6\text{CO} \quad (10)$$

The electrical conductivity of all these neutral organometallic polymers follows the same temperature dependence as the TTN compounds (Equation 7). The T_o values and the room temperature resistivity ρ_{300}, however, are significantly different for different tetrathiolene ligands. For the same TTL ligand, the T_o and the ρ_{300} values are very similar despite the fact that different metals of varying local geometries (nickel vs. dicobalt dicarbonyl moieties) are involved. In going from TTT to TTN to TCTTN, the slope-related quantity $T_o = 1.1$–5.6×10^5 K increases only slightly (less than or equal to a factor of two) whereas the resistivity ρ_{300} increases by one order of magnitude in each step, going from 10^5 to 10^6 to 10^7 ohm-cm (*5*). We believe that this trend is related to the oxidation potential of the TTL ligand. That is, since the oxidation potentials of the free ligands increase along the series TTT < TTN < TCTTN, and since it has been shown previously for the discrete metal tetrathiolene clusters that the resulting metal complexes exhibit trends of oxidation potential parallel that of the ligands (vide supra), it is conceivable that as the ligand becomes harder to oxidize (TTT → TTN → TCTTN), the negative charges localized on the ligand increase progressively. This implies that the number of effective charge carriers (or the density of the disordered localized states) decreases as one goes from TTT to TTN to TCTTN. This then causes a slight increase of approximately less than or equal to a factor of two in T_o but an exponential increase in ρ_{300} along the series TTT < TTN < TCTTN.

Both $[Ni(TTL)]_x$ and $[Co_2(CO)_2(TTL)]_x$ react with oxidizing agents such as iodine either in solution–suspension or in solid–gas phases, according to Equations 11 and 12, respectively (5). A suspension of the

$$\frac{1}{x}[Ni(TTL)]_x + \frac{y}{2}I_2 \rightleftarrows \frac{1}{x}[Ni(TTL)I_y]_x \qquad (11)$$

$$\frac{1}{x}[Co_2(CO)_2(TTL)]_x + \frac{y}{2}I_2 \rightleftarrows \frac{1}{x}[Co_2(CO)_2(TTL)I_y]_x \qquad (12)$$

complexes in a benzene (or other organic solvents) solution of iodine will absorb molecular iodine to form $[Ni(TTL)I_y]_x$ and $[Co_2(CO)_2(TTL)I_y]_x$ (where y is the molar ratio of iodine to $[Ni(TTL)]_x$ and $[Co_2(CO)_2(TTL)]_x$, respectively). The resulting complexes are also insoluble in most organic solvents. The same reaction can be carried out in solid–gas phase. The solid $[Ni(TTL)]_x$ or $[Co_2(CO)_2(TTL)]_x$ complexes absorb iodine vapor slowly at room temperature to form $[Ni(TTL)I_y]_x$ or $[Co_2(CO)_2(TTL)I_y]_x$. The rate of these solid–gas reactions can be accelerated by increasing temperature. Furthermore, these reactions are totally reversible in the sense that the absorbed iodine can be removed by pumping $[Ni(TTL)I_y]_x$ and $[Co_2(CO)_2(TTL)I_y]_x$ under vacuum at elevated temperatures. The kinetics of these solid–gas reactions can be monitored by measuring the weight gain (forward reactions, Equations 11 and 12) and weight loss (reversed reactions, Equations 11 and 12) of the solid complexes in an atmosphere saturated with iodine vapor and in vacuo, respectively (5).

The IR spectra of these oxidized polymers are very similar to those of the neutral polymers except for a new band at 1050 cm^{-1} which can reasonably be assigned to the formation of new C=S bonds in the oxidized polymer chains.

The reversibility of these reactions suggests that they are topotactic or intercalation reactions. We believe that these reactions represent oxidation of the polymeric $[Ni(TTL)]_x$ or $[Co_2(CO)_2TTL]_x$ chains with molecular iodine which is concomitantly reduced to iodide or polyiodide anions which then "intercalate" into the solid matrix.

To investigate the oxidation states of $[Ni(TTL)I_y]_x$ and $[Co_2(CO)_2(TTL)I_y]_x$, we measured the resonant Raman spectra of $[Ni(TTT)I_y]_x$ where $y = 1,2,3$. For $[Ni(TTT)I_3]_x$, intense resonance–enhanced totally symmetric I–I–I stretching frequency of I_3^- was observed at 107 cm^{-1} along with the expected overtone progressions (65, 66). This is consistent with the formulation of $[\overline{Ni(TTT)}\cdot I_3^-]_x$. For $[Ni(TTT)I]_x$ on the other hand, no I_3^- frequency was observed. Though the failure to observe the expected frequencies in Raman spectroscopy does not constitute a proof of the absence of the species, it is our belief that the result is consistent with the formulation $[\overline{Ni(TTT)}^{y+} \cdot yI^-]_x$ for $y \leq 1$ since we expect the $[Ni(TTT)]_x$ chain chromophore to be relatively little affected by the oxidation, and hence the presence of $y/3$ mole equivalents of I_3^- would have been easily detectable. For $[Ni(TTT)I_2]_x$, the resonance Raman spectrum shows not only the intense fundamental and overtone progression of I_3^- but also the "antisymmetric" stretch at 143 cm^{-1}, suggesting either some distortion of the triiodide ion from the idealized $D_{\infty h}$ symmetry and/or the presence of both symmetric and asymmetric triiodide ions (65, 66). We propose that $[Ni(TTT)I_2]_x$ can be formulated as $(\overline{NiTTT^+} \cdot (1/2)I^- \cdot (1/2)I_3^-)_x$ in which the coexistence of I^- and I_3^- in the channels provided by the oxidized $[Ni(TTT)]_x$ chains causes some or all of the triiodide ions to be distorted. Preliminary extended x-ray absorption fine structure (EXAFS) spectroscopic measurements of the oxidized polymeric species also revealed no distinct Ni–I and Co–I bonds in $[Ni(TTL)I_y]_x$ (nickel K edge) and $[Co_2(CO)_2(TTL)I_y]_x$ (cobalt K edge), respectively, which is consistent with the formulation that the iodides or polyiodides are not directly bonded to the metal tetrathiolene polymer chains (67).

The electrical conductivity data of the oxidized species $[Ni(TTT)I_y]_x$ (where $y = 1,2,3$) follow the same temperature dependence (Equation 7) as the neutral species with, however, a significant decrease in T_o by a factor of two to three and a dramatic decrease in room temperature resistivity ρ_{300} by three orders of magnitude in going from the neutral to the oxidized polymers. This is consistent with the increase in the number of effective charge carriers on the organometallic chain upon oxidation. An initially puzzling observation that the electrical conductivities of $[Ni(TTT)I_y]_x$ are virtually invariant to the degree of oxidation with $y = 1, 2$, and 3 can now be explained by the above formulation that for $y > 1$, the extra iodines go in as neutral iodine molecules, converting part ($y = 2$) or all ($y = 3$) of the iodide (I^-) into the triiodide (I_3^-) ions such that the oxidation states of the nickel tetrathiolene chains remain essentially the same for $y \geq 1$.

Applications

There are several plausible technological applications of these new metal tetrathiolene polymers. A prime example is the use of these organometallic polymers as reversible anode materials in rechargeable (secondary) batteries. We take advantage of the following facts: (1) these polymers can be oxidized with oxidants such as molecular iodine reversibly; (2) both the neutral and the oxidized species are insoluble in common solvents; and (3) both the neutral and the oxidized species are thermally, air, and moisture stable. We reason that, in the discharging process of a battery with metal tetrathiolene polymers as reversible anodes, electrons flow through the outer load circuit; the oxidized polymers then pick up an equivalent amount of iodide or polyiodide ions from the electrolyte solution. In the recharging process, an opposite potential is applied which reduces the oxidized species to the neutral polymers with the concomitant ejection of an equivalent amount of the iodide or polyiodide ions back to the electrolyte.

A typical rechargeable battery based on this idea has been constructed. It uses the $[Ni(TTL)]_x$ polymer as the anode, poly-2-vinylpyridine-iodine $(P2VP \cdot (x/2)I_2)$ (68) complex as the cathode, and aqueous KI solution as the electrolyte solution (Equation 13). In the discharging process, electrons flow from the anode $[Ni(TTL)]_x$ to the cathode $P2VP \cdot (x/2)I_2$ through the load circuit; the iodide (or polyiodide) ions formed at the cathode then enter the electrolyte while an equivalent amount of iodide ions from the electrolyte solution intercalate into the oxidized metal tetrathiolene polymer (anode). The electrolyte concentration is therefore conserved. Upon recharging with an opposite

$$[Ni(TTL)]_x / KI(H_2O) / P2VP \cdot \frac{x}{2} I_2 \qquad (13)$$

potential, these processes are reversed. Since both the anode and the cathode are insoluble in the electrolyte, the overall process amounts to transporting iodine from P2VP polymer to metal tetrathiolene polymer (a redox reaction) in the discharging process and vice versa in the recharging process. The measured voltage for such a battery ranges from 0.5 to 0.8 V, depending upon the type of metal tetrathiolene polymer chosen. It is conceivable that higher voltages can be achieved by changing the metal and/or the ligand(s) of the anode or by replacing the cathode with more powerful oxidizing materials.

Conclusions

In conclusion, we have demonstrated that the organochalcogen compounds discussed in the section on "Organochalcogen Ligands" are excellent ligands for organometallic syntheses. The resulting metal complexes, either as discrete oligomeric clusters or infinite-chain polymers, exhibit novel stereochemistry and rich electrochemistry. The pseudo-one-dimensional semiconductivities and the reversible topotactic oxidation reactions of the insoluble metal tetrathiolene polymers open up a new dimension of potential technological applications. We continue to develop the chemistry, stereochemistry, and electrochemistry of these (and related) materials.

Acknowledgment

I thank J. J. Hauser and P. K. Gallagher at Bell Laboratories (Murray Hill) for permission to quote the conductivity and thermogravimetric results (Ref. 5), respectively. I am also grateful to J. San Filippo at Rutgers University (New Brunswick) for laser Raman measurements. Special thanks go to P. A. Snyder for her skillful technical assistance. I also enjoyed helpful discussions with D. W. Murphy and J. N. Carides. Other contributors are acknowledged in Ref. *1, 2, 3, 4,* and *5.*

Literature Cited

1. Teo, B. K., Wudl, F., Marshall, J. H., Kruger, A., *J. Am. Chem. Soc.* (1977) **99**, 2349.
2. Teo, B. K., Wudl, F., Hauser, J. J., Kruger, A., *J. Am. Chem. Soc.* (1977) **99**, 4862.
3. Teo, B. K., Snyder-Robinson, P. A., *J. Chem. Soc., Chem. Commun.,* submitted for publication.
4. Teo, B. K., Snyder-Robinson, P. A., *Inorg. Chem.,* accepted for publication.
5. Teo, B. K., Snyder-Robinson, P. A., Hauser, J. J., Gallagher, P. K., *Inorg. Chem.,* submitted for publication.
6. Girgis, A. Y., Sohn, Y. S., Balch, A. L., *Inorg. Chem.* (1975) **14**, 2327.
7. Kaiser, S. W., Saillant, R. B., Butler, W. M., Rasmussen, P. G., *Inorg. Chem.* (1976) **15**, 2681, 2688.
8. Mueller-Westerhoff, U. T., Heinrich, F., "Extended Interactions Between Metal Ions," L. V. Interrante, Ed., ACS Symp. Ser. (1974) **5**, 396.
9. Mueller-Westerhoff, U. T., "Inorganic Compounds with Unusual Properties," R. B. King, Ed., ADV. CHEM. SER. (1976) **150**, 31–45.
10. Leitherser, M., Coucouvanis, D., *Inorg. Chem.* (1977) **16**, 1611.
11. Hollander, F. J., Leitheiser, M., Coucouvanis, D., *Inorg. Chem.* (1977) **16**, 1615.
12. Hollander, F. J., Coucouvanis, D., *Inorg. Chem.* (1974) **13**, 2381.
13. Coucouvanis, D., Baenziger, N. C., Johnson, S. M., *J. Am. Chem. Soc.* (1973) **95**, 3875.
14. Coucouvanis, D., Piltingsrud, D., *J. Am. Chem. Soc.* (1973) **95**, 5556.
15. Hollander, F. J., Coucouvanis, D., *J. Am. Chem. Soc.* (1977) **99**, 6268.
16. Callahan, R. W., Keene, F. R., Meyer, T. J., Salmon, D. J., *J. Am. Chem. Soc.* (1977) **99**, 1064.
17. Taube, H., *Surv. Prog. Chem.* (1974) **6**, 1.
18. Krentzien, H., Taube, H., *J. Am. Chem. Soc.* (1976) **98**, 6379.
19. Meyer, T. J., *Prog. Inorg. Chem.* (1975) **19**, 1.

20. Hunziker, M., Ludi, A., *J. Am. Chem. Soc.* (1977) **99**, 7370.
21. Trofimenko, S. J., *J. Org. Chem.* (1964) **29**, 3046.
22. Cunningham, J. A., Sievers, R. E., *J. Am. Chem. Soc.* (1973) **95**, 7183.
23. Cetinkaya, B., Hitchcock, P. B., Lappert, M. F., Pye, P. L., *J. Chem. Soc., Chem. Commun.* (1975) 683.
24. Interrante, L. V., "Inorganic Compounds with Unusual Properties," R. B. King, Ed., ADV. CHEM. SER. (1976) **150**, 1–17.
25. Interrante, L. V., Ed., "Extended Interactions Between Metal Ions," ACS Symp. Ser. (1974) **5**.
26. Miller, J. S., Epstein, A. J., *Prog. Inorg. Chem.* (1976) **20**, 1.
27. Schrauzer, G. N., Prakash, H., *Inorg. Chem.* (1975) **14**, 1200.
28. Rosa, E. J., Schrauzer, G. N., *J. Phys. Chem.* (1969) **73**, 3132.
29. Laufrey, M., *Compt. Rend.* (1911) **152**, 92.
30. Zweig, A., Hoffmann, A. K., *J. Org. Chem.* (1965) **30**, 3997.
31. Meinwald, J., Dauplaise, D., Wudl, F., Hauser, J. J., *J. Am. Chem Soc..* (1977) **99**, 255.
32. Meinwald, J., Dauplaise, D., Clardy, J., *J. Am. Chem. Soc.* (1977) **99**, 7743.
33. Wudl, F., Schafer, D. E., Miller, B., *J. Am. Chem. Soc.* (1976) **98**, 252.
34. Marschalk, C., Stumm, C., *Bull. Soc. Chim. Fr.* (1948) **15**, 418.
35. Marschalk, C., *Bull. Soc. Chim. Fr.* (1952) **19**, 800.
36. Klingsberg, E., *Tetrahedron* (1972) **28**, 963.
37. Bonamico, M., Dessy, G., Fares, V., *Chem. Commun.* (1969) 324.
38. Inokuchi, H., Kochi, M., Harada, Y., *Bull. Chem. Soc. Jpn.* (1967) **40**, 2695.
39. Perez-Albuerne, E. A., Johnson, H., Jr., Trevoy, D. J., *J. Chem. Phys.* (1971) **55**, 1547, and references cited therein.
40. Bray, J. W., Hart, H. R., Jr., Interrante, L. V., Jacobs, I. S., Kasper, J. S., Piacente, P. A., Watkins, G. D., *Phys. Rev. B* (1977) **16**, 1359.
41. Wudl, F., Wobschall, D., Hufnagel, E. J., *J. Am. Chem. Soc.* (1972) **94**, 670.
42. Ferraris, J. P., Cowan, D. O., Walatka, V., Perlstein, J. A., *J. Am. Chem. Soc.* (1973) **95**, 948.
43. Schafer, D. E., Wudl, F., Thomas, G. A., Ferraris, J. P., Cowan, D. O., *Solid State Commun.* (1974) **14**, 347.
44. Wudl, F., *J. Am. Chem. Soc.* (1975) **97**, 1962.
45. Bloch, A. N., Weisman, R. B., Varma, C. M., *Phys. Rev. Lett.* (1972) **28**, 753.
46. Melby, L. R., *Can. J. Chem.* (1965) **43**, 1448.
47. Masuda, K., Silver, M., Eds., "Energy and Charge Transfer in Organic Semiconductors," Plenum, New York, 1974.
48. Fenske, R. F., *Pure Appl. Chem.* (1971) **27**, 61.
49. Geiger, W. E., Jr., *J. Phys. Chem.* (1973) **77**, 1862.
50. Teo, B. K., Snyder-Robinson, P. A., submitted for publication.
51. Phillips, T. E., Kistenmacher, T. J., Ferraris, J. P., Cowan, D. O., *J. Chem. Soc., Chem. Commun.* (1973) 471.
52. Kistenmacher, T. J., Phillips, T. E., Cowan, D. O., *Acta Crystallogr. Sect. B* (1974) **30**, 763.
53. Peters, C., Eagen, C. F., *Inorg. Chem.* (1976) **15**, 782.
54. Williams, J. M., Ross, F. K., Iwata, M., Petersen, J. L., Peterson, S. W., Lin, S. C., Keefer, K., *Solid State Commun.* (1975) **17**, 45.
55. Williams, J. M., Iwata, M., Peterson, S. W., Leslie, K. A., Guggenheim, H. J., *Phys. Rev. Lett.* (1975) **34**, 1653.
56. Shante, V. K. S., Varma, C. M., Bloch, A. N., *Phys. Rev. B* (1973) **8**, 4885.
57. Bloch, A. N., Varma, C. M., *J. Phys. C* (1973) **6**, 1849.
58. Hamilton, E. M., *Philos. Mag.* (1972) **26**, 1043.

59. Keller, H. J., Ed., "Low-Dimensional Cooperative Phenomena," Plenum, New York, 1975.
60. Thomas, T. W., Underhill, A. E., *Chem. Soc. Rev.* (1972) **1**, 99.
61. Zeller, H. R., Beck, A., *J. Phys. Chem. Solids* (1974) **35**, 77.
62. Thomas, T. W., Hsu, C. H., Labes, M. M., Gomm, P. S., Underhill, A. E., Watkins, D. M., *J. Chem. Soc.* (1972) 2050.
63. Ginsberg, A. P., Koepke, J. W., Hauser, J. J., West, K. W., Di Salvo, F. J., Sprinkle, C. R., Cohen, R. L., *Inorg. Chem.* (1976) **15**, 514.
64. Mott, N., "Metal-Insulator Transitions," Chaps. 4, 6, Taylor and Francis, London, 1974.
65. Kiefer, W., Bernstein, H. J., *Chem. Phys. Lett.* (1972) **16**, 5.
66. Kaya, K., Mikami, N., Udagawa, Y., Ito, M., *Chem. Phys. Lett.* (1972) **16**, 151.
67. Teo, B. K., unpublished data.
68. Schneider, A. A., Greatbatch, W., Mead, R., "Power Sources 5," D. H. Collins, Ed., p. 651, Academic, London, 1975.

RECEIVED April 10, 1978.

30

Templates in Zeolite Crystallization

LOUIS D. ROLLMANN

Central Research Division, Mobil Research and Development Corp., P.O. Box 1025, Princeton, NJ 08540

Crystallization of zeolites is often a nucleation-controlled process occurring from molecularly inhomogeneous, aqueous gels. The product is strongly dependent on the cation distribution in these mixtures. A methodology is developed for interpreting crystallization data and for identifying template effects by added quaternary ammonium cations. These techniques are then examined, by way of example, with a series of quaternary ammonium polymers. It is shown that such polymers can force crystallization of large-pore zeolites (where small-pore structures would otherwise result), and that they can preserve the integrity of a pore system during crystallization of fault-prone zeolite structures.

Zeolites are three-dimensional, crystalline networks of AlO_4 and SiO_4 tetrahedra, a unit negative charge being associated with each AlO_4 tetrahedron in the framework. Balancing that charge must be some intracrystalline cation—such as sodium, potassium, or a quaternary ammonium ion—so that the resultant zeolite phase has an idealized formula:

$$M_2O \cdot Al_2O_3 \cdot x\,SiO_2 \cdot y\,H_2O$$

wherein M represents a cation (mono-valent in this example). Synthesis of a zeolite structure is commonly effected by heating an alkaline, aluminosilicate "gel," and the structure obtained is strongly dependent on the cation distribution.

More than 130 natural and synthetic zeolites are already described in the literature (*1*), and several reviews have detailed the rapid progress that has been made in synthesis efforts (*2–7*). A correlation exists

Table I. Comparison of Hydrocarbon and Zeolite Pore Dimensions

Pore	Diameter	Examples
12-Ring	6–8 Å	Mordenite, Y, gmelinite
10-Ring	5–7 Å	Ferrierite, dachiardite
8-Ring	4–6 Å	Erionite, A
6-Ring	2–3 Å	Sodalite, sodalite cages in A and in Y

Hydrocarbon	Minimum Dimensions
n-Hexane	3.5×4.2 Å
Cyclohexane	4.8×6.4 Å
Benzene	2.2×5.6 Å

between the smaller cations, sodium, potassium, and possibly tetramethylammonium (TMA), and the polyhedral building units of several zeolite structures (5), suggesting that these cations template or direct crystallization of specific zeolite phases. In general however, the role of organic cations in directing crystallization is a topic open to debate (5).

Once formed, many zeolite frameworks possess a pore system accessible to hydrocarbons and a large, internal surface area which is potentially the locus of catalytic activity. Alternatively, the accessibility of a zeolite pore system to hydrocarbons of differing size can be a probe of the dimensions of those pores. Each channel in a zeolite is defined by a ring of SiO_4 and AlO_4 tetrahedra, and it is useful to group zeolites according to the number of tetrahedra which constitute that ring, for example, 6-, 8-, 10- or 12-ring zeolites. The last three groups comprise the most common catalyst components, and the diameters of various examples of each pore size are compared with minimum dimensions of several C_6 hydrocarbons in Table I.

In the following pages, the potential effects of organic cations in directing crystallization of specific zeolite structures and techniques for identifying such directing functions will be reviewed briefly. Then, by way of example, the influence of selected cationic polymers on zeolite crystallization will be described.

Template Effects

In general, addition of a quaternary ammonium cation to a reaction mixture can effect changes of three types: (a) a different zeolite structure is obtained; (b) a zeolite crystallizes where the reaction mixture would otherwise remain amorphous indefinitely; (c) the same zeolite is obtained as without quaternary, but it possesses an altered chemical composition. Unless crystallization was markedly accelerated, only (a) and (b) would represent "template effects" attributable to the quaternary cation, and (a) is by far the more common of the two effects.

Identification of such effects is made difficult, however, by the complexity of a crystallization experiment. Crystallization typically occurs from a molecularly inhomogeneous, aqueous gel, prepared by combination of a silica and an alumina source together with varying amounts of hydroxide ion. Since the product obtained is often nucleation-controlled, variables in an experiment can include not only the silica source (gel, sol, sodium silicate) and the alumina source (sodium aluminate, aluminum sulfate) but also the detailed mixing and crystallizing procedures (temperatures, aging, stirring rate, etc.).

A crystallization experiment is best described by mole ratios of reaction mixture components, as listed in Table II. Sodium silicate, for example, is treated as a mixture of NaOH and SiO_2, sodium aluminate as NaOH and Al_2O_3, and aluminum sulfate as Al_2O_3 and H_2SO_4. It is recognized in the calculation that Al_2O_3 is incorporated into a zeolite framework as AlO_2^-, i.e., each mole consumes two moles of hydroxide: $Al_2O_3 + 2OH^- \rightarrow 2AlO_2^- + H_2O$.

Most simply, the ratios in Table II can be divided by function. $SiO_2:Al_2O_3$ ratio defines a constraint on the overall framework composition of the product. Since many zeolite structures can be synthesized in only a limited range of composition, $SiO_2:Al_2O_3$ ratio can also restrict the number of zeolite phases possible from a given reaction mixture. (The actual number will be, of course, unknown since new phases continue to be discovered.) In most cases (the well-known Linde zeolite 4A being an exception), the $SiO_2:Al_2O_3$ ratio of a product zeolite will be lower than that of the reaction mixture, zeolites frequently incorporating all of the aluminum in a mixture and leaving varying amounts of silicate in solution. In the examples below, $SiO_2:Al_2O_3$ ratios of the reaction mixtures are in the range 10–100.

Formation of an ordered aluminosilicate framework requires hydrolysis and rearrangement of Si–O–Si (and Si–O–Al) bonds, the rate of that formation being dependent on hydroxide concentration (19). The interrelated values of $H_2O:SiO_2$ and $OH^-:SiO_2$ define both the molecular

Table II. Reaction Mixture Composition

Mole Ratio	Influence
$SiO_2:Al_2O_3$	Framework composition
$H_2O:SiO_2$	Viscosity ⎫ Hydroxide
$OH^-:SiO_2$	Silicate molec. wt. ⎬ concentration
$Na^+:SiO_2$	Cation distribution
$R_4N^+:SiO_2$	Templating / Framework Al content (20, 21)

weight distribution of the silicate species in a reaction mixture and the rate of change of that distribution. As the $OH^-:SiO_2$ ratio increases, more silicate remains in solution and lower $SiO_2:Al_2O_3$ products result. Thus the ratios $H_2O:SiO_2$ and $OH^-:SiO_2$ exert a second, important influence on the course of a crystallization and on the product obtained.

At high hydroxide levels, more dense, more thermodynamically stable phases result, either as a result of the facile redistribution possible with low molecular weight silicate species or as a result of product dissolution and recrystallization. For example, zeolite A (an eight-ring structure capable of sorbing normal paraffins) converts to zeolite P (thermally unstable, sorbing only molecules smaller than H_2O) in dilute sodium hydroxide and, in more concentrated solution, to the dense, six-ring sodalite structure (1). In the examples below, $H_2O:SiO_2$ and $OH^-:SiO_2$ have values in the ranges 10–50 and 0.1–2.0, respectively.

If templating is important, key ratios will define the cation distribution, $Na^+:SiO_2$, $K^+:SiO_2$, $R_4N^+:SiO_2$, and so forth, ratios which can range (individually) from zero to more than two. In Table III, examples are selected from the literature to illustrate cation effects in crystallization experiments. The Na-series examples show a shift in product from Y to mazzite-related structures (ZSM-4, Omega) on addition of only small amounts of TMA ion. When potassium replaces sodium, the L structure results but limited addition of TMA produces yet another structural type, the offretite-type zeolites.

Experiments such as those with TMA strongly suggest template effects, but they are not of themselves conclusive. As noted earlier, crystallization products are often determined by nucleation, the first species to nucleate being the product (12). Reaction mixtures are normally inhomogeneous, an inhomogenity which could be influenced by added quaternary ammonium cations. Additional evidence for templating by TMA in these examples is provided by elemental analysis.

The offretite structure is known (13). Offretite has a two-dimensional pore system, large 12-ring channels interconnected by an eight-ring network. It is in the detailed structure that evidence for templating

Table III. Product Dependence on Cation Distribution
$SiO_2:Al_2O_3 = 16$–20, $H_2O:SiO_2 = 14$–25 100°C, Static

Zeolite	$OH^-:SiO_2$	$Na^+:SiO_2$	$TMA^+:SiO_2$	$K^+:SiO_2$	Ref.
Y	0.8	0.8	0	0	1
ZSM-4	0.8	0.8	0.04	0	9, 10
Omega	0.7	0.6	0.14	0	1
L	0.8	0	0	0.8	1
TMA-O	0.9	0	0.09	0.8	1
TMA-Offretite	1.1	0.4	0.10	0.7	11

should be sought. A unit cell of offretite contains two aluminosilicate cages, an 11-hedron (called an ϵ- or cancrinite cage) accessible only via six rings and a 14-hedron restricted by eight rings (a γ- or gmelinite cage). When synthesized, TMA-O and TMA-offretite contain one TMA per unit cell that cannot be exchanged and that is located in each of the gmelinite cages (1, 13). Thus the TMA must have been inside the cage when it was formed. (It is interesting that these TMA samples also contain one non-exchangeable potassium ion per unit cell, located inside the cancrinite cage.)

Omega and mazzite are probably isostructural (14). Omega then would be a large-pore, 12-ring zeolite constructed of the same 14-hedron or gmelinite cages found in the offretite structure but with two cages per unit cell. When synthesized, omega typically contains 1.6 TMA ions per unit cell, almost one per gmelinite cage (1). Thus both structural examples from Table III show that TMA was incorporated into the zeolite framework as it was formed. It is the combination of this fact with the crystallization data in Table III that constitutes strong evidence for templating by TMA in these synthesis experiments.

Polymeric Templates

Until recently (8), quaternary ammonium polymers represented an unexplored special class of organic cations. With polymers, a growing crystal had to admit not just a single cation but a complete, linked chain of defined structure.

For the present example, polymers were synthesized by reaction of 1,4-diaza[2,2,2]bicyclooctane (dabco) with the compounds $Br-(CH_2)_n-Br$, where $n = 3, 4, 5, 6,$ and 10 (15, 16, 17). The compounds had a structure as follows,

$$\left[-N \bigcirc N^+ - (CH_2)_n - \right]_x$$

and were designated by the bromide used. With 1,4-dibromobutane, for example, the polymer was designated "Dab-4 Br." Molecular weights of the polymers, estimated from the concentration dependence of the kinematic viscosity in aqueous solution, were in the range 2000–15,000. The Dab-4 Br, for example, had an average molecular weight of about 10,000 (8).

A polymer must, by its very nature, be accommodated in the pores rather than in the small cages of a crystallizing zeolite structure. In the following results two zeolites will be important, gmelinite and mordenite, both large-pore, 12-ring structures as noted in Table I. Gmelinites, however, both natural and synthetic, exhibit the sorptive properties of small- rather than large-pore zeolites. Such behavior is attributed to chabazite stacking faults, faults that form in a plane perpendicular to and blocking access to the large gmelinite channels and that can be observed by x-ray diffraction. If a fault-free gmelinite could be prepared, then, substantial changes in sorptive properties should result from those found with current samples.

Mordenite is a second 12-ring zeolite whose sorptive properties do not always correspond to expectations based on structure. Two synthetic types of mordenite have been reported by Sand, "large-port" and "small-port" mordenite (18). The former, synthesized from only a narrow range of reaction mixture compositions, exhibits the sorptive properties expected of a 12-ring channel (having a diameter of 6–8 Å). The latter has an adsorption diameter of only about 4 Å. It is indistinguishable from large-port mordenite by x-ray diffraction and may co-exist with dense structures such as analcite.

Two series of experiments were conducted to explore the effects of cationic polymers in crystallizing mixtures; one at 90°C and the other at 180°C. The composition of the reaction mixtures was varied about a base case of $SiO_2:Al_2O_3 = 30$, $H_2O:SiO_2 = 20$, $OH^-:SiO_2 = 1.2$, and $Na^+:SiO_2 = 1.2$. Time was varied in the range of 3–13 days (or more) to ensure that the results obtained were not artifacts of a sequential crystallization process.

In the absence of polyelectrolyte, the low-temperature experiments produced the large-pore zeolite Y, contaminated with varying amounts of the thermally unstable phase P. That two zeolites, Y and P, were produced was expected since it is well known that both can form in the same overall composition field, their respective amounts depending on relative nucleation and growth rates (12). The significant point is that only these two zeolites were observed in these polymer-free, low-temperature experiments.

As shown in Table IV, even very small amounts of the Dab-4 Br polymer ($R_4N^+:SiO_2 = 0.01$) effected a complete change in the product. Gmelinite was produced (although x-ray diffraction showed it to be faulted by chabazite). As the amount of polymer was increased, the product shifted to pure gmelinite. Eventually crystallization was inhibited by the polymer, probably because the forming zeolite could accommodate organic cations only in amounts corresponding to its pore volume.

If templating is involved, one would expect the effectiveness of a polyelectrolyte to depend on its structure. The data in Table V were obtained at the composition optimum for Dab-4 Br in the synthesis of

Table IV. Crystallization Results at Low Temperature Static, 90°C

Polymer	$N^+:SiO_2$	Product
None	0	Y(+P)
Dab-4 Br	0.01	Gmelinite (faulted)
Dab-4 Br	0.14	Gmelinite (faulted)
Dab-4 Br	0.23	Pure gmelinite
Dab-4 Br	0.43	Amorphous

pure gmelinite, and three polymers produced gmelinite, Dab-4, -5, and -6. Neither Dab-3 or -10 nor the dipropyl and dibutyl monomers were effective.

Product analysis provides supporting evidence for the directing function of the polymer. The pure gmelinite samples sorbed 7.3% cyclohexane (25°C, $p:p_o = 0.67$) after calcination, compared with only 1.0% for a natural, faulted sample. The C:N ratio in the gmelinite samples averaged 5.4, compared with 5.1 in the Dab-4 Br used.

The structure of gmelinite is known and it is instructive to compare its unit cell dimensions with the length of a polymer repeating unit. A unit cell of gmelinite contains one large pore which is 10.0 Å long in the direction of that pore. From Courtauld models one can estimate the length of a polymer repeating unit as follows:

Dab-3	7.5 Å
-4	8.7
-5	9.9
-6	11.0
-10	14.5

Only those polymers that had repeating units 9–11 Å long were effective, an excellent correlation with the dimensions of the gmelinite unit cell. Furthermore, one then would expect two quaternary nitrogen atoms per unit cell. Elemental analysis averaged 2.3.

Table V. Effect of Polymer Structure on Crystallization at 90°C

Polymer	Product
Dab-3 Br	P
Dab-4 Br	Pure gmelinite
Dab-5 Br	Gmelinite (faulted)
Dab-6 Br	Gmelinite (faulted)
Dab-10 Br	Y + P
Dab-Pr$_2$	Y + P
Dab-Bu$_2$	Y + P

Table VI. Crystallization Results at High Temperature Static, 180°C

Polymer	$N^+:SiO_2$	Product
None	0	Analcite
Dab-4 Br	0.11	Analcite
Dab-4 Br	0.23	Analcite + Mordenite
Dab-4 Br	0.43	Pure mordenite

In Table VI are data showing that these results are not an artifact specific to the gmelinite structure. At high temperature the reaction mixture originally producing Y yielded a dense phase, analcite; addition of polymer shifted the product to a large-pore structure, mordenite.

That this mordenite was of the "large-port" variety was shown by cyclohexane sorption (6.3%, 25°C, $p:p_o = 0.67$). Correlation was again found between the unit cell dimension along a mordenite pore (7.5 Å) and the length of a polymer repeating unit (8.7 Å). Based on these dimensions one would expect somewhat less than two nitrogen atoms per unit cell and analysis shows 1.7, in excellent agreement. Thus a templating and directing function of these polymers can be shown for a range of temperatures, compositions, and resultant structures.

Acknowledgment

A survey paper in the area of zeolite crystallization can only be written from the legacy and the accomplishments of a large number of talented researchers. I am particularly indebted to G. T. Kerr, R. H. Daniels, E. W. Valyocsik, and D. H. Olson, who have directly contributed to these results, and to the many people in our Paulsboro laboratory whose work has provided the basis for summary.

Literature Cited

1. Breck, D. W., "Zeolite Molecular Sieves," John Wiley and Sons, New York, 1974.
2. Barrer, R. M., "Molecular Sieves," p. 39, Society of the Chemical Industry, London, 1968.
3. Zhdanov, S. P., Adv. Chem. Ser. (1971) **101**, 20.
4. Senderov, E. E., Khitarov, N. E., "Zeolites, Their Synthesis and Conditions of Formation in Nature," Nauka Publishing House, Moscow, 1970.
5. Flanigen, E. M., Adv. Chem. Ser. (1973) **121**, 119.
6. Schwochow, F., Puppe, L., Angew. Chem., Int. Ed. Engl. (1975) **14**, 620.
7. Robson, H. E., Prepr., Div. Pet. Chem., Am. Chem. Soc. (1977) **22(2)**, 500.
8. Daniels, R. H., Kerr, G. T., Rollmann, L. D., J. Am. Chem. Soc. (1978) **100**, 3097.
9. Ciric, J., French Patent **1,502,289** (1966).
10. Plank, C. J., Rosinski, E. J., Rubin, M. K., British Patent **1,117,568** (1968).

11. Jenkins, E. E., U.S. Patent **3,578,398** (1971).
12. Kerr, G. T., *J .Phys. Chem.* (1968) **72**, 1385.
13. Gard, J. A., Tait, J. M., *Acta Crystallogr.* (1972) **B28**, 825.
14. Rinaldi, R., Pluth, J. J., Smith, J. V., *Acta Crystallogr.* (1975) **B31**, 1603.
15. Rembaum, A., Baumgartner, W., Eisenberg, A., *J. Polym. Sci. B* (1968) **6**, 159.
16. Noguchi, H., Rembaum, A., *J. Polym. Sci. B* (1969) **7**, 383.
17. Salamone, J. C., Snider, B., *J. Polym. Sci. A-1* (1970) **8**, 3495.
18. Sand, L. B., "Molecular Sieves," p. 71, Society of the Chemical Industry, London, 1968.
19. Kerr, G. T., *J. Phys. Chem.* (1966) **70**, 1047.
20. Barrer, R. M., Denny, P. J., *J. Chem. Soc.* (1961) 971.
21. Kerr, G. T., *Inorg. Chem.* (1966) **5**, 1537.

RECEIVED February 2, 1978.

31

Reaction Schemes for Dinuclear Compounds Containing Metal–Metal Triple Bonds Illustrated by Recent Findings in the Chemistry of Molybdenum and Tungsten

MALCOLM H. CHISHOLM

Department of Chemistry, Princeton University, Princeton, NJ 08540

The following generalized reactions are proposed for compounds containing metal–metal triple bonds (M = Mo, W). (1) Carbene-like addition across the M–M bond; (2) oxidative addition with reduction in M–M bond order (three to two); (3) reductive elimination with an increase in M–M bond order (three to four); (4) reversible Lewis base association which may or may not change the bond order, depending upon the electronic configuration of the M_2 moiety; (5) metal–metal triple bond cleavage by a carbyne-like reagent; and (6) oligomerization of the $M \equiv M$ unit to form a cluster or polynuclear complex. These generalized reactions are discussed in the light of recent experimental observations in the reactivity patterns of $M_2(OR)_6$ and $Cp_2M_2(CO)_4$ compounds.

The ability of transition metals to form multiple bonds with themselves is well recognized, and over the past decade a number of such compounds have received detailed examination by a variety of spectroscopic and structural techniques. For a recent review of M–M quadruple bonds see Ref. 1. More recently, certain compounds containing M–M quadruple and triple bonds have been the subject of theoretical treatments (2). However, the reactivity patterns of these compounds remains to be explored; this should prove a rich and exciting new area of transition metal chemistry. It is possible that organometallic reaction schemes evolved (3) for mononuclear transition metal complexes can be applicable to dinuclear systems and, furthermore, that dinuclear compounds can provide building blocks for the much desired systematic syntheses of new polynuclear and cluster compounds (4, 5, 6). In this account, a number of general modes of reaction are proposed for compounds containing metal–metal triple bonds. These are then discussed in light of recent experimental observations.

The notation M≡M is used to represent any compound containing a homonuclear metal–metal triple bond in which the metal atoms are in very similar, if not equivalent, environments; they have the same number of valence shell electrons, the same coordination number, and the same formal oxidation state. Fitting these requirements are two classes of molybdenum and tungsten compounds. Class I are M_2X_6 and $M_2X_{6-n}Y_n$ compounds, where $Y = R(alkyl)$, NR_2, OR, O_2CNR_2, O_2COR, and halide (7). Class II are $Cp_2M_2(CO)_4$ compounds ($M = Cr$, Mo, and W) in which the metal atoms are formally in the $+1$ oxidation state and attain an 18-valence shell electronic configuration by the formation of the metal–metal triple bond (8, 9). In class I the metals are formally trivalent (M^{3+}) and, even after forming a metal–metal triple bond, do not attain an 18-valence shell electronic configuration. Both oxidation state and valence shell electronic configuration are expected to influence the reactivity of the metal–metal triple bond. The proposed reactions involve the symmetrical addition/elimination of substrate molecules to M≡M compounds; the products are considered to have equivalent metal atoms.

Reactions which might lead to an odd number of electrons in the products are not considered. This is not meant to imply that such reactions cannot occur nor to imply that odd electron intermediates are not involved in some of the proposed reactions. (Compounds containing M–M bonds of fractional order are well documented. E.g., as in (i) $K_3Mo_2(SO_4)_4 \cdot 3.5H_2O$ (10), (ii) $MoW(O_2CBu^{tert})_4(CH_3CN)(I)$ (11), (iii) $[Cp_2Co_2(CO)_2]^-$ (12).) However, thus far in our studies we have neither obtained as products nor detected as intermediates odd electron dinuclear species.

Addition of X: to M≡M in Reaction 1 represents a carbene-like addition to a triple bond. The moiety X: could indeed be a carbene or

$$M{\equiv}M + X: \rightleftharpoons M\overset{X}{\underset{}{\diagup\diagdown}}M \qquad (1)$$

an organic molecule capable of reacting with a metal-to-metal triple bond in this way, e.g., carbon monoxide or an iso-nitrile. X: could also be an inorganic/organometallic substrate such as $Fe(CO)_4$, Cp_2NbH, SnR_2, or a d^8 square planar transition metal complex. The requirement of X: is merely that it is capable of expanding its coordination number and oxidation state by two.

An interesting example of Reaction 1 has just been discovered in a study of the reaction between $Mo_2(OR)_6$ compounds and carbon monoxide (13). The compound $Mo_2(OBu^{tert})_6$ reacts reversibly with carbon monoxide in hydrocarbon solvents at room temperature and 1 atm to give a deep purple crystalline compound $Mo_2(OBu^{tert})_6CO$, $\nu(CO) = 1670$ cm^{-1}. The molecular structure deduced from x-ray studies (13) is shown in Figure 1. The molecule has virtual C_{2v} symmetry. The coordination polyhedron about each metal atom is approximately a square pyramid with the bridging carbonyl carbon at the common apex. The short metal–metal distance, 2.498(1) Å (cf. Mo—Mo = 2.222(1) Å in $Mo_2(OR)_6$), the diamagnetic nature of the compound, and electron counting require the existence of a metal–metal double bond.

Reactions 2 and 3 represent oxidative addition and reductive elimination sequences. Reactions 1, 2, and 3 all involve a reversible addition/elimination of a substrate molecule that contributes two electrons to a dinuclear center. However, since the proposed reactions proceed with a change in M–M bond order, the number of metal valence shell electrons

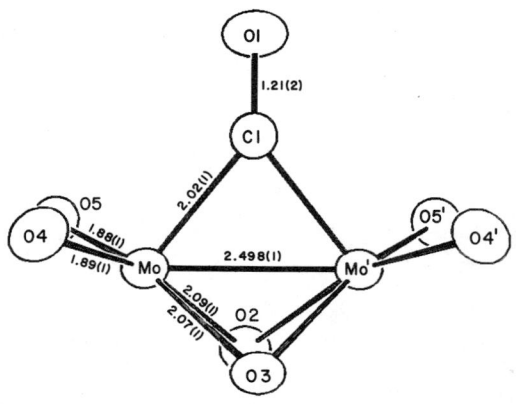

$Mo_2O_6(CO)$ Skeleton of

$Mo_2(O\text{-}t\text{-}Bu)_6(CO)$

Figure 1. *A view of the coordination geometry of $Mo_2(OBu^{tert})_6CO$ showing the main internuclear distances. Each atom is represented by its thermal ellipsoid of vibration, scaled to enclose 40% of the electron density. The tertiary butyl groups are omitted for clarity.*

$$M\equiv M + X-X \rightleftarrows X-M=M-X \quad (2)$$

$$X-M\equiv M-X \rightleftarrows M\equiv M + X-X \quad (3)$$

is not increased. This contrasts with the reactions of mononuclear transition metal complexes where the metals change their number of valence shell electrons by two.

At present there are no well-documented examples of Reactions 2 and 3 although several known compounds could serve as excellent models for these types of reactions. For example, the compounds $M_2Me_2(O_2CNR_2)_4$, when heated to > 150°C in vacuo, eliminate ethane and yield residues which, by elemental analyses, can be formulated as $M_2(O_2CNR_2)_4$ compounds. Both $Mo_2(CH_2SiMe_3)_6$ and $Mo_2(OPr^i)_6$ have been found to react with acetic acid to yield, upon vacuum sublimation (200°C, 10^{-4} cmHg), $Mo_2(OAc)_4$. Here a M–M triple-to-quadruple bond transformation is achieved, Reaction 3, but the detailed reaction pathway and the nature of the eliminated organic compounds are not known.

The simple oxidative addition of X–X across a M–M triple bond to yield an unbridged M–M double bond has yet to be structurally established although there are a number of reactions in which this might occur, e.g. (8), $Cp_2M_2(CO)_4 + I_2 \rightarrow Cp_2M_2(CO)_4I_2$. There are, however, known examples of where an X–X addition to a compound containing a multiple bond occurs with the formation of metal–ligand bridges. The reaction of $Mo_2(OPr^i)_6$ to give $Mo_2(OPr^i)_8$, which is discussed later, is representative of this type of M–M triple-to-double bond transformation since in the product, $Mo_2(OPr^i)_8$, there are bridging alkoxy ligands. The addition of X_2 (X = I or Br) to $Mo_2(S_2COEt)_4$, which contains a M–M quadruple bond, yields $Mo_2X_2(S_2COEt)_4$ compounds having Mo–Mo single bonds (Mo–Mo = 2.72 Å) as a result of a surprising rearrangement in the bonding mode of the xanthate ligand (14). Clearly the reactivity of compounds containing M–M multiple bonds towards oxidative addition/reductive elimination reactions is going to be as complex and even less predictable than analogous reactions involving mononuclear transition metal complexes (15, 16).

There are several examples of Lewis base association reactions of type 4.

$$M\equiv M + 2L: \rightleftarrows L-M\equiv M-L \quad (4)$$

Here four electrons are donated to the M_2 center with retention of the M–M triple bond. Lewis base association should be applicable only to metal-to-metal triple-bonded compounds in which the metal atoms have 16 or less valence shell electronic configurations. This is the case for

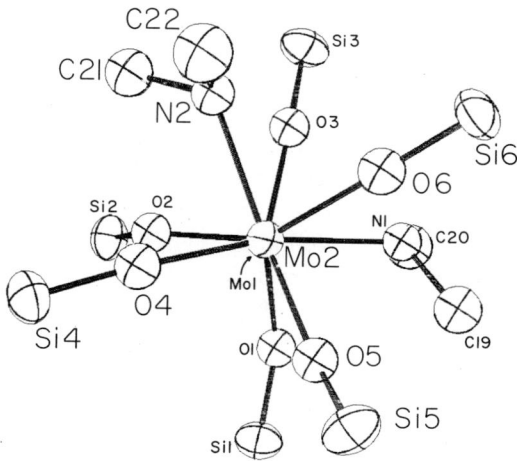

Figure 2. An ORTEP view of the $Mo_2(OSi)_6$-$(NC_2)_2$ portion of the $Mo_2(OSiMe_3)_6(HNMe_2)_2$ molecule looking directly down the Mo–Mo bond with Mo(1) eclipsed by Mo(2). Atoms labelled in smaller print are bonded to Mo(1). All atoms are represented by 50% probability ellipsoids. Some important interatomic distances and angles are: Mo–Mo = 2.242(1) Å, Mo–O(av) = 1.95 Å, Mo–N(av) = 2.28 Å, Mo–Mo–O(av) = 102°, Mo–Mo–N(av) = 95°.

$Mo_2(OR)_6$ compounds, and these react reversibly with amines to give adducts $Mo_2(OR)_6(amine)_2$ (17). A view of the central core of the $Mo_2(OSiMe_3)_6(HNMe_2)_6$ molecule is shown in Figure 2. The Mo–Mo distance is 2.242(1) Å.

Other examples in which metal atoms in M≡M compounds expand their coordination number and number of valence shell electrons are seen in the reactions of $Mo_2(OR)_6$, (18), $W_2Me_2(NEt_2)_4$ (19), and $W_2(NMe_2)_6$ (19) compounds with CO_2. The products $Mo_2(OR)_4(O_2COR)_2$, $W_2Me_2(O_2CNEt_2)_4$, and $W_2(O_2CNMe_2)_6$ provide examples of compounds containing metal–metal triple bonds between metal atoms that are coordinated to four, five, and six ligand atoms, respectively.

In contrast, Lewis base association to a M≡M compound in which the metal atoms have an 18-valence shell electronic configuration will proceed with reduction in M–M bond order as shown in Reaction 5. The reversible reaction between $Cp_2Mo_2(CO)_4$ (Mo–Mo = 2.40 Å) and CO which gives $Cp_2Mo_2(CO)_6$ (Mo–Mo = 3.27 Å) provides a good example of Reaction 5 (8).

$$M \equiv M + 2L: \rightleftarrows L\text{---}M\text{---}M\text{---}L \qquad (5)$$

The compounds $Cp_2M_2(CO)_4$ also have been found to be reactive towards a number of unsaturated molecules, un, giving simple addition products $Cp_2M_2(CO)_4(un)$. The compounds where M = Mo and un = $PhC \equiv CPh$, $EtC \equiv CEt$, $HC \equiv CH$ (20, 21), $CH_2=C=CH_2$ (22, 23), and Me_2NCN (24) have been structurally characterized. In all cases, the unsaturated organic molecule spans the Mo_2 bond (see Figure 3) which increases in length from 2.40 Å in $Cp_2Mo_2(CO)_4$ to 2.974, 3.015, and 3.117 Å where un = $HC \equiv CH$, Me_2NCN, and $CH_2=C=CH_2$, respectively. The organic molecules act as four electron donors to the M_2 group and can be considered as further examples of products formed in reactions of type 5. The compound $Cp_2Mo_2(CO)_4(allene)$ has C_2 symmetry and thus equivalent molybdenum atoms. However, the compounds $Cp_2Mo_2(CO)_4(RC_2R)$ and $Cp_2Mo_2(CO)_4(NCNMe_2)$ adopt structures in which the molybdenum atoms are inequivalent. In $Cp_2Mo_2(CO)_4(RC \equiv CR)$ compounds, the asymmetry is associated with the carbonyl bonding and presumably arises from internal crowding. In $Cp_2Mo_2(CO)_4(NCNMe_2)$,

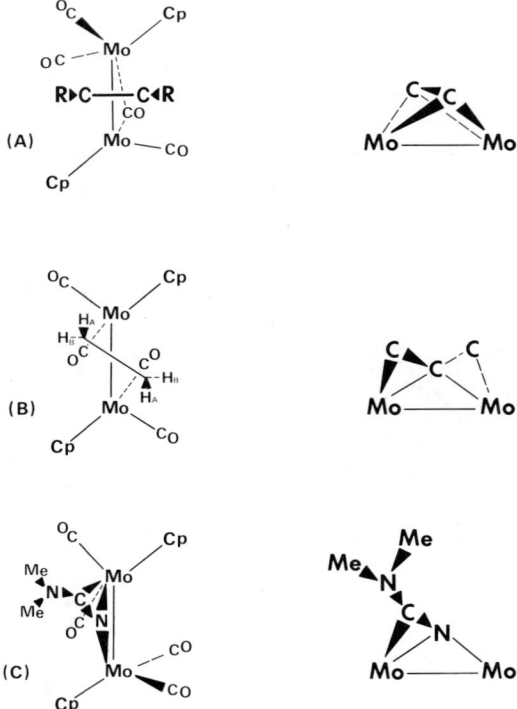

Figure 3. Representations for the molecular structures of $Cp_2Mo_2(CO)_4(un)$ compounds: (a) un=$RC \equiv CR$, (b) un=$CH_2=C=CH_2$, and (c) un=Me_2NCN

the bridging Me$_2$NCN group donates a nitrogen lone pair to one molybdenum atom and a CN π-electron pair to the other. ^{13}C NMR studies indicate that Cp$_2$Mo$_2$(CO)$_4$(PhC$_2$Ph) and Cp$_2$Mo$_2$(CO)$_4$(NCNMe$_2$) compounds adopt structures in solution akin to those found in the solid state and that low energy processes cause the two metal centers to become equivalent on the NMR time scale above $-$ 40°C.

In Reaction 6 the metal–metal triple bond is cleaved and replaced by a metal–ligand triple bond.

$$M{\equiv}M + 2X\colon \to 2M{\equiv}X \qquad (6)$$

There is therefore, no overall change in the number of metal valence shell electrons. Although not many substrates meet the requirement of being carbyne-like, the reactions between nitric oxide and a metal-to-metal triple bond can be viewed as examples of Reaction 6.

Cp$_2$Mo$_2$(CO)$_4$ compounds react readily with NO(2 equivalents) to give the mononuclear complexes CpMo(CO)$_2$(NO) (25). Similarly, Mo$_2$(OR)$_6$ compounds react with NO(2 equivalents) to give [Mo(OR)$_3$(NO)]$_2$ compounds (26). Here there is a pair of bridging alkoxide ligands, which

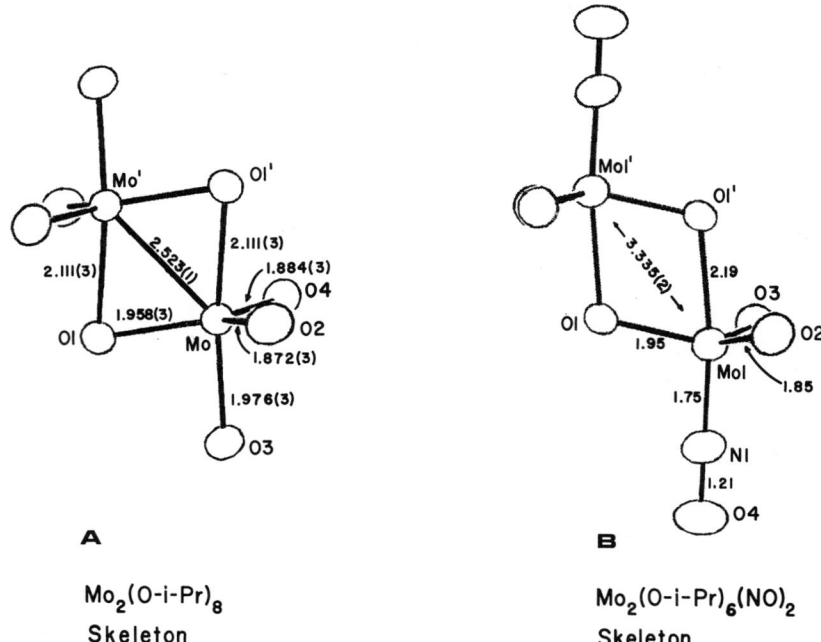

Figure 4. Coordination geometries of (A) Mo$_2$(OPr^i)$_8$ and (B) Mo$_2$(OPr^i)$_6$-(NO)$_2$, showing some pertinent bond distances. Distances shown for B are averaged over two independent molecules. In both A and B the molecules possess rigorous C$_i$ and virtual C$_{2h}$ symmetry.

leads to a 14-valence shell electronic configuration for molybdenum. The Mo–Mo distance is 3.325 Å, which precludes any direct metal–metal bond. The dimer can be cleaved by the addition of a donor ligand such as pyridine, and a mononuclear compound $W(OBu^{tert})_3(NO)(py)$ has recently been structurally characterized (27). The structure of [Mo-$(OPr^i)_3NO]_2$ is shown in Figure 4. In $W(OBu^{tert})_3(NO)(py)$ there is also a linear M–N–O moiety in an axial position of a trigonal bipyramid; the pyridine ligand is in the other axial position (27). The value of the NO stretching frequency, 1555 cm^{-1}, is low for a linear M–N–O group (for a recent review of metal–nitrosyl complexes see Ref. 28), which indicates very extensive W-to-NOπ* bonding and the significance of the resonance form M $\equiv$ N–O: (the two other resonance structures for a linear M–NO group are M̈ = N = Ö and N–N $\equiv$ O:).

A potential source of an X: substrate is, of course, an X $\equiv$ X type of molecule. Reaction 6 would then simply represent a metathesis reaction. Since transition metal carbyne complexes are well known, it is not inconceivable that Reaction 7 can occur. Alternatively an X $\equiv$ X or

$$M\equiv M + RC\equiv CR \rightleftarrows 2M\equiv CR \qquad (7)$$

2X: substrate could react to form a planar M_2X_2 moiety of the type shown below.

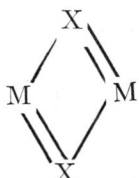

Compounds containing bridging carbyne ligands, e.g., $[Me_3SiCH_2)_2M$-$(\mu$-$CSiMe_3)]_2$ where M = Nb, Ta (29), and W (30), contain planar M_2C_2 moieties of this type.

Another mode of reaction for a three-electron donor substrate is shown in Reaction 18. This is closely related to the simple oxidative-addition reaction shown in Reaction 2 and can be expected to occur whenever the X moiety has one or more lone pairs of electrons, providing that the formation of the two M–X–M bridges does not require rupture

$$M\equiv M + 2X: \rightleftarrows \quad \begin{array}{c} X \\ M \diagdown M \\ X \end{array} \qquad (8)$$

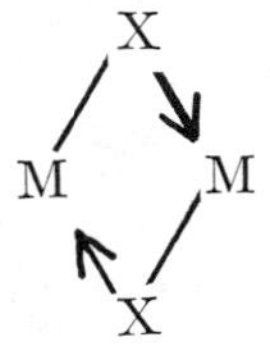

of the M–M double bond as in Reaction 5. Alternatively the addition of 2X · or X–X across a M–M triple bond can occur as indicated in Reaction 2, and one of the other ligands can then take up a bridging position, as seen in Ref. *14*.

The formation of $Mo_2(OPr^i)_8$ in the reaction between $Mo_2(OPr^i)_6$, $AgPF_6$(2 equivalents), and proton sponge (2 equivalents) in isopropyl alcohol can be viewed as an example of Reaction 8 (*31*). The structure of $Mo_2(OPr^i)_8$ is shown in Figure 4, where a simple comparison is made with the related compound $Mo_2(OPr^i)_6(NO)_2$ (*26*). In both compounds there is essentially trigonal bipyramidal coordination about each molybdenum atom, and there is a pair of Pr^iO bridging ligands which form alternately long (axial) and short (equatorial) Mo–O bonds. The most striking differences between the two structures are: (i) the Mo–Mo distances, which are 3.335(2) and 2.525(1) Å for $Mo_2(OPr^i)_6(NO)_2$ and $Mo_2(OPr^i)_8$, respectively; and (ii) the angles of the $Mo_2(\mu\text{-}O)_2$ moiety. These differences are readily accounted for by simple ligand field considerations. A trigonal bipyramidal field splits the metal d orbitals into three sets $e'(d_{x^2-y^2}, d_{xy})$, $e''(d_{xz}, d_{yz})$, and $a'(d_{z^2})$ with the degenerate pair, d_{xz} and d_{yz}, lying lowest in energy. In $Mo_2(OPr^i)_6(NO)_2$, each molybdenum atom can be assumed, formally, to have four $4d$ electrons after the formation of σ-bonds to each of the five ligands. This form of electron counting uses the conventional, though purely formal, description of the linear Mo–N–O group as $M^- \leftarrow NO^+$. These four electrons then occupy the $e''(d_{xz}, d_{yz})$ orbitals where they can very effectively participate in Mo–NO π^* back bonding, thus explaining the very low value (1632 cm^{-1}) of $\nu(NO)$ in $Mo_2(OPr^i)_6(NO)_2$. The bonding in the dimeric compound $Cr_2(OPr^i)_6(NO)_2(\nu(NO) = 1720$ cm$^{-1})$ and the mononuclear compound $W(OBu^{tert})_3(NO)(pyridine)(\nu(NO) = 1555$ cm$^{-1})$ must be essentially the same. In all of these compounds there is extensive metal e''-to-NO π^* bonding which, based on the values of $\nu(NO)$, follows the order $W > Mo > Cr$.

In the compound $Mo_2(OPr^i)_8$, the formal oxidation state of molybdenum is $+4$, and each molybdenum atom has two $4d$ electrons. It is thus possible to envision the formation of the metal–metal double bond as the result of d_{xz}–d_{xz} and d_{yz}–d_{yz} interactions. It should be noted that the compounds $M_2(OR)_6(NO)_2$, $M(OR)_3(NO)L$, $Mo_2(OPr^i)_8$, and $Mo_2(OBu^{tert})_6CO$ provide a new class of group VI transition metal complexes in which the metal atoms are five coordinate having 14 valence shell electronic configurations.

The factors which lead to the formation of dinuclear compounds containing M–M bonds of multiple order n rather than to the formation of polynuclear or cluster compounds in which the metal atoms form n σ-bonds with each other are not well understood. The size of the ligands

$$2M\equiv M \rightleftarrows M_4 \quad\quad (9)$$

is one important factor, and, in principle, a reversible association reaction, represented by Reaction 9 above, is to be expected for certain metal–ligand combinations. The possible geometries for the M_4 moiety are many and include tetrahedral, square planar, and open chain structures. We are not presently in a position to make predictions concerning the preferred geometries of M_4 compounds formed in Reaction 9, but we do note that this type of oligomerization is found in the chemistry of tri-valent molybdenum and tungsten alkoxides. For molybdenum, the neo-pentoxide exists in both dinuclear and polynuclear forms (29). The ethoxide is tetrameric and diamagnetic in benzene and shows $Mo_4(OEt)_{12}{}^+$, $Mo_3(OEt)_9{}^+$, and $Mo_2(OEt)_6{}^+$ ions in the mass spectrometer (32). For tungsten, only the very bulky tri-ethylsiloxy and tertiary butoxy ligands give dinuclear compounds. The less bulky iso-propoxy and neo-pentoxy groups give tetranuclear complexes. A black crystalline tetranuclear compound $W_4(OPr^i)_{12}(HOPr^i)_2$ has been structurally characterized (see Figure 5) and is believed to have one of the two Pr^iOH ligands coordinated at each terminal tungsten (33). Formation of $W_4(OPr^i)_{12}(HOPr^i)_2$ can be viewed as the first step in a polymerization of $W_2(OPr^i)_6$ (an M≡M compound), which is halted in this instance by the coordination of the Pr^iOH ligands.

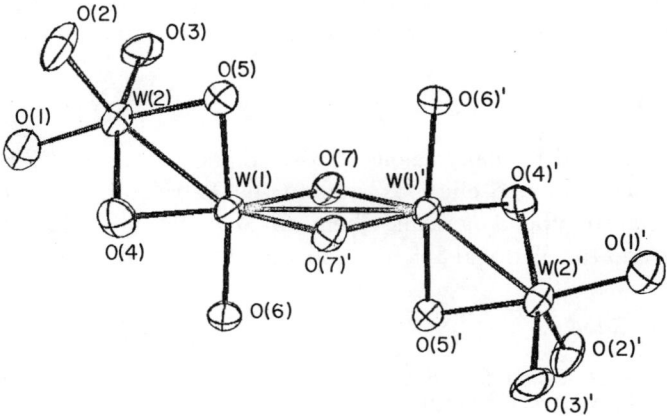

Figure 5. An ORTEP view of the $W_4(OPr^i)_{12}$ $(HOPr^i)_2$ molecule showing only the W_4O_{14} skeleton. The molecule has C_i symmetry. Some important parameters are: W(1)–W(2) = 2.46 Å; W(1)–W(1)' = 3.30 Å; W(2)–W(1)–W(1)' angle = 140°.

Conclusions

(1) The reactions of symmetrical M ≡ M compounds with symmetrical substrates are expected to yield products in which the metal atoms are in equivalent environments. Whenever an exception is found, low energy pathways will readily interconvert the two ends of the dinuclear compound.

(2) Reactions leading to stepwise changes in M–M bond order are possible and may or may not be accompanied by a formal valence change of the metal atoms. Predictions with regard to M–M bond order changes presently are not possible because of uncertainties regarding metal–ligand rearrangements.

(3) The potential for carrying out dinuclear hydrocarbon catalysis should be recognized. One catalytic sequence leading to selective hydrogenation is already suggested by the ability of $Cp_2M_2(CO)_4$ compounds to coordinate unsaturated molecules that are four- but not two-electron donors:

$$Cp_2M_2(CO)_4 + un \rightarrow Cp_2M_2(CO)_4(un)$$
$$Cp_2M_2(CO)_4(un) + H_2 \rightarrow Cp_2M_2(CO)_4 + un\, H_2$$

(4) The general reactions proposed herein for M ≡ M compounds with symmetrical substrates are not exhaustive but merely pertinent to some recent experimental observations. The reaction schemes involving unsymmetrical substrates and heteronuclear M–M^1 multiple bonded compounds are virtually unlimited, all of which indicates the growth potential of this area of transition metal chemistry.

Acknowledgment

I am grateful to many talented co-workers, whose names are to be found in the reference citations, and to the Petroleum Research Fund administered by the American Chemical Society, the Office of Naval Research, and the National Science Foundation for their financial support.

Literature Cited

1. Cotton, F. A., *Chem. Soc. Rev.* (1975) **4**, 27.
2. Cotton, F. A., *Acc. Chem. Res.* (1978) **11**, 225.
3. Tolman, C. A., *Chem. Soc. Rev.* (1972) **1**, 337.
4. Muetterties, E. L., *Bull. Soc. Chim. Belg.* (1975) **84**, 959.
5. Muetterties, E. L., *Science* (1977) **196**, 839.
6. Ugo, R., *Catal. Rev.* (1975) **11**, 225.
7. Chisholm, M. H., Cotton, F. A., *Acc. Chem. Res.* (1978) **10**, 000.
8. Klinger, R. J., Butler, W., Curtis, M. D., *J. Am. Chem. Soc.* (1975) **97**, 3535.

9. Ginley, D. S., Bock, C. R., Wrighton, M. S., *Inorg. Chim. Acta* (1977) **23**, 85.
10. Cotton, F. A., Frenz, B. A., Pederson, E., Webb, T. R., *Inorg. Chem.* (1975) **14**, 391.
11. Katoric, V., Templeton, J. J., Hoxmeier, R. J., McCarley, R. E., *J. Am. Chem. Soc.* (1975) **97**, 5300.
12. Bergman, R. G., Ilendin, C. S., Schore, N. E., *J. Am. Chem. Soc.* (1976) **98**, 255, 256.
13. Chisholm, M. H., Cotton, F. A., Extine, M. W., Kelly, R. L., *J. Am. Chem. Soc.* (1978) **100**, 2256.
14. Cotton, F. A., Extine, M. W., Niswander, R. H., *Inorg. Chem.* (1978) **17**, 692.
15. Halpern, J., *Acc. Chem. Res.* (1970) **3**, 386.
16. Collman, J. P., *Acc. Chem. Res.* (1968) **1**, 136.
17. Chisholm, M. H., Cotton, F. A., Extine, M. W., Reichert, W. W., *J. Am. Chem. Soc.* (1978) **100**, 153.
18. Chisholm, M. H., Cotton, F. A., Extine, M. W., Reichert, W. W., *J. Am. Chem. Soc.* (1978) **100**, 1727.
19. Chisholm, M. H., Cotton, F. A., Stults, B. R., *Inorg. Chem.* (1977) **16**, 603.
20. Bailey, W. I., Cotton, F. A., Jameson, J. D., Kolb, J. R., *J. Organomet. Chem.* (1976) **121**, C23.
21. Bailey, W. I., Chisholm, M. H., Cotton, F. A., Rankel, L. A., *J. Am. Chem. Soc.* (1978) **100**, 000.
22. Bailey, W. I., Chisholm, M. H., Cotton, F. A., Murillo, C. A., Rankel, L. A., *J. Am. Chem. Soc.* (1977) **99**, 1261.
23. *Ibid.* (1978) **100**, 802.
24. Chisholm, M. H., Cotton, F. A., Extine, M. W., Rankel, L. A., *J. Am. Chem. Soc.* (1978) **100**, 807.
25. King, R. B., Efraty, A., Douglas, W. M., *J. Organomet. Chem.* (1973) **60**, 125.
26. Chisholm, M. H., Cotton, F. A., Extine, M. W., Kelly, R. L., *J. Am. Chem. Soc.* (1978) **100**, 3354.
27. Chisholm, M. H., Cotton, F. A., Extine, M. W., Kelly, R. L., results to be published.
28. Eisenberg, R., Meyer, C. D., *Acc. Chem. Res.* (1975) **8**, 26.
29. Huq, F., Mowat, W., Skapski, A. C., Wilkinson, G., *J. Chem. Soc., Chem. Commun.* (1971) 1477.
30. Chisholm, M. H., Cotton, F. A., Extine, M. W., Murillo, C. A., *Inorg. Chem.* (1978) **17**, 696.
31. Chisholm, M. H., Kelly, R. L., results to be published.
32. Chisholm, M. H., Cotton, F. A., Murillo, C. A., Reichert, W. W., *Inorg. Chem.* (1977) **16**, 1801.
33. Akiyama, M., Chisholm, M. H., Cotton, F. A., Extine, M. W., results to be published.

RECEIVED February 22, 1978.

INDEX

INDEX

A

Acetylenes, oxidative coupling of . 179
Acetonitrile and
    $Fe(DMPE)_2D(C_6D_5)$ 74f
Acetonitrile–2,6-lutidine mixture .. 247f
Acetophenone and
    $Fe(DMPE)_2D(C_6D_5)$ 79f
Acid esters 169f
Acyl complex, mechanism for
    formation of 154f
Alkaline earth oxides, thermally
    activated as catalysts 144
Alkylbenzenes 40
Alkylcyclohexanes 40
Alkynes and nickel(0) complexes . 196
Alloy–hydrogen systems 281
Alloy interstitial sites 291t
Aluminum, effect on thermodynamics and structure of
    $AlNi_{5-x}Al_x$ hydrides 297
Amide ligands 180, 189, 190t
Amidocinnamic acids 22
Amine
    catalysts 26, 28t
    ligands 180, 187, 188t
    reduction of aromatic nitro
        compounds to 126
Aniline as reaction product 27
Arene hydrogenation 31, 39, 40
Aromatic nitro compounds 26
    hydrogenation of 28t
    reduction to amines 126, 128t
Asymmetric
    catalysis 51
    hydrosilylation 50, 57
    synthesis vs. resolution 51f
Autoxidation rate 175

B

Back-electron-transfer reactions ... 238
Benzophenones, reduction of
    prochiral 56
Bicyclo[1.1.0] butanes 317
Bicyclo[2.1.0] pentane 310
Bidentate ligand WL_4 complexes . 252
Bidentate-phosphorus ligand
    complexes of iron 67
Binuclear
    compounds, metal–metal
        triple bonds 396
    metal complexes 162

Binuclear (*continued*)
    rhodium(I) isocyanide complexes 225
    photochemistry 232
    structural and spectroscopic
        properties 226
    thermal reactions 230
Bioinorganic models 162
Biphenylenes, metallocyclopentadienes from 208
BIPY(COD)Ni 203, 208
1,8-Bis-homocubane system 315
    metal catalysis 316f

C

$Cp_2Mo_2(CO)_4(un)$ compounds,
    molecular structures of 401f
$Cr(CO)_6$, isotopic species
    with $C^{18}O$ 111t
$Cr(CO)_6$, oxygen exchange
    reactions of 109, 110f
CuCl and NBD in ethanol 334f
$Cu[HB(pz)_3]CO$, structure of ... 328f
$[CuH(PPh_3)]_6$, structure of 341f
"CuO" iniator species,
    pyridine-stabilized 181
$Cu(PPh_3)_2BH_4$, structure of 336f
$Cu_2(SR)_2$ core 362t
$Cu_4X_4L_4$ cluster compounds 327f
Carbene, regioselectivity 317
Carbon
  –carbon σ bond and a transition
    metal 314f
  –hydrogen bond activation 67
    aromatic 75
    bidentate–phosphorus ligand
        complexes of iron 67
    kinetics and thermodynamics . 78
    nonaromatic sp^2 77
    sp 75
    sp^3 71
  –hydrogen bond formation 52f
  monoxide
    catalytic reduction using 121
    electron transfer to 143
    –metal oxide interactions ... 140
    reduction to hydrocarbons ... 131
    surface structures 144
  –oxygen bonds 199
Carbonyl ligands, activation
    barriers 273t
Carbonyl nickel(II)
    complexes 153, 154t, 156t

411

Carbonylation reactions 152
Catalysis
  in acidic solutions 89
  in alkaline solutions 82
  in amine solutions 90, 91t
  asymmetric 51
  binuclear hydrocarbon 406
  iridium complexes 38, 40
    poisons to 6
  possible mechanisms for 84
  by ruthenium carbonyl 82
  tri-μ-hydrido complex 38
  of the water gas shift
    reaction, homogeneous 81, 94
Catalyst(s)
  in alkaline solutions 81, 86
  amine 28t
  carbon monoxide and water 121
  cobalt(II) tetraphenylporphyrin,
    supported 345, 346t, 348f
  group VIII metals 32
  heterogeneous 11, 12t, 82
  homogeneous 43, 351
    oxidative coupling 178
  hydrogenation 16, 19, 21f
  iridium, catalytically inactivated 38
  iron oxides 128t
  for isomerization of
    quadricyclane 344
  K$_3$[Co(CN)$_5$H] 43, 45–46t
  metal 131
    carbonyls 81
    dithiolenes 351, 354t, 355
  mono-μ-hydrides 39
  nickel(0) 307
  nickel(II) phosphine complexes 152
  olefin hydrogenation 37
  palladium chloride, supported .. 349
  pentamethylcyclopentadienyl–
    rhodium–iridium complexes 31
  phase transfer 106
  phosphine 28t
  in photochemical energy storage 344
  polymeric 12t
  for polyunsaturated olefins 2
  pyridine 179
  rhodium 52
    –phosphine complex 26, 27t
  ruthenium carbonyl 82
  -substrate intermediates 22
  supported chiral 60
  thermally activated alkaline
    earth oxides 144
  transition metals 43
  triphenylcyclopropenylnickel
    halide 351, 352t, 353f
  for the water gas shift reaction . 86
Catalytic
  activity 6
  reductions, carbon monoxide
    and water 121
  reductions with Fe(CO)$_5$ 128t
  system derived from Fe(CO)$_5$. 96

Cation distribution, product
  dependence on 390
Cation radicals of Mg–Mg
  diporphyrin 172
  EPR spectra of 172f
Chain length, carbon
  hydrogenation 10
Charge-transfer bands, WL$_4$
  low-energy 259t, 260t
Charge-transfer transitions,
  low-energy 258
Chelate
  diphosphine ligands 16
  hydrogenation 20
  rings 256f
  tungsten IV 252
Chemical heat pump, two
  metal hydride 294f
Chiral
  catalysts 60
  chelating 53
  ligands 53
  phosphines 53
Chlorophyll a dihydrate polycrystals, platinized ... 210, 212–213f
Chromophore, type 3 copper 358
Cleavage of allyl ethers 208
Cleavage of nickel(0) complexes,
  novel 195
Coal liquefaction 149
Cobalt(II) tetraphenylporphyrin
  catalysts, supported 345
  polystyrene-anchored 346t
  recycling study 348f, 349t
Cofacial diporphyrins 162, 166–167t
Coordination chemistry, cationic
  rhodium–phosphine complexes 16
Coordination geometries 402f
Copper
  complexes, bridged 360t
  chromophore, type 3 358
    structural constraint 359
  metalloproteins 358
  in redox enzymes, entatic
    state for 358
Copper(I)
  chemistry ... 326, 327–328f, 336f, 341f
  chloride, stoichiometry
    with oxygen 180
  compounds 325
    excited state of 329
    spectral characteristics ... 328t, 334f
  oxidation, aprotic 178
  sensitization processes,
    classification of 338, 339t
  -sensitized olefin photoreactions 330
Copper(II)
  complexes 182
    direct synthesis 183
    ligand exchange 183
    in situ generation 184
    ligand/solvent combinations . 184
Crystallographic data,
  LaNi$_{5-x}$Al$_x$ 289t, 293t

INDEX
413

Crystallization
  effects of polymer structure on . 393t
  results at low temperature 393t
  results at high temperature 394t
  templates in zeolite 387
    cationic polymers 388
Cycloaddition, nickel(0)-catalyzed 310
Cycloaddition, stereochemistry ... 313
Cyclohexane hydrogenation,
    kinetics 37
Cyclooctadiene 4
    hydrogenation 6f
1,5-Cyclooctadiene, catalytic
    hydrogenation of 11f
1,5-Cyclooctadiene, catalytic
    isomerization of 10f
Cyclopentadiene chromium
    tricarbonyl 4

D

Desulfurization 195
  of sulfur heterocycles 203
Desulfurizing process 208
Dicarbonyl bis(N,N-dimethyldi-
    thiocarbamato)tungsten(II) . 265
Dichalocolene ligands 365f
o-Dichlorobenzene 184
1,1-Dicyanocyclopropane 323f
Dienes 11, 12t, 34
  hydrogenation to monenes . 44, 45–46t
Diffusion pyrolysis apparatus 219f
Diffusion pyrolysis, quantitative
    determination by 219
Dihydrosilanes 54
Dimer, exciton splitting in a 170f
Dimerization, reductive 196
2,3-Dimethyl-1,3-butadiene 182f
1,1-Dimethylcyclopropane 68
2,6-Dimethylphenol, oxidative
    coupling of 6
Dinuclear compounds, reaction
    schemes for 396
Diolefins 6
DIOP, asymmetric hydrosilylation . 54
Direct synthesis, copper(II)
    complexes 183
Dithionaphthalene, oxidation
    potential of 368f, 369f
DMPE 68
Dodecahedral d^2 complexes,
    low-spin eight-coordinate 254
Double bonds, terminal 6
Dyes 236

E

Et$_4$N$^+$ μ-H[Cr(CO)$_5$]$_2^-$, reaction
    with CO 114t
Et$_4$N$^+$ μ-H[M(CO)$_5$]$^-$ species,
    electronic spectra of 117f
Electrochemistry of metal
    tetrathiolenes 364, 375

Electrodes 302
Electron-deficient olefins,
    trapping with 317
Electron transfer 140
  to CO from MgO and ThO$_2$ ... 143
  complexes, copper 362
  complexes, potential 252
  for conversion and storage 236
  excited state 239
  reactions of metal complexes,
    light-induced 236
  thermal activation 143
  transition state 362
Energy levels of a strained molecule
    and a transition metal 314f
Entatic state for copper in
    redox enzymes 358
Enthalpy of LaNi$_{5-x}$Al$_x$
    hydrides 279, 281
Entropy of LaNi$_{5-x}$Al$_x$ hydrides .. 279
Enyl complexes 34
EPR spectra
  of CO-derived radical 148f
  cation radicals 172f
  of Cu-Cu diporphyrin 167, 168f
  oxygen-containing samples of
    diporphyrins 174f
ESR signal 144
ESR spectra 191f
Ester, soybean methyl 1
Excited-state electron transfer 239
Exciton interaction 167, 170f
Exciton splitting in a dimer 170f

F

Fe(CO)$_5$, catalytic reductions with 128t
Fe(CO)$_5$-catalyzed water gas
    shift reaction 96–97t
Fe(DMPE)$_2$, adducts to 76f
Fe(DMPE)$_2$Cl$_2$, reduction of 68
Fe(DPPE)$_2$ 67
Fischer–Tropsch process 106, 121
Five-coordinate, zero-valent
    species formation 69
Fluxional process 275

G

Gaseous evolution of molecular
    hydrogen and oxygen 69
  diffusion pyrolysis 219–220f
  mass spectrometric
    analysis 214, 215–218f
Gel-permeation resins 184
Group VI metal carbonyls 101t

H

μ-H[Cr(CO)$_5$]$_2^-$, ligand substi-
    tution reactions of 113
HFe(CO)$_4^-$ as source of electrons 129
HRh$_2$Cl^{2+}, spectra of 231f, 233f

High pressure spectroscopy 102
^{2}H-NMR of *trans*-2-pentene 48
HOMO 167f, 315, 322–323f
Homogeneous
  catalysis, quadricyclane
    to norbornadiene 351
  catalysis, water gas shift reactions   81
  catalysts 26, 32, 43, 81, 178, 351
  deoxygenation 195
Hydrides, metal 34
Hydrides, stoichimetric reactions .. 36
Hydrocarbons 2
Hydrocarbons from CO 131
Hydrodehalogenation 30
Hydrodesulfurization, deoxygenation of opoxides 203
Hydrodesulfurization, of sulfur
  heterocycles 203
Hydroformylation reaction
  catalysts 124
  of propane 126t
  Reppe modification of the 122
Hydrogen
  dissociation pressures 280
  gaseous evolution of molecular 210, 214
  production rate 98–99f
  reaction of
    [Rh(DIPHOS)(NOR]$^+$ with   17
  transfer reactions 34
Hydrogenation
  of arenes 31, 39
  aromatic nitro compounds .. 26, 27–28t
  carbon chain length 10
  catalysts
    of aromatic nitro compounds . 26
    asymmetric 16, 19, 21f, 26
    homogeneous 26, 27–28t
    for olefins 37
    of prochiral olefins 22
  catalytic homogeneous 43
  catalytic, of monoenes 10t
  cyclohexane 37
  dienes 44, 45–46t
  double bonds 1
  ethylenic bonds 5f
  of metal carbene complexes ... 137
  micellar conditions 43
  of olefins 31, 37, 39, 40
  pentamethylcyclophenadienyl–
    rhodium and –iridium
    complexes 31
  phase transfer conditions 43
  polyolefins 2
  of propylene 4t
  rate 48
  reactions 8
  selective 406
    of polyunsaturated olefins ... 1
  solvent 6t
  soybean methyl ester 1, 12t
  of 2,3,3-trimethyl-1,4-pentene .. 9
  using excess substrate 46t
Hydrogenolysis of strained
  molecules, catalytic 321

Hydrophobic ruthenium(II)
  complex 240
Hydrosilylation 50
  of acetophenone 54, 56f
  heterogeneous 63
  in homogeneous media 64
  of ketones 50, 57t, 58f
  mechanism 57, 58t
  of Ph–CO–CH$_3$ 55t
  by supported rhodium catalyst .. 64

I

Induction periods 181, 183f
Initiator polymerization 181, 189
Iron, bidentate phosphorus ligand
  complexes of 67
Iron–thiazine photogalvanic cells . 296
  charge carriers 300
  decay of 300
  diffusion of 302
  formation of 302
  electrode transfer 302
  light absorption 299
  processes 298
Iron–thionine photogalvanic cell .. 304f
Iron–thionine photoredox reaction
  in acid solution 297
Irradiation
  quantum efficiencies 249
  spectral changes 247f
Isomerization of quadricyclane,
  catalysts for 344

K

K$_3$[Co(CN)$_5$H] 43
  -catalyzed hydrogenation 45–46t
Ketones, hydrosilylation 54, 58f

L

LaNi$_4$Al, model of 290f
LaNi$_{5-x}$Al$_x$ alloys, hysteresis in ... 288t
LaNi$_{5-x}$Al$_x$ hydrides, thermodynamic and structural
  properties 279
LaNi$_{4.5}$Al$_{0.5}$, desorption
  isotherms for 285f
LaNi$_{4.5}$Al$_{0.5}$, pressure–composition data for 286t
LaNi$_{4.6}$Al$_{0.4}$, desorption
  isotherms for 284f
LaNi$_{4.6}$Al$_{0.4}$, pressure–composition data for 283t
Lactam ligands 180, 189, 190t
Ligand(s)
  abbreviations 253t
  alkoxide 402
  amine 180, 187, 188t
  binuclear metal 162
  carbonyl 273
    activation barriers 273t

INDEX 415

Ligand(s) (continued)
  carbyne 403
  copper complexes 360t
  dichalcolene 365f
  exchange 187
    copper(II) complexes 183
    initiator-stabilizing 180
    isomerization 308
    liquid 183
    organochalcogen 366
    oxidation 185
    /solvent systems 183
  substitution
    in Cr(CO)₆ and
      μ-H[Cr(CO)₅]₂⁻ 106, 113
    rate 113, 114t, 115f
    tetrathiolene 365f, 371f
  triorgano-phosphine and
    -phosphite 31
Light, absorption of 299
Low-energy charge-transfer
  transitions 258
LUMO 315, 322–323f

M

M(CO)₆, catalytic systems
  derived from 102
Me₄Si, chemical shifts relative to.. 276t
MgO, paramagnetic centers
  generated on 141f
Mo₂(OBu^{tert})₆CO, coordination
  geometry of 398f
Mo₂(OSiMe₃)₆(HNMe₂)₂
  molecule 400f
Metal
  carbene complexes, hydrogena-
    tion of 137
  carbonyls 81
    catalytic systems 102
    group VI 101t
    hexacarbonyls 94
  complexes
    of cofacial diporphyrins,
      binuclear 162
    for energy conversion and
      storage 236
    formyl 131
      isolation of 132
      kinetic stability of 132
      and metal hydrides,
        equilibrium between .. 132
      synthesis and NMR
        observation of 132
  hexacarbonyls 94
  iridium 31
  iron 67
  light-induced electron-
    transfer reactions of 236
  rhodium 16, 31
  rhodium–phosphine 26
  transition 60, 344
  tungsten 263
  dithiolenes as catalysts 354–355t

Metal (continued)
  hydrides 34
  –metal triple bonds 396
  oxide–carbon monoxide inter-
    actions 140
  oxides, relationship to catalysis,
    coal liquefaction, and
    electron movement 149
  tetrathiolene(s) 364, 366f
    clusters, molecular 372
Metallocyclopentadienes from
  biphenylenes 208
Metalloproteins, copper 358
Z-Methyl-α-acetamidocinnamate .. 23f
2-Methyl-1,3-butadiene 45t
Methylene chloride 184
2,6-Methylphenol 185
Micelles 44
  excited-state electron transfer in 239
MO energy levels 116
MO theory 313, 314f
Monodentate tertiary
  phosphine:CO fixation 152
Monenes 12t
  catalytic hydrogenation of 10t
  diene hydrogenation to 44
Molecular metal tetrathiolene
  clusters 372
Molecular orbital control, metal
  catalysis 315
Molybdenum, metal–metal triple
  bonds 396
Mono-μ-hydrides 39
Multielectron reduction of
  molecular oxygen 173
Multielectron transfer 173f

N

NiI₂(CO)(PMe₃)₂, x-ray
  structure of 160
Naphthalene, oxidation potential
  of 368f, 369f
Naphthalene systems,
  1,8-substituted 365f
Natural oils 2
Nickel(0)
  -CH₂ complex 320f
  into carbon-oxygen bonds 199
  -catalyzed cycloaddition 310
  -catalyzed reactions of strained
    ring systems 307
  -catalyzed valence isomerization. 307
  complexes with alkynes 196
Nickel(II) complexes
  carbonyl 156–157t
  -d^8 152
  molecular and cationic
    carbonyl 154–155t
Nitro compounds, hydrogenation
  of aromatic 26
Nitrobenzene, hydrogenation of ..28–29t
Nitrobenzene, reduction of 128t
Nonspecific oxidative processes ... 183

Norbornadiene, catalytic trapping
of organometallic intermediate 307
Norbornadiene, photosensitized
conversion to quadricyclane ... 344
Novel cleavage of nickel(0)
complexes 195

O

Octahedral d^6 complexes 254
Olefins
hydrogenation 31
catalysts 37
rates 39, 40
photoreactions by Cu(I)
compounds 325
polyunsaturated, hydrogenation . 1
Oligomerization reactions of
nickel(0) complexes 173, 195
Optical yield, influence of
silane structure 54
Organochalcogen ligands 366
Organometallic intermediate 307
Organometallic polymers,
semiconducting 378
Orgel's rule 252
Oxidation 183
aprotic 178
of $C_{10}H_{8-n}S_n$ 367, 368–369f
of Cu(I) 178
ligand 185
reactions of excited states 236
Oxidative coupling catalysts
of acetylenes 179
homogeneous 178
of phenols 179
Oxidative coupling of 2,6-
dimethylphenol 182f
Oxygen
exchange in Cr(CO)$_6$ and
μ-H[Cr(CO)$_5$]$_2^-$ 106, 109
exchange in the water gas
shift reaction 106
gaseous evolution of molecular . 210
multielectron reduction of ... 173
reduction potentials of 173f
reduction, voltammograms of .. 175f
uptake rate 185f

P

^{31}P ^{1}H FT NMR data 158f
Pt(PO$_3$)$_2$Cl$_2$ + SnCl$_2$,
PMR spectrum 9f
Palladium chloride catalysts,
polystyrene-anchored 351t
Palladium chloride catalysts,
supported 349
Paramagnetic species,
formation of 145, 147t
Pentacarbonyliron, catalyst
precursor 94
Pentacoordinate carbonyl
nickel(II) 152
1,3-Pentadiene 46t

Pentamethylcyclopentadienyl–
iridium complexes 31
Pentamethylcyclopentadienyl–
rhodium complexes 31
Phase-transfer reaction
conditions 44, 47
Phenols, oxidative coupling of ... 179
Phosphine(s)
CO fixation 152
chiral 53t
-rhodium complexes 16, 26
Phosphorus ligand complexes of
iron, bidentate 67
Photochemical energy storage
system 344
Photochemical splitting of water . 210
Photochemistry, rhodium(I)
isocyanide complexes 225
Photoelectrolytic process 215f
Photogalvanic cells 296
processes in 298
Photolytic products of
D$_2$O–H$_2$O 216–217f
Photolytic products using
H$_2$O^{16}O–H$_2^{18}$O 218f
Photoreactions, olefin 325
Photoreduction by triethylamine .. 245
Photosensitivity, quinolinolato-
tungsten(IV), chelates 261
Photosynthesis, in vitro 210
Photovoltaic activity, effect of
oxygen on 211
Platinized Chl a cell 212f
Platinized electrode 211, 213f
Platinum catalyst, electronic
structure 4, 7t
Polymeric species 180
Polymerization, initiator 181, 189
Polymerization, tungsten(IV)
chelates 261
Polymers, cationic 388
Polypyridyl complexes 237
Polystyrene 11
Pore dimensions, hydrocarbon
and zeolite 388t
Porphyrins 162
Pressure–composition data for
LaNi$_{4.6}$Al$_{0.4}$ 283t
Pressure data 282f
Primary light reaction, in vitro ... 210
Probability ellipsoids 361f
Prochiral benzophenones,
reduction of 56
Prochiral olefins, hydrogenation of 22
Pyridine 179
derivatives 180, 183, 186t
ligand exchange 183
synthesis 183
ligands, substituted 185
-stabilized "CuO" initiator
species 181f
Pyrolysis apparatus, diffusion 219f
Pyrolysis, quantitative determination by diffusion 219
Pyrolytic analyses 220f

Q

Quadricyclane isomerization	307
Quenching	238

R

$RhCl(PO)_3$	2
$[Rh(DIPHOS)]^+$	18
arene and alkene adducts of	19
-catalyzed hydrogenation	19, 21f
$[Rh(DIPHOS)(BENZENE)]BF_4$	24
$[Rh(DIPHOS)(NOR)]^+$, reaction with hydrogen	17f
$[Rh(DIPHOS)(NOR)]BF_4$	24
$[Rh(DIPHOS)(Z$-methyl-α-acetamidocinnamate$)]BF_4$	24
$Rh_2(bridge)_4^{2+}$, spectral data	229t
$Rh_2(bridge)_4^{2+}$, structure of	226f
$Rh_2(bridge)_4(BPh_4)_2$, spectra of	228f
$Rh_2(bridge)_4Cl_2^{2+}$, structure of	230f
$[Rh_2(DIPHOS)_2BF_4]_2$	24
$Rh_2(TM4$-bridge$)_4^{2+}$, spectral data	229t
$Rh_2(TM4$-bridge$)_4^{2+}$, structure of	229f
$[Rh_3(DIPHOS)_3(OCH_3)_2]PF_6$	24
Radical concentrations	146f
Redox enzymes, entatic state for copper in	358
Redox, excited state	236
Redox relationships	237
Reduction	
of catalytic hydrogenolysis of strained molecules	321
of $Fe(DMPE)_2Cl_2$	68
of nitrobenzene	128t
of prochiral benzophenones	56
reactions of excited states	236
Reppe modification of hydroformylation reaction	122
Resin, metal-free	60t
Rhodium	
–phosphine complexes, cationic	16
catalysts	26, 27t
hydrogenation of nitro compounds	26
transfer	62
experiments	64
Rhodium(I)	
complexes, binuclear	225
energies of molecular orbitals in	227f
photooxidation of	232t
-tertiary phosphine ligands	16
Rhus vernicifera	359
Ring-contracting process	203, 208
Ring systems, Ni(0)-catalyzed reactions of strained	307
Ruthenium carbonyl	82
Ruthenium(II) complex	
excited states, oxidative quenching of	246
excited states, reductive quenching of	241
hydrophobic	240
Ruthenium(III) species	248

S

SnO_2 anode	304f
Semiconductivity of metal tetrathiolenes	364, 378
Seven-coordinate compounds, tungsten	263
Silanes	54
S_{N2} type mechanism, nickel(II)-d^8	152, 159
Solar energy storage	344
Solvent, hydrogenation	6t, 13t
Soybean	
ester, hydrogenaiton of	12t
methyl ester	1
oil	1, 2t
Spectra, IR	103f
Spectroscopic studies, high pressure	102
Spectroscopy, binuclear rhodium(I) isocyanide complexes	225
Spin trap	58
Stereochemistry	
of the $[(C_6H_5)_3P]_4Pt_2$-$(C_{10}H_4S_4)$ molecule	376f
catalyzed cycloaddition	313
of expanded coordination spheres	263
of $LaNi_{5-x}Al_x$	288, 289t, 290f, 291t, 293t
of metal tetra-thiolenes	364, 374, 375–376f
seven-coordinate tungsten complexes	263
Stereoselectivity $[Rh_2(DIPHOS)_2]^{2+}$ and $[Rh_3(DIPHOS)_3(OMe)_2]^+$	22
Stereospecificity, hydrogenation	40
Stereospecificity, hydrosilylation	50
Stoichiometric reactions of hydrides	36
Strained ring systems, nickel(0) catalysis	307
Substituted pyridine ligands	185, 186t
Substrate, excess in hydrogenation	46t
Substrate, structure and optical yield	55
Sulfur complexes, bridged	362t
Sulfur heterocycles, desulfurization and hydrodesulfurization of	203
Sunlight engineering efficiency	296
μ-Superoxo dicobalt complex	174
Supported catalysts, chiral	60t
Surface-induced back reaction in thin-layer iron–thiazine photogalvanic cells	296
Surfactant complex, monolayer-bound	240
Surfactants, micelle-forming	45–46t

T

Templates	
in zeolite crystallization	387
effects	388
polymeric	391

Terminal double bonds 6
Ternary alloys, of $LaNi_{5-x}Al_x$ 280
Tetramic cluster 193
Tetrathiolene clusters, molecular
   metal 372, 375–376f
Tetrathiolene ligands 365f
   oxidation potentials of 371f
Tetrathionaphthalene, oxidation
   potential of 368f, 369f
Thermodynamic data of
   $LaNi_{5-x}Al_x$ 281, 287t
Transformylation of metal formyl
   complexes 136
Transition metal complexes 60
Tricarbonylbis(N,N-dimethyldithio-
   carbamato)tungsten(II) ..265, 266
Trienes 12t
Triethyl amine 26, 28
   conversion by irradiation 244
   photoreduction of 245
Tri-μ-hydride complex 38
Trimerization 196
2,3,3-Trimethyl-1,4-pentene
   hydrogenation 9t
Triphenylcyclopropenylnickel
   halide catalysts 351t, 353t
Triphenylphosphinedicarbonylbis-
   (N,N-dimethyldithiocarba-
   mato)tungsten(II) 265
Tungsten
   complexes, seven-coordinate ... 263
     IR spectra 266, 267f
     NMR 268, 269f, 272f
   metal–metal triple bonds 396
Tungsten(IV) chelates 252
   photosensitivity 261
   polymerization 261
   potential energy transfer
     complexes 252

U

USDA 4

V

Voltammograms of oxygen
   reduction 175f

W

$W(CO)_3(dmtc)_2$, spectrum
   of 267f, 269f, 272f
$W_4(OPr^i)_{12}(HOPr^i)_2$ molecule .. 405f
WL_4 complexes 252
WL_4 isomers
   chelate rings in 256f
   low-energy charge-transfer
     bands 259t, 260t
   synthesis 257
Water
   catalytic reductions using
     CO and 121
   gas shift reaction
     $Fe(CO)_5$-catalyzed 96–97t
     metal carbonyl catalysis 81, 86t
     metal hexacarbonyls 94
     pentacarbonyliron 94
     phase transfer catalysis 106
   photochemical splitting of 210
Wilkinson catalyst 2, 41

Z

Zeolite crystallization
   cationic polymer influence ..388, 390t
   temperature 393–394t
   templates in 388, 391, 393t
Zeolite synthesis 387

The text of this book is set in 10 point Caledonia with two points of leading. The chapter numerals are set in 30 Point Garamond; the chapter titles are set in 18 point Garamond Bold.

The book is printed offset on Text White Opaque 50-pound. The cover is Joanna Book Binding blue linen.

*Jacket design by Linda Mattingly.
Editing and production by Saundra Goss.*

The book was composed by Service Composition Co., Baltimore, MD., printed and bound by The Maple Press Co., York, PA.

QD
1
A355
no.173

APR 26 1979